The United Nations World Water Development Report 3

WATER IN A CHANGING WORLD

World Water Assessment Programme

UN WATER

UNESCO Publishing

United Nations Educational, Scientific and Cultural Organization

earthscan from Routledge

First published 2009 by Earthscan

2 Park Square, Milton Park, Abingdon, Oxfordshire OX14 4RN
605 Third Avenue, New York, NY 10017

Routledge is an imprint of the Taylor & Francis Group, an informa business

First issued in hardback 2020

UNESCO ISBN: 978-9-23104-095-5
Earthscan ISBN: 978-1-84407-839-4 (hbk)
978-1-84407-840-0 (pbk)

UNESCO Publishing: http://publishing.unesco.org/

Earthscan: www.earthscan.co.uk

Suggested citation
World Water Assessment Programme. 2009. *The United Nations World Water Development Report 3: Water in a Changing World.* Paris: UNESCO, and London: Earthscan.

Cover design
Peter Grundy, London, UK

Book design and typesetting
Communications Development Incorporated, Washington, DC

A catalogue record for this book is available from the British Library.

Library of Congress Cataloging-in-Publication data has been applied for.

This Report has been published on behalf of the United Nations World Water Assessment Programme (WWAP), with the support of the following organizations:

United Nations Funds and Programmes
United Nations Centre for Human Settlements (UN-HABITAT)
United Nations Children's Fund (UNICEF)
United Nations Conference on Trade and Development (UNCTAD)
United Nations Department of Economic and Social Affairs (UNDESA)
United Nations Development Programme (UNDP)
United Nations Environment Programme (UNEP)
United Nations High Commissioner for Refugees (UNHCR)
United Nations University (UNU)

Specialized UN Agencies
Food and Agriculture Organization (FAO)
International Atomic Energy Agency (IAEA)
International Bank for Reconstruction and Development (World Bank)
World Health Organization (WHO)
World Meteorological Organization (WMO)
United Nations Educational, Scientific and Cultural Organization (UNESCO)
International Fund for Agricultural Development (IFAD)
United Nations Industrial Development Organization (UNIDO)

United Nations Regional Commissions
Economic Commission for Europe (ECE)
Economic and Social Commission for Asia and the Pacific (ESCAP)
Economic Commission for Africa (ECA)
Economic Commission for Latin America and the Caribbean (ECLAC)
Economic and Social Commission for Western Asia (ESCWA)

Secretariats of United Nations Conventions and Decades
Secretariat of the Convention to Combat Desertification (CCD)
Secretariat of the Convention on Biological Diversity (CBD)
Secretariat of the United Nations Framework Convention on Climate Change (FCCC)
Secretariat of the International Strategy for Disaster Reduction (ISDR)

Table of contents

Foreword by Ban Ki-moon, Secretary-General, United Nations

It is well known that water is life; what this Report shows is that water also means livelihoods. It is the route out of poverty for individuals and communities. Managing water is essential if the world is to achieve sustainable development.

This challenge is even more pressing as the world confronts the triple threats of climate change, rising food and energy costs, and the global economic crisis. All three are exacerbating poverty, inequality and underdevelopment.

The United Nations has responded by consolidating our work and joining with partners who can make a difference through UN-Water, which brings together more than two dozen UN agencies and other stakeholders. The initiative's World Water Assessment Programme is setting an example of system-wide cooperation based on the understanding that water is such a central consideration that it must be an integral part of all planning and investments.

Developing countries and countries in transition are striving to manage their water resources more effectively. I call on the bilateral donors to support those efforts by increasing water's share of official development assistance above the current level of 5.4%.

This is important not only for development; it is a matter of security, too. Lack of basic services can contribute to political instability. Armed conflicts can further disrupt these services.

There has been a widespread failure to recognize water's vital role in providing food, energy, sanitation, disaster relief, environmental sustainability and other benefits. This has left hundreds of millions of people suffering from poverty and ill health and exposed to the risks of water-related diseases.

This situation is unconscionable. Governments and the international development community must make more and immediate investments in water management and related infrastructure. We must all work together to address this matter of life and livelihoods. This Report is meant to spur such action, and I commend it to a wide global audience.

Ban Ki-moon
Secretary-General
United Nations

Foreword by Koïchiro Matsuura, Director-General, United Nations Educational, Scientific and Cultural Organization

With the release of this third edition of *The United Nations World Water Development Report,* it is clear that urgent action is needed if we are to avoid a global water crisis. Despite the vital importance of water to all aspects of human life, the sector has been plagued by a chronic lack of political support, poor governance and underinvestment. As a result, hundreds of millions of people around the world remain trapped in poverty and ill health and exposed to the risks of water-related disasters, environmental degradation and even political instability and conflict. Population growth, increasing consumption and climate change are among the factors that threaten to exacerbate these problems, with grave implications for human security and development.

The current Report provides a comprehensive analysis of the state of the world's freshwater resources. It also, for the first time, shows how changes in water demand and supply are affected by and affect other global dynamics. It represents a considerable collaborative achievement for the 26 UN agencies that make up UN-Water and are engaged in the World Water Assessment Programme (WWAP), which leads the monitoring and evaluation behind the Report. UNESCO is very proud to have played a pivotal role in the launch of this flagship programme and to continue to support its work by housing the WWAP Secretariat. I am confident that this third volume will prove crucial as a working tool for policy-makers and other stakeholders, providing solid evidence from which to develop an effective and sustainable approach to water issues.

The Report could not come at a more important time. We have passed the halfway point towards the 2015 target date for achieving the Millennium Development Goals, and despite progress, massive challenges remain. Millennium Development Goal 7 calls for halving the proportion of people without sustainable access to safe drinking water and basic sanitation. While the world is on track to achieve the water target globally, large regions of the world and many countries lag behind, and some risk backsliding. This is particularly the case in sub-Saharan Africa and low-income Arab states. On current trends the sanitation target will be missed by a wide margin in the majority of developing countries. But water is linked not only to Millennium Development Goal 7. It also directly affects, as this Report establishes, the achievement of all eight Millennium Development Goals, including, notably, the first goal, the eradication of extreme poverty and hunger.

Water is a cross-cutting issue that demands a coordinated approach. Our success in avoiding a global water crisis is directly linked to our ability to address other global challenges, from poverty eradication and environmental sustainability to fluctuating food and energy costs and financial turmoil in world economies. It is therefore imperative that global risks, including those associated with water, be dealt with in an integrated manner. We must develop interdisciplinary tools that can take into account different drivers such as climate change and financial markets to achieve sustainable water management. This requires the engagement of all stakeholders, particularly government leaders, as well as global coordination through the UN system.

Water is essential to facing today's global challenges and achieving the Millennium Development Goals. As such, it should be a priority for the United Nations and the global community as a whole. Be assured that UNESCO stands ready to play its part in this process.

Koïchiro Matsuura
Director-General
United Nations Educational, Scientific and Cultural Organization

Preface

In 1999 the United Nations system resolved to issue regular editions of *The United Nations World Water Development Report*. An expert group, convened by the United Nations Department of Economic and Social Affairs, developed recommendations for the objectives and targeted audience of the report (box 1).

The first edition, *The United Nations World Water Development Report: Water for People, Water for Life,* was released in March 2003 at the 3rd World Water Forum in Kyoto, Japan. The second, *Water, a Shared Responsibility,* was released in March 2006 at the 4th World Water Forum in Mexico City. The first report provided an inaugural assessment of progress since the 1992 United Nations Conference on Environment and Development in Rio de Janeiro. Both reports were based on key challenge areas (such as water for food, water for energy, and challenges for governance). Stand-alone assessments were prepared by UN agencies. The assessments included pilot case studies on which the Report drew in developing appropriate assessment methodologies and lessons learned.

This third edition embraces a holistic structure and focuses on the second objective established by the expert group – to accelerate coverage and investments for basic human water needs (drinking water supply, sanitation and health, food security, mitigation of floods and droughts and prevention of conflicts), giving priority to developing countries.

Contents of the Report

A major theme of this Report is that important decisions affecting water management are made outside the water sector and are driven by external, largely unpredictable forces – forces of demography, climate change, the global economy, changing societal values and norms, technological innovation, laws and customs and financial markets. Many of these external drivers are dynamic, and changes are accelerating. The conceptual framework

Box 1 Objectives and targeted audience of *The United Nations World Water Development Report*

It is recommended that *The United Nations World Water Development Report* be targeted for national decision-makers and water resources managers, with two complementary objectives:

- To strengthen and stimulate national capacities and cross-sector institutions in integrated water development planning and in sustainable management of water resources at river basin and aquifer levels.
- To stimulate an acceleration of coverage and investments, in priority, for basic human water needs (drinking water supply, sanitation and health, food security, mitigation of floods and droughts and prevention of conflicts), giving a priority to developing countries.

A more effective and targeted support of the international community for such local and national efforts would also be an important objective of this awareness-raising and action-oriented report.

Source: United Nations Expert Group Meeting to Examine Methodologies for the Preparation of a Biennial 'World Water Development Report', convened and organized by the UN Department of Economic and Social Affairs, New York, 11-14 January 2000.

that evolved for the Report is on the inside front cover of the Report and in figure 1.1 in chapter 1. The figure illustrates how developments outside the water domain influence water management strategies and policies. The Report emphasizes that decisions in other sectors and those related to development, growth and livelihoods should incorporate water as an integral component, including responses to climate change, food and energy challenges and disaster management.

At the same time, the Report's analysis of the state of the world's water resources is imbedded in a more expansive context of what can be accomplished through water management. The analysis leads to a set of responses and recommendations for action that differ from those that have emerged from more introspective analyses of the water sector because they incorporate the contribution of water to sustainable development.

This Report offers a holistic approach to links between water and climate change, food, energy, health and human security. Human security, broadly conceived, includes basic needs for food, water, health, livelihoods and a place to live – issues addressed in the Millennium Development Goals. As the second part of the Fourth Assessment Report of the Intergovernmental Panel on Climate Change (IPCC), released in April 2007, demonstrates, poor people are likely to suffer most from the effects of climate change.[1]

New processes

In keeping with the Report's broader view on policy options, new processes were applied in preparing this Report. Fuller treatment is given to such topics as climate change, business and trade, financing, the role of the private sector, water transport and innovations and new technologies.

The United Nations Expert Group recommendation to involve countries in preparing the reports was reflected in the first edition in case studies based on 10 countries (including 10 national river basins) with different physical, climate and socioeconomic conditions. This method was followed in the second edition and in this Report, which presents the case studies in a companion volume to the main report. The World Water Assessment Programme is also launching a series of supporting publications that include scientific side papers, topic and sector reports and dialogue reports, taking the programme out of its rigid three-year cycle.

The preparatory process for this Report has followed an inclusive, participatory approach benefiting from opinion and feedback from the scientific, professional and decision-making communities from within and outside the water sector.

Broader input to the Report and the World Water Assessment Programme processes in general has been achieved through four mechanisms:

- A Technical Advisory Committee of 11 prominent individuals from around the world with water sector expertise and broader policy-making experience in their countries and internationally.

- Expert groups on indicators, monitoring and data/metadata bases; scenarios; climate change and water; policy relevance; business, trade, finance and the private sector; legal issues and water storage.

- A Report team composed of UN-Water member agencies, their professional and non-governmental organization partners and the broader community of water and water-related sectors.

- Stakeholder engagement through the World Water Assessment Programme Website and review processes, including public as well as solicited input and feedback from hundreds of individuals and organizations.

1. 'Poor communities can be especially vulnerable, in particular those concentrated in high risk areas. They tend to have more limited adaptive capacities and are more dependent on climate-sensitive resources such as local water and food supplies.' (IPCC, 2007, Summary for Policymakers. In *Climate Change 2007: Impacts, Adaptation and Vulnerability. Contribution of Working Group II to the Fourth Assessment Report of the Intergovernmental Panel on Climate Change,* eds., M. L. Parry, O. F. Canziani, J. P. Palutikof, P. J. van der Linden and C. E. Hanson, Cambridge, UK: Cambridge University Press, p. 9).

This Report marks a transition from the first two reports – a transition from being a report primarily for water managers to being a report for leaders at all levels of government, the private sector and civil society, whose decisions depend on the availability of water resources and make demands on water management. The Expert Group on Policy Relevance consulted hundreds of such leaders to obtain their views on policy issues relevant to the water sector. At the same time, the Report continues to provide useful data for water managers on the state and use of this precious resource. Past reports have looked at trends based on historical data. It is clear that change is accelerating and that the effects of change are not easily projected from trends. To help us understand possible futures and how to cope with their impact on water resources, the World Water Assessment Programme process looks at the development of scenarios that will serve the fourth *World Water Development Report*. This scenario effort takes into account the main drivers of water, including demographics, climate change, social and economic processes and technology, along with their interactions.

In preparing this Report new data were available to update only a third of the 60-plus indicators that were reported in the second edition. And some indicators were found to be no longer valid. The lack of data was echoed by the coordinators and authors of this Report, who found that indicators and data were often not available for analysing and reporting on issues considered important. As a consequence, a new process was developed for indicators and monitoring that aims at a better understanding of the trends and developments, including changes, in the state of water resources, their uses and the interface between the state and water uses and between water and other sectors. This reflects a recommendation of Agenda 21 – a comprehensive plan of action agreed at the Rio Summit for all areas of human impact on the environment – that a detailed data collection for both fluxes of 'exploitable water resources' and of 'associated costs and finances' be conducted within a comprehensive plan for water development at the basin level.[2]

To this end, the World Water Assessment Programme established an Expert Group on Indicators, Monitoring and Data/Metadata Bases, and UN-Water established a Task Force on Indicators, Monitoring and Reporting, which is coordinated by the World Water Assessment Programme. Their results will be reported by the World Water Assessment Programme in a process leading to the fourth *World Water Development Report* and by UN-Water. A table showing the status of indicators reported on in this Report is presented in appendix 1. More detailed information may be found at www.unesco.org/water/wwap.

Few countries know how much water is being used and for what purposes, the quantity and quality of water that is available and that can be withdrawn without serious environmental consequences and how much is being invested in water management and infrastructure. Despite the availability of new remote sensing and geographic information system technologies that can simplify monitoring and reporting and despite the growing need for such information in an increasingly complex and rapidly changing world, less is known with each passing decade. Strengthening such information systems is vital not only at a national scale but also at a global scale – to inform the construction of global models of the hydrologic cycle and decisions on where interventions, including external aid, would be most useful. Chapters 10 and 13 of the Report, in particular, treat this subject.

Challenges remain in managing water resources for development

The contribution of sustainable access to safe drinking water and adequate sanitation to achieving the Millennium Development Goals is well established. Largely ignored, however, is the fact that water resources are at the core of many of the Millennium Development Goals on which progress is lagging. This Report and others elaborate the direct and indirect contributions of water management across all the Millennium Development Goals.

It is not enough to hope that the trickle-down effects of economic growth will result in equitable distribution that includes the poor. The economic growth and poverty-reducing contributions of water resources must be made explicit and specific at the country level. Intergovernmental efforts must support such actions and maintain the momentum of the global commitments made since the Millennium Declaration in 2000.

2. United Nations, 1992, Agenda 21, Chapter 18, Protection of the Quality and Supply of Freshwater Resources: Application of Integrated Approaches to the Development, Management and Use of Water Resources, New York: Department of Economic and Social Affairs, United Nations.

While mitigation of anthropogenic climate change is vital, the blunt reality is that all countries – particularly developing countries that will be hit hardest and earliest – and business sectors must also adapt to climate change. Even if greenhouse gas concentrations stabilize in the coming years, some impacts from climate change are unavoidable. These include increasing water stress in many regions, more extreme weather events, the potential for large population migration and the disruption of international markets. These challenges cannot be separated from the challenges of sustainable development in a complex global context.

This report provides evidence of the need for public investments in water resources infrastructure and implementation capacity. It also provides evidence of the vital importance of water resources and environmental sustainability to engage the private sector, civil society and communities to invest and become involved, offering examples of how this can be done.

Bilateral donors, important in funding water investments, must avoid the temptation to reduce their aid budgets during the current global financial and economic crises. Multilateral aid could be an important source of financing for many years to come. Yet both bilateral and multilateral donors appear not to recognize the contribution of the water sector to growth: the water sector's share of official development assistance has remained below 6% for some time. This said, the flow of official development assistance has increased in recent years and so has the water component in dollar terms. But most of the increase has gone to water supply (and sanitation, to a lesser degree), while aid flows to other water sectors have stagnated in dollar terms and fell as a percentage of total assistance.

Like other physical infrastructure, water infrastructure deteriorates over time and needs repair and replacement. Investment is also required in operation and maintenance and in developing the capacity of the sector so that infrastructure meets appropriate standards and functions efficiently.

The case of sub-Saharan Africa

Sub-Saharan Africa, in particular, remains mired in poverty. Its progress towards achieving the Millennium Development Goals lags behind that of other regions. The percentage of the population living in absolute poverty is essentially the same as it was 25 years ago. About 340 million Africans lack access to safe drinking water, and almost 500 million lack access to adequate sanitation. Countries in sub-Saharan Africa store only about 4% of their annual renewable flows, compared with 70%-90% in many developed countries, yet water storage is essential to ensure reliable sources of water for irrigation, water supply and hydropower and to provide a buffer for flood management.

The need to act now

The challenges that face decision-makers are numerous. The context in which they must make decisions is not well defined. This Report does not attempt to provide a full set of answers. But it identifies the key issues that must be faced. It describes some of the ways that decision-makers have dealt with these challenges, providing options for consideration across levels of government and sectors.

Despite the many unknowns, we need to act now – with decisions about investments in water infrastructure and in implementation capacity to enable environmentally sustainable economic growth and social development and with decisions on safety nets to ensure basic services that protect the poor.

We hope that this third *United Nations World Water Development Report* will stimulate decision-makers in government, the private sector and civil society to act.

Olcay Ünver
World Water Assessment Programme Coordinator
United Nations Educational, Scientific and Cultural Organization

William Cosgrove
United Nations World Water Development Report 3
Content Coordinator

Acknowledgements

This Report would not have been possible without the essential and gracious support of many individuals. The personal support and interest of Koïchiro Matsuura, Director-General of the United Nations Educational, Scientific and Cultural Organization (UNESCO), created an enabling environment. The leadership and guidance of Andras Szöllösi-Nagy, director of the Division of Water Sciences of UNESCO, allowed the World Water Assessment Programme (WWAP) team to mobilize its collective energy and capacity in the best way possible. Pasquale Steduto, chair of UN-Water and chief of the Land and Water Division of the Food and Agriculture Organization of the United Nations (FAO), graciously extended his catalytic support during a challenging process of preparation and production.

The leadership of Uri Shamir, chair of the Report's Technical Advisory Committee and professor of engineering at Technion-Israeli Institute of Technology, and the expertise of the committee members helped create a product of sound scientific basis and superior quality, which was further enhanced by the work of WWAP's Expert Groups. We acknowledge the efforts of Gerald Galloway, professor of engineering at the University of Maryland, in helping to reach out to hundreds of decision-makers worldwide, enabling the Report to be policy relevant to its primary intended audience.

We acknowledge the support of the World Bank, FAO and Organisation for Economic Co-operation and Development in providing the most recent data and information available, sometimes before they were published. We especially thank the Italian Ministry of Environment, Land and Sea for its generous provision of funds; the Italian Ministry of Foreign Affairs for its support; and the government of the Umbria Region of Italy for the beautiful new premises that now house the WWAP at the Villa La Colombella, Perugia.

We thank the Report's editors, Bruce Ross-Larson and Meta de Coquereaumont, and their production team at Communications Development Incorporated – Joseph Caponio, Amye Kenall, Allison Kerns, Christopher Trott and Elaine Wilson – for the extraordinary support they provided.

The WWAP also thanks the following individuals and organizations from around the world for their generous and varied contributions to the Report.

UN-Water

Pasquale Steduto, chair; Aslam Chaudhry, Johan Kuylenstierna and Frederik Pischke

UNESCO

Alice Aureli, Jonathan Baker, Jeanne Damlamian, Siegfried Demuth, Walter Erdelen, Rosanna Karam, Shahbaz Khan, Anil Mishra, Djaffar Moussa-Elkadhum, Anna Movsisyan, Mohan Perera, Amale Reinholt-Gauthier, Léna Salamé and Alberto Tejada-Guibert

World Water Development Report 3 teams

Chapter coordinators

Tim Kasten (UNEP) with the support of Thomas Chiramba (UNEP); Andras Szöllösi-Nagy (UNESCO) and Wolfgang Grabs (WMO), associate coordinator, with the support of Siegfried Demuth (UNESCO) and Anil Mishra (UNESCO); Jean-Marc Faurès (FAO); Håkan Tropp (UNDP Water Governance Facility); Olcay Ünver (WWAP)

Acknowledgements

Content coordinator
William Cosgrove

Process manager
George de Gooijer

Chapter facilitators
Richard Connor, William Cosgrove, George de Gooijer, Denis Hughes and Domitille Vallée

Graphics coordinator
Akif Altundaş

WWAP publications coordinator
Samantha Wauchope

United Nations World Water Assessment Programme
Technical Advisory Committee
Uri Shamir, chair; Dipak Gyawali, deputy chair; Fatma Attia, Anders Berntell, Elias Fereres, M. Gopalakrishnan, Daniel Pete Loucks, Laszlo Somlyody, Lucio Ubertini, Henk van Schaik, Albert Wright

Sponsors and donors
Italian Ministry of Environment, Land and Sea; Government of Region of Umbria, Italy; Government of Japan; UNESCO Etxea, Basque Water Agency (URA) and Danish International Development Assistance; and US Army Corps of Engineers, Institute for Water Resources

Secretariat
Olcay Ünver, coordinator; Michela Miletto, deputy coordinator; Akif Altundaş, Floriana Barcaioli, Adriana Fusco, Lisa Gastaldin, Georgette Gobina, Simone Grego, Shaukat Hakim, Rosanna Karam, Engin Koncagül, Lucilla Minelli, Stéfanie Néno, Abigail Parish, Daniel Perna, Jean-Baptiste Poncelet, Astrid Schmitz, Marina Solecki, Toshihiro Sonoda, Jair Torres, Domitille Vallée, Casey Walther and Samantha Wauchope

Expert groups
Indicators, monitoring and databases
Mike Muller and Roland Schulze, co-chairs; Joseph Alcamo, Amithirigala Jayawardena, Torkil Jønch-Clausen, Peter C. Letitre, Aaron Salzberg, Charles Vörösmarty, Albert Wright and Daniel Zimmer

Business, trade, finance and involvement of private sector
Ger Bergkamp and Jack Moss, co-chairs; Margaret Catley-Carlson, Joppe Cramwinckel, Mai Flor, Richard Franceys, Jürg Gerber, Gustavo Heredia, Karin Krchnak, Neil McLeoud, Herbert Oberhansli, Jeremy Pelczer and Robin Simpson

Climate change and water
Pierre Baril and BertJan Heij, co-chairs; Bryson Bates, Filippo Giorgi, Fekri Hassan, Daniela Jacob, Pavel Kabat, Levent Kavvas, Zbigniew Kundzewicz, Zekai Şen and Roland Shulze

Legal issues
Stefano Burchi and Patricia Wouters, co-chairs; Rutgerd Boelens, Carl Bruch, Salman M. A. Salman, Miguel Solanes, Raya Stephan and Jessica Troell

Policy relevance
Gerry Galloway and Dipak Gyawali, co-chairs; Adnan Badran, Qiu Baoxing, Antonio Bernardini, Benito Braga, Max Campos, Peter Gleick, Rajiv Gupta, Mohammed Ait Kadi, Celalettin Kart, Juliette Biao Koudenoukpo, Juan Mayr, Jack Moss, Mike Muller, Hideaki Oda, Marc Overmars, Victor Pochat, Jerome Delli Priscoli, Cletus Springer, Carel de Villeneuve, Zhang Xiangwei and Jiao Yong

Scenarios
Joseph Alcamo and Gilberto Gallopin, co-chairs; Vahid Alavian, Nadezhda Gaponenko, Allen Hammond, Kejun Jiang, Emilio Lebre la Rovere, Robert Martin, David Molden,

Mike Muller, Mark Rosegrant, Igor Shiklomanov, Jill Slinger, Narasingarao Sreenath, Ken Strzepek, Isabel Valencia and Wang Rusong

Storage
Luis Berga and Johan Rockström, co-chairs; Alison Bartle, Jean-Pierre Chabal, William Critchley, Nuhu Hatibu, Theib Oweis, Michel de Vivo, Arthur Walz and Carissa Wong

Contributing and partner organizations
AquaFed; Conservation International; Global Water Partnership; International Centre for Water Hazard and Risk Management; International Institute for Advanced Systems Analysis; International Research and Training Center on Erosion and Sedimentation; International Water Association; Organisation for Economic Co-operation and Development; Stockholm Environment Institute; UNDP Water Governance Facility at Stockholm International Water Institute; UN-Water Decade Programme on Capacity Development; UNEP-DHI Centre for Water and Environment; University of Dundee Centre for Water Law, Policy and Science; UNESCO-IHE Institute for Water Education; World Business Council on Sustainable Development and World Water Council

Case studies
Editor
Engin Koncagül (WWAP)

Editorial team
Rebecca Brite and Alison McKelvey Clayson

Maps
AFDEC

Case study contributors
Argentina, Bolivia, Brazil, Paraguay and Uruguay (La Plata River basin)
Miguel Ángel López Arzamendia, Silvia González, Verónica Luquich, Victor Pochat and staff of the Intergovernmental Coordinating Committee of the La Plata River Basin

Bangladesh
Saiful Alam, Mozaddad Faruque, Azizul Haque, Md. Anwarul Hoque, Jalaluddin Md. Abdul Hye, Md. Azharul Islam, Andrew Jenkins, A. H. M. Kausher, Hosne Rabbi, Md. Mustafizur Rahman, Md. Shahjahan and the Bangladesh Ministry of Water Resources

Brazil and Uruguay (Lake Merín Basin)
Gerardo Amaral, José Luis Fay de Azambuja, Ambrosio Barreiro, Artigas Barrios, Jorge Luiz Cardozo, Daniel Corsino, Adolfo Hax Franz, Henrique Knorr, Fiona Mathy, Juan José Mazzeo, Joao Menegheti, Claudio Pereira, Jussara Beatriz Pereira, Martha Petrocelli, Carlos María Prigioni, Hamilton Rodrigues, Aldyr Garcia Schlee, Carlos María Serrentino, Manoel de Souza Maia and Silvio Steinmetz

Cameroon
Kodwo Andah and Mathias Fru Fonteh

China
Dong Wu, Hao Zhao, Jin Hai, Ramasamy Jayakumar, Liu Ke, Pang Hui, Shang Hongqi, Song Ruipeng, Sun Feng, Sun Yangbo and Xu Jing

Estonia
Erki Endjärv, Harry Liiv, Peeter Marksoo and Karin Pachel

Finland and Russian Federation (Vuoksi River basin)
Natalia Alexeeva, Sari Mitikka, Raimo Peltola, Bertel Vehviläinen, Noora Veijalainen and Riitta-Sisko Wirkkala

Italy
Beatrice Bertolo and Francesco Tornatore

Republic of Korea
Republic of Korea Ministry of Land, Transport and Maritime Affairs

Netherlands
Marcel E. Boomgaard, Joost J. Buntsma, Michelle J. A. Hendriks, Olivier Hoes, Rens L. M. Huisman, Jan Koedood, Ed R. Kramer, Eric Kuindersma, Cathelijn Peters, Jan Strijker, Sonja Timmer, Frans A. N. van Baardwijk, Tim van Hattum and Hans Waals

Pacific Islands
Marc Overmars, Hans Thulstrup and Ian White

Pakistan
Ch. Muhammad Akram, Mi Hua and Zamir Somroo

Spain (Autonomous Community of the Basque Country)
Fernando Díaz Alpuente, Ana Oregi Bastarrika, Iñaki Urrutia Garayo, Mikel Mancisidor, Sabin Intxaurraga Mendibil, Josu Sanz and Tomás Epalza Solano

Sri Lanka
M. M. M. Aheeyar, Sanath Fernando, K. A. U. S. Imbulana, V. K. Nanayakkara, B. V. R. Punyawardena, Uditha Ratnayake, Anoja Seneviratne, H. S. Somatilake, P. Thalagala and K. D. N. Weerasinghe

Sudan
Gamal Abdo, Abdalla Abdelsalam Ahmed, Kodwo Andah, Abdin Salih,

Swaziland
Kodwo Andah, E. J. Mwendera and the Swaziland Department of Water Affairs

Tunisia
Mustapha Besbes, Jamel Chaded, Abdelkader Hamdane and Mekki Hamza

Turkey (Istanbul)
Gülçin Aşkın, Zeynep Eynur, Canan Gökçen, Canan Hastürk, S. Erkan Kaçmaz, Selami Oğuz, Gürcan Özkan, Vildan Şahin, Turgut Berk Sezgin, Aynur Uluğtekin and Aynur Züran

Uzbekistan
Abdi Kadir Ergashev, Eh. Dj. Makhmudov, Anna Paolini and Sh. I. Salikhov

Zambia
Osward M. Chanda, Hastings Chibuye, Christopher Chileshe, Peter Chola, Ben Chundu, Adam Hussen, Joseph Kanyanga, Peter Lubambo, Andrew Mondoka, Peter Mumba, Mumbuwa Munumi, Priscilla Musonda, Christopher Mwasile, Kenneth Nkhowani, Imasiku A. Nyambe, Liswaniso Pelekelo, Zebediah Phiri, Friday Shisala, Lovemore Sievu and George W. Sikuleka

Participants at preparatory meetings and workshops

Inception meeting – 7-11 November 2007 – Paris, France

Virginie Aimard, Guy Alaerts, Joseph Alcamo, Reza Ardakanian, Pierre Baril, Francesca Bernardini, Gunilla Björklund, Janos Bogardi, Rudolph Cleveringa, James Dorsey, Elias Fereres, M. Gopalakrishnan, Wolfgang Grabs, Dipak Gyawali, BertJan Heij, Molly Hellmuth, Denis Hughes, Tim Kasten, Henrik Larsen, Peter C. Letitre, Daniel Pete Loucks, Jan Luijendijk, Robert Martin, Michel Meybeck, Jack Moss, Yuichi Ono, Léna Salamé, Monica Scatasta, Uri Shamir, Laszlo Somlyody, Manfred Spreafico, Alberto Tejada-Guibert, Lucio Ubertini, Henk van Schaik, Charles Vörösmarty, James Winpenny, Junichi Yoshitani and Daniel Zimmer

Integration meeting – 19-25 April 2008 – Perugia, Italy

Daniel Adom, Virginie Aimard, Guy Alaerts, Joseph Alcamo, Youssef Al-Mooji, Fatma Attia, Pierre Baril, Luis Berga, Anders Berntell, Gunilla Björklund, Robert Bos, Andrew Bullock, Stefano Burchi, Thomas Chiramba, Engin Çitak, Rudolph Cleveringa, Elias Fereres, Carlos Fernandez, Gilberto Gallopin, Gerry Galloway, M. Gopalakrishanan, Wolfgang Grabs, Dipak Gyawali, Joakim Harlin, BertJan Heij, Molly Hellmuth, Sarah Hendry, Denis Hughes, Niels Ipsen, Tim Kasten, Yanikoglu Kubra, Kshitij M. Kulkarni, Johan Kuylenstierna, Jon Lane, Henrik Larsen, Peter C. Letitre, Dennis Lettenmaier, Daniel Pete Loucks, Robert Martin, Anil Mishra, Jack Moss, Mike Muller, Yuichi Ono, Walter Rast,

Ahmet Saatci, Léna Salamé, Darren Saywell, Roland Schulze, Uri Shamir, Laszlo Somlyody, Toshihiro Sonoda, Alberto Tejada-Guibert, Jon Martin Trondalen, Duygu Tuna, Lucio Ubertini, Stefan Uhlenbrook, Wim van der Hoek, Pieter van der Zaag, Henk van Schaik, Charles Vörösmarty, James Winpenny, Albert Wright, Adikari Yoganath and Daniel Zimmer

Workshop on indicators, monitoring and databases – 18-20 June 2008 – Perugia, Italy
Karen Frenken, George de Gooijer, Jan Hassing, Engin Koncagül, Mike Muller, Stéfanie Néno, Gerard Payen, Roland Schulze, Charles Vörösmarty and Casey Walther

Workshop on policy relevance – 28 July – 1 August 2008 – Perugia, Italy
Michael Abebe, Altay Altinors, Kodwo Andah, Ger Bergkamp, Thanade Dawasuwan, Gerry Galloway, Dipak Gyawali, Saadou Ebih Mohamed, Jack Moss, Stéfanie Néno, Joshua Newton, Jerome Delli Priscoli, Khomoatsana Tau and Håkan Tropp

Solicited consultations

Real-time Delphi survey on scenarios, October 2007
Joseph Alcamo, Fatma Attia, Pierre Baril, Bryon Bates, Anders Berntell, Elias Fereres Castiel, Gilberto Gallopin, Nadezhda Gaponenko, Filipo Giorgi, Jerome Glenn, Stela Goldenstein, M. Gopalakrishnan, Wolfgang Grabs, Dipak Gyawali, BertJan Heij, Danielle Jacob, Pavel Kabat, Tim Kasten, Zbigniew Kundzewicz, Peter Loucks, David Molden, David Seckler, Uri Shamir, Zekai Şen, Igor Shiklomanov, Roland Shulze, Lazslo Somlyody, Ken Strzepek, Lucio Ubertini, Isabel Valencia, Henk van Schaik, Wang Rusong and Albert Wright

Real-time Delphi survey on storage, February 2008
Alison Bartle, Luis Berga, Jean-Pierre Chabal, Imo Efiong Ekpo, John Gowing, Robert T. Heath, Jia Jinsheng, Marna de Lange, Peter Stuart Lee, Jan Lundqvist, Maimbo Mabanga Malesu, Norihisa Matsumoto, Adama Nombre, Alberto Marulanda Posada, Johan Rockström, Herman E. Roo, Giovanni Ruggeri, Bernard Tardieu, Richard M. Taylor, Barbara van Koppen, Arthur Walz, Martin Wieland, Qiang Zhu and Przemyslaw Zielinski

Public online consultation on the table of contents, March 2008
Diepeveen Aleid, Abdullatif Al-Mugrin, Elfadil Azrag, Nick Blazquez, Marcia M. Brewster, Olga Daguia, Binayak Das, Orock Tanyi Fidelis, Mikkel Funder, Cristy Gallano, Andreas Grohmann, Alfred Heuperman, Peter Kabongo, Tom McAuley, F. H. Mughal, Farhad Mukhtarov, Kefah Naom, N. Parasuraman Ngappan, Cyprien Ntahomvukiye, Gerd Odenwaelder, Gbenga Olatunji, Michaela Oldfield, Ramadhan, Friederike Schubert, Paulo de Tarso Castro, Mase Toru, Nicola Tynan, Etiosa Uyigue, Hideo Watanabe, Maya Wolfensberger, Nayyer Alam Zaigham and the Gender and Water Alliance

Real-time Delphi survey on policy relevance, March 2008
Emaduddin Ahmad, Natalia Alexeeva, Ali Al-Jabbari, Elena Isabel Benitez Alonso, Miguel Angel, Lina Sergie Atassi, Manuel Rodríguez Becerra, Charlie Bepapa, Benedito Braga, Martina Bussettini, Mokhtar Bzioui, Adrian Cashman, Sharif Uzzaman Choudhury, Betsy A. Cody, Christopher Cox, Basandorj Davaagiin, Dwarika Dhugnel, Francis Flynn, Bertha Cruz Forero, Gerald Galloway, Iñaki Urrutia Garayo, Zaheer Hussain Shah Gardezi, Peter Gleick, Biksham Gujja, Handagama, Islam-ul-Haque, Kocou Armand Houanye, Mukdad Hussein, Upali Senarath Imbulana, Abbasgholi Jahani, Ananda Jayasinghe, Mohamed Ait Kadi, Badra Kamaladasa, Ville Keskisarja, Julio Thadeu S. Kettelhut, Arzel Hossain Khan, Juliette Biao Koudenoukpo, Latu S. Kupa, Juan Mayr Maldonado, Olga Marecos, Jurado Marquez, Polioptro F. Martínez-Austria, Miguel A. Medina, Jr., G. Tracy Mehan, A. M. Muller, Jadambaa Namjilin, Gustavo Victor Necco, Visa Niittyniemi, Ali Noorzad, Michel Ouellet, Marc Overmars, Mauri Cesar Barbosa Pereira, Claudia Patricia Mora Pineda, Giorgio Pineschi, Victor Pochat, Syed Ayub Qutub, Walid Abed Rabboh, Hifza Rasheed, Josu Sanz, Henk van Schaik, Carlos María Serrentino, Cletus Springer, Steven L. Stockton, Sumitha Sumanaweera, Vincent D. Sweeney, Muhammad Aslam Tahir, Sonja Timmer, Francesco Tornatore, Robert Reece Twilley, Carel de Villeneuve, Erik K. Webb, Cevat Yaman and Farhad Yazdandoost

Electronic survey for water leaders and water experts, July 2008
Sameh Mohamed Abdel-Gawad, Florence Grace Adongo, Emaduddin Ahmad, Abdalla A. Ahmed, Fernando Alberto, Sibel Algan, Daouda Aliou, Mirtha Almada, Hugo Pablo Amicarelli, Paula Antunes, Bayoumi Bayoumi Attia, Van Baardwijk, Banadda, Jayanta Bandyopadhyay, Elena Benitez, Emilia Bocanegra, Lisa Bourget, John Carey,

Adrian Cashman, Roberto Torres Castro, Lucas Chamorro, Xu Cheng, Mourad Choyekh, Murray Clamen, Michael J. Clark, Betsy A. Cody, Ken Conca, Filiz Demirayak, Carlos Diaz, Kayembe Ditanta, Ajaya Dixit, Ould Mohamed El Hacen Saadou Ebih, Omar Elbadawy, Evens Emmanuel, Loic Fauchon, Miriam Feilberg, Bertha Cruz Forero, Iñaki Urrutia Garayo, Roberto Galan Garcia, Elda Guadalupe Vasquez de Godoy, Elizabeth Granados, Norman Grannemann, Pilar Cornejo R. de Grunauer, Sylvain Guebanda, Guero, Adrian Ortega Guerrero, Biksham Gujja, G. J. C. Gunatilake, Carlos Gutiérrez-Ojeda, Dipak Gyawali, Charles Hakizimana, Azizul Haque, Islam-ul-Haque, Liu Heng, Oda Hideaki, Eduardo Zamudio Huertas, Magda Amin Idris, Upali S. Imbulana, Mulipola Pologa Ioane, Vijay Jagannathan, Jahani, Santiago Jara, H. M. Jayatillake, Gerald Jean-Batiste, Badra Kamaladasa, Vakup Karaaslan, Ville Keskisarja, Wael M. Khairy, Arzel Hossain Khan, Nguyen Hong Khanh, Abdelaziz H. Konsowa, Juan Jose Ledesma, Peter Letitre, Mark Limbaugh, Ana Deisy López Ramos, Lutfi Ali Madi, Yvon Maranda, Darysbeth Martinez, Andrés Pérez Mattiauda, Marcus Moench, Ekhlas Gamal Eldin Mohamed, David Molden, Sadí Laporte Molina, Isaìas Montoya B., Mike Muller, Hamza Ozguler, Gürcan Özkan, Eddy Gabriel Baldellón Pedraza, Amataga Penaia, Ralph Pentland, Mauri Cesar Barbosa Pereira, Andrés Pérez, Odalis Perez, Mathieu Pinkers, Syed Ayub Qutub, Walid Abed Rabboh, Santiago Maria Reyna, Decarli Rodríguez, Jorge Rucks, Jayampathy Samarakoon, Monica Elizabeth Urbieta Sanabria, João Bosco Senra, Carlos Maria Serrentino, José Joaquín Chacón Solano, Toshihiro Sonoda, Guido Soto, Hugo Herrera Soto, Steven L. Stockton, Sumitha Sumanaweera, Veronica Tarbaeva, U. Tsedendamba, Aynur Uluğtekin, Kishor Uprety, Jeroen van der Sommen, Ximena Vargas, Celso Velazquez, Ingrid Verstraeten, Carel de Villeneuve, Carissa Wong, Jorge Montaño Xavier, Alaa Yassin and Farhad Yazdandoost

UN-Water consultation, August-September 2008
Secretariat of the Convention on Biological Diversity, Food and Agriculture Organization of the United Nations, United Nations Environment Programme and World Bank

WWAP side publications series coordinator
Marwa Daoudy

Partners in WWDR Side Publications Series
Zafar Adeel, Yoganath Adikari, Joseph Alcamo, Maite Martinez Aldaya, Reza Ardakanian, Pierre Baril, Dominique Berteaux, Harriet Bigas, David Bird, Gunilla Björklund, Sylvie de Blois, Amadou Idrissa Bokoye, Sobhanlal Bonnerjee, Leon Braat, Marco Braun, Anne Cann, Diane Chaumont, Torkil-Jønch Clausen, David Coates, Jean-François Cyr, Claude Desjarlais, Paris Edwards, Marie-Joëlle Fluet, Louis-Guillaume Fortin, Gilberto Gallopín, Jerome Glenn, Matt Hare, Joakim Harlin, Jan Hassing, BertJan Heij, Andrew Hudson, Niels Ipsen, Harald Koethe, David Lammie, Henrik Larsen, Jan Leentvaar, Geerinck Lieven, Palle Lindgaard-Jørgensen, Manuel Ramon Llamas, Ralf Ludwig, Wolfram Mauser, Alastair Morrison, Jasna Muskatirovic, André Musy, Benjamin Ndala, Gernot Pauli, Alain Rousseau, René Roy, Brigitte Schuster, Lynette de Silva, Lucia De Stefano, Jon Martin Trondalen, Håkan Tropp, Richard Turcotte, Wim van der Hoek, Charlotte van der Schaaf, Luc Vescovi, Ruth Vollmer, Ian White, James Winpenny, Lars Wirkus, Aaron T. Wolf and Junichi Yoshitani

***World Water Development Report 3* Messages Series**
Coordinator
George de Gooijer

Contributors
Altay Altinörs, Ger Bergkamp, Claire Furlong, George de Gooijer, Dipak Gyawali, Jack Moss, Joshua Newton, Sharon Velasquez Orta, Darren Saywell, Alberto Tejada-Guibert and James Winpenny

Special thanks
The Turkish Ministry of Foreign Affairs, the Secretariat of the 5th World Water Forum (Istanbul), the Greater Municipality of Istanbul Water and Sewerage Administration and the General Directorate of State Hydraulic Works of Turkey

We apologize for any inadvertent errors or omissions of contributors to the Report. Some names may be incomplete because they come from participants' own online registration information, which may have been incomplete.

Overview of key messages

The amount of freshwater on Earth is finite, but its distribution has varied considerably, driven mainly by natural cycles of freezing and thawing and fluctuations in precipitation, water runoff patterns and evapotranspiration levels. That situation has changed, however. Alongside natural causes are new and continuing human activities that have become primary 'drivers' of the pressures affecting our planet's water systems. These pressures are most often related to human development and economic growth.

History shows a strong link between economic development and water resources development. There are abundant examples of how water has contributed to economic development and how development has demanded increased harnessing of water. Such benefits came at a cost and in some places led to increasing pressure on the environment and increasing competition among users. Our requirements for water to meet our fundamental needs and our collective pursuit of higher living standards, coupled with the need for water to sustain our planet's fragile ecosystems, make water unique among our planet's natural resources.

Important decisions affecting water management are made outside the water sector and are driven by external, largely unpredictable drivers – demography, climate change, the global economy, changing societal values and norms, technological innovation, laws and customs, and financial markets. Many of these external drivers are dynamic and changing at a faster pace. Developments outside the water domain influence water management strategies and policies. Decisions in other sectors and those related to development, growth and livelihoods need to incorporate water as an integral component, including responses to climate change, food and energy challenges and disaster management. The analysis of these issues leads to a set of responses and recommendations for action that incorporate the contribution of water to sustainable development.

Chapter 1. Getting out of the box – linking water to decisions for sustainable development

The news media today are full of talk of crises – in climate change, energy and food supplies and prices, and troubled financial markets. These global crises are linked to each other and to water resources management. They arise against a background of continuing poverty for a large part of the world. Unless resolved, they may lead to increasing political insecurity and conflict at local and national levels.

- The 'water box' dilemma must be resolved. Leaders in the water sector – in water supply and sanitation, hydropower, irrigation and flood control – have long been aware that water is essential to sustainable development, but they do not make the decisions on development objectives and the allocation of human and financial resources to meet them. These decisions are made or influenced by leaders in government, the private sector and civil society, who must learn to recognize water's role in obtaining their objectives.
- Water is essential for achieving sustainable development and the Millennium Development Goals. Properly managing water resources is an essential component of growth, social and economic development, poverty reduction and equity – all essential for achieving the Millennium Development Goals.

Leaders in the water sector have long been aware that water is essential to sustainable development, but they do not make the decisions on development objectives and the allocation of human and financial resources to meet them. These decisions are made by leaders in government, the private sector and civil society

Alongside the natural forces affecting water resources are new human activities that have become the primary 'drivers' of the pressures affecting our planet's water systems

- Water is linked to the crises of climate change, energy and food supplies and prices, and troubled financial markets. Unless their links with water are addressed and water crises around the world are resolved, these other crises may intensify and local water crises may worsen, converging into a global water crisis and leading to political insecurity and conflict at various levels.

Specialists and managers in water supply and sanitation, hydropower, irrigation and flood control have long been aware of this. But they often have a narrow, sectoral perspective that blinds many decisions on water. And they do not make the decisions on development objectives and financial resources needed to meet these broader objectives.

Action is required now. Lives and livelihoods depend on water for development. After decades of inaction, the problems are enormous. And they will worsen if left unattended. But while the challenges are substantial, they are not insurmountable. The Report has examples of how some countries and regional and local governments have solved similar challenges. Recognizing the links between water resources and other crises around the world and between water resources and development, leaders in the water domain and decision-makers outside it must act together now to meet these challenges.

Part 1. Understanding what drives the pressures on water

Alongside the natural forces affecting water resources are new human activities that have become the primary 'drivers' of the pressures affecting our planet's water systems. These pressures are most often related to human activities and economic growth. Our requirements for water to meet our fundamental needs and our collective pursuit of higher living standards, coupled with the need for water to sustain our planet's fragile ecosystems, make water unique among natural resources.

Drivers should not be considered in isolation of related socioeconomic and political factors or of other drivers. Many natural links also influence how drivers affect changes, directly and indirectly. Water properties are governed by biological, chemical and physical laws that define the quantity and quality of water resources, regardless of human influences, and that are linked in various ways. Superimposed on these natural processes are human activities that intensify these processes and disrupt the natural balance of water systems.

Economic growth, a principal driver of water use, is affected by a wide range of policy decisions, from international trade to education and public health, while the potential rate of economic growth can be affected by demographic variables such as population distribution (local workforce availability) and social characteristics (workforce capacity and the role of women) and by the availability of new technologies. Water availability is also directly subject to the impacts of climate change, which also can exert additional pressures on the other drivers.

The result of these combined and interacting forces is a continuously increasing demand for finite water resources for which there are no substitutes. When water resources of acceptable quality can no longer be provided in sustainable quantities, the outcome can be overexploitation of aquatic ecosystems. The ultimate losers are the exploited aquatic ecosystems and the organisms (including humans) dependent on them for survival and well-being.

Chapter 2. Demographic, economic and social drivers

Human activities and processes of all types – demographic, economic and social – can exert pressures on water resources and need to be managed. These pressures are in turn affected by a range of factors such as technological innovation, institutional and financial conditions and climate change.

Demographic drivers. Population dynamics (growth, gender and age distribution, migration) create pressures on freshwater resources through increased water demands and pollution. Changes in the natural landscape associated with population dynamics (migration, urbanization) can create additional pressures on local water resources and the need for more water-related services.

Economic drivers. Growth and changes in the global economy are having far-reaching impacts on water resources and their use. Growing international trade in goods and services can aggravate water stress in some countries while relieving it in others through flows of 'virtual water' (water embedded in products and used in their production, particularly in the form of imported agricultural commodities).

Social drivers. Social drivers are mainly about individual rather than collective actions and about the way people think and act on a day-to-day basis. Social drivers influence human perceptions and attitudes about the environment, including water resources, in turn influencing the pressures people exert on water through water demands and uses. Changes in lifestyles represent one of the principal drivers of change. They reflect human needs, desires and attitudes (as illustrated in consumption and production patterns), which are influenced by such social drivers as culture and education and by economic drivers and technological innovation; the rapid global rise in living standards combined with population growth presents the major threat to the sustainability of water resources and the environment.

Chapter 3. Technological innovation
Technological innovation is driven largely by both human wants and needs. It can create both positive and negative pressures, sometimes simultaneously, resulting in increased or decreased water demand, supply and quality. One of the most unpredictable drivers, technological innovation can create rapid, dramatic and unexpected changes, both in pressures and solutions. Impediments to the dissemination of technology must be overcome for developing countries to benefit from innovations developed in richer countries.

Chapter 4. Policies, laws and finance
Efforts to implement water management effectively and efficiently and to properly inform the decision-making process are facilitated by the adoption of water resources management laws, policies and strategies that reflect links between water and the social and economic sectors. Good examples can be found in many countries.

But even if all the necessary policies and laws are in place, development of water resources will not take place without adequate funding of infrastructure and the institutional and human capacity of the sector.

Policies and laws. Effective policies and legal frameworks are necessary to develop, carry out and enforce the rules and regulations that govern water use and protect the resource. Water policy operates within a context of local, national, regional and global policy and legal frameworks that must all support sound water management goals.

Legitimate, transparent and participatory processes can effectively mobilize input for designing and implementing water resources policy and create a strong deterrent to corruption. Corruption remains a poorly addressed governance issue in the water domain. It can lead to uncontrolled pollution of water sources, overpumping and depletion of groundwater, lack of planning, degradation of ecosystems, weakened flood protection, urban expansion leading to heightened water tensions, and other harmful effects.

Finance. Although water is often described as a 'gift of nature', harnessing and managing it for the wide variety of human and ecological needs entail financial costs. While there may appear to be many financing options for water resources development, governments still have only three basic means of financing them: tariffs, taxes and transfers through external aid and philanthropy.

Although water is often described as a 'gift of nature', harnessing and managing it for the wide variety of human and ecological needs entail financial costs

Policy-makers need to make political decisions on socially and environmentally acceptable trade-offs among different objectives and on who bears the costs of such compromise. Commitments have been made by the donor community to increase assistance to the broad water sector, but this has led mainly to an increase in allocations for water supply and sanitation in dollar terms (although its share of total official development assistance has stagnated at 4%), and the percentage of total aid allocated to the water sector remains below 6% and has been declining.

Chapter 5. Climate change and possible futures
The external drivers of change, strongly connected, create complex challenges and opportunities for water managers and decision-makers in government, the private sector and civil society. Climate change and variability, while seldom the main stressors on sustainable development, can impede or even reverse development gains.

Climate change. There is evidence that the global climate is changing and that some of the change is human-induced. The main impacts of climate change on humans and the environment occur through water. Climate change is a fundamental driver of changes in water resources and an additional stressor through its effects on external drivers. Policies and practices for mitigating climate change or adapting to it can have impacts on water resources, and the way we manage water can affect the climate.

Public policy, so far dominated by mitigation, could benefit from a better balance between mitigation and adaptation. Carbon is a measure of the anthropogenic causes of climate change – water is a measure of its impacts. The international community also has to balance investing for tomorrow's likely problems of greater climate variability and global warming against investing for today's problems of climate variability to

Steadily increasing demand for agricultural products to satisfy the needs of a growing population, and the desire for a more varied diet, continues to be the main driver behind water use

prevent losses from droughts and floods. While both are vital, focusing on today's problems can also create greater resilience for dealing with the problems of tomorrow.

Possible futures. Each of the external water drivers is dynamic and continues to evolve, as do the direct and indirect pressures they exert on water resources. Thus, it is difficult to draw a comprehensive picture of the future by examining each driver independently. Because the drivers can have even more of an impact on future water resources collectively than they can individually, future scenarios that consider these interactions offer a more holistic picture. Existing global water scenarios are outdated, incomplete or sectoral and do not fully incorporate each of the external drivers. The evolution of the drivers and the logic behind their storylines need to be examined and possibly redefined in view of developments both inside and outside the water sector that have occurred over the past decade.

Part 2. Using water

History shows a strong link between economic development and water resources development. There are abundant examples of how water has contributed to economic development and how development has demanded increased harnessing of water. Steadily rising demand for agricultural products to satisfy the diverse needs of growing populations (for food, fibre and now fuel) has been the main driver behind agricultural water use.

The effects of water-depleting and water-polluting activities on human and ecosystem health remain largely unreported or difficult to measure, and the need grows stronger for effective protection of ecosystems and the goods and services they produce – on which life and livelihoods depend. As competition among demands on water increases, society will need to respond with improved water management, more effective policies and transparent and efficient water allocation mechanisms.

Chapter 6. Water's many benefits

Water has always played a key role in economic development, and economic development has always been accompanied by water development. Investment in water management has been repaid through livelihood security and reductions in health risks, vulnerability and ultimately poverty. Water contributes to poverty alleviation in many ways – through sanitation services, water supply, affordable food and enhanced resilience of poor communities faced with disease, climate shocks and environmental degradation. Water of the right quality can improve health through better sanitation and hygiene and, when applied at the right time, can enhance the productivity of land, labour and other productive inputs. In addition, healthy freshwater ecosystems provide multiple goods and services essential to life and livelihood.

The importance of water services is especially apparent in societies where normal social life and political structures have broken down. In these fragile states the government cannot or will not deliver core functions to most of its people, including the poor. While each fragile state is fragile in different ways and for different reasons – war, post-conflict recovery, major natural catastrophe, prolonged mismanagement and political repression – a striking commonality in reports from aid agencies is the prominence of water and sanitation in relief and reconstruction programmes. The rapid restoration of viable water services is often a crucial ingredient of nation-building in these fragile states.

Chapter 7. Evolution of water use

While most of the old challenges of water supply, sanitation and environmental sustainability remain, new challenges such as adaptation to climate change, rising food and energy prices, and ageing infrastructure are increasing the complexity and financial burden of water management. Population growth and rapid economic development have led to accelerated freshwater withdrawals.

Trends in access to domestic water supply indicate substantial improvement in the past decade, putting most countries on track to achieve the water supply target of the Millennium Development Goals. However, sanitation is lagging well behind, and most sub-Saharan African countries and many rural areas still show unsatisfactory records for both water supply and sanitation.

Steadily increasing demand for agricultural products to satisfy the needs of a growing population continues to be the main driver behind water use. While world population growth has slowed since the 1970s and is expected to continue its downward trend, steady economic development, in particular in emerging market economies, has translated into demand for a more varied diet, including meat and dairy products, putting additional pressure on water resources.

After agriculture, the two major users of water for development are industry and energy (20% of total water withdrawals), which are transforming the patterns of water use in emerging market economies. Water and energy share the same drivers: demographic, economic, social and technological processes put pressure on both energy and water. The recent acceleration in the production of biofuel and the impacts of climate change bring new challenges and add to the pressures on land and water resources.

Freshwater ecosystems provide an extensive array of vital services to support human well-being. A variety of economic and recreational activities such as navigation, fisheries and pastoral activities depend on direct use of water in healthy ecosystems. Yet some environmental services receive inadequate policy attention and are endangered by the way development sectors use water.

Chapter 8. Impacts of water use on water systems and the environment
The pattern and intensity of human activity have disrupted – through impacts on quantity and quality – the role of water as the prime environmental agent. In some areas depletion and pollution of economically important river basins and associated aquifers have gone beyond the point of no-return, and coping with a future without reliable water resources systems is now a real prospect in parts of the world.

While the intensity of groundwater use, partly encouraged by subsidized rural electrification, has led to the emergence of many groundwater-dependent economies, their future is now threatened by aquifer depletion and pollution. Prospects for relaxing use of these key aquifers, remediating water quality and restoring groundwater services to ecosystems look remote unless alternative management approaches are developed.

Our ability to maintain the environmental services we depend on has improved but remains constrained by an incomplete understanding of the magnitude and impact of pollution, the resilience of affected ecosystems and the social institutions that use and manage water resources systems. A failure to monitor the negative impacts of water use on the environment and institutional weaknesses in many developing countries prevent effective enforcement of regulatory provisions.

Relevant information about pollution loads and changes in water quality is lacking precisely where water use is most intense – in densely populated developing countries. As a result, the often serious impacts of polluting activities on the health of people and ecosystems remain largely unreported. Still, there are signs of progress in how pollution and the risks of pollution can be mitigated and trends in environmental degradation reversed.

Chapter 9. Managing competition for water and the pressure on ecosystems
Competition for water and shortcomings in managing it to meet the needs of society and the environment call for enhanced societal responses through improved management, better legislation and more effective and transparent allocation mechanisms.

Challenges include wise planning for water resources, evaluation of availability and needs in a watershed, possible reallocation or storage expansion in existing reservoirs, more emphasis on water demand management, a better balance between equity and efficiency in water use, inadequate legislative and institutional frameworks and the rising financial burden of ageing infrastructure.

Water management choices should emerge from informed consultation and negotiation on the costs and benefits of all options after considering basin interconnectedness, relationships between land and water resources, and the consistency and coherence of decisions with other government policies.

Part 3. State of the resource

The uneven distribution over time and space of water resources and their modification through human use and abuse are sources of water crises in many parts of the world. In many areas hydrologic extremes have increased. Deaths and material damage from extreme floods can be high, and more intense droughts, affecting increasing numbers of people, have been observed in the 21st century. Worldwide, water observation networks are inadequate for current and future management needs and risk further decline. There are insufficient data to understand and predict the current and future quantity and quality of water resources, and political protocols and imperatives for sharing data are inadequate.

Chapter 10. The Earth's natural water cycles
Water resources are made up of many components associated with water in its three physical states (liquid, solid and gas). The components of the water cycle (rainfall,

Water and energy share the same drivers: demographic, economic, social and technological processes put pressure on both energy and water

> **Most climate scientists agree that global warming will result in an intensification, acceleration or enhancement of the global hydrologic cycle, and there is some observational evidence that this is already happening**

evaporation, runoff, groundwater, storage and others) therefore all differ in their chemical and biochemical qualities, spatial and temporal variability, resilience, vulnerability to pressures (including land use and climate change), susceptibility to pollution and capacity to provide useful services and to be used sustainably. A consequence of this variability is that while human pressures have resulted in large modifications to the global hydrologic cycle, the directions and degrees of change are complex and difficult to ascertain. The uneven distribution of water resources over time and space and the way human activity is affecting that distribution today are fundamental sources of water crises in many parts of the world. Adding complexity, climate change and variability also influence the water supply, demand and buffering system, although their precise impacts can be difficult to isolate.

Chapter 11. Changes in the global water cycle

Most climate scientists agree that global warming will result in an intensification, acceleration or enhancement of the global hydrologic cycle, and there is some observational evidence that this is already happening. While trends in precipitation have been noted in some parts of the world, in other areas precipitation patterns have remained about the same within the period of observed data. Changes have been observed in snow cover extent and snow water equivalent and in the frequency with which precipitation falls as snow. More than 15% of the world's population live where water resources availability depends heavily on snowmelt from ephemeral snowpacks or perennial glaciers. Despite the evidence of temperature changes, there is little evidence of detectable changes in evaporation and evapotranspiration.

Climate change is being superimposed on an already complex hydrologic landscape, making its signal difficult to isolate, and yet making its influence felt throughout the water supply, demand and buffering system. Data limitations in record length, continuity and spatial coverage contribute to the uncertainty, while natural climate variability and multiyear variability associated with large-scale atmospheric circulation patterns influence the interpretation of many trends in ways that are not yet fully understood.

Despite the limitations of global datasets, many studies have shown changes in runoff and streamflow. Many have focused on low (drought) or high (flood) extremes. Except in regions with flows affected by glacier meltwater, the general conclusion is that global trends are not present or cannot be detected at this stage, although climate change-related trends are evident in some regions. Groundwater resources have been heavily used for human supply and agriculture for many years. While many groundwater abstraction schemes access fossil water (water unrelated to current conditions), renewable groundwater resources depend on highly variable recharge volumes.

It is thus realistic to expect future recharge regimes to reflect changes in the driving hydrologic processes (such as precipitation and evapotranspiration) that might result from anticipated climate changes. It is increasingly clear that the assumption of statistical stationarity is no longer a defensible basis for water planning.

Among the consequences of a changing hydrologic cycle is its interaction with the terrestrial carbon cycle. The terrestrial biosphere may have taken up roughly 25% of anthropogenic carbon emissions during the last century; it is unclear how long this can continue.

Chapter 12. Evolving hazards – and emerging opportunities

Water-related hazards can be naturally occurring or anthropogenic. Hazards can result from too much water (floods, erosion, landslides and so on) or too little (droughts and loss of wetlands or habitat) and from the effects of chemical and biological pollution on water quality and in-stream ecosystems. The natural variability of water resources and changes, whatever the cause, can provide opportunities for management strategies to respond to potential climate change threats by implementing more resource-sustainable policies and practices.

In many places climate-related water events have become more frequent and more extreme. In developing countries extreme floods can result in many deaths, while in developed countries they can result in billions of dollars in damages. More intense droughts in the past decade, affecting an increasing number of people, have been linked to higher temperatures and decreased precipitation but are also frequently a consequence of the mismanagement of resources and the neglect of risk management. The increased exposure to potential climate change hazards has led to more awareness of water resources management.

Changes in flow and inputs of chemical and biological waste from human activity have altered the water quality and ecological functioning of many of the world's rivers. Global warming is expected to have substantial effects on energy flows and matter recycling through its impact on water temperature, resulting in algal blooms, increases in toxic cyanobacteria bloom and reductions in biodiversity.

In areas of increasing water stress groundwater is an important buffer resource, capable of responding to increased water demands or of compensating for the declining availability of surface water.

Chapter 13. Bridging the observational gap

Worldwide, water observation networks provide incomplete and incompatible data on water quantity and quality for properly managing water resources and predicting future needs – and these networks are in jeopardy of further decline. Also, no comprehensive information exists on wastewater generation and treatment and receiving water quality on a regional or global scale. While new technologies based on satellite remote sensing and modelling present opportunities, their value is limited by our ability to ground-truth and validate the simulated information.

Management of the world's water resources requires reliable information about the state of the resource and how it is changing in response to external drivers such as climate change and water and land use. There is little sharing of hydrologic data, due largely to limited physical access to data, policy and security issues; lack of agreed protocols for sharing; and commercial considerations. This hampers regional and global projects that have to build on shared datasets for scientific and applications-oriented purposes, such as seasonal regional hydrologic outlooks, forecasting, disaster warning and prevention, and integrated water resources management in transboundary basins.

Improving water resources management requires investments in monitoring and more efficient use of existing data, including traditional ground-based observations and newer satellite-based data products. Most countries, developed and developing, need to give greater attention and more resources to monitoring, observations and continual assessments of the status of water resources.

Part 4. Responses and choices

We have many of the answers. Across the planet we have already shown that it can be done! But there is no one-size-fits-all solution. The best mix of responses to a country's development objectives and policy priorities to meet its water challenges depends on the availability of water in space and in time and the country's technical, financial, institutional and human capacities – its culture, political and regulatory frameworks, and markets.

Options within the water domain are distinct from those outside it. Leaders in the water domain can inform the processes outside their domain and implement decisions for the water domain; but it is the leaders in government, the private sector and civil society who determine the directions that will be taken. Responses outside the water domain strongly affect the macro changes that influence how water is used and allocated. They also make water adaptation measures more (or less) effective and less (or more) costly.

Many countries face multiple challenges but have limited financial and natural resources and implementation capacities. Countries need to fully use synergy opportunities and to make trade-offs and difficult decisions on how to allocate among uses and users to protect their water resources. To achieve results, many actors need to participate in these decisions.

Chapter 14. Options inside the water box

There are many practical examples of solutions within the water domain. Some options show particular promise. Preparing institutions to deal with current and future challenges requires support for institutional development through such reforms as decentralization, stakeholder participation and transparency, increased corporatization where feasible and fair, partnerships and coordination (public-private, public-public, public-civil society), and new administrative systems based on shared benefits of water, including when water crosses borders. Decision-makers need to consider the influence of water law, both formal and customary, including regulations in other sectors that influence the management of water resources.

Decision-making is improved by consulting with stakeholders and ensuring accountability in planning, implementation and management as well as building trust within the water and related sectors and fighting corruption and mismanagement. Strengthening organization structures and improving the operating efficiency of water supply utilities will help to improve service

Worldwide, water observation networks provide incomplete and incompatible data on water quantity and quality for properly managing water resources and predicting future needs – and these networks are in jeopardy of further decline

Unsustainable management and inequitable access to water resources cannot continue. We might not have all the information we would like to have before acting, but we do know enough now to begin to take significant steps

quality and increase the coverage and density of connections, while also boosting revenues and creating a more viable financial base to attract further investment.

Innovation and research are critical for developing appropriate solutions. And greater institutional capacity and human capacity are needed, both within the water domain and in areas or sectors outside the water domain. Capacity development can occur through traditional forms of education, on-the-job training, e-learning, public awareness raising, knowledge management and professional networks.

Sound management accountability and good governance within the water sector contribute to creating a favourable investment climate. This should include new approaches such as payment for environmental services.

Chapter 15. Options from beyond the water box

Dealing with risk and uncertainty has long been a routine challenge for water resources managers and policy-makers across sectors and the world. However, issues like climate change and demographic dynamics have made the risks greater and the task more complex. Risk management is now much more important – indeed essential – to analysis and decision-making.

Drivers and policies outside the water sector have more impact on water management than do many policies championed and implemented by water-related ministries. Identifying trade-offs and synergies between water and other policy sectors can enhance policy impacts in all sectors and avoid some adverse effects on water. Because governments, civil society and business leaders make decisions every day that can affect water, it is important to identify where such decisions can also lead to improvements in water sector management and in water sector and environmental services.

Examples of win-win situations abound – whether created by governments, communities or businesses – that point to promoting deliberate cooperation between water and non-water actors and integrating water issues into external decisions. International organizations, notably the UN system, can provide support and expertise to governments, help civil society build capacity and catalyse leadership in the private sector.

Chapter 16. The way forward

Water and water systems must be managed to achieve social and economic development objectives and to sustain development. Water resources, properly managed, are critical to the survival and well-being of individuals. They can ensure equity and security in water and sanitation for families, businesses and communities. And they can ensure adequate water for food, energy and the environment as well as protection from floods and droughts.

Decision-making on water requires seeking synergies and selecting appropriate trade-offs. It also requires distinguishing between short-term 'fire-fighting' – responding to the urgent issues of the day – and long-term strategic development. Developing multipurpose water schemes and reusing water wherever feasible can lessen the need for trade-offs by enabling the same volumes of scarce water to deliver multiple outcomes.

The donor community can incorporate water into the broader frameworks of development aid and focus assistance on areas where it is needed most – in sub-Saharan Africa, in Asian and Latin American slums and in states recovering from conflict. Recent G-8 efforts in this direction are promising.

The chief executives of the UN agencies, following the example of their joint discussions of and collective responses to climate change, can convene to examine the role of water, water systems and water management in development and environmental services, providing direction to agencies and advice to member countries.

The World Water Assessment Programme and its partners are working to help reduce uncertainty, facilitate decision-making and accelerate investment by highlighting the links between socioeconomic development and investment in water management capacity and infrastructure in other sectors.

The challenges are great, but unsustainable management and inequitable access to water resources cannot continue. We might not have all the information we would like to have before acting, but we do know enough now to begin to take significant steps. Actions must include increased investment in water infrastructure and capacity development. Leaders in the water domain can inform the processes outside their domain and manage water resources to achieve agreed socioeconomic objectives and environmental integrity. But leaders in government, the private sector and civil society will determine the direction that actions take. Recognizing this responsibility, they must act now!

Water in a changing world

Chapter

1 Getting out of the box – linking water to decisions for sustainable development

Chapter 1 Coordinator
Olcay Ünver
(WWAP)

Facilitator
William Cosgrove

Chapter 1

Getting out of the box – linking water to decisions for sustainable development

Authors: Andy Bullock, William Cosgrove, Wim van der Hoek, James Winpenny
Contributors: Gerry Galloway, BertJan Heij, Molly Hellmuth, Jack Moss, Monica Scatasta
Coordinator: Olcay Ünver (WWAP)
Facilitator: William Cosgrove

Key messages

- The 'water box' dilemma must be resolved. Leaders in the water sector – in water supply and sanitation, hydropower, irrigation and flood control – have long been aware that water is essential to sustainable development, but they do not make the decisions on development objectives and the allocation of human and financial resources to meet them. These decisions are made or influenced by leaders in government, the private sector and civil society, who must learn to recognize water's role in obtaining their objectives.

- Water is essential for achieving sustainable development and the Millennium Development Goals. Properly managing water resources is an essential component of growth, social and economic development, poverty reduction and equity, and sustainable environmental services – all essential for achieving the Millennium Development Goals.

- Water is linked to the crises of climate change, energy and food supplies and prices, and troubled financial markets. Unless their links with water are addressed and water crises around the world are resolved, these other crises may intensify and local water crises may worsen, converging into a global water crisis and leading to political insecurity and conflict at various levels.

The media today are full of talk of crises – in climate change, energy and food supplies and prices, and troubled financial markets. These global crises are linked to each other and to water resources. Unless resolved, they may lead to increasing political insecurity and conflict at local and national levels.

These crises arise against a background of continuing poverty for much of the world. Managing water resources is essential to social and economic development, poverty reduction and equity and to achieving the Millennium Development Goals. Sustainable development depends on managing the costs of service provision using existing infrastructure along with additional investments in new water infrastructure and rehabilitation, both physical and institutional.

Specialists and managers in water supply and sanitation, hydropower, irrigation and flood control have long been aware that water is essential to sustainable development. But they often have a narrow, sectoral perspective that blinds many decisions on water. And they do not make the decisions on development objectives and the allocation of human and financial resources needed to meet these broader objectives. These decisions are made or influenced by leaders in government, the private sector and civil society. These leaders must learn to recognize water's role in attaining their objectives and act accordingly.

An understanding of water issues and of the support needed for investments, institutions, incentives, information and capacity inside the 'water sector' requires partnerships between those responsible for the economy-wide benefits of water and those responsible for managing water

And they must act in a changing world, a world driven by forces that they often do not control – forces of demography, the global economy, changing societal values and norms, technological innovation, international law, financial markets and climate change.

Opening the water box

Until the 1990s (and continuing in some countries) water subsectors generally worked independently, with specialists in water supply and sanitation, hydropower, irrigation, flood control and so on interacting very little.[1] As population growth and other pressures on water ('water drivers') brought more and more basins near closure (the allocation of all of the water in a basin), the need to manage water across subsectors at the basin level became evident. Water management was expanded during the 1990s to incorporate efficient water use, equitable sharing of benefits, and environmental sustainability – what came to be called integrated water resources management. And in 2002 the World Summit on Sustainable Development in Johannesburg set for all countries the goal to develop integrated water resources management plans by 2005.

Many countries are applying integrated water resources management at the basin level. But management is still largely confined to the water sector, where it is well understood that water is essential to all life on the planet (human and other species) and to human livelihoods. The sector is beginning to recognize that decisions by people outside the water sector determine how water will be used, but the other sectors are seen as cross-cutting in water management. The approach within the sector has been to invite those working in other socioeconomic sectors to join in integrated water resources management. But the societal and political questions that determine the real allocation and management of water resources also need to take into account the technological aspects of integrated water resources management.

The sphere of decision-making and the water box

Within government, water use is decided by the interaction of decision-makers in the main socioeconomic sectors – health, education, agriculture, housing, industry, energy, economic development and environment. In many countries this interaction occurs through a cabinet of ministers presided over by the prime minister or president. Parallel mechanisms may exist at a regional, state (provincial) or local (municipal) government level. The role of these government structures is critical in water management.

In many countries government directly controls only a small fraction of investments in the economy, but it determines the conditions that will attract or discourage investment. To be most effective, decisions should be taken through an interactive process that involves leaders in business (finance, industries, commerce) and civil society (community-based organizations and other non-governmental organizations).

Ideally, government, business and civil society leaders would work together in the interest of society. Because of the implications of their decisions for water use, an understanding of water issues and of the support needed for investments, institutions, incentives, information and capacity inside what has traditionally been considered the 'water sector' requires partnerships between those responsible for the economy-wide benefits of water and those responsible for managing water. Leaders in the water sector must thus ensure that these leaders outside the 'water box' know the constraints and options for water resources and help them implement their decisions efficiently and effectively.

Among the decisions that affect water the most are those relating to how a country meets its objectives for energy and food security, employment, disaster preparedness, environmental sustainability and other societal goals. These decisions are made in broader political frameworks and not by water managers, who subsequently deal with their implications for water and with other outcomes that touch on water. Figure 1.1 illustrates this process.

Outside the water sector is an area of synergy, tradeoffs, coordination and integration, involving higher-level, multisectoral decision-making processes. Water professionals, stakeholders and individuals can inform and influence decisions in this area, affecting outcomes. But they need to have a seat at the decision-making table and to respond by implementing water management effectively and efficiently and by properly informing the decision-making process. These efforts are facilitated in the many countries that have adopted water resources management laws, policies or strategies that reflect links between water and the social and economic sectors.

Figure 1.1 **Decision-making affecting water**

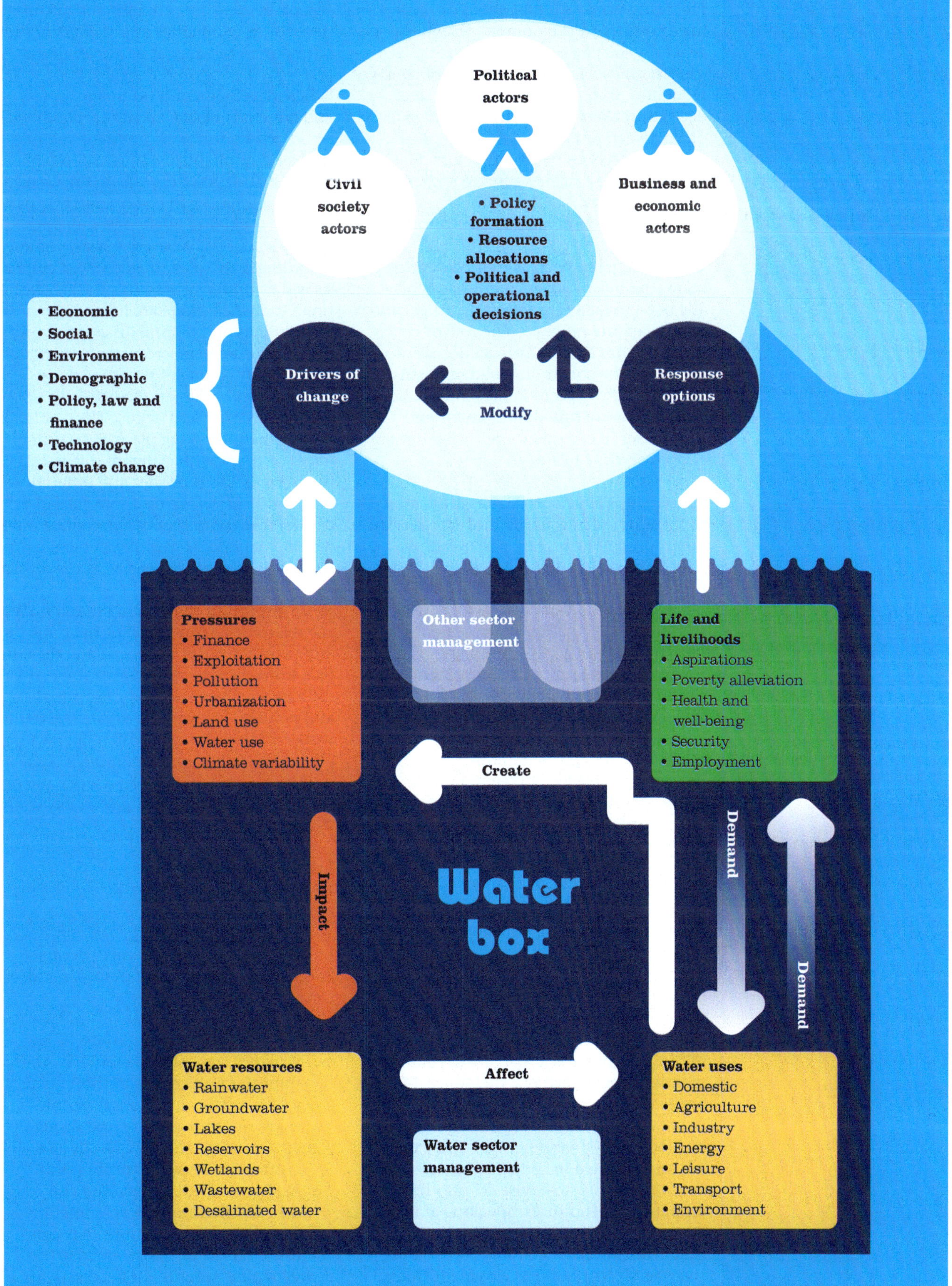

Source: Authors' construction.

Everywhere decisions related to development of necessity incorporate water development decisions, whether explicitly recognized or not

Decision-makers and water management

Providing water is but a means of achieving a country's development objectives – generally job creation, food security, GDP growth and social goals including poverty reduction. In pursuing these objectives, decision-makers are challenged by trade-offs between possible investments and possible synergies between sectors. Making trade-offs and searching for synergies require cooperation between those responsible for different sectors of the economy.

Where there has been sustained development, the role of government has generally been to facilitate action by others and to regulate the process.[2] The role of water managers has been to inform decision-makers of the constraints and opportunities of water resources management and water infrastructure development and then to act in accordance with the national development strategy.

Partnerships have been strongly promoted in the water sector, particularly for service provision. Public-private partnerships have been the predominant model, some functioning as intended, and some with mixed impacts. Water user associations in participatory irrigation management have become widespread in a number of countries, with some success in improving irrigation scheme management. But whether the operator is a private company, a public corporation or a municipal service, the successes have clearly demonstrated the importance of the complementary roles of public decision-makers and authorities on the one side and service operators on the other. In the long-term neither can succeed without the other.

Other types of partnerships include civil society organizations, municipalities and the private sector. A recent study on Latin America concluded that proper institutional frameworks, incentives and mutual trust are keys to successful partnerships.[3] River basin organizations are increasingly playing an important role. Broad coalitions of development partners, including different levels of government; donors; multinational, international and regional agencies; and local non-governmental organizations are being created in some countries, such as Mozambique,[4] to advise on priorities for public expenditures. Speaking at the Davos economic summit in January 2008, U.K. Prime Minister Gordon Brown said that the Millennium Development Goals will not be met 'unless there is a private, voluntary and government partnership'.[5] He added that 'governments have to understand that they have to make it possible for companies to affect change' and at times have to see companies as providers not just of resources but also of resourcefulness.

Where development is occurring rapidly and growth is viable, greater emphasis will be on private sector engagement and market-based mechanisms. Where development is slower and growth prospects are weaker, greater emphasis will be on providing basic services, including safety nets targeting society's poorest. Where governments and institutions are weak (fragile states) emphasis will be on reconstruction and rehabilitation. And where there are humanitarian crises, conflicts and natural disasters, emphasis will be on emergency responses. Working across many countries simultaneously, regional approaches emphasize integration, regional security and equity. Thus, although development is taking place in very different settings, with different integrating frameworks and processes and different sets of actors, everywhere decisions related to development of necessity incorporate water development decisions, whether explicitly recognized or not.

More important than trying to quantify the relative 'market share' of the public and private sectors is recognizing that they face similar challenges, constraints and difficulties. The task for decision-makers and political leaders is to create the framework conditions under which operators of all kinds – public, private, mixed, community providers and others – can provide services and investments effectively over the long term.

Sustainable development as the framework for water management

In the overview of *The Growth Report* of 2008 the Commission on Growth and Development argues that

> Growth is not an end in itself. But it makes it possible to achieve other important objectives of individuals and societies. It can spare people *en masse* from poverty and drudgery. Nothing else ever has. It also creates the resources to support health care, education, and the other Millennium Development Goals to which the world has committed itself. In short, we take the view that growth is a necessary, if not sufficient, condition for broader development, enlarging the scope

> for individuals to be productive and creative.[6]

Sustained growth requires water

Growth requires access to natural resources. *The Growth Report* acknowledges that we may be entering a period in which natural resources, broadly defined, impose new limits on growth. But the report makes no major reference to the essential role of water resources. *World Water Development Report 3*, which places more emphasis on development than its predecessors, makes the case that the availability of water resources and their management are determinants of a country's growth strategy.

Africa provides a good example because both growth and water are major challenges there. The African heads of state recognized the importance of water to development when they gathered in Sharm el-Sheikh, Egypt, in mid-2008 and adopted a declaration explicitly noting the role of water as a key to sustainable development in the region (box 1.1).

Societies do not become wealthy first and then invest in water management; they find ways to manage water and risk first, which then leads to wealth. If they are wise, they do this in a way that avoids pollution, cares for equity and otherwise ensures the sustainability of the resource.

Investment in water infrastructure is required to meet basic needs in rural areas and to enhance agricultural productivity through better management of water. As development proceeds, with the shift to commercial and industrial activities in urban areas, water has to be managed for energy and food production, transportation, flood control, and drinking water and sanitation, as well as for industrial and commercial activities.

Asian Water Development Outlook 2007 highlights the significant global development challenge this represents.[7] That report emphasizes a 'multidisciplinary and multi-sector perspective [on water] around the Asia and Pacific region' in facing the challenges of sustaining growth. It highlights

> important topics that have been neglected or are being inadequately considered in most countries of the region. Among these is the urgent need to address the inherent interrelationships between water and other important development-related sectors, like energy, food, and the environment.

It has little in the way of a detailed roadmap for water resources development, however.

Box 1.1 Commitment of African heads of state to water as a key to sustainable development

WE, the Heads of State and Government of the African Union, meeting at the 11th Ordinary Session of our Assembly in Sharm el-Sheikh, Arab Republic of Egypt, from 30 June to 1 July 2008,

Recognizing the importance of water and sanitation for social, economic and environmental development of our countries and Continent; . . .

Recognizing that water is and must remain a key to sustainable development in Africa and that water supply and sanitation are prerequisites for Africa's human capital development;

Concerned that there is an under-utilization and uneven sharing of water resources in Africa, and that remains a growing challenge in the achievement of food and energy securities. . . .

WE COMMIT OURSELVES TO:

(a) *Increase* our efforts to implement our past declarations related to water and sanitation.

(b) *Raise* the profile of sanitation by addressing the gaps in the context of the 2008 eThekwini Ministerial Declaration on sanitation in Africa adopted by [the African Ministers Council on Water].

(c) *Address* issues pertaining to agricultural water use for food security as provided for in the Ministerial Declaration and outcomes of the first African Water Week.

And particularly;

(d) *Develop and/or update* national water management policies, regulatory frameworks, and programmes, and prepare national strategies and action plans for achieving the [Millennium Development Goal] targets for water and sanitation over the next seven (7) years;

(e) *Create* conducive environment to enhance the effective engagement of local authorities and the private sector;

(f) *Ensure* the equitable and sustainable use, as well as promote integrated management and development, of national and shared water resources in Africa;

(g) *Build* institutional and human resources capacity at all levels including the decentralized local government level for programme implementation, enhance information and knowledge management as well as strengthen monitoring and evaluation;

(h) *Put in place* adaptation measures to improve the resilience of our countries to the increasing threat of climate change and variability to our water resources and our capacity to meet the water and sanitation targets;

(i) *Significantly increase* domestic financial resources allocated for implementing national and regional water and sanitation development activities and call upon Ministers of water and finance to develop appropriate investment plans;

(j) *Develop* local financial instruments and markets for investments in the water and sanitation sectors;

(k) *Mobilize* increased donor and other financing for the water and sanitation initiatives. . . .

Source: African Union 2008.

Benefits from investing in water

Many water investments have been evaluated by the rate of return of single-purpose schemes without considering the additional benefits possible from multipurpose projects.[8] Increasingly, evidence is emerging of the direct economy-wide benefits of investments in water (see chapter 6). For example, there is evidence that local action on water management in China has delivered measurable improvements in local GDP.[9] In the 335 counties in China with primary electrification from hydropower, annual average income per farmer rose 8.1% a year, nearly 3 percentage points more than the national average. In those communities 30 million people upgraded their livelihoods from marginalized farming to off-farm labourers in the industrial and services sector without any negative impact on agricultural production.

Evidence is also growing of the macroeconomic returns to investments in water management – and the costs of failures to invest. Disasters such as floods (resulting from typhoons and hurricanes and from rainfall exceeding the carrying capacity of channels) and droughts hurt poor economies more than wealthy ones, which are better prepared to cope with such disasters (figure 1.2).

Investments in environmental sustainability and water management to prevent water-related disasters can have large payoffs, so countries need not wait to invest until they have achieved middle- or high-income status. Investments in water infrastructure by the US Army Corps of Engineers between 1930 and 1999, for example, yielded returns of $6 for each $1 spent and controlled flood damage despite rising population numbers and property value at risk over the period (figure 1.3). The World Health Organization (WHO) estimates returns of $3-$34, depending on the region and technology, for each $1 invested in safe drinking water and basic sanitation.[10] There is thus a strong case that improved coverage of drinking water and sanitation contributes to economic growth. Policy-makers can use these data to justify their actions, identify areas of deficiency and better prioritize actions.[11]

Policy-makers also need to better understand the benefits for national development that result from sustainable water management and provision of safe water. Expanding safe drinking water and sanitation services would drastically cut the loss of life from water-related illness and free up scarce health resources in developing countries. Five thousand children die each day from diarrhoea alone – one every 17 seconds.[12] Upgrading water supply and sanitation services can also improve education, allowing more girls to attend school instead of spending hours each day collecting water. Improved access would also save millions of work days. The overall economic loss in Africa alone due to lack of access to safe water and basic sanitation is estimated at $28.4 billion a year, or around 5% of GDP.[13] Box 1.2 estimates the

Figure 1.2 The costs of disasters as a share of GDP are much higher in poor countries than rich countries

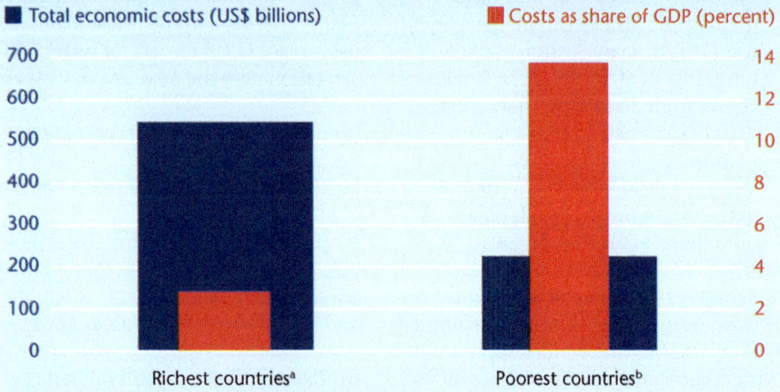

a. Annual GDP per capita above $9,361.
b. Annual GDP per capita below $760.
Source: Delli Priscoli and Wolf 2009.

Figure 1.3 US government investments in water infrastructure during 1930-96 yielded $6 in damages averted for each $1 invested

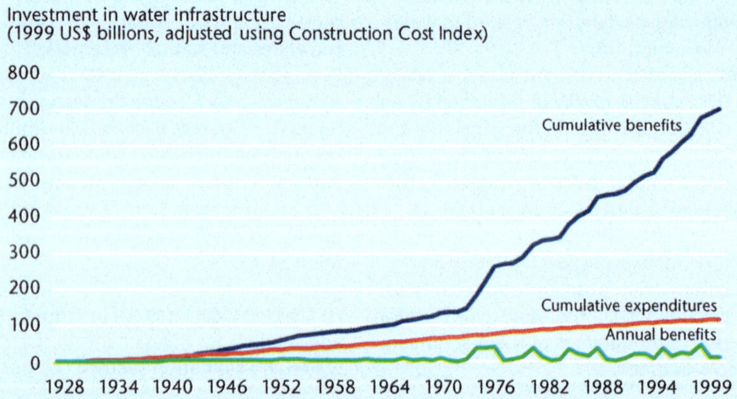

Source: Based on Delli Priscoli and Wolf 2009.

Box 1.2 Economic impacts of lack of adequate sanitation facilities in South-East Asia

Cambodia, Indonesia, the Philippines and Viet Nam lose an estimated $9 billion a year because of poor sanitation (based on 2005 prices), or approximately 2% of their combined GDP, according to the first regional study on the economic impacts of poor sanitation, undertaken in South-East Asia by the World Bank Water and Sanitation Project. The highest economic costs ($4.8 billion for the four countries combined) are from sanitation- and hygiene-related diseases. Poor sanitation also contributes substantially to water pollution, adding to the cost of providing safe water for households and reducing the production of fish in rivers and lakes ($2.3 billion). There are also environmental losses (loss of productive land, $220 million) and tourism losses ($350 million). Universal sanitation would lead to an annual gain of $6.3 billion in the four countries. Implementing ecological sanitation approaches (latrines separating urine and fæces for use as fertilizer) would be worth an estimated $270 million annually.

Source: Hutton, Haller, and Bartram 2007.

costs of lack of access to adequate sanitation facilities for four South-East Asian countries.

Environmental degradation from water pollution and excessive withdrawals also has negative economic impacts. For example, the damage cost of environmental degradation in the Middle East and North Africa has been estimated at some $9 billion a year, or 2.1%-7.4% of GDP.[14] Industrial countries are learning the enormous costs associated with restoring essential ecosystems. In the United States the costs have been estimated at more than $60 billion and continue to rise as more becomes known (box 1.3).

Box 1.3 Estimated costs of restoring essential ecosystems in the United States

The following are estimates for restoring major essential ecosystems in the United States. The cost exceeds $60 billion, and the total is likely to be higher still as more information becomes available.

- Everglades Restoration: $10.9 billion. Groundwork laid for Everglades restoration, but projects are experiencing delays (www8.nationalacademies.org/onpinews/newsitem.aspx?RecordID=11754).
- Restoration of the Upper Mississippi River: $5.3 billion for a 50-year ecosystem restoration plan (www.nationalaglawcenter.org/assets/crs/RL32470.pdf).
- Restoration of Coastal Louisiana: $14 billion towards a Sustainable Coastal Louisiana by 2050 (www.coast2050.gov/2050reports.htm).
- Restoration of Chesapeake Bay: $19 billion for the Chesapeake Bay Program (www.chesapeakebay.net/fundingandfinancing.aspx?menuitem=14907).
- Restoration of Great Lakes: $8 billion for Great Lakes restoration and protection priorities (www.cglg.org/projects/priorities/PolicySolutionsReport12-10-04.pdf).
- Restoration of California Bay Delta: $8.5 billion (first seven years) for large-scale ecosystem restoration initiatives (www.nemw.org/calfed.htm).
- Restoration of Missouri River – to be determined.

Figure 1.4 Water investment requires a holistic approach – links between pricing, financing and stakeholders

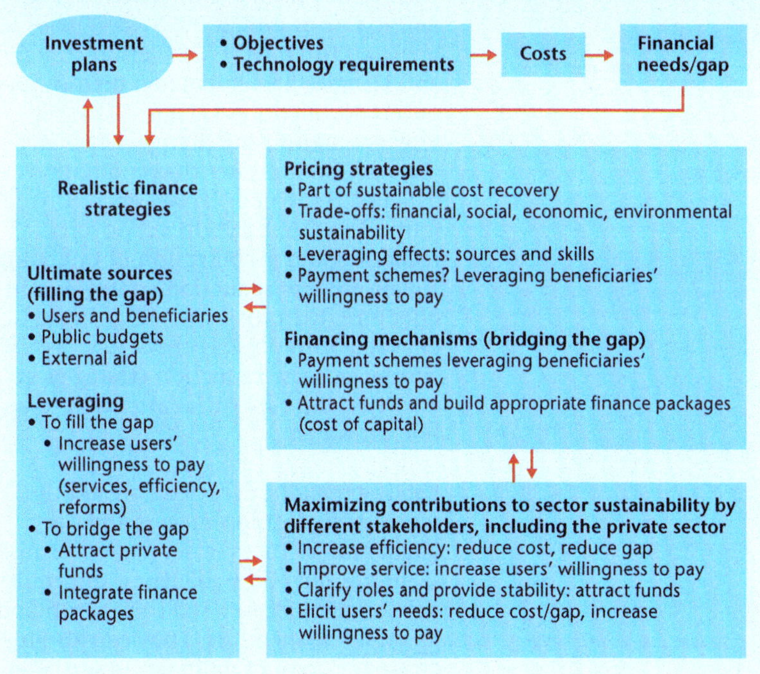

Source: Authors' construction.

Investing in water

Investment flows to uses with the highest economic rate of returns. Currently, water often gives very low returns for very long payback periods primarily because of the way it is governed (see chapter 4). Much political interaction in the water sector drives operations to 'structural bankruptcy'. It is not surprising that new investors are not eager to enter the water sector. Yet public investment in infrastructure is declining. And so the needs of the water sector go unmet.

The challenges in financing water services have been well described in recent years. Proposed solutions and innovative responses are presented in the reports of the World Panel on Financing Water Infrastructure[15] and the Task Force on Financing Water for All.[16] Ultimately, there are only three sources of financing: user tariffs, public expenditure and external aid (official or philanthropical). Recourse to these sources should be preceded and accompanied by efficiency measures to control operating costs and by careful project selection and design to ensure the best return to scarce resources.

Many studies have attempted to estimate the total investments that would be required to provide adequate infrastructure for water supply and sanitation. Typically presented as global or regional estimates, they often ignore the essential precondition of investments in institutions, reform, and implementation and management capacities and in replacement of ageing infrastructure. Because water can be managed only locally, investments must also be managed locally. Investing in water requires a holistic approach (figure 1.4). Sound financial management, as illustrated in figure 1.4, will make it possible for water authorities and governments to attract loans or external aid to supplement their own sources of capital.

Nonetheless, many developing countries, having applied all of the measures implied by such a process, will still lack the capital required to meet basic needs through

Today, poverty reduction strategies still offer only the prospect of aligning action on water with poverty reduction, as few current poverty reduction strategies give anything but superficial attention to action on water

water resources development and service delivery. In those cases it is relevant to question how much external aid is available, where it is applied and whether the amount can or should be increased.

Distributing the benefits of growth

The 2007 U.K. Department for International Development policy paper 'Growth and Infrastructure' stated that 'Growth is the single most important way of pulling people out of poverty'.[17] It cites empirical literature attributing more than 80% of recent poverty reduction worldwide to growth and less than 20% to redistribution (social protection). It gives the examples of China, where 450 million people have been lifted out of poverty since 1979, helped by exceptionally high growth rates, and Viet Nam, which experienced the most rapid reduction in poverty rates on record, from 75% in the late 1980s to less than a third in 2002, thanks to high growth rates.

That poverty reduction is the overriding policy concern is evidenced by the primacy of poverty reduction strategies and national development plans as the governing mechanisms for partnerships and finance from the international community. As of mid-2008, 59 countries had prepared full poverty reduction strategies and 11 more had completed preliminary poverty reduction strategies. This represents a significant change. For many years action on water that could deliver benefits to the poor lacked government frameworks that prioritized poverty reduction and mobilization of financing. Today, poverty reduction strategies still offer only the prospect of aligning action on water with poverty reduction, as few current poverty reduction strategies give anything but superficial attention to action on water.

Public expenditure reviews are another tool to help decision-makers allocate public funds. These reviews of government spending can boost efficiency and equity, development impact and the accountability of public spending. They can also increase the accountability and transparency of results and support governance reforms and anticorruption programs.

Economic justification for water investments come from their translation into economy-wide growth through employment, capital and labour productivity, taxes, government expenditure, revenue control, debt, purchasing power, balance of payments, foreign exchange reserves, trade balances, accelerator impacts on capital investment, business confidence and the stock market.

In India water development evened out the seasonal demand for labour, resulting in major gains for the country.[18] Forecasts by the New Partnership for Africa's Development concerning African agriculture's contribution to growth and poverty reduction are founded on the economic justifications of reduced food import bills, more predictable import profiles, increased export revenues and reduced poverty at the household level.[19]

To attract development-oriented finance, the growth-increasing and poverty-reducing contributions of water resources must be made explicit and specific at the country level. Such specifics will influence the sources, costs, viability, sustainability and instruments of finance. National, basin and local action plans are needed to align water resources, economic growth and poverty reduction. Making such alignments and other essential connections will be more successful within frameworks such as a round of poverty reduction strategies, public expenditure reviews and national development plans.

Reducing poverty, which limits access to water

The world must acknowledge the crisis of persistent underdevelopment and poverty. Since the end of the Second World War more than 3 billion people have benefited from economic development, but at least 2 billion people remain in need. Some 1.4 billion people lived in 'absolute poverty' in 2005,[20] a number that does not take into account the recent wave of increases in energy and food prices.[21] These women, men and children daily face the consequences of poverty – disease, malnutrition and hunger. They have no capacity to prepare for natural disasters, such as earthquakes and floods, or to respond when they strike. The world community has set the Millennium Development Goal target of halving the proportion of people living in poverty by 2015. But we are far from being on track, particularly in regions where the need is highest.

Human Development Report 2006 considers the experience of water and sanitation as reinforcing the 'long-standing human development lesson' that rates of coverage in access to water and sanitation rise with income on average (figure 1.5).[22] *Global Monitoring Report 2005* notes that in South Asia an improving investment climate and stronger policies, along with gains in basic service delivery, have sustained rapid economic growth since 1990 and contributed significantly to poverty reduction and to reaching the Millennium Development Goals in some countries.[23]

The case for investing in Africa

Where investment in water has been weak, GDP growth has been constrained – by as much as 10% where the effects of droughts, floods and natural hydrologic variability are compounded in less developed economies. Where weak economic growth has been accompanied by inadequate investment in social protection, the gap in achieving the Millennium Development Goals has worsened in many countries, with devastating social impacts.

Africa, in particular, remains mired in poverty (figure 1.6) despite recent economic growth trends in some countries. In developed countries water storage ensures reliable sources of water for irrigation, water supply and hydropower as well as a buffer for flood management. Countries in Africa store only about 4% of annual renewable flows, compared with 70%-90% in many developed countries. About 340 million Africans lack access to safe drinking water, and almost 500 million lack access to improved sanitation facilities. The First African Water Week, convened in Tunis in March 2008, opened with a call for greater efforts to ensure water security nationally and regionally. Donald Kaberuka, president of the African Development Bank Group, emphasized that

> it is no longer acceptable that the African continent continues to utilize only 4% of its water resources, when a huge proportion of the people do not have access to safe water, and when large populations are faced with frequent floods and drought, in addition to food and energy shortages. Action is urgently needed.[24]

In June 2008 the MDG Africa Steering Group published a number of concrete recommendations for scaling up opportunities to address poverty in Africa.[25] Their recommendations related to achieving the Millennium Development Goals in Africa are summarized in table 1.1.

Investing in water to reach the Millennium Development Goals

This third edition of the United Nations *World Water Development Report* is being published just beyond the half-way point along the timeline from the Millennium Summit of 2000 and the 2015 target date for attaining the Millennium Development Goals. Making progress towards those goals will rise even higher on political agendas within the next six years.

The Millennium Declaration placed safe drinking water and basic sanitation firmly among the development objectives, making it a target of Millennium Development Goal 7. But while adequate progress is

Figure 1.5 Access to water and sanitation rises with income

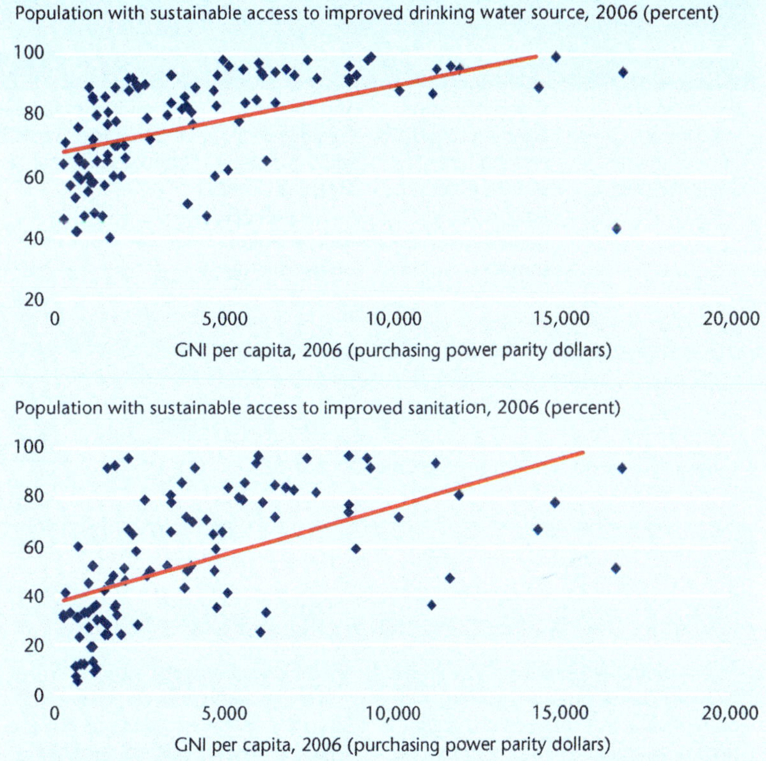

Source: Based on data from WHO Statistical Information System (www.who.int/whosis/en/).

Figure 1.6 Poverty remains high in sub-Saharan Africa

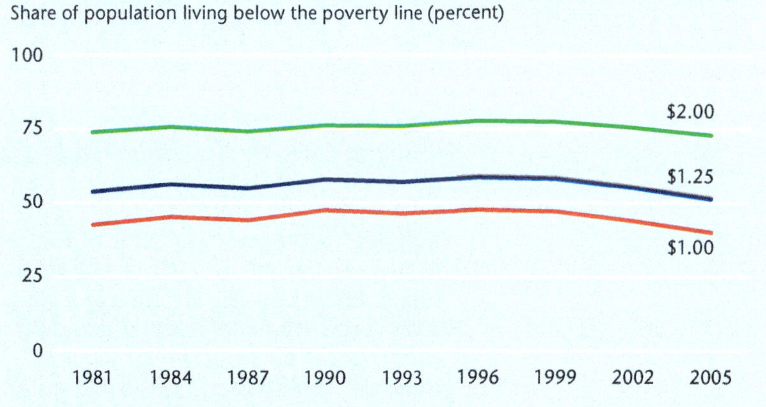

Note: Poverty lines in 2005 prices.
Source: Based on Chen and Ravallion 2008, p. 41.

Table 1.1 **Summary of scaling-up opportunities related to achieving the Millennium Development Goals in Africa**

Scaling-up opportunity	Summary of key results	Policy leadership	Key multilateral financing mechanisms (among several funding sources)	Estimated public external financing needs by 2010 from all funding sources
Achieving the Millennium Development Goals in Africa	Comprehensive cross-sector public expenditure programmes against clear quantitative targets	Secretary-General and MDG Africa Steering Group, G-8 leadership, African Union, private sector, foundations	All multilateral, bilateral and private mechanisms providing high-quality, predictable financing	Some $72 billion a year, of which $62 billion (in 2007 terms) from Development Assistance Committee members (following the Gleneagles G-8 meeting, Monterrey Consensus and EU official development assistance targets), with additional financing from non-Development Assistance Committee donors, developing country collaboration, private foundations and innovative private co-financing

Source: Based on MDG Africa Steering Group 2008, p. 32.

Box 1.4 **Progress in meeting the Millennium Development Goal target on water supply and sanitation**

The world is on track to meet the Millennium Development Goal target on drinking water. Current trends suggest that more than 90% of the global population will use improved drinking water sources by 2015.

The world is not on track to meet the Millennium Development Goal sanitation target. Between 1990 and 2006 the proportion of people without improved sanitation decreased by only 8 percentage points. Without an immediate acceleration in progress, the world will not achieve even half the sanitation target by 2015. Based on current trends, the total population without improved sanitation in 2015 will have decreased only slightly, from 2.5 billion to 2.4 billion.

Source: WHO and UNICEF Joint Monitoring Programme 2008, pp. 8 and 23.

being made towards the provision of safe drinking water, the sanitation target is far from being met (box 1.4).

And despite progress, the scale of the challenge remains massive. While the water supply target is being attained at a global level, large regions of the world and many countries are far from the target, and some risk backsliding. This is particularly the case in sub-Saharan Africa and low-income Arab states. In many places the sanitation targets will be missed by a wide margin.

Both the drinking water and sanitation targets are vitally important. The contribution of improved drinking water and sanitation to the achievement of all the Millennium Development Goals is now well established.[26] This report demonstrates this link throughout; others have elaborated the direct and indirect contributions of water management across all the Millennium Development Goals.[27] Figure 1.7 depicts these links graphically. These links served as an important advocacy instrument during the International Year of Sanitation in 2008. High-profile international attention has focused on basic services in recent years, including declarations at Brasilia (2003), Beppu (2007), eThekwini (2008), Tunis (2008) and Sharm el-Sheik (2008). Gaps in drinking water and sanitation, in particular, have attracted political attention at the highest levels.

Development partnerships are helping countries that are off track for achieving the Millennium Development Goals get back on track. Intergovernmental efforts are working to maintain the momentum of the global commitments made since the Millennium Declaration and of water-specific processes such as the G-8 Evian Action Plan[28](box 1.5). New initiatives, such as the 2007 launch of the Millennium Development Goal Africa Initiative by the UN system, have sought to reinvigorate the efforts of countries that are off track in their progress towards achieving the Millennium Development Goals.

Sustaining the environment

Environmental sustainability, broadly, refers to the ability of the environment to continue to support progressive social and economic development and to provide many types of ecosystem services (table 1.2). Multistakeholder processes, such as the World Commission on Dams, have seen environmental sustainability rise in prominence as a factor influencing water development decisions. And such international conventions as the United Nations Convention to Combat Desertification and the United Nations Convention on Biodiversity have made water a global issue.

Figure 1.7 **Cause-effect chains and links between water and the Millennium Development Goals**

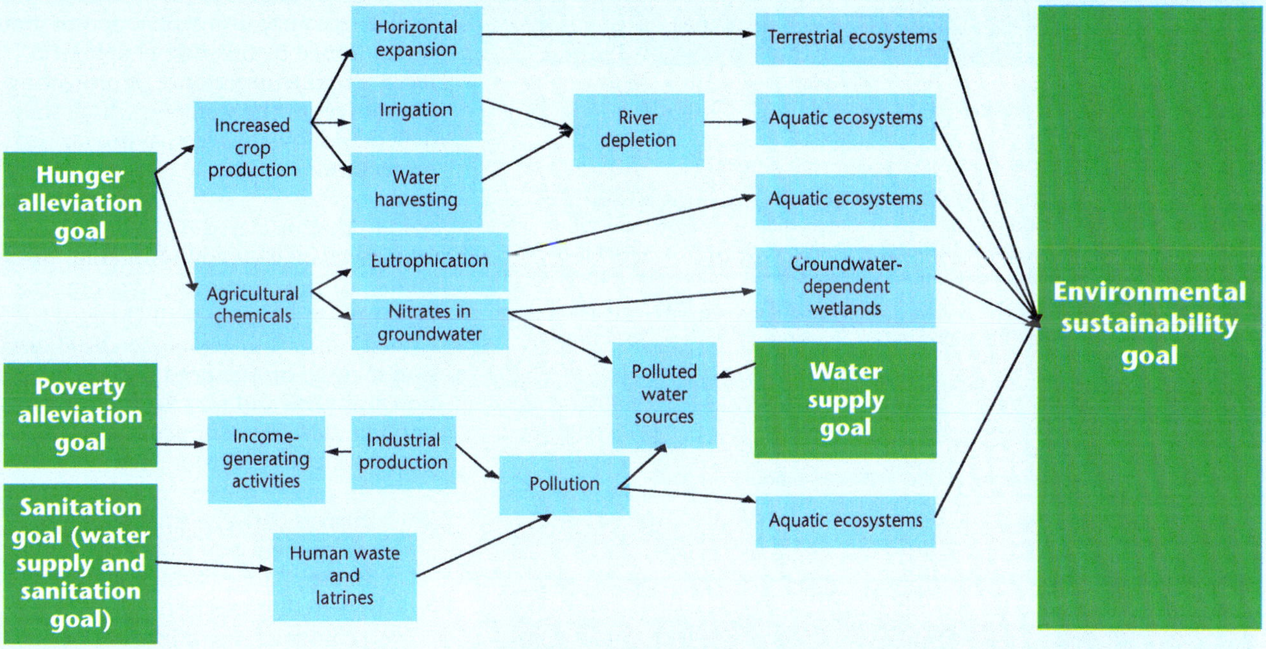

Source: Based on Cosgrove 2006, p. 38.

Today, water management crises are developing in most of the world. UN-Water reports that in just one week in mid-November 2006 national media sources reported local but high-profile shortages in parts of Australia, Botswana, Canada, China, Fiji, Kuwait, Liberia, Malawi, Pakistan, Philippines, South Africa, Uganda, the United Arab Emirates and the United States.[29]

Generally regional phenomena, water crises can emerge as water shortages and droughts, floods or both, now aggravated by the consequences of climate change. They may be natural or caused by demands that exceed supply, lack of infrastructure or poor water management. They may be the result of waste or abuse resulting in pollution. Together they threaten the lives and livelihoods of billions of people and risk irrevocably altering the planet's ecosystems.

Every year in developing countries an estimated 3 million people die prematurely from water-related diseases. The largest proportion of these deaths are among infants and young children, followed by women, from poor rural families who lack access to safe water and improved sanitation (box 1.6).[30] More than 1 million people die annually from malaria, the vast majority in poverty-stricken Africa. Another 1 million people die from air pollution in urban areas. And everywhere the poor suffer most.

Box 1.5 **High-Level Event on the Millennium Development Goals, United Nations, New York, 25 September 2008: Extract from compilation of partnership events and commitments**

The event [Water and Sanitation for All] reiterated the strong political and diplomatic support for international efforts needed to address the water and sanitation issues and enhance human security. It promoted good water cycle management and the application of Integrated Water Resources Management. It reaffirmed the importance of formulation and implementation of national assistance strategies building on the 'Paris Declaration on Aid Effectiveness', while considering the specific needs and resources of the recipient countries.

The event emphasized the importance of mobilizing adequate international and national financial resources for the implementation of the national strategies and the need to strive towards using sector-wide approaches; and developed partnerships with civil society organizations, local authorities and the private sector to implement national strategies and action plans to improve the accessibility and quality of water and sanitation services as well as initiatives to establish a 'Framework for Action' to focus on the off-track countries, including the possible consideration for a 'Fast Track Initiative' with catalytic funding to install a High-Level 'Task Force' to reach [Millennium Development Goal 7], and to make one annual global progress report and to hold one annual high-level review meeting.

Japan committed to establish a Water Security Action Team for Africa to provide safe drinking water for 6.5 million people and implement a water supply capacity-building program that would train 5,000 people over the next five years. Tajikistan said it would host the International Freshwater Forum in 2010 as a venue for a preliminary discussion of achievements, challenges and experiences within the International Decade Water for Life, 2005-15.

The Netherlands said it would help provide access to safe drinking water and sanitation for at least 50 million people by 2015 having already signed various agreements that will benefit almost 30 million people, at a cost of around €1.3 billion. Germany will continue to train Central Asian water experts. The Netherlands and the United Kingdom committed €106 million in joint funding for water and sanitation initiatives in developing countries over the next five years.

Source: UN 2008.

Table 1.2	Types of ecosystem services		
	Forests	**Oceans**	**Cultivated/ agricultural lands**
Environmental goods	• Food • Freshwater • Fuel • Fibre	• Food • Fuel	• Food • Fuel • Fibre
Regulating services	• Climate regulation • Flood regulation • Disease regulation • Water purification	• Climate regulation • Disease regulation	• Climate regulation • Water purification
Supporting services	• Nutrient cycling • Soil formation	• Nutrient cycling • Primary production	• Nutrient cycling • Soil formation
Cultural services	• Aesthetic • Spiritual • Educational • Recreational	• Aesthetic • Spiritual • Educational • Recreational	• Aesthetic • Educational

Source: Based on MEA 2005.

Box 1.6 Malnutrition attributable to environmental risks

Experts estimate that poor water and sanitation services and hygiene practices and inadequate water resources management contribute to half of all cases of infant and child underweight, an estimate corroborated by a World Bank technical review of 38 recent cohort studies (confidence interval of 39%-61%). Evidence from several of those studies demonstrates that exposure to environmental health risks in early childhood leads to permanent growth faltering, lowered immunity and increased mortality. A recent large study from Bangladesh reveals that dysentery and watery diarrhoea together can retard weight gain by 20%-25% compared with periods of no infections.

Source: Prüss-Üstün and Corvalán 2006; World Bank 2008; Alam et al. 2000.

Box 1.7 Water as capital

Classical economists recognized land (all natural resources), labour and produced capital as the basic sources of wealth. Neoclassical economists focused only on labour and capital, treating 'land' as another interchangeable form of capital. Natural resources were considered abundant relative to demand and therefore not an important focus for economics, whose task was to allocate scarce resources – those whose use constrained alternative economic opportunities. There was little consideration of the environment's dual role as a source of valuable inputs and as a sink for the economy's waste and pollution. Nor was much thought given to the possibility that the world might reach a scale of resource exploitation at which the capacity of both the source and sink functions of the environment could become binding constraints on economic growth.

The focus on produced rather than natural capital is particularly misleading for water. Prices are typically related to the capital outlays required to deliver water (that is, for the infrastructure and operations and maintenance charges), with little or no value attributed to the resource itself. Not only do undervalued water resources tend to be overused, but undervaluation also induces distorted prices that provide poor information about whether investments make sense. Focusing only on capital costs provides no insight into whether economic activities are creating value or whether the resource is running out and needs to be conserved.

Water delivery is highly capital-intensive, so produced capital will remain a crucial focus for financial and economic analyses of water investments. But the value of water resources also matters, and water's availability, quality and timing cannot simply be assumed.

Source: Bergkamp and Sadoff 2008.

The value of water goes well beyond its productive value (box 1.7). Citizens who realize this are calling for action to protect water, joined by business people who recognize the importance of protecting the sources of the water on which they depend. Many are even paying for such protection.[31]

Also to be considered is the impact of climate change on environmental sustainability. At the High-Level Event on the Millennium Development Goals at the United Nations in September 2008 discussion focused on the need for new adaptation strategies and for climate-resilient national development plans, especially for the least developed countries:

> Linkages between financing for development and international climate change financing were discussed. It was also agreed that all countries, including donor countries, the UN system and the Bretton Woods institutions, need to clarify the budgetary implications of adaptation; ensure that adequate finance mechanisms are in place; and help meet the additional costs that climate-resilient development will entail.[32]

Global crises and water

While climate change will create important pressures on water, it is not currently the most important driver of these pressures outside the water sector. The most important drivers – forces and processes generated by human activities – are demographics and the increasing consumption that comes with rising per capita incomes (see chapter 2).

In the early stages of development population growth is the most important driver. But most of the projected growth in demand comes not from high-population-growth countries but from countries with high rates of economic growth and large current populations. As incomes permit, people consume more. To start with, there will be a requirement for more water to produce food for tens of millions of people moving from one meal to two meals a day. Later, still more water will be needed for food production as people include more meat in their diets. Changes in lifestyles will require large amounts of water to produce and process non-food goods and services (virtual water), further increasing pressures on the quantity and quality of water resources. Other demographic drivers include rural-urban migration and migration in response to political conflict and environmental crises.

Other external forces that may create either positive or negative pressures on water resources include pricing policies and subsidies for water and water-related goods, trade patterns, developments in science and technology, consumption patterns, evolution of policies and laws, social movements and global and national politics.

Except for climate change, these forces will not create pressures directly (or only) on water management. The pressures will be felt first at the level of sector ministers, whose responses will translate into strategies that affect the water sector. These ministers will have to make decisions under conditions of risk and uncertainty. The better informed they are, the more likely they are to make the right decisions. For water managers this means being able to provide reliable information about where and when water is available, of what quality, where and how it is used, what happens to wastewater, how much water leaves the country in exports of goods that use water in their production (virtual water) and how much enters the country in imports. This will be a challenge for water managers in most countries, which lack the necessary measurements and do not systematically collect the necessary data. But when the information is available, it will be possible to calculate the country's water balance and the water footprints (volume of water used) of various users. Using this information, water managers can advise decision-makers in other sectors of the feasibility of their plans and the implications for water.

The Report provides ample evidence of all these facts. This is not the first time that professionals in the water sector have attempted to bring them to the world's attention. But this time the effort may be more successful, because this time the world is facing other global crises – in energy, food and climate change and global warming – that cannot adequately be addressed without considering the role of water.

The world is facing global crises in energy, food, and climate change and global warming that cannot adequately be addressed without considering the role of water

Water for energy

Demand for energy – for heat, light, power and transportation – is increasing rapidly (see chapter 7). The price of energy commodities has been rising as well. Volatile, the nominal price of oil – the benchmark commodity – rose from less than $25 a barrel eight years ago to about $100 early in 2008 and more than $140 in June 2008. Within two months it fell below levels projected for the longer term by the Energy Information Administration of the U.S. Department of Energy and was at $35 a barrel on December 19, 2008 (figure 1.8). Energy

Figure 1.8 **Historical and projected energy demand and oil prices show steadily rising demand and rapidly rising prices**

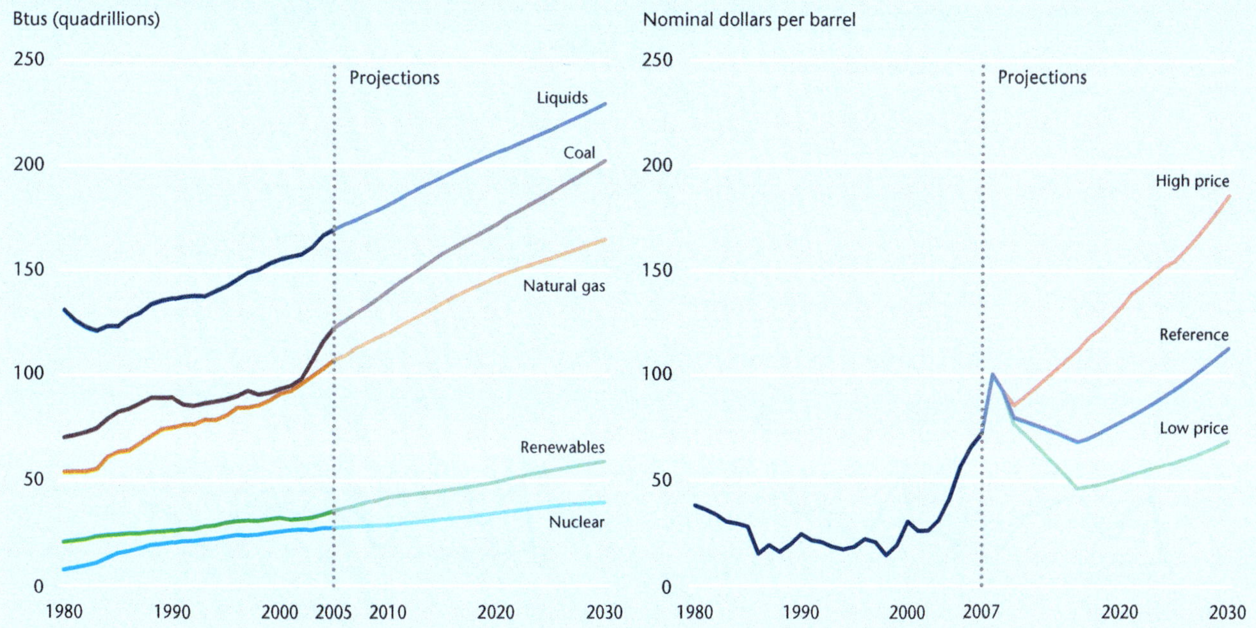

Note: The reference case assumes average GDP growth of 2.4% a year, the high case assumes 3.0% a year, and the low case assumes 1.8% a year.
Source: Based on EIA 2005, 2008a.

The number of countries without enough water to produce their food is rising. The situation can be remedied by investing in water infrastructure, markets, credit, agricultural technology and extension services

prices, particularly the oil prices that drive them, earlier reflected rising world demand and constraints. The recent financial crisis, which has slowed economic growth throughout the world, reducing anticipated demand, was largely responsible for the low price of oil at the end of 2008.

The combination of high prices and a desire to substitute other sources of fuel led to the recent increase in the production of bioenergy, which has potentially important impacts on water quality and availability. Hydropower may be a renewable and non-polluting source of energy in some countries. Water for cooling is needed for all thermal sources of power, including nuclear. In the United States water withdrawn for cooling (39%) equals agriculture's share of water use. At the same time energy is required to lift groundwater, pump it through pipes and treat both groundwater and wastewater. An estimated 7% of all energy produced is used for such purposes.[33] Increased demand for water through desalination may increase energy demand in some countries, although marginally on a global scale.

Water for food

Agriculture is by far the largest consumer of freshwater – about 70% of all freshwater withdrawals go to irrigated agriculture (see chapter 7). The recent steep rise in food prices (figure 1.9) has severely hurt many food-importing countries. Rising demand for food caused by growing populations and shifting diets, production shortfall in some countries, increased costs for key agricultural inputs such as fertilizers (driven in turn by energy costs), bioenergy-related incentives in some countries and possible financial speculation have all contributed to the problem. The High-Level Conference on World Food Security: The Challenges of Climate Change and Bioenergy, a Food and Agriculture Organization summit in Rome on 3 June 2008, adopted a declaration acknowledging 'an urgent need to help developing countries and countries in transition expand agriculture and food production, and to increase investment in agriculture, agribusiness and rural development, from both public and private sources'.[34] It calls on donors to provide balance of payments and budget support to low-income food-importing countries.

At the summit Robert B. Zoellick, president of the World Bank, said that the Bank recognizes that the energy-food nexus means that food prices will stay high and that the 'task is two-fold, to handle today's danger to those for whom securing food has become a daily struggle, and turn higher food prices into an opportunity for developing world agriculture, and

Figure 1.9 **Wheat and rice prices have risen sharply in recent years**

Historical and projected prices of wheat and rice, 1970-2017

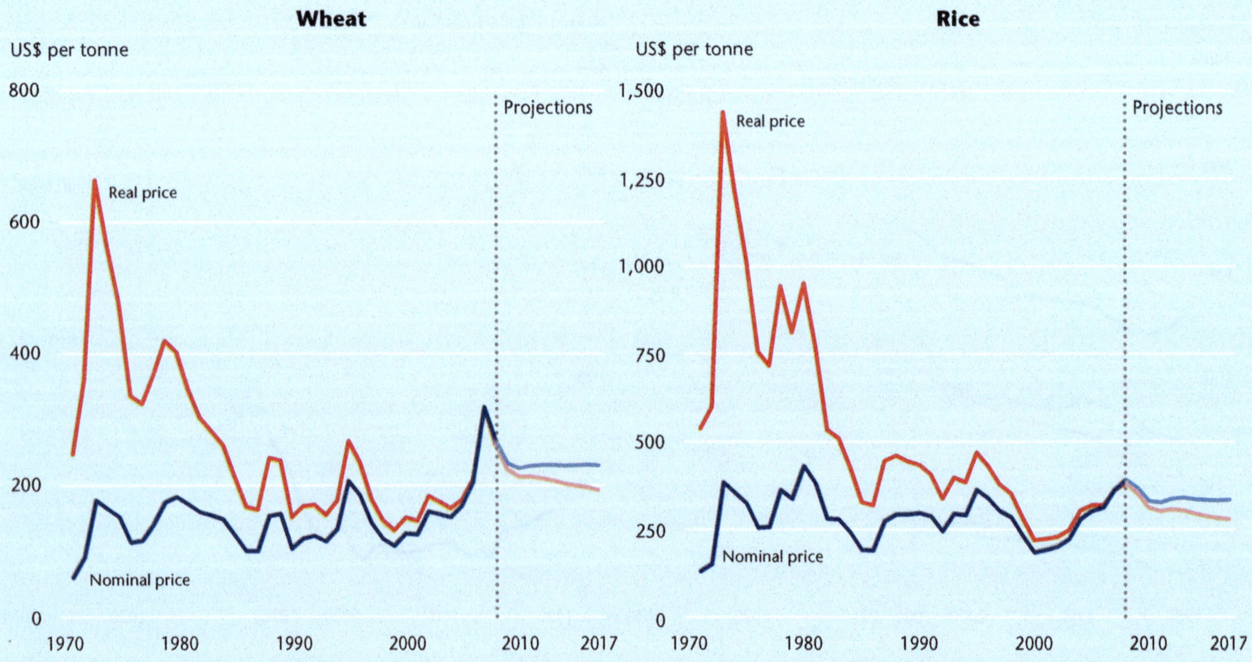

Source: Based on OECD and FAO 2008.

for farmers in developing countries'.[35] The summit highlighted the strong links among food security, economic development, climate change, markets, development assistance and energy and how actions have implications for other sectors. While the role of water in agriculture was discussed at the summit, the final declaration did not mention water and water's strong links with these and many other issues.

Water scarcity may limit food production and supply, putting pressure on food prices and increasing countries' dependence on food imports. The number of countries and regions without enough water to produce their food is rising as populations increase. The situation can be remedied in many developing countries by investing in water infrastructure, markets, credit, agricultural technology and extension services.

Underinvestment in water

The energy and food crises are taking place during a time of global financial crisis. A credit crunch has followed the financial crises that began in the United States and Europe in 2007 and spread around the globe. The credit crunch has resulted in a slowdown in economic growth around the world. The International Monetary Fund forecast in January 2009 that all industrial countries would face a period of recession and that some developing countries are more at risk than others (box 1.8).[36] According to the Commission on Growth and Development 'developing countries are most vulnerable to sudden stoppages of credit and sudden switches of international custom or supply.'[37]

Developing countries most at risk include those exporting directly to crisis-affected countries, those whose exports are experiencing falling world prices and those whose exports have high income elasticity (luxury goods, including tourism). Declining tourism revenues and employment will directly affect the poor. Countries dependent on foreign direct investment, remittances and development funds to finance the current account deficit will also be at risk. Oil-importing countries have already been hard hit by the period of high oil prices.

The high rates of global savings and strong productivity growth in the three decades before the financial crisis – when the stock of financial assets grew three times faster than GDP – were not accompanied by investments in physical assets, and their levels are below those in the last decade. While other factors may also have contributed to these lower levels of investment, economic uncertainty is a major factor. Uncertainty about the policy environment in developing and emerging market economies has always been a concern, but its influence has strengthened in the currently highly competitive global markets.[38]

The impact on developing countries will vary. Budgetary spending on infrastructure is often cut during periods of financial tightening, although for governments that can afford it, investing in infrastructure can help counter an economic slowdown. Private investment may also suffer, but since the private sector's contribution to the water sector has been relatively small, the sector is less exposed to any financial tightening. Countries dependent on aid face uncertain times. Bilateral donors, important in funding water investments, may be tempted to reduce their aid budgets. Multilateral aid could be an important source of financing for the next few years, especially following recent record multiyear replenishments of the International Development Association, African Development Fund and European Development Fund. Yet both bilateral and multilateral aid donors still appear not to recognize the contribution of the water sector to growth, as indicated by the sector's small share of total official development assistance in recent years (less than 4%; see table 4.4 in chapter 4).

Inadequate information on water and water crises

Managing water is made more difficult by the lack of knowledge and information required for decision-making and long-term planning. Few countries know how much water is being used and for what purposes, the quantity and quality of water that is available and that can be withdrawn without serious environmental consequences

Few countries know how much water is being used and for what purposes, the quantity and quality of water that is available and that can be withdrawn without serious environmental consequences and how much is being invested in water management and infrastructure

Box 1.8 International Monetary Fund updated economic forecast for 2009

World growth is projected to fall to 0.5 % in 2009, its lowest rate since World War II. Despite wide-ranging policy actions, financial strains remain acute, pulling down the real economy. A sustained economic recovery will not be possible until the financial sector's functionality is restored and credit markets are unclogged.

Financial markets are expected to remain strained during 2009. In the advanced economies, market conditions will likely continue to be difficult until forceful policy actions are implemented to restructure the financial sector, resolve the uncertainty about losses, and break the adverse feedback loop with the slowing real economy. In emerging economies, financing conditions will likely remain acute for some time – especially for corporate sectors that have very high rollover requirements.

Source : IMF 2009, pp. 1-2.

Scarcity – low available water per capita – is forecast to worsen where population growth is still high, as in sub-Saharan Africa, South Asia and some countries in South America and the Middle East

and how much is being invested in water management and infrastructure (see chapter 13).

Underfunding of observation, monitoring and information systems leads to weaknesses in infrastructure, research and development, and training and to reduced efficiencies. Less is known with each passing decade, despite the availability of new remote sensing and geographic information system technologies that can simplify monitoring and reporting and despite the growing need for such information in an increasingly complex and rapidly changing world. Such information is vital not only at a national scale but also at a global scale – to inform the construction of global models of the hydrologic cycle and decisions on where interventions, including external aid, would be most useful. One move in that direction is the United Nations Economic Commission for Europe Convention on the Protection and Use of Transboundary Watercourses and International Lakes, which requires signatories to exchange data on water quality and quantity and pollution sources and the environmental conditions of transboundary waters.

Climate change and water

Some parts of the world have no shortage of water. Others, such as North and Southern Africa, the Middle East and parts of South Asia, South-East Asia and South America, suffer scarcity because of low annual rainfall. Others suffer seasonal scarcity. Yet others suffer from extreme rainfall, causing floods. Some suffer from both low and extreme rainfall, at different times. In some large countries, such as Mozambique and the United States, parts of the country may experience damaging intensive rainfalls while other parts suffer prolonged drought. These variations matter most where they affect large populations. Scarcity – low available water per capita – is forecast to worsen where population growth is still high, as in sub-Saharan Africa, South Asia and some countries in South America and the Middle East.

Adapting to climate change adds a critical challenge to this picture for all countries, particularly for developing countries, whose capacity to adapt is low, and for cities in coastal areas (see chapter 5). Even if greenhouse gas concentrations stabilize in the coming years, some impacts from climate change are unavoidable. These include growing water stress, more extreme weather events, higher levels of migration and the disruption of international markets. Climate models show that extremes of rainfall are likely to worsen, resulting in more floods and droughts in regions already affected – often regions with low income levels per capita, widespread absolute poverty, high population growth and rapid urbanization. If climate change brings significant shifts in the availability of water resources, patterns of human migration could be affected.

These challenges cannot be separated from the challenges of sustainable development. For some developing countries the incremental costs of adapting to climate change will soon approach the current value of aid inflows. The leaders of the G-8, meeting in Hokkaido, Japan, in July 2008, committed to accelerating action on technology development, transfer, financing and capacity building to support adaptation (box 1.9). Such action must include water resources, which will be most affected by climate change. A recent United Nations Framework Convention on Climate Change document on adaptation noted that:

> sector-specific adaptation planning and practices were discussed in the areas of agriculture and food security, water resources, coastal zones and health. Those sectors were selected based on their importance to Parties and organizations as highlighted in their submissions.[39]

The world is right to be concerned about climate change, which poses major threats to humans and ecosystems. The 2007 United Nations Climate Change Conference in Bali, Indonesia, acknowledged that

Box 1.9 Extracts from Declaration of Leaders Meeting of Major Economies on Energy Security and Climate Change at the G-8 Hokkaido, Toyako, summit, 9 July 2008

Climate change is one of the great global challenges of our time. Conscious of our leadership role in meeting such challenges, we, the leaders of the world's major economies, both developed and developing, commit to combat climate change in accordance with our common but differentiated responsibilities and respective capabilities and confront the interlinked challenges of sustainable development, including energy and food security, and human health.

We will work together in accordance with our Convention commitments to strengthen the ability of developing countries, particularly the most vulnerable ones, to adapt to climate change. This includes the development and dissemination of tools and methodologies to improve vulnerability and adaptation assessments, the integration of climate change adaptation into overall development strategies, increased implementation of adaptation strategies, increased emphasis on adaptation technologies, strengthening resilience and reducing vulnerability, and consideration of means to stimulate investment and increased availability of financial and technical assistance.

Source: G-8 2008.

even the minimum predicted shifts in climate for the 21st century, at more than twice the 0.6° Celsius increase that has occurred since 1900, would be significant and disruptive. The intergovernmental response has focused primarily on mitigation of climate change, embracing wide-ranging measures, including reducing greenhouse gas emissions, transferring clean technologies and protecting forests. These measures may slow climate change. They will not halt or reverse it.

It will be two generations before these measures begin to have an effect. And even if successful, they imply a considerably changed future climate. (They are not aimed at reversing changes already under way.) In the meantime people must be protected from the consequences of global climate change through adaptation measures. Adaptation, as embodied in the Nairobi Work Programme of the United Nations Framework Convention on Climate Change, is based on gaining a better understanding of the impacts of climate change and making informed decisions on practical measures.[40]

The water situation and the vulnerability of poor communities present a strong case for action on climate change. Projections warn of changes in water availability and quality that could have disastrous consequences. Water is the principal medium through which climate change will affect economic, social and environmental conditions. Changes in water availability will have economy-wide impacts.

Yet while the world appears motivated to respond to the impacts of future climate change, it remains unmotivated to act on the water crises that are with us today. Even without climate change, development is threatened in many regions by factors that we have already failed to address time and again. The Intergovernmental Panel on Climate Change's April 2008 report on water points this out clearly (box 1.10).

Security and water

Climate change, especially its implications for scarce water resources, is a matter of collective security in a fragile and increasingly interdependent world. At a 2007 UN Security Council debate on the impact of climate change on peace and security UN Secretary-General Ban Ki-moon noted that climate change has implications for peace and security, as well as serious environmental, social and economic implications, especially 'in vulnerable regions that face multiple stresses at the same time – pre-existing conflict, poverty and unequal access to resources, weak institutions, food insecurity and incidence of diseases such as HIV/AIDS.'[41] He outlined 'alarming, though not alarmist' scenarios, including limited or threatened access to energy increasing the risk of conflict, a scarcity of food and water transforming peaceful competition into violence, and floods and droughts sparking massive human migrations, polarizing societies and weakening the ability of countries to resolve conflicts peacefully.

In Africa alone by 2020, 75-250 million people may be exposed to increased water stress due to climate change. If coupled with increased demand, this will hurt livelihoods and exacerbate water-related problems.[42] Research centres such as the Oxford Research Group[43] are underpinning the security concerns of the United Nations, the European Union[44] and national governments[45] about climate change and its impacts on water. The forces at work are global in scale, the aggregate result of the behaviour of all countries. Dealing with them will require international cooperation and coordination. Yet at the same time national leaders must continue to act and make decisions at a national level.

As climate change and adverse water impacts increase in politically charged areas, conflicts will likely intensify, requiring new and rapid adaptive security strategies. Hydrologic shocks that may occur through climate change increase the risk of major national and international security threats, especially in unstable areas (box 1.11). Adverse changes in internal, interjurisdictional and transboundary waters can put food, social, health, economic, political and military security at risk.

Some fragile states (map 1.1) have experienced widespread conflict that has resulted in the destruction of economic

While the world appears motivated to respond to the impacts of future climate change, it remains unmotivated to act on water crises that are with us today

Box 1.10 Intergovernmental Panel on Climate Change Technical Report on Water and Climate Change

Current water management practices may not be robust enough to cope with the impacts of climate change on water supply reliability, flood risk, health, agriculture, energy and aquatic ecosystems. In many locations, water management cannot satisfactorily cope even with current climate variability, so that large flood and drought damages occur. As a first step, improved incorporation of information about current climate variability into water-related management would assist adaptation to longer-term climate change impacts. Climatic and non-climatic factors, such as growth of population and damage potential, would exacerbate problems in the future. (*very high confidence*)

Source: IPCC 2008.

Box 1.11 UN Secretary General Ban Ki-moon warns that water shortages are increasingly driving conflicts

'The challenge of securing safe and plentiful water for all is one of the most daunting challenges faced by the world today.

'Until only recently, we generally assumed that water trends do not pose much risk to our businesses. While many countries have engaged in waste-water treatment and some conservation efforts, the notion of water sustainability in a broad sense has not been seriously examined.

'Our experiences tell us that environmental stress due to lack of water may lead to conflict and would be greater in poor nations.

'Ten years ago – even five years ago – few people paid much attention to the arid regions of western Sudan. Not many noticed when fighting broke out between farmers and herders, after the rains failed and water became scarce.

'Today everyone knows Darfur. More than 200,000 people have died. Several million have fled their homes.

'There are many factors at work in this conflict, of course. But almost forgotten is the event that touched it off – drought. A shortage of life's vital resource.

'We can change the names in this sad story. Somalia. Chad. Israel. The occupied Palestinian territories. Nigeria. Sri Lanka. Haiti. Colombia. Kazakhstan. All are places where shortages of water contribute to poverty. They cause social hardship and impede development. They create tensions in conflict-prone regions. Too often, where we need water we find guns. . . .'

Source: Ban Ki-moon 2008.

infrastructure. The vulnerability of affected populations is worsened by the state's loss of control over the forces of law and order and ultimately by its loss of political legitimacy. Installing infrastructure and renewing institutional capacity following conflict have the potential to set post-conflict nations on a path to recovery. For example, rehabilitation of damaged irrigation infrastructure and expansion of water supply and sanitation formed a significant part of the 2006 Somali Rehabilitation and Reconstruction Plan.[46] Similarly, rehabilitation of infrastructure after major natural disasters provides an opportunity to address long-standing infrastructure deficits.

The need for action – now

Water has remained too low on the list of political priorities for too long, a situation that cannot be allowed to continue. Action is required now. Lives and livelihoods depend on water for development. Changes in human behaviour and activity are accelerating, affecting demand for water and its supply. Because investments have been neglected, development is lagging, people are suffering and the environment is deteriorating. The resources needed to address the problems of water management are minuscule compared with the financial resources that have been pledged and secured to deal with carbon emissions or the current financial crisis. After decades of inaction, the problems are enormous. And they will worsen if left unattended.

Although substantial, the challenges are not insurmountable. In part 4 the Report shows how some countries and regional and local governments have solved similar challenges. The decisions on development objectives and the allocation of human and financial resources needed to meet

Map 1.1 Fragile states as defined by the International Development Association

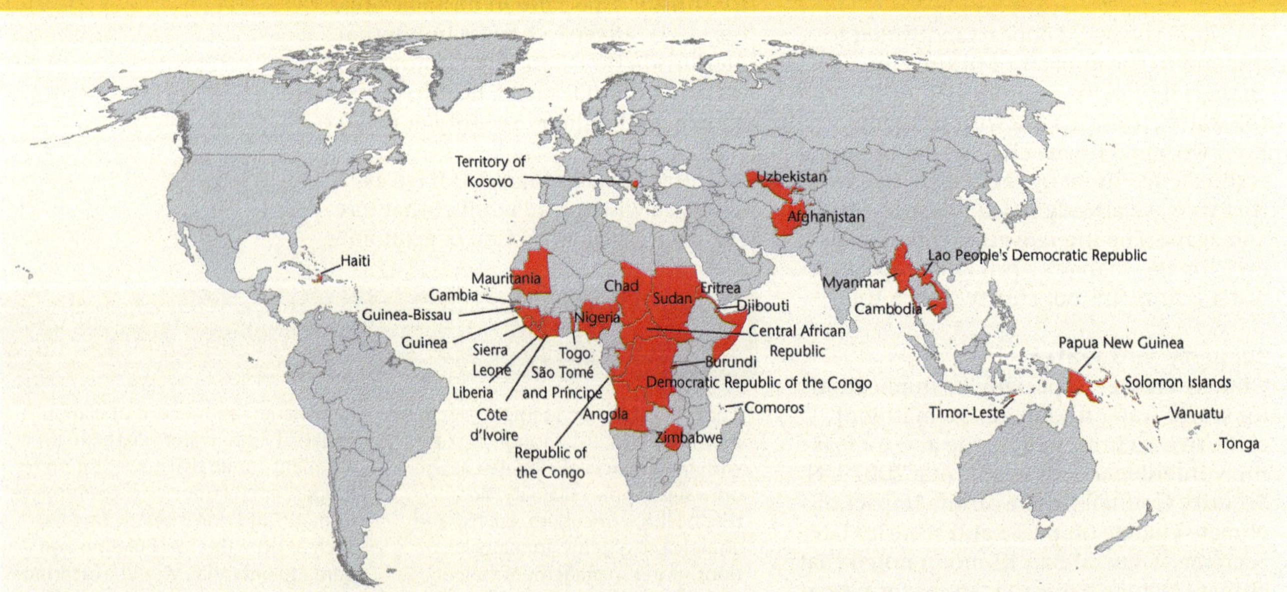

Note: Fragile states are low-income countries that score below a threshold on the International Development Association's Country Policy and Institutional Assessment, a tool used to assess the quality of country policies. The list is prepared annually.
Source: Based on IDA 2007.

them are made or influenced by leaders in government, the private sector and civil society. They are the ones who must recognize the role of water in attaining their objectives – and demonstrate the will to act now.

Structure of the Report

The Report has four parts. Part 1 examines water drivers – or what drives the pressures on water. Externalities, mostly human-induced, create pressures on water. Human activities and processes of all types – demographic, economic and social – can exert pressures on water resources that need to be managed. These pressures are affected by a range of factors such as technological innovation, climate change, and policies, laws and financial conditions

Part 2 is about using water. History shows strong, mutual links between economic development and water development. Steadily increasing demand for agricultural products to satisfy the diverse needs of a growing population (for food, fibre and now fuel) has long been the main driver behind agricultural water use. In a situation of tight balance between food supply and demand, climate events – droughts in particular – have an increasingly strong impact on food price volatility. There is a growing need to protect ecosystems and the goods and services they produce and on which life and livelihoods depend. As competition among demands on water increases, society will need to respond more effectively through improved water management, policies and transparent and efficient water allocation mechanisms.

Part 3 explores the state of water resources. The uneven distribution over time and space of water resources, and how that distribution is being modified, are fundamental sources of the water crisis. Global warming is expected to result in an intensification, acceleration or enhancement of the global hydrologic cycle. There is some observational evidence that this is already happening. In many places climate extremes have become more frequent or more intense, with droughts and floods affecting increasing numbers of people. Worldwide, water observation networks are inadequate for current needs and are at risk of further decline. The data to understand and predict water quantity and quality are lacking.

Part 4 is on responses and choices. It shows that we can do what it takes to manage water resources properly to avert crises and promote sustainable socioeconomic development. Others have already shown the way. But there is no one-size-fits-all solution. The best mix of responses to a specific country's development objectives and policy priorities to meet various water challenges depends on the availability of water of acceptable quality for its intended use and the country's technical, financial, institutional and human capacities and its culture, political and regulatory frameworks and markets.

Leaders within the water domain can inform the processes outside their domain and manage water resources to achieve agreed socioeconomic objectives. But it is the leaders in government, the private sector and civil society who determine the directions that development will take. Recognizing this, they must act now!

The decisions on development objectives and the allocation of human and financial resources needed to meet them are made or influenced by leaders in government, the private sector and civil society – not by water managers or specialists

Notes

1. There were exceptions, such as the development of the Tennessee River in the United States beginning in the 1930s under the Tennessee Valley Authority.
2. Commission on Growth and Development 2008.
3. Phumpiu and Gustafsson 2007.
4. See www.pap.org.mz.
5. Speaking at the session Re-Thinking Social Responsibility on 25 January 2008, as cited in Maidmont 2008.
6. Commission on Growth and Development 2008, p. 1.
7. ADB 2007, p. vi.
8. The benefits of investing in water are presented in greater detail in chapter 6.
9. SIWI 2005.
10. Hutton and Haller 2004.
11. Schuster-Wallace et al. 2008.
12. UN-Water 2008.
13. WHO 2006.
14. Hussein 2008.
15. Winpenny 2003.
16. van Hofwegen 2006.
17. DFID 2007, p. 2.
18. World Bank 2003.
19. NEPAD 2002.
20. Originally defined as $1.00 per day and revised to $1.25 in 2005 to reflect evolving purchasing power parity
21. Chen and Ravallion 2008.
22. UNDP 2006, p. 6.
23. World Bank 2005.
24. Kaberuka 2008.
25. MDG Africa Steering Group 2008.
26. WELL 2005.
27. Poverty-Environment Partnership 2006.
28. G-8 2003.
29. UN-Water 2007.
30. World Bank 2008.
31. Worldwatch Institute 2008, pp. 117-21.
32. United Nations 2008.
33. Hoffman 2004.
34. FAO 2008.
35. Zoellnick 2008.
36. IMF 2009.
37. Commission on Growth and Development 2008, p. 103.
38. Rajan 2006.

39. UNFCCC 2007.
40. UNFCCC 2005.
41. UN Security Council 2007.
42. IPCC 2008.
43. The Oxford Research Group, in a briefing paper on sustainable security, argues that the effects of climate change – displacement of peoples, food shortages, social unrest – have long-term security implications far greater than those of terrorism and notes that the U.S. Department of Defense's Office of Net Assessment takes the same view (Abott, Rogers, and Sloboda 2006, p. 7).
44. Such as the statement by the European Commission and the Secretary General/ High Representative for Foreign and Security Policy Javier Solana (2006, p. 2): 'Investment in mitigation . . . as well as ways to adjust to the unavoidable should go hand in hand with addressing the international security threats created by climate change.'
45. Such as U.K. Foreign Secretary Margaret Beckett's statement in the 2007 UN Security Council debate on the impact of climate change on peace and security that climate change exacerbates many threats (UN Security Council 2007) and the testimony of Deputy Director of National Intelligence for Analysis (NIA) Thomas Finger before a Joint House committee that an NIA assessment found that sub-Saharan Africa, the Middle East and Central and South-East Asia are most vulnerable to warming-related drought, flooding, extreme weather and hunger (House Permanent Select Committee on Intelligence and U.S. House Select Committee on Energy Independence and Global Warming 2008, p.13).
46. UNDP and World Bank 2007.

References

Abott, Chris, Paul Rogers, and John Sloboda. 2006. *Global Responses to Global Threats Sustainable Security for the 21st Century.* Oxford, UK: Oxford Research Group.

ADB (Asian Development Bank). 2007. *Asian Water Development Outlook 2007: Achieving Water Security for Asia.* Manila: Asian Development Bank.

African Union. 2008. Sharm El-Sheikh Commitments for Accelerating the Achievement of Water and Sanitation Goals in Africa. Declaration I. Assembly of the African Union, Eleventh Ordinary Session, 30 June-1 July, Sharm el-Sheikh, Egypt. www.africa-union.org/root/au/Conferences/2008/june/summit/dec/ASSEMBLY%20DECISIONS%20193%20-%20207%20(XI).pdf.

Alam, Dewan S., Geoffrey C. Marks, Abdullah H. Baqui, M. Yunus, and George J. Fuchs. 2000. Association between Clinical Type of Diarrhoea and Growth of Children under 5 Years in Rural Bangladesh. *International Journal of Epidemiology* 29 (5): 916-21.

Ban Ki-moon. 2008. Address by the UN Secretary-General to the session "Time is Running Out on Water," of the Davos World Economic Forum, 24 January 2008, in Davos, Switzerland. www.un.org/apps/news/infocus/sgspeeches/search_full.asp?statID=177.

Bergkamp, G., and C. W. Sadoff. 2008. Water in a Sustainable Economy. In *State of the World: Innovations for a Sustainable Economy.* Washington, DC: Worldwatch Institute.

Chen, Shaohua, and Martin Ravallion. 2008. *The Developing World Is Poorer than We Thought, but No Less Successful in the Fight against Poverty.* Policy Research Working Paper 4703. World Bank, Washington, DC.

Commission on Growth and Development. 2008. *The Growth Report: Strategies for Sustained Growth and Inclusive Development.* Conference Edition. Washington, DC: World Bank.

Comprehensive Assessment of Water Management in Agriculture. 2007. *Water for Food, Water for Life: A Comprehensive Assessment of Water Management in Agriculture.* London: Earthscan, and Colombo: International Water Management Institute.

Cosgrove, W. J. 2006. Water for Growth and Security. In *Water Crisis: Myth or Reality?: Marcelino Botin Water Forum 2004.* Peter Rogers, M. Ramon Llamas, and Louis Martinez-Cortina, eds. London: Taylor and Francis.

Delli Priscoli, J., and A. T. Wolf. 2009. *Managing and Transforming Water Conflicts.* International Hydrology Series. Cambridge, UK: Cambridge University Press.

DFID (Department for International Development). 2007. Growth and Infrastructure Policy Paper. Department for International Development, London.

EIA (Energy Information Administration). 2005. *International Energy Annual 2005.* Washington, DC: U.S. Department of Energy. www.eia.doe.gov/iea.

———. **2008a.** *Annual Energy Outlook 2008* (DOE/EIA 0383/2008). Washington, DC: U.S. Department of Energy.

———. **2008b.** *World Energy Projections Plus.* Washington, DC: U.S. Department of Energy.

European Commission and the Secretary-General/High Representative for Foreign and Security Policy. 2006. Report on Climate Change and International Security to the European Council, 3 March 2006. Council of the European Union, Brussels. www.envirosecurity.org/activities/diplomacy/gfsp/documents/Solana_security_report.pdf.

FAO (Food and Agriculture Organization of the United Nations). 2006. *Rapid Growth of Selected Asian Economies: Lessons and Implications for Agriculture and Food Security. Synthesis Report.* Bangkok: Food and Agriculture Organization of the United Nations.

———. **2008.** Declaration of the High-Level Conference on World Food Security: The Challenges of Climate Change and Bioenergy. High-Level Conference on World Food Security, 3-5 June, Rome. www.un.org/issues/food/taskforce/declaration-E.pdf.

G-8. 2003. Water: A G8 Action Plan. G8 Summit, 1-3 June, Evian, France.

———. **2008.** Declaration of Leaders Meeting of Major Economies on Energy Security and Climate Change. Hokkaido Toyako Summit, 9 July, Toyako, Hokkaido, Japan. www.mofa.go.jp/policy/economy/summit/2008/doc/doc080709_10_en.html.

Hoffman, Allan R. 2004. The Connection: Water and Energy Security. *Energy Security*, 13 August. Institute for Analysis of Global Security. www.iags.org/n0813043.htm.

House Permanent Select Committee on Intelligence and House Select Committee on Energy Independence and Global Warming. 2008. National Intelligence Assessment on the National Security Implications of Global Climate Change to 2030: Statement for the Record of Dr. Thomas Fingar, Deputy Director of National Intelligence for Analysis and Chairman of the National Intelligence Council, 25 June 2008. U.S. Congress, Washington, DC.

Hussein, M. A. 2008. Costs of Environmental Degradation: An Analysis in the Middle East and North Africa Region. *Management of Environmental Quality* 19 (3): 305-17.

Hutton, Guy, and Laurence Haller. 2004. *Evaluation of the Costs and Benefits of Water and Sanitation Improvements at the Global Level.* Geneva: World Health Organization.

Hutton, Guy, Laurence Haller, and Jamie Bartram. 2007. Economic and Health Effects of Increasing Coverage of Low Cost Household Drinking-water Supply and Sanitation Interventions to Countries Off-track to Meet MDG Target 10. Background paper for *Human Development Report 2006*, World Health Organization, Public Health and the Environment, Geneva.

IDA (International Development Association). 2007. List of Fragile States. The International Development Association of the World Bank, Washington, DC. http://web.worldbank.org/.

IMF (International Monetary Fund). 2009. *World Economic Outlook Update. January.* Washington, DC: International Monetary Fund. www.imf.org/external/pubs/ft/weo/2009/update/01/index.htm.

IPCC (Intergovernmental Panel on Climate Change). 2008. Technical Paper on Climate Change and Water. IPCC-XXVIII/Doc. 13, Intergovernmental Panel on Climate Change, Geneva. www.ipcc.ch/meetings/session28/doc13.pdf.

Kaberuka, Donald. 2008. Opening Statement. African Development Group First African Water Week: Accelerating Water Security for Socio-Economic Development of Africa, 26-28 March, Tunis.

Maidmont, Paul. 2008. Re-thinking Social Responsibility. *Forbes* 25 (January). www.forbes.com/leadership/citizenship/2008/01/25/davos-corporate-responsibility-lead-cx_pm_0125notes.html.

Manning, Nick. 1999. *Strategic Decision-making in Cabinet Government: Institutional Underpinnings and Obstacles.* Washington, DC: World Bank.

MDG Africa Steering Group. 2008. *Achieving the Millennium Development Goals in Africa: Recommendations of the MDG Africa Steering Group.* New York: United Nations.

MEA (Millennium Ecosystem Assessment). 2005. *Ecosystems and Human Well-Being: Wetlands and Water Synthesis.* Washington, DC: World Resources Institute.

NEPAD (New Partnership for Africa's Development). 2002. *Comprehensive Africa Agriculture Development Programme.* New Partnership for Africa's Development, Midrand, South Africa.

OECD (Organisation for Economic Co-operation and Development) and FAO (Food and Agriculture Organization of the United Nations). 2008. *OECD-FAO Agricultural Outlook 2008-2017.* Paris: Organisation for Economic Co-operation and Development.

Phumpiu, P., and J. E. Gustafsson. 2007. Reform or Adjustment? From the Liberal and Neo-liberal Policy Reform Process in Honduras. Paper from 5th International Water History Association Conference: Pasts and Futures of Water, 13-17 June, Tampere, Finland.

Poverty-Environment Partnership. 2006. *Linking Poverty Reduction and Water Management.* Poverty-Environment Partnership.

Prüss-Üstün, A., and C. Corvalán. 2006. *Preventing Disease through Healthy Environments: Towards an Estimate of the Environmental Burden of Disease.* Geneva: World Health Organization. www.who.int/quantifying_ehimpacts/publications/preventingdisease/en/index.html.

Rajan, Raghuram G. 2006. Investment Restraint, the Liquidity Glut and Global Imbalances. Remarks by the Economic Counsellor and Director of Research, International Monetary Fund, at the Conference on Global Imbalances organized by the Bank of Indonesia in Bali, Indonesia, November 16, 2006. www.imf.org/external/np/speeches/2006/111506.htm.

Schuster-Wallace, Corinne J., Velma I. Grover, Zafar Adeel, Ulisses Confalonieri, and Susan Elliott. 2008. *Safe Water as the Key to Global Health.* Hamilton, Canada: United Nations University International Network on Water, Environment and Health.

SIWI (Stockholm International Water Institute). 2005. *Making Water a Part of Economic Development: The Economic Benefits of Improved Water Management and Services.* Stockholm: Stockholm International Water Institute.

UN Security Council. 2007. Security Council Holds First-ever Debate on Impact of Climate Change on Peace, Security, Hearing over 50 Speakers. UN Security Council 5663rd meeting, 17 April 2007. United Nations Security Council, Department of Public Information, News and Media Division, New York, 17 April. www.un.org/News/Press/docs/2007/sc9000.doc.htm.

UNDP (United Nations Development Programme). 2006. *Human Development Report 2006. Beyond Scarcity. Power, Poverty and the Global Water Crisis.* New York: Palgrave Macmillan.

UNDP (United Nations Development Programme) and World Bank. 2007. Somalia Reconstruction and Development Programme: Deepening Peace and Reducing Poverty. Somalia Joint Needs Assessment, United Nations and World Bank Coordination Secretariat, Nairobi. www.somali-jna.org/downloads/ACF7C9C.pdf.

UNFCCC (United Nations Framework Convention on Climate Change). 2005. Five-year Programme of Work on the Subsidiary Body for Scientific and Technological Advice on Impacts, Vulnerability and Adaptation to Climate Change. FCCC/CP/2005/5/Add.1, Bonn, Germany.

———. **2007.** Report on the Workshop on Adaptation Planning and Practices. FCCC/SBSTA/2007/15, United Nations Framework Convention on Climate Change.

United Nations. 2008. Compilation of Initiatives and Commitments relating to the High-Level Event on the Millennium Development Goals. High-Level Event on the Millennium Development Goals: Committing to Action: Achieving the Millennium Development Goals, 25 September, New York.

UN-Water. 2007. Outcome of Seminar Coping with Water Scarcity. Outcome of the Seminar on Coping with Water Security, 23 August, Stockholm. www.unwater.org/www-seminar2.html.

———. **2008.** Sanitation Is Vital for Health. Factsheet 1. International Year for Sanitation, New York.

van Hofwegen, Paul, and Task Force on Financing Water for All. 2006. *Enhancing Access to Finance for Local Governments. Financing Water for Agriculture.* Chaired by Angel Gurría. Marseille, France: World Water Council.

WELL. 2005. Supporting the Achievement of the MDG Sanitation Target. WELL Briefing Note for CSD-13, WELL, Leicestershire.

WHO (World Health Organization). 2006. Economic and Health Effects of Increasing Coverage of Low Cost Water and Sanitation Interventions. UNHDR Occasional Paper, World Health Organization, Geneva.

WHO (World Health Organization) and UNICEF (United Nations Children's Fund) Joint Monitoring Programme. 2008. *Progress on Drinking Water and Sanitation: Special Focus on Sanitation.* New York and Geneva: World Health Organization.

Winpenny, James. 2003. *Financing Water for All.* Report of The World Panel on Financing Water Infrastructure, chaired by Michel Camdessus. Kyoto: World Water Council, 3rd World Water Forum, and Global Water Partnership.

World Bank. 2003. Water Resources Sector Strategy: Strategic Directions for World Bank Engagement. World Bank, Washington, DC.

———. **2005.** *Global Monitoring Report. Millennium Development Goals: From Consensus to Momentum.* Washington, DC: World Bank.

———. **2008.** *Environmental Health and Child Survival: Epidemiology, Economics Experiences.* Washington, DC: World Bank.

Worldwatch Institute. 2008. *State of the World: Innovations for a Sustainable Economy.* New York: Norton Institute.

Zoellnick, Robert B. 2008. Speech by President of World Bank at United Nations' High-Level Conference on World Food Security: The Challenges of Climate Change and Bioenergy, 3 June 2008, Rome, Italy.

Understanding what drives the pressures on water

PART 1

Chapters 2-5 Coordinator
Tim Kasten (UNEP) with the support of Thomas Chiramba (UNEP)

Facilitator
Richard Connor

PART

1

Authors: Richard Connor and Walter Rast

The amount of freshwater on Earth is finite, but its distribution has varied considerably, driven mainly by natural cycles of freezing and thawing and fluctuations in precipitation, water runoff patterns and evapotranspiration levels.

That situation has changed, however. Alongside natural causes are new and continuing human activities that have become primary 'drivers' of the pressures affecting our planet's water systems. These pressures are most often related to human development and economic growth. Our requirements for water to meet our fundamental needs and our collective pursuit of higher living standards, coupled with the need for water to sustain our planet's fragile ecosystems, make water unique among our planet's natural resources.

Chapters 2-5 describe these water drivers and their interactions as they relate to the sustainability of water resources and systems. They also examine how to make reasonable predictions about the future. Such forecasts are relevant for policy-making directed to water resources and for development activities, investment planning and other activities generally considered to be outside the domain of the water sector – or 'outside the water box'.

Part 1 examines the processes behind the rising pressures on our water supplies, identify the ones most likely to have the greatest impact on the world's water resources in the coming decades and describe the context within which water will be managed. These chapters describe what we know about the current situation and recent trends and forecast possible futures related to processes that we refer to as drivers of change and define as:

> a set of fundamental processes that are external to the water sector and that directly or indirectly co-determine the evolution of the water system in terms of the quality, quantity and spatial distribution of the resource.

At the turn of the century the World Water Vision exercise of the World Water Council – the first and largest international effort to develop global water scenarios – identified a series of 'driving forces' that 'represent key factors, trends or processes which influence the situation, focal issues or decisions, and actually propel the system forward and determine the story's outcome'.[1] Using this definition, the Vision team selected major drivers and organized them into six clusters: demographic, economic, technological, social, governance and environmental. This part of the Report draws on these clusters, with the exception of the environment, which is defined as a use and is covered extensively in part 2. To this list we have added climate change, discussed in chapter 5 and throughout the Report. This part of the Report also describes many of the complex links between the drivers, which can cause both positive and negative feedback impacts.

In describing drivers 'external to the water sector', we have sought to identify key forces or processes of change over which water sector users, managers and decision-makers have little direct influence. Thus, water use sectors (agriculture, energy, domestic and industrial) are not drivers even though they have a major impact on the resource because they are not external to the water sector. The drivers of agriculture – and its demand for water – are such fundamental processes as population growth, changes in dietary preferences as living standards rise, and increasing demand for non-food agricultural products such as bioenergy. The drivers of change are the demographic, economic and social forces that, in combination, exert pressures on the agriculture sector. This leads to an evolution in agriculture practices, which can also be influenced by technological innovation and agricultural and trade policies, all of which eventually affect the quality and quantity of water.

These drivers should not be considered in isolation of related socioeconomic or political factors and other drivers. Many natural links influence how drivers affect changes, directly and indirectly. Water properties are governed by biological, chemical and physical laws that define the quantity and quality of water resources and that are linked in various ways. Temperature, a physical factor, can affect the metabolism of aquatic organisms, a biological process. The excessive biological production (such as excessive algal growth) associated with increased temperature can degrade water quality, a chemical property.

Superimposed on these natural processes are human activities that exacerbate these processes, disrupting the natural balance of water systems. The growth of algae or aquatic plants in a lake, for example, is stimulated by excessive nutrients and minerals washed into the lake as a result of human activities, accelerating natural growth processes to levels that can cause water quality degradation and interfere with beneficial water uses.

Drivers are thus the forces and processes generated by human activities. Consider governments' efforts to improve citizens' livelihoods and standards of living by increasing economic growth. Economic growth is affected by a wide range of policy decisions, from international trade to education and public health, while the potential rate of economic growth can be affected by demographic variables such as population distribution (local workforce availability) and social characteristics (workforce capacity) and by the availability of new technologies. Economic activity also requires adequate quantities of natural resources, including freshwater. And water availability is directly subject to the impacts of climate change, which can exert additional pressures on other drivers.

A rising standard of living is typically accompanied by increased consumption and production of goods, along with rising demands for water-related household services and water resources to facilitate economic growth and related activities. Rising demand for meat and fish in urbanized and emerging market economies, for example, has increased fishery activities and livestock production, generally a water-intensive activity. The feedback loop of degraded water quality from livestock feedlot runoff can diminish fish production or alter its quality. There is also sociological evidence that urbanization shifts fishing pressures from natural water systems to artificial systems. Thus, urbanization and globalization, with changes in diets and lifestyles, are strong drivers of water use, even though decisions made outside the water sector are driving them.

The result is a continuously increasing demand for finite water resources for which there are no substitutes. When water resources of acceptable quality can no longer be provided in sustainable quantities to meet such demands, aquatic ecosystems can be overexploited as each sector or user group tries to satisfy its own water needs at the expense of others. The ultimate loser is the sustainability of the exploited aquatic ecosystems and the organisms (including humans) dependent on them for survival and well-being.

Note

1. Gallopín and Rijsberman 2000, p. 18.

Chapter 2

Demographic, economic and social drivers

Authors: Gunilla Björklund, Richard Connor, Anne Goujon, Molly Hellmuth, Patrick Moriarty, Walter Rast, Koko Warner and James Winpenny

Contributors: Arjen Hoekstra, Walter Rast and David Wiberg

Coordinator: Tim Kasten (UNEP)

Facilitator: Richard Connor

Key messages

- Human activities and processes of all types – demographic, economic and social – can exert pressures on water resources and need to be managed.
- These pressures are in turn affected by a range of factors such as technological innovation, institutional and financial conditions and climate change.
- The rapid global rise in living standards combined with population growth presents the major threat to the sustainability of water resources and environmental services.

Demographic drivers

Authors: Richard Connor, Anne Goujon, Molly Hellmuth and Koko Warner

Contributors: Walter Rast and David Wiberg

Key messages

- Population dynamics (growth, age distribution, urbanization and migration) create pressures on freshwater resources through increased water demands and pollution.
- Changes in the natural landscape associated with population dynamics (migration, urbanization) can create additional pressures on local freshwater resources and the need for more water-related services.

Demographic processes such as population growth, age distribution, urbanization and migration create some of the greatest pressures on water resources quantity and quality. These demographic processes directly affect water availability and quality through increased water demands and consumption and through pollution resulting from water use. They affect water resources indirectly through changes in land use and water use patterns, with significant implications at local, regional and global levels. And the availability and quality of water as well as trends in water use can influence demographic processes.

The world's population is growing by about 80 million people a year, implying increased freshwater demand of about 64 billion cubic metres a year.[1] An estimated 90% of the 3 billion people who are expected to be added to the population by 2050 will be in developing countries, many in regions where the current population does not have sustainable access to safe drinking water and adequate sanitation.[2] Many governments lack the financial resources and institutional capacity to provide for these needs, while countries that have experienced gains in the number of people with access to water supply and sanitation services since 1990 may see these gains eroded by population growth.

The demographics of the global population are changing, with important implications for water resources. By 2050, 22% of the world's population is expected to be 60 years old or older, up from 10% in 2005. At the same time, the world has more young

The world will have substantially more people in vulnerable urban and coastal areas in the next 20 years. In areas with already-scarce water resources water managers will have to look beyond the water sector for solutions

people than ever, with nearly half the world population being under the age of 25.

While the world's urban population grew rapidly during the 20th century (from 220 million to 2.8 billion), the next few decades will see an unprecedented scale of urban growth across developing countries. In Africa and Asia the urban population is expected to double between 2000 and 2030. By 2030 the towns and cities of the developing world will make up an estimated 81% of urban humanity.[3]

Today, there are an estimated 192 million migrants worldwide, up from 176 million in 2000.[4] Coastal areas, with 18 of the world's 27 megacities (populations of 10 million or greater), are thought to face the largest migration pressures.[5] About 75% of people residing in low-lying areas are in Asia, with the most vulnerable being poor people. International migration is increasing as a result of such factors as demographic changes, economic disparities, trade liberalization, environmental changes and new communication technologies. Impacts of climate change can substantially accelerate migration (see chapter 5). Demographic changes affect international migration in two ways. Rapid population growth, combined with economic difficulties, push people to cities, while a declining and ageing population induces countries to accept migrants, who are typically willing to work at much lower wages than native workers. Water shortages and hazards, particularly where people are directly dependent on the environment for their livelihood, can also induce migration.

The net implication of these demographic processes is clear: the world will have substantially more people in vulnerable urban and coastal areas in the next 20 years. The rate of slum formation is nearly the same as the rate of urban growth. In areas with already-scarce water resources water managers will have to look beyond the water sector for solutions. They will have to work closely with leaders in other sectors, such as education, health, social services and agriculture, to respond effectively to the demographic challenge.

Population growth

We live in a demographically divided world, with population still growing rapidly in some regions (Africa and the Middle East), ageing rapidly in others (Europe and East Asia) and already declining in others (Europe; map 2.1).

Besides Eastern Europe and the former Soviet Union, where annual population growth is already negative, Australia, China, Japan, New Zealand and Western Europe will also soon see shrinking populations. Around 2060 South Asia and Pacific will also experience negative population growth rates. Other regions are less susceptible to negative population growth forces. Sub-Saharan Africa and the Middle East will continue to experience high rates of population growth well into the future. This timing will characterize most of the problems of water scarcity.

Map 2.1 **Expected areas of population growth and decline, 2000-2080**

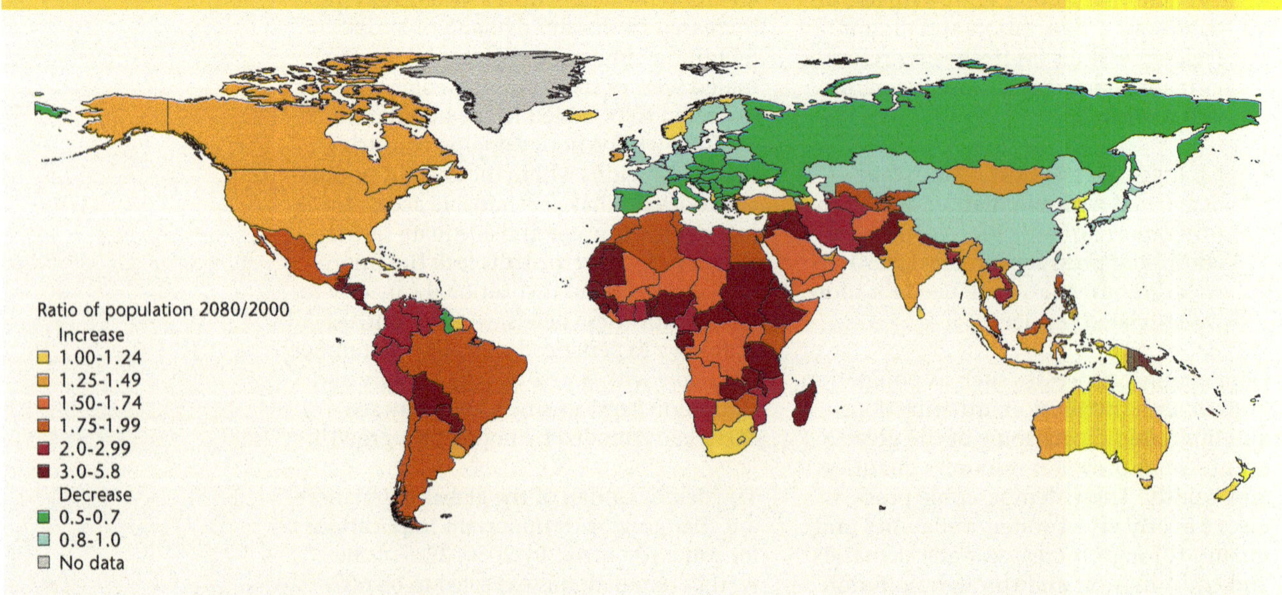

Source: Lutz, Sanderson, and Scherbov 2008.

Most population growth will occur in developing countries, mainly in regions that are already in water stress and in areas with limited access to safe drinking water and adequate sanitation facilities. More than 60% of the world's population growth between 2008 and 2100 will be in sub-Saharan Africa (32%) and South Asia (30%). Together, these regions are expected to account for half of world population in 2100. Such rates of population growth will have major social and environmental impacts, given the level of economic development in many affected countries.

Most population growth will occur in developing countries and mainly in regions that are already in water stress, in areas with limited access to safe drinking water and adequate sanitation facilities

Age distribution

The age of the population will influence consumption and production patterns, with attendant impacts on natural resource needs, including freshwater. The resource needs and services associated with increasing longevity will include greater provision of medicines, medical facilities and health-care providers. For younger people the globalization of trade and advertising tempts those in developing countries to want more and those in developed countries who already have more to want even more. These needs and wants translate into higher consumption and production patterns, requiring additional resources, including freshwater.

Urbanization and the growth of informal human settlements

In 2008 world population was estimated to be equally split between urban and rural areas, marking the transition from a rural dominated to an urban dominated world. By 2030 the number of urban dwellers is expected to be about 1.8 billion more than in 2005 and to constitute about 60% of the world's population (figure 2.1), while the number of rural inhabitants is expected to decline slightly from 3.3 billion to 3.2 billion. Almost all (95%) of the increase in urban populations is expected in developing countries, especially in Africa and Asia, where the urban population is projected to double between 2000 and 2030.[6] Urbanization rates are much lower in developed countries and are even declining in some countries.

Despite the continuing growth of megacities – which require natural resources and create waste in quantities not seen in human history – most of the world's urban populations live in cities with fewer than 500,000 inhabitants. The growth of small and mid-size cities will have significant impacts on water resources. In most established or formal urban areas access to water supply and sanitation services is believed to be better than in rural areas. But in informal urban areas residents have little access to safe drinking water or adequate sanitation services, increasing the danger of water- and sanitation-related diseases. It is through such informal urban areas that most urban growth occurs.

In addition to the sociological and health implications of increased population density in urban settlements, urbanization has unique environmental impacts. Urbanization is accompanied by the transformation of natural land surfaces into impervious surfaces, such as streets, parking lots, roofs and other types of structures that block the percolation of rainwater and snowmelt into soil. Such construction increases the flow velocity of water over the land surface, carrying polluting materials into receiving water systems, degrading water quality and causing local pollution problems. This urban drainage effect has increased the frequency of flash floods, causing casualties and infrastructure damage.

Migration

Migratory populations include traditional groups of subsistence-level pastoralists and agriculturalists, as well as family groups and individuals seeking greater opportunities and refugees fleeing the consequences of war, conflict or natural disasters. Refugees often pass through camps or informal settlements that may be artificially sustained by aid agencies or governments. The result is the rapid denuding of the surrounding area as people search for water and fuel wood in order to survive – leading to soil degradation, deforestation, land clearing and a scarcity of potable water. Migratory pastoralists

Figure 2.1 **By 2030 about 60% of the world's population is expected to live in urban areas**

Share of population residing in urban areas, 2005 and 2030 (percent)

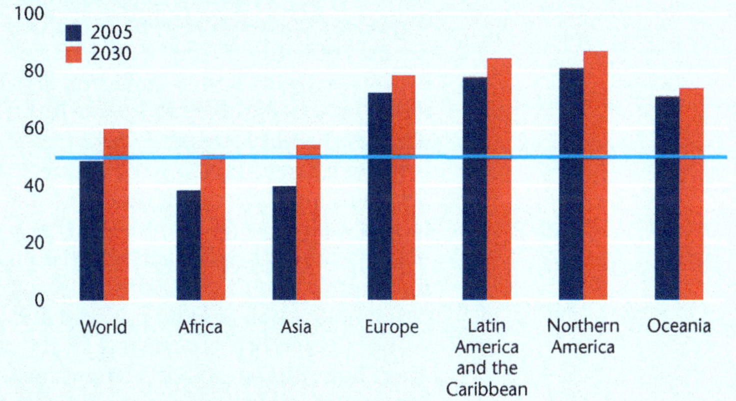

Note: Regions are official UN regions.
Source: United Nations 2006b.

The relation between water and migration is two-way: water stressors drive migration, and migration contributes to water stress

and agriculturalists can also have significant localized impacts on the surrounding landscape through such practices as overgrazing of livestock and slash and burn agriculture. While generally considered rural dwellers, these migrants frequently constitute a large share of those seeking better economic opportunities and better access to water and sanitation services, shelter, health services and food stocks in established urban areas, leading to the proliferation of informal communities on the fringes of cities.

The relation between water and migration is two-way: water stressors drive migration, and migration contributes to water stress. Water stressors, such as water scarcity and flooding, can trigger migration decisions. The social, economic and political context in which water stresses occur will influence the migration response. And if the natural environment becomes inhospitable, people are motivated to move to areas where their locally specific knowledge may no longer apply. Once people move, their places of destination must provide them with water resources, which can lead to further environmental stresses.

In these situations the arrival of additional people can worsen existing water crises and strain the capacity of the urban infrastructure. Water conflicts can be exacerbated through migration or the presence of refugees, and the fragile balance of human populations and water resources can be upset. Increasingly, links between environmental issues, including water, and security issues, including migration, have become a topic of scientific research and policy debate. Climate change, which is predicted to lead to greater frequency and intensity of extreme weather events, is likely to result in an overall increase in the displacement of people in the future (see chapter 5).

Estimates of potential environmentally displaced people range from 24 million to almost 700 million who could be displaced by water-related factors, including development projects designed to relieve future water availability stresses.[7] Part of the complexity in unraveling the connection between migration and environmental factors such as water resources is that people rely indirectly or directly on the environment for their livelihoods. In addition, development policies and political and economic stability – or the lack of it – can affect both migration and water resources. Given these complexities, it is difficult to estimate the magnitude of potential migration as a result of environmental factors.

One positive outcome of migration is the lessening of the pressures on the vacated lands, which may allow some ecosystems to recover. In Europe and North America the rural exodus has resulted in the growth of new parklands in some locations.

Challenges

With rapidly ageing populations in some places and rapidly ageing populations combined with a shrinking population in others, it is important to consider the quality dimensions of education and health as well as the quantitative dimensions of population size and age structure in addressing the water needs of evolving communities. To meet the challenges of rapid urban population growth, decision-makers can focus on positive factors that affect fertility decline – social development, investments in health and education, empowerment of women and better access to reproductive health services – in contrast to antimigration approaches.

Economic drivers

Authors: Richard Connor, Walter Rast and James Winpenny

Contributor: Arjen Hoekstra

Key messages

- Growth and changes in the global economy are having far-reaching impacts on water resources and their use.

- Growing international trade in goods and services can aggravate water stress in some countries while relieving it in others through flows of 'virtual water', particularly in the form of imported agricultural commodities.

Global economic expansion affects water through growth in the number of consumers and through changes in their consumption habits, in the way goods and services are produced and in the location of activities, all of which affect international trade. Growth in global output is currently estimated to slow to 2.2% in 2009, though this will likely be less because of the economic volatility arising from the global financial crisis.[8] The growth output is also unevenly distributed. Several emerging market economies are registering continuously high growth rates, transforming them into major global economic forces. Brazil, China, India and the Russian Federation are, on Goldman Sachs' latest forecast, expected to overtake the combined economic strength of the G-8 by 2032.[9] Even sub-Saharan Africa, long a growth straggler, is experiencing growth rates of 6% or more, fuelled largely by oil and commodities.

Water is affected by economic forces, while the state of water resources has a strong feedback to the economy. In periods of water shortages public authorities are likely to close factories and divert water from farmers to release water supplies for households. Water contamination from industrial effluents may result in factory closures and relocation, while the depletion and contamination of groundwater may compel industries to relocate. Lack of water storage infrastructure may cause heavy economic losses from flooding and drought. Polluted water has high costs for human health. In short, adequate investments in water management, infrastructure and services can yield a high economic return by avoiding such related costs.[10]

Globalization – used here as shorthand for the increasing international flows of goods and services, people, investments and finance – may make the situation worse, but it can also provide solutions. Producing and exporting goods and services with a large water footprint (the volume of water used in producing the goods and services consumed) could aggravate the problems of a water-scarce economy. Yet such an economy could gain from importing goods with a high water content (importing virtual water). Companies can escape their local water problems by relocating to other countries. However, growing corporate awareness of a firm's water footprint is leading to greater transparency about the impact of a firm's supply chain on its water environment. Globalization is also enabling the spread of water expertise provided by international firms and through global communications of service providers in other countries. These companies are a key part of water solutions through the desalination, re-use and wastewater treatment technologies they bring with them.

The following sections focus on economic processes that have exerted pressures on water resources and how they are managed. In addition to globalization, these processes include the global food and fuel crises and international trade (virtual water and increasing awareness of the water footprints of production and services).

Globalization

While economic integration is a dominant feature of globalization, social, cultural, political and institutional aspects are also important. Changes in consumption patterns through growing demands and easier access to goods and services, increased transport and energy needs and global access to innovation and knowledge all play a role in globalization – and all have an impact on water resources and the environment.

Globalization has raised the productivity and living standards of people in the countries that have opened themselves to the global marketplace. However, the gains from globalization have not been evenly distributed. Many people remain on the fringes, and some have fallen further behind. Exclusion, grinding poverty and environmental damage create dangers. An estimated 1.4 billion people – often referred to as the 'bottom billion' – live on just $1.25 a day.[11] Those who suffer the most usually have the least to start with – indigenous peoples, women in developing countries, the rural poor and their children.

In many cases rapid economic growth has failed to provide opportunities for these poorest of the poor. Social services remain severely unfunded, and environmental and energy problems, including water quality and lack of service delivery, are acute. In advanced economies increased economic insecurity has been associated with rising inequality and the squeezing of social provisioning. In middle-income countries economic shocks, accelerated trade liberalization and premature deindustrialization have constrained economic diversification and formal job creation. Elsewhere, intractable poverty has fed a vicious circle of economic insecurity and political instability and, on occasion, communal violence.[12] Such situations increase the threat of degrading water resources and reducing environmental services.

In addition to these indirect pressures are the direct pressures, such as proliferation of invasive species. Related to the increasing exchange of goods through international shipping, invasive species have caused enormous environmental damage to aquatic and terrestrial ecosystems.

The global food crises and the rising cost of fuel and energy

Reversing decades of low prices, the two-year period 2006-08 has seen sharp, and largely unanticipated, increases in food prices. Because poor people spend one-half to three-quarters of their income on food, a steep increase in the price of rice, grains and edible oils is tantamount to a large reduction in income. While in the long run higher food prices are an opportunity for those who live and work in rural areas (especially if they have the technology and

The gains from globalization have not been evenly distributed: many people remain on the fringes, and some have fallen further behind

The world will need almost 60% more energy in 2030 than in 2002. Water is needed for the production of energy of all types, so expansion of energy supply will affect water resources

the inputs – including water – needed to raise their productivity to its full potential), in the short run higher prices create a crisis for the urban and rural poor. Although Africa and other low-income countries are particularly vulnerable, even middle-income countries are at risk if they lack well developed safety nets.

According to the Commission on Growth and Development, there are many potential causes for the steep food price increases. Contributing factors include rising demand, shifting diets, droughts, increased costs of agricultural inputs (such as fertilizers) and policies that encourage the use of agricultural land and output for bioenergy production. Although there is no consensus yet on the relative importance of these factors, many believe that policies favouring bioenergy over food need to be reviewed.[13] The 2008 Declaration of the High-Level Conference on World Food Security: The Challenges of Climate Change and Bioenergy cautions:

> We are convinced that in-depth studies are necessary to ensure that production and use of biofuels is sustainable in accordance with the three pillars of sustainable development and takes into account the need to achieve and maintain global food security.[14]

Other longer-term factors may also have been at play. The low agricultural prices prevailing until recently may have led governments to neglect investments in rural infrastructure, research and development, storage and food security programmes that were once a priority. In parallel, agricultural policies in many countries encouraged non-food commodities such as bioenergy, fibres and narcotics over food commodities. Many major food-producing countries have reacted to the crisis by restricting exports to help contain prices at home, driving international prices still higher. Global markets in food have become temporarily fragmented. The recent food crisis has encouraged countries to re-consider food self-sufficiency, giving it prominence over purely economic considerations. This will likely have an impact on national food and agriculture policies for several years, with implications for water resources management.

A drive towards food self-sufficiency would have undesirable consequences for national water security, especially for countries in arid regions. Such policies, though beneficial for rural development, increase a country's national water footprint and forfeit growth in higher-income, less water-intensive sectors.

Crude oil prices have also risen sharply in recent years – from under $25 a barrel in 2002 to more than $150 in July 2008 before dropping back to just under $40 in early January 2009. Among the likely contributing factors to the rise is increased demand linked to economic growth in emerging market economies. This growing demand has also increased pressure to exploit new sources of oil. Many of these, such as the tar sands in Western Canada, have a very high water – and environmental – footprint (see chapter 3). Increasing oil prices are also likely linked to the overall increase in the cost of energy, which has been rising steadily since the early 1970s (figure 2.2).

Like food security, energy security is important for GDP growth. According to the International Energy Agency, the world will need almost 60% more energy in 2030 than in 2002, with economic growth in developing countries driving most of the increase.[15] Development of hydropower is one energy strategy to reduce dependence on fossil fuels and limit greenhouse gas emissions, and developing countries possess significant hydropower potential. Water is needed for the production of energy of all types (see chapter 7), so expansion of energy supply will affect water resources and related environmental services. Energy to support growth within urban centres will depend largely on water resources management responses to centralized power production. Growth in small towns will likely rely more on off-grid renewable energy sources.

High prices can provide incentives for greater efficiency in fuel consumption and agricultural production and can generate

Figure 2.2 The cost of energy to consumers has been rising since the 1970s

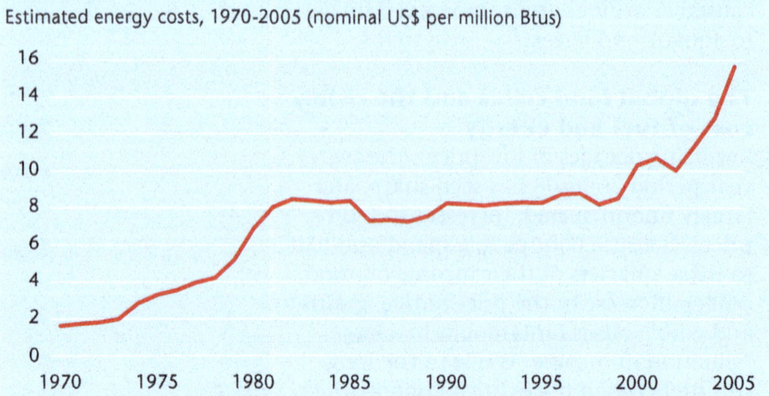

Source: Based on EIA 2008.

more income for people in rural areas. High fuel prices are likely to spur the development of alternative energy types like wind and solar, which require little water, and many countries also benefit from higher tax proceeds when energy prices rise – resources that could be used for further investments in efficiency and development.

Water and trade: virtual water and growing awareness of water footprints

The concepts of water footprints and virtual water are used to describe the relations among water management, international trade and politics and policies, and water resources use as it pertains to human consumption. Water footprints measure how much water is used in the production and consumption of goods and services (as well as how much pollution is generated), while virtual water is a tool for determining the movement of water through international trade.

Because water is heavy relative to its value, it is not feasible to transport it in bulk over long distances, with the exception of limited schemes for drinking water. Thus, water is predominantly a local concern, although it becomes a regional issue where rivers or lakes cross national boundaries. What transforms water into a global issue is trade in goods and services with a substantial water content either in the finished product or in its production (so-called virtual water). Providing water and wastewater services to households, industries and farmers can also have implications for international trade.

Countries with water shortages can import water-intensive goods and services, while water-abundant countries can take advantage of their bountiful water supplies through exports. While this beneficial trade happens broadly at a regional level (box 2.1), many countries have trade patterns that do not promote or benefit from this advantage. Through patterns of consumption and imports, countries can aggravate water shortages and pollution of their water supplies. Trade distortions and failure to properly price water resources may worsen the water-related problems of trading partners (see map 7.3 in chapter 7 and map 8.1 in chapter 8).

Many companies are beginning to understand the need to measure their water footprint, including that of their supply chains, and to relieve water stress in the communities where they operate.

Countries with water shortages can import water-intensive goods and services, while water-abundant countries can take advantage of their bountiful water supplies through exports

Box 2.1 Virtual water

Water-intensive products are heavily traded over large distances, as countries import and export water in virtual form as agricultural and industrial commodities. The global volume of virtual water flows in commodities is 1,625 billion cubic metres (m^3) a year, accounting for about 40% of total water consumption. About 80% of these virtual water flows relate to agricultural products trade, and the remainder to industrial products trade.

Global virtual water trade can save water if products are traded from countries with high water productivity to countries with low productivity. For example, Mexico imports wheat, maize and sorghum from the United States, which requires 7.1 billion m^3 of water a year in the United States to produce. If Mexico produced the imported crops domestically, it would require 15.6 billion m^3 a year. From a global perspective this trade in cereals saves 8.5 billion m^3 of water a year. Despite some trade from countries with low water productivity to countries with high productivity, global water savings through international trade of agricultural products has been estimated at about 350 billion m^3 a year, equivalent to 6% of the global volume of water used for agricultural production.

Many countries, including Japan, Mexico and most countries in Europe, the Middle East and North Africa, have net virtual water imports (see map). Water security in many countries thus strongly depends on external water resources (see chapter 7).

Regional virtual water balances and net interregional virtual water flows related to trade in agricultural products, 1997-2001

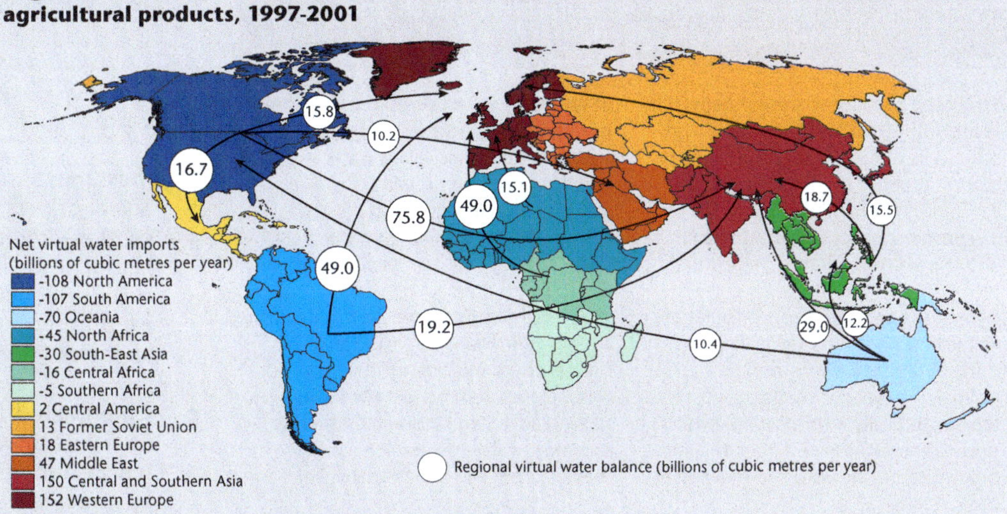

Source: Based on Hoekstra and Chapagain 2008.

> **A growing middle class is consuming much more milk, bread, eggs, chicken and beef, the production of which is more water-intensive than the simpler diets they are replacing**

Motivating companies to assess their water footprints is the desire to gain the goodwill of customers and potential customers and the need for cost control and risk management, including safeguarding access to the water essential for their operations. Recent business initiatives to support sustainable water management include the CEO Water Mandate launched at the 2007 UN Global Leadership Forum, the World Economic Forum's call for a 'coalition' of businesses to engage in water management partnerships and the World Business Council for Sustainable Development's creation of a water diagnostic tool and water scenario planning supports.[16]

Water is increasingly viewed as a potential threat and constraint to economic growth. As an example, China's remarkable economic growth has been accompanied by serious environmental problems, most notably water shortages in the north and pollution from wastewater effluents across the country (box 2.2). Massive projects begun to divert extensive water resources from the south to its more populated north will doubtless result in major environmental and social issues.

Trade and investment patterns are ultimately driven by demand, and changes in consumption and lifestyle accompany rising income levels in all countries. 'How much water do people drink?' (on average, 2-5 litres a day in developed countries) is much less relevant than 'How much water do people eat?' (3,000 litres a day in developed countries, according to one estimate).[17] Economic growth in emerging market economies is driving the growth of a middle class that is consuming much more milk, bread, eggs, chicken and beef, the production of which is more water-intensive than the simpler diets they are replacing.[18] Likewise in the services sector, tourism and recreation are creating an increasingly large water footprint in host societies.

The concepts of virtual water and water footprints are useful in illustrating the true influence of economic activity on water. With greater awareness should come measures to improve water productivity ('output per drop') in water-stressed environments and to reduce the polluting side effects of production.

Challenges

Globalization is bringing increasing economic opportunities to many, while leaving behind some who need them most: the world's poorest people living in the least developed countries. The first challenge is to shift this balance so that the less fortunate can have access to basic products and services, including sustainable access to safe drinking water and adequate sanitation services.

A second major challenge is to ensure that the cumulative action of economic activities and all other water drivers does not overwhelm nature's ability to provide for human needs. The expansion and growth of the global economy, and the resulting increases in human consumption, drive human demands to use more natural resources, including freshwater. However, the goods and services provided by ecosystems (such as water, biodiversity, fibre, food, feed and climate) are finite and vulnerable. Balancing economic development and environmental sustainability – and all the drivers influencing these links – remains a core requirement for sustainable development.

Social drivers

Authors: Gunilla Björklund, Richard Connor, Anne Goujon, Patrick Moriarty, Walter Rast and James Winpenny

Key messages

- Social drivers influence human perceptions and attitudes about the environment, including water resources, in turn influencing the pressures people exert on water through water demands and uses.

- Changes in lifestyles are one of the principal drivers of change. They reflect human needs, desires and attitudes (as illustrated in consumption and production patterns), which are influenced by such social drivers

---

Box 2.2 Water: a brake on economic growth and corporate prospects

While the scarcity of freshwater is felt acutely in Africa and West Asia, water scarcity is already an economic constraint in major growth markets such as China, India and Indonesia, as well as commercial centres in Australia and the western United States. If current consumption patterns continue, two-thirds of the world's population will live in water-stressed conditions by 2025. Compounding – and politicizing – these challenges is the reality that fully a third of the world's population lacks access to sufficient quantities of safe water to meet their basic needs.

In the next two to five years many companies will need to adapt to water availability concerns, including water stress and flooding; water quality concerns, including increasingly contaminated surface and groundwater supplies; and water access concerns, specifically competition with other water users. Corporate leaders who prepare careful water strategies for managing medium-term business risks and opportunities will not only be prepared to meet the future – gaining advantage in some of the key, and most water-constrained, global markets – but can also help shape it.

Source: Pacific Institute 2007.

---

as culture and education and by economic drivers and technological innovation; the rapid global rise in living standards combined with population growth presents the major threat to the sustainability of water resources and the environment.

Social drivers are mainly about individual rather than collective actions and about the way people think and act on a day-to-day basis. The four social drivers considered here are poverty, education, cultures and value systems, and lifestyles and consumption patterns.

Poverty

Poverty leaves people with few choices. They must do what is necessary for their survival, whatever the environmental consequences. Slash-and-burn agricultural practices, overexploitation of inland fisheries and the proliferation of informal settlements around urban areas in developing countries attest to this reality. And even as many developing countries have addressed problems of hunger and malnutrition, water quality has been degraded and per capita water availability has worsened. The poorest communities are also commonly in areas most vulnerable to the impacts of climate change and variability, including unstable hillsides and low-lying coastal areas, and lack the capacity to cope with natural disasters.

The inadequate water resources and sanitation facilities associated with poverty result in such environmental consequences as water pollution and degraded aquatic ecosystems, often the source of poor people's livelihoods. High levels of water-associated disease (such as schistosomiasis, malaria, trachoma, cholera and typhoid) are also common. And many people living in poverty engage in artisanal activities, such as metal working, that can generate large quantities of water pollutants.

Poor people often pay the highest relative prices for water. Inhabitants of informal settlements, for example, do not normally receive water delivery services from central water supply agencies, but typically pay exorbitant prices for drinking water (sometimes of dubious quality) from local water dealers. And in rural areas in developing countries people can spend hours each day fetching water.

History suggests that some initial level of economic development may be necessary before attention is given to environmental sustainability. However, there is sufficient evidence to believe that the two may not necessarily be in conflict. First, some processes are irreversible (for example aquifer depletion and contamination) and need to be halted now. Second, the state of water resources – and the environment in general – affects the poor disproportionately, so attention to environmental sustainability must especially recognize their urgent needs. And third, investments in environmental protection, water management and water supply and sanitation services, among others, can have high pay-offs in economic benefits.

But whatever actions are taken to reduce poverty, it must also be recognized that increasing the economic well-being of the very poor will ultimately translate into higher demand for natural resources, including water. This will require trade-offs, especially where these resources are lacking or over-exploited.

Education

An educated populace typically has a better understanding of the need for sustainable use of aquatic ecosystems and the important environmental goods and services they provide. Education can also lead to greater water use efficiency. For example, knowledge of water systems, new materials and emerging technologies (such as package treatment plants) can help extend water services to informal areas. Knowledge of water conservation practices also facilitates improved water use efficiency in these areas.

More education enables people to improve their economic circumstances, leading to empowerment, better health and longer life expectancy. At the community level the education of broad segments of society can accelerate the demographic transition, through declines in fertility and infant mortality rates (figure 2.3). An educated

Whatever actions are taken to reduce poverty, it must also be recognized that increasing the economic well-being of the very poor will ultimately translate into higher demand for natural resources, including water

Figure 2.3 **The fertility rate declines with rising female literacy, 1990**

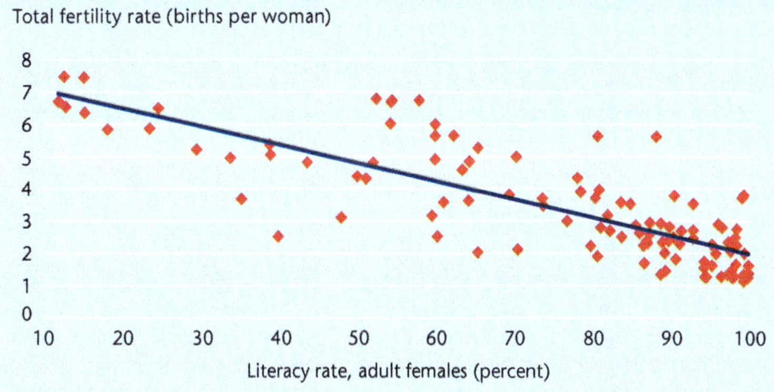

Source: Institute for Statistics 2006; World Population Prospects Database.

An educated populace typically has a better understanding of the need for sustainable use of aquatic ecosystems and the important environmental goods and services they provide. Education can also lead to greater water use efficiency

society is also more likely to be democratic and politically stable, working to reduce inequity and promote the acceptance of cultural diversity. Thus, education not only fosters economic growth, but also increases expectations of a better quality of life for individuals, their families and society.

While education is fundamental to improved economic and social well-being, in many fast-growing countries in Africa, South Asia and elsewhere, the predicted growth in population is likely to depress school enrolment. Because of budget and capacity constraints, schools may not be able to cope with the growth in the number of children to be enrolled.

In many settings access to schooling also is linked to improved access to safe drinking water and sanitation facilities. Separate sanitation facilities in schools for boys and girls have been shown to increase the attendance of girls and are also important for maintaining a minimum comfort level for female teachers. Improving access to water and sanitation facilities, by increasing family incomes, enables households to pay for school fees and equipment. And a reduced incidence of water- and sanitation-related diseases contributes to less absenteeism and better performance.

Culture and values

Culture describes the patterns of human activities and the symbolic structures that impart significance and importance to these activities (such as art, institutions, science, beliefs and moral systems). Because such structures are passed from generation to generation, culture can be defined as the way of life for an entire society.

In several regions the empowerment of women has emerged as an important driver, particularly at the household and community levels. As described in box 2.3, this ongoing process brings about social, environmental and health benefits that can have positive repercussions for the community as a whole in how water services are received and managed.

The perceived values of natural resources reflect cultural perspectives as well as economic perspectives. Lakes and reservoirs, for example, provide many valuable services, including water for drinking and sanitation, agriculture, industry and livestock uses and, in the case of reservoirs, for electricity generation. They serve as buffers against water shortages and excesses and as contaminant sinks for their drainage basins. They provide food and economic livelihoods through fisheries, aquaculture and environmental tourism. They are important aquatic ecosystems and provide habitat for rare and threatened species. And they can possess important cultural and religious values that emphasize humanity's connections to the natural world. Which of these uses are pursued or emphasized depends largely on the cultural perspectives and economic values assigned to them by society.

One of the most powerful manifestations of cultural values is religious belief. Many religions describe the role of humanity as both a moulder and a steward of the environment. Virtually all of the world's major religions see a spiritual challenge in the ecological crises evident today.[19] Religious beliefs that highlight humanity as a steward, rather than master, of the environment can be a powerful influence in developing and sustaining the awareness of societies and communities of their roles in using and conserving natural resources, including water.

Box 2.3 The role of women within the water sector and the importance of gender mainstreaming

In most developing countries gender inequity persists in access to and control of a range of productive, human and social capital assets. Consequently, the core components of poverty (capability, opportunity, security and empowerment) differ along gender lines.

In the water sector women labour to provide water for household needs while men make decisions about water resources management and development at both the local and national levels. Women draw water for household use, transport it home and store it until it is used for cooking, cleaning and washing. In areas of low water coverage women collect water from drains, ditches or streams that are often infected with pathogens and bacteria, causing severe illness or even death. In addition, women spend considerable time collecting water at the expense of income-generating activities. This also exposes them to sexual abuse and other forms of violence and leaves less time for girls to attend school.

Lessons from Africa and the rest of the world have demonstrated that increased participation by women in decision-making leads to better operation and maintenance of water facilities, better health for the community, greater privacy and dignity for women, more girls attending school and increased income opportunities for women.

The immediate action by water sector participants is to ensure gender mainstreaming in any planned action, including legislation, policies and programmes in all areas and at all levels. This will ensure that the voices of marginalized and disadvantaged women and men are integrated in design, implementation, monitoring and evaluation of policies and programmes and therefore help to achieve sustainable water provision for all.

Source: Adapted from Mutagamba 2008.

Religious beliefs can also sometimes accelerate the degradation of these resources. One example is the Hindu practice of cremating their deceased family members in funeral pyres and placing their ashes into the Ganges River, which is considered holy. However, incomplete cremation results in incompletely burned human remains being put into the river, causing degraded water quality and increasing the potential for the transmission of waterborne diseases. The custom is deeply rooted in religious beliefs, making it difficult to address with a strictly scientific rationale. Religious significance has been observed for water systems in other societies around the world.

Lifestyles and consumption patterns

Lifestyles and associated consumption choices are increasingly considered the most important drivers affecting water resources, along with population growth. And the pressures these drivers generate can be transmitted through trade and investment activities to other regions. As standards of living rise in developing countries and countries undergoing economic transition, the demand for larger homes and for 'luxury' items such as kitchen appliances, cars and other vehicles and the energy to run, heat and or cool them is increasing the demand for the resources required to produce, generate and operate them. Thus, humanity's environmental footprint is expanding dramatically. And despite some laudable efforts to develop cleaner technologies to shrink this footprint (see chapter 3), population growth and the changing lifestyles and consumption choices associated with rising living standards will continue to threaten the sustainability of water resources and the environment.

The evolution of eating habits and changes in diets as living standards rise are among the most important drivers of agricultural water use for several crops in many countries. The quantity of water used per person for food production depends on a society's dietary habits, in particular on the relative importance of meat and dairy products in diets. Massive social and economic changes taking place in many developing countries are lifting millions of people out of poverty and creating a new middle class with increasing demands for such food as milk, bread, eggs, chicken and beef to complement their traditional and less water-intensive diets.[20]

A simple calculation illustrates the impacts of changing food habits on water resources. It is estimated that the Chinese consumer who ate 20 kilograms (kg) of meat in 1985 will eat more than 50 kg in 2009,[21] increasing demand for grain to feed livestock. Assuming that 1 kg of grain requires 1,000 liters of water to produce, the annual water footprint of this change in diet for some 1.3 billion Chinese will translate into a need for 390 cubic kilometres (km^3) of water. Similar changes are taking place in other countries with growing economies. For the extremely poor, eating even two meals a day instead of one can substantially increase per capita water consumption (see box 7.4 in chapter 7).

As this example suggests, lifestyles and consumption patterns are, in essence, the sum of all drivers. They bring together economic growth, technological innovation, the evolution of culture and values, population dynamics (population growth and the number of people who have reached a certain standard of living) and governance (how wealth is distributed).

Challenges

Once people's survival needs are met, their wants become more prominent. These wants usually focus on increasing human comfort and convenience and are generally associated with rising consumption of material goods and non-essential services such as travel and leisure. The desire for a better lifestyle is arguably one of the most powerful human motivations, and the rapid global rise in living standards, combined with population growth, poses the major threat to the sustainability of water resources and the environment. The production of goods to satisfy these growing human wants is often not possible without the overuse of natural resources. Further, it is accompanied by the production of wastes and other non-useful by-products. Unrestrained fulfillment of the desire for a better lifestyle will be accompanied by environmental stresses, many of them unprecedented.

The major challenge is to reconcile human needs and human wants with the ability of nature to provide or replenish the resources to produce them. Global society must address the dual goal of enhancing human well-being and lifestyles while ensuring the sustainability of the ecosystems and environmental conditions that provide the desired goods and services. Achieving this goal will prove impossible unless humans recognize and better understand the links between their actions and the condition and sustainability of the natural environment. Raising awareness to bring about behavioural change is one approach, but a still elusive goal.

The desire for a better lifestyle is arguably one of the most powerful human motivations, and the production of goods to satisfy these growing human wants is often not possible without the overuse of natural resources

Notes

1. Hinrichsen, Robey, and Upadhyay 1997.
2. United Nations 2007.
3. UNFPA 2007.
4. United Nations 2006a.
5. Morton, Boncour, and Laczko 2008.
6. UNFPA 2007.
7. Klaus Töpfer, former head of the United Nations Environment Programme, talks of 22-24 million environmental migrants (Biermann 2001), whereas Norman Myers (2005) reports 'at least' 25 million in 1995 (latest date for a comprehensive assessment), especially in the African Southern Sahara, Central America, China and South Asia. Myers expects the number to reach 50 million by 2010. The United Nations Refugee Agency (UNHCR 2002, p. 12) estimated that there were approximately 24 million people around the world who fled their homes because of floods, famine and other environmental factors. Christian Aid released a report in 2007 estimating that up to 685 million people were forced to move because of environmental factors, including development projects such as dams that inundated large areas of inhabited land. All of these estimates are from OSCE (2007).
8. More information on the revised forecast can be found in IMF 2008a.
9. Poddar and Yi 2007.
10. SIWI 2005.
11. World Bank 2008.
12. United Nations 2008.
13. Commission on Growth and Development 2008.
14. High Level Conference on World Food Security 2008, article 7.f.
15. IEA 2006.
16. WBCSD 2006.
17. World Economic Forum 2008.
18. Wiggins 2008.
19. Bassett, Brinkman, and Pedersen 2000.
20. Wiggins 2008.
21. Wiggins 2008.

References

Bassett, Libby, John T. Brinkman, and Kusumita P. Pedersen. 2000. *Earth and Faith. A Book of Reflection for Action.* Nairobi: Interfaith Partnership for the Environment and United Nations Environment Programme.

Commission on Growth and Development. 2008. *The Growth Report: Strategies for Sustained Growth and Inclusive Development.* Conference Edition. Washington, DC: World Bank.

EIA (Energy Information Administration). 2008. *Annual Energy Review 2007.* Washington, DC: Government Printing Office. www.eia.doe.gov/emeu/aer/pdf/aer.pdf.

FAO (Food and Agriculture Organization of the United Nations). 2008. Declaration of the High-Level Conference on World Food Security: The Challenges of Climate Change and Bioenergy. High Level Conference on World Food Security, 3-5 June, Rome.

Gallopín, G. C., and F. Rijsberman. 2000. Three Global Water Scenarios. *International Journal of Water* 1 (1): 16-40.

Hinrichsen, D., B. Robey, and U. D. Upadhyay. 1997. *Solutions for a Water-Short World.* Population Reports Series M, no. 14. Baltimore, MD: Population Information Program, Johns Hopkins School of Public Health.

Hoekstra, A. Y., and A. L. Chapagain. 2008. *Globalization of Water: Sharing the Planet's Freshwater Resources.* Oxford: Blackwell Publishing.

IEA (International Energy Agency). 2006. *World Energy Outlook 2006.* Paris: Organisation for Economic Co-operation and Development, and International Energy Agency.

IMF (International Monetary Fund). 2008a. World Economic Outlook Update: Rapidly Weakening Prospects Call for New Policy Stimulus. Washington, DC: International Monetary Fund. www.imf.org/external/pubs/ft/weo/2008/update/03/.

———. **2008b.** *World Economic Outlook: Financial Stress, Downturns, and Recoveries.* Washington, DC: International Monetary Fund.

Institute for Statistics. 2006. *World Education Indicators Data Centre Literacy Statistics.* Paris: United Nations Educational, Scientific and Cultural Organization.

Lutz, W., W. Sanderson, and S. Scherbov. 2008. The Coming Acceleration of Global Population Ageing. *Nature* 451 (20): 716-19.

Morton, A., P. Boncour, and F. Laczko. 2008. Human Security Policy Challenges. *Forced Migration Review: Climate Change and Displacement.* Issue 31. Oxford, United Kingdom: Refugee Studies Centre, University of Oxford.

Mutagamba, Maria Lubega. 2008. The Role of Women within the Water Sector and The Importance of Gender Mainstreaming. *The 5th World Water Forum Newsletter 4.*

Pacific Institute. 2007. *At the Crest of a Wave: A Proactive Approach to Corporate Water Strategy.* Oakland, CA: Business for Social Responsibility and the Pacific Institute.

Poddar, Tushar, and Eva Yi. 2007. *Global Economics Paper Issue no. 152: India's Rising Growth Potential.* Goldman Sachs Global Economic Website, Goldman Sachs. http://usindiafriendship.net/viewpoints1/Indias_Rising_Growth_Potential.pdf.

SIWI (Stockholm International Water Institute). 2005. Making Water a Part of Economic Development: The Economic Benefits of Improved Water Management and Services. Stockholm: Stockholm International Water Institute.

United Nations. 2006a. *Trends in Total Migrant Stock: The 2005 Revision.* New York: Population Division, Department of Economic and Social Affairs, United Nations.

———. **2006b.** *World Urbanisation Prospects: The 2005 Revision. Fact Sheet 3.* New York: Population Division, Department of Economic and Social Affairs, United Nations. www.un.org/esa/population/publications/WUP2005/2005WUP_FS3.pdf.

———. **2007.** *World Population Prospects: The 2006 Revision.* New York: Population Division, Department of Economic and Social Affairs, United Nations.

———. **2008.** *World Economic and Social Survey 2008: Overcoming Economic Insecurity.* New York: Department of Economic and Social Affairs, United Nations.

UN-HABITAT (United Nations Human Settlements Programme). 2006. *State of the World's Cities 2006/7.* London: Earthscan.

UNFPA (United Nations Population Fund). 2007. *State of World Population 2007: Unleashing the Potential of Urban Growth.* New York: United Nations Population Fund.

UNICEF (United Nations Children's Fund). 2006a. Childinfo: Monitoring the Situation of Children and Women, Child Mortality Info Database. www.childmortality.org/.

———. **2006b.** *The State of the World's Children 2007: The Double Dividend of Gender Equality.* New York: United Nations Children's Fund.

WBCSD (World Business Council on Sustainable Development). 2006. *Business in the World of Water: WBCSD Water Scenarios to 2025.* Washington, DC: World Business Council on Sustainable Development.

Wiggins, Jenny. 2008. Feature: Developing Tastes. *Financial Times Magazine,* 27-28 January. www.ft.com/cms/s/0/8e606e1e-cbb2-11dc-97ff-000077b07658.html?nclick_check=1

World Bank. 2008. *World Development Indicators 2008.* Washington, DC: World Bank.

World Economic Forum. 2008. Managing Our Future Water Needs for Agriculture, Industry, Human Health, and the Environment. Discussion Document for the World Economic Forum Annual Meeting 2008. www.european-waternews.com/download/whitepaper_uploadfile_2.pdf.

Demographic, economic and social drivers

Chapter 3

Technological innovation

Authors: Richard Connor and Walter Rast
Contributor: Gunilla Björklund
Coordinator: Tim Kasten (UNEP)
Facilitator: Richard Connor

Key messages

- Technological innovation is driven largely by both human wants and needs.
- Technological innovation can create both positive and negative pressures, sometimes simultaneously, resulting in increased or decreased water demand, supply and quality.
- Technological innovation is one of the most unpredictable drivers. It can create rapid, dramatic and unexpected changes, both in pressures and solutions.
- Impediments to the dissemination of technology must be overcome if developing countries are to benefit from innovations developed in richer countries.

Technological change takes different forms, each with different potential impacts on the environment. Some innovations reduce environmental pressures (by lowering emissions or using water resources more efficiently, for example), while others increase them (by increasing water demands for their production, for example). Most innovations create both positive and negative pressures on the environment, while the main purpose of technology is to make processes (production, transformation and communication, for example) more efficient, which generally means more cost-effective, the environmental benefits of some technologies have also yielded broader economic benefits. In recent decades, for example, greater environmental regulation and corporate social responsibility, combined with pressures from society, have prompted cleaner and more environmentally friendly technologies and increased their overall value.

Disseminating technology is as important as developing it. Controls on the dissemination of technology, especially from developed countries (where much of the technology is generated) to developing countries (which are less able to afford or generate it), inhibit the ability of developing countries to stay economically and environmentally competitive.

In the water sector the expansion of scientific knowledge and technological applications is changing the way water is used, cleaned and reused to meet human, economic and environmental needs. Industries are investing in new technologies and processes that reduce water use and wastewater discharges. Household consumers are being offered water-saving technologies such as low-flush toilets, low-flow showers and faucet aerators. Agricultural productivity is being leveraged by drip irrigation and maintained by soil fertility and conservation techniques. Water supplies are being enhanced in many countries through innovative wastewater treatment and reuse techniques. And breakthroughs continue in desalination: advances in technologies and energy efficiency in the past decade have made

Water supplies are being enhanced in many countries through innovative wastewater treatment and reuse techniques

desalination an economic option for water supplies in coastal cities (see figure 9.3 and box 9.5 in chapter 9).[1]

This chapter looks at six areas – in which water-related technologies are emerging rapidly – that are likely to exert strong pressures on the supply, use and management of water resources: environmental research and development, renewable energy, information and communications technology, biotechnology, bioenergy and nanotechnology. It also describes the challenges and difficulties associated with the dissemination of technology, which is especially important for developing countries.

Recent trends and advances in science and technology

Key message

- Technological innovation is driven largely by both human wants and needs.

People are the ultimate drivers of change on a global scale, through both their needs (their requirements for survival) and their wants (their desires for products and services that enhance safety, comfort and well-being). Although not true everywhere or to the same degree, technological advances that address these wants and needs are a major reason why many people enjoy a standard of living that includes access to safe drinking water and adequate sanitation.

It is sometimes difficult to determine whether technology development drives water demands or whether increasing water demands associated with human activities drive technology. Some new technologies can have positive benefits – reducing water demand and increasing water availability (for example, rainwater harvesting) – while others can increase water demands (such as using crops to produce bioenergy). In analysing technological advances and interventions, it is also useful to distinguish their structural elements (such as construction of a plant, dam or irrigation system) and their non-structural elements (including public awareness campaigns, educational programmes and information sharing). This section outlines some key technology areas and provides some insight on how new developments can affect water resources.

Environmental research and development

Many developed countries have increased their investment in environmental research and development (R&D) to encourage new technologies to improve environmental quality (figure 3.1). Perhaps more important, developed countries also encourage research by the private sector through subsidies and tax incentives for specific types of research. This has been much less the case in most developing countries, however, because of the many competing claims on their limited financial resources. Thus, the main path of technology transfer is from developed countries to developing countries.

The focus of the R&D activities varies with national sustainable development priorities and interests and available funding. Germany, for example, has focused on clean processes and production technologies, Norway on energy and the environment and the United States on climate, water and hydrogen as an energy source.

There also appears to be a correlation between environmental regulations and environmental technology, with regulation spurring industries and water use sectors to address water availability and water quality. Environmental regulations may be counterproductive in facilitating environmental technology in some situations, however, since once required standards are met, incentives to engage in further technology development may dissipate.

Renewable energy

The renewable energy sector has seen remarkable innovation over the past two decades. Innovation has accelerated in response to recent public and political pressure to reduce greenhouse gas emissions thought to be contributing to global climate change. First-generation

Figure 3.1 **Many developed countries have increased their investment in environmental research and development**

Share of environmental R&D in total government R&D, 1981-2005 (percent)

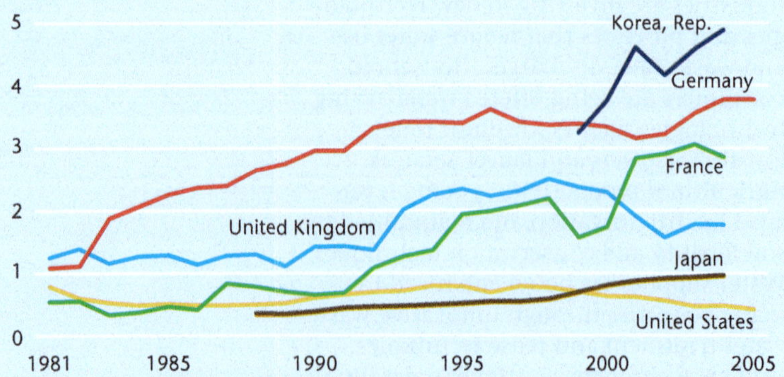

Source: Based on OECD 2008.

(hydropower and biomass combustion) and second-generation (solar heating and wind power) technologies are now being followed by third-generation technologies such as concentrated solar power, ocean energy, enhanced geothermal systems and integrated bioenergy systems. As these innovations have lowered relative costs, the use of renewable energy resources has risen worldwide (figure 3.2).

If current policies are maintained, global energy demands are expected to grow by as much as 55% through 2030, according to the International Energy Agency.[2] China and India alone would account for about 45% of this projected increase (based on conservative economic growth figures), and developing countries overall for 74%. Electricity generation from hydroelectric and other renewable energy resources is projected to increase at an average annual rate of 1.7% between 2004 and 2030, for an overall increase of 60%. Although renewable energy would still account for only a small part of total energy demand, the increase in renewable energy production could have a large impact on water resources, especially increases in hydropower generation.

Future development of hydropower will be limited by two main factors. One is the spatial and geophysical potential for new hydropower installations. In many developed countries, including Australia, the United States and much of Western Europe, most of the suitable sites for hydropower installations have already been developed (see map 7.6 in chapter 7). The second limiting factor is financial investment capacity, which has been the primary constraint in developing countries, including in most of Africa. Pressure from environmental groups opposed to dams, particularly to large dams, may also constrain future hydropower development.

Since renewable energy resources alone are not sufficient to meet the predicted dramatic increase in energy demands through 2030, fossil fuel extraction and development of nuclear energy will continue to increase, as will their impacts on water resources and the environment. Coal consumes about 2 cubic metres (m^3) of water per megawatt hour of electricity generated, nuclear power about 2.5 m^3 and petroleum about 4 m^3. Extracting petroleum from Canada's tar sands, which have received much criticism as an 'unclean' source of oil, consumes an estimated 20-45 m^3 of water per megawatt hour, nearly 10 times that for conventional oil extraction. Thus, as fossil fuel sources become increasingly inaccessible, the water footprint of oil tar sands is likely to increase dramatically.

Figure 3.2 **The use of renewable energy sources rose worldwide between 1990 and 2004**

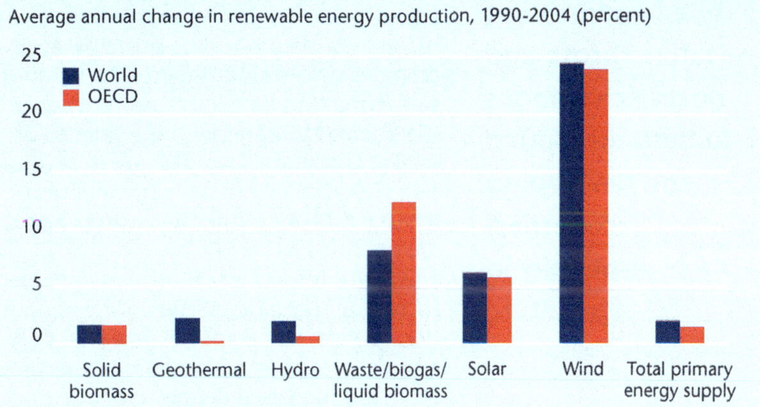

Source: Based on OECD 2008.

Information and communications technology

Advances in information and communications technology can affect the cost and effectiveness of monitoring ecosystem health and quality. Reductions in the costs of sensors, coupled with satellite-based wireless data transfer, have greatly facilitated the monitoring of water resources (water quality, water levels, flow rates and so on) and the delivery of water-related services, all in real time.

Improved monitoring through advanced information and communications technology can intensify the environmental effectiveness of policy measures, from the improved tracking of potentially hazardous materials to the monitoring of emissions from large stationary and smaller diffuse (non-point) and mobile sources. The greatest number of patents for monitoring environmental impacts between 1978 and 2002 was granted for water pollution treatment, attesting to the importance of information and communications technology innovations in the sustainable management of water resources. Still lacking, however, are adequate original field data required for ground-proofing, monitoring and forecasting data and for informed decision-making (see also chapter 14).

Biotechnology and genetically modified organisms

Plant and animal breeding has increased agricultural productivity and therefore affected water productivity. Progress has been concentrated in crop and animal productivity and resistance to pests, disease and weather extremes.

Biotechnology can have a valuable role in addressing water scarcity and quality challenges in both developed and developing countries, especially in agriculture

The green revolution of the 1970s and 1980s is an example of the dramatic effects of how taking advantage of technological advances can improve the livelihoods and incomes of the poor. The principal technologies involved in the green revolution were irrigation, fertilizer and pest control, together with high-yielding varieties of maize, wheat and rice. The green revolution in Asia doubled cereal production during 1970-95, while increasing the land area devoted to cereals by only 4%. By the late-1990s it was clear that many people, including segments of the poorest population groups, had reaped substantial benefits from higher incomes, less expensive food and increased demand for their labour associated with the green revolution.

The green revolution also demonstrates that unintended consequences can accompany new technologies. The excessive use of agrochemicals has polluted waterways, while wasteful irrigation has contributed to water scarcity in some areas and to water logging and soil salinization in others. High livestock concentrations have contributed to the spread of disease. As monoculture of crops for export or for use as animal feed replaced traditional polyculture techniques, the economic outcomes for some small farmers deteriorated as increased production of cereal crops caused prices to fall and crop susceptibility to pests and plant diseases spread. Increased agricultural production also led to higher water demands, exacerbating water scarcity problems in some arid and semi-arid regions (see chapter 8).

Genetically modified organisms are a more recent agricultural advance. A genetically modified organism is an organism whose genome has been altered through genetic engineering. A large share of food crops, such as corn, cotton and soybeans, have been genetically modified to increase yields and resistance to pests and chemical herbicides. Although this technology offers the potential for developing drought-resistant crops, with obvious advantages for water-scarce regions, little progress has been made towards this goal, and no breakthrough is expected in the near future.

Micro-organisms are an especially promising avenue, since there is considerable knowledge and experience in genetic experimentation with them. As decomposers of organic material, they are capable of breaking down or otherwise neutralizing many types of polluting materials in the environment. Micro-organisms are currently used, for example, in biological processes in municipal wastewater treatment plants to treat or break down organic materials in wastewater. Micro-organisms that can more efficiently break down oil pollution in aquatic ecosystems and soils following oil spills or other industrial accidents are receiving attention. Similar avenues may become evident for research into the treatment of other types of water pollutants.

Bioenergy

Bioenergy, derived most commonly from plant materials, is a renewable energy source that is less likely to increase carbon dioxide emissions that contribute to global warming (in contrast to fossil fuels, which return long-stored carbon to the atmosphere). Cellulose, including agricultural residues, waste products and woody biomass, is also showing promise as a bioenergy source (see chapter 7).

This new technology is not without problems. For maize and sugarcane used to create bioenergy, a major problem is the need for large quantities of water to grow the crops (see box 7.2 in chapter 7) and for considerable quantities of fossil fuel energy for tillage, fertilizers, pesticides, irrigation, harvesting and transport machinery, and processing.[3] Research is currently focused on the development of second-generation bioenergy, converting wood, crop residues and other biomass sources into liquid biofuel. Non-food crops such as jatropha do not require the intensive management and soil quality that food crops need and therefore may not compete directly with food crops for resources (water and good agricultural land). Second-generation bioenergy technology has the potential to increase energy yields significantly, but may not be commercially viable for 5-10 years.

Producing bioenergy from crops traditionally grown as food will require additional agricultural production to make up for the lost food sources, and more water as well. Increased bioenergy production has also resulted in a significant increase in some food prices by diverting grain traditionally grown for food.[4] More than one-third of maize production in the United States in 2008 was being used to produce ethanol[5] and about half the vegetable oils produced in the European Union were being used for biodiesel fuel.[6] Although the impact is extremely difficult to assess, bioenergy production is estimated to have caused up to 70%-75% of the rise in the global prices of some food stocks, including approximately 70% of the increase in maize prices.[7] Higher energy prices worldwide and a weak US dollar are believed to have caused the remainder.[8]

Bioenergy production also causes environmental impacts unrelated to climate, particularly impacts arising from agricultural practices (see chapter 7). Examples include tillage-based soil erosion, eutrophication from fertilizer runoff, increased pesticide loads to aquatic habitats and biodiversity loss from land use changes. Further, the use of bioenergy could spawn other problems, as reductions in greenhouse gas emissions (from switching from fossil fuels to biofuels) could be offset by the clearing of new land to make room for more crop production. Cutting down forests could release carbon dioxide and reduce biodiversity. Under conditions of water scarcity, producing fuel for automobiles instead of producing food to feed a growing population becomes less socially acceptable, especially in developing countries.

Nanotechnology

Nanotechnology, the design and manufacture of extremely small electronic circuits and mechanical devices built at the molecular level of matter, shows particular promise for water resources. Key areas are desalinization (see box 9.5 in chapter 9), water purification, wastewater treatment and monitoring. The first three areas involve the use of nanofiltration technology, nanomaterials and nanoparticles to remove or reduce water contaminants. Monitoring involves the use of nanosensors.

Many nanotechnology-based approaches are less a major departure from traditional methods of addressing such issues than a means of improving existing applications and devices.[9] Seawater desalinization plants are already in operation around the world, and many technologies can effectively remove microbes and other contaminants from water. And although operation efficiencies vary, wastewater treatment plants also exist in many developed and developing countries.

Nanotechnology has the potential to greatly improve water quality and quantity through water treatment or remediation. Nanofiltration membranes and other advanced filtration materials can facilitate water desalinization and increased water reuse and recycling, improving desalinization efficiency and reducing associated costs (especially for energy). Another emerging area is the development of nanomaterials, which can act as a 'sponge' to enhance the removal of specific heavy metals from water supplies. Research is exploring the use of nanoparticles as catalysts for chemical reactions of other materials as a means of degrading them and for removing salts and heavy metals. Such treatments could be targeted to chemicals for which existing technologies are inefficient or costly and could eventually permit human use of heavily polluted and saline water for drinking, sanitation and irrigation.

For water monitoring nanotechnology encompasses new and enhanced sensors for detecting biological and chemical water contaminants present in very low concentrations. New sensor technology, coupled with micro- and nanofabrication technology, may eventually lead to the development of highly accurate and portable sensors.

There are also impediments to the large-scale use of nanotechnologies to address water resources issues. While many nanotechnologies are already in use, many are still at varying stages of research or development. Thus, although such technology could help developing countries increase water treatment or remediation efficiency and reduce costs associated with traditional treatment methodologies, it is unclear when nanotechnology-based applications will be ready for wide-scale use. And even though nanotechnologies may prove very efficient and cost-effective over the long term, initial acquisition and application costs are high in many cases. Using such technologies also will require the technical capacity to maintain and operate them.

There also are some risks associated with nanotechnology-based approaches, specifically the possibility that engineered nanoparticles used to catalyse chemical reactions may end up in water systems. Little is yet known about how such materials may interact with biological organisms, so the possibility of toxicity to humans and ecosystems must be considered.

The technology dissemination challenge

Key message

- Technology is constantly evolving, and the availability of technologies can differ widely between developed and developing countries because of impediments to dissemination of research and adaptation to local conditions.

Nanotechnology shows particular promise for desalinization, water purification, wastewater treatment and monitoring

With the bulk of technological innovation originating in developed countries, introducing appropriate technologies into developing countries is a key challenge of development

Technological progress is both a determinant and an outcome of rising incomes. At the national level it can occur through invention and innovation, the adoption and adaptation of existing but new-to-the-market technologies and the spread of technologies across individuals, firms and the public sector within a country.

With the bulk of technological innovation originating in developed countries, introducing appropriate technologies into developing countries is a key challenge of development. It requires both the willingness to transfer the technology and the capacity to pay for, absorb, adapt and use the technology so that it generates long-term benefits.

Exporting technology to developing countries

The number of patents and scientific journal articles focusing on technology is strongly correlated with GDP per capita.[10] Most developing countries lack the ability to generate innovations at the technological frontier. Moreover, relatively undeveloped domestic technology sectors and the lure of better economic and scientific opportunities abroad draw highly educated nationals from many developing countries to cutting-edge research sectors in high-income countries.

The lack of advanced technological competence in developing countries means that technological progress occurs there mainly through the adoption and adaptation of existing technologies. The penetration of older technologies, such as fixed-line telephones, electric power networks, transportation, health care and water services – many ultimately provided by governments – has tended to lag behind that of newer technologies (figure 3.3). Many of these older technologies require infrastructure that is expensive to create and maintain and that relies on large numbers of people with scarce technical skills. In addition, the diffusion of older technologies today depends on the intensity and efficiency with which government services were delivered in the past, many of which have a poor record.

The rates of acceptance and application of newer technologies have been higher than those of older technologies because rates of acceptance and application are more directly correlated with income. The infrastructure for newer technologies such as mobile phones and the Internet is generally less expensive to create and requires fewer (although more skilled) workers to operate and maintain. Moreover, with regulatory reform in many countries the private sector now offers these services in a competitive environment, rather than in the state-owned, monopolistic environments of the past. Supplying such new technologies has thus been more responsive to market demands and less constrained by the stringencies of government budgets or state enterprises. Furthermore, demands for such products have been boosted by low end-user costs, resulting from competitive pricing strategies, and the characteristics of some newer technologies that lend themselves to sharing more readily than do some older technologies.

Absorptive capacity for technology

Most technological progress in developing countries has been achieved through absorption and adaptation. A country's ability to absorb, adapt and apply foreign technologies depends mainly on its exposure to foreign technologies (the pace at

Figure 3.3 The absorption of older and more recent technologies depends on more than income

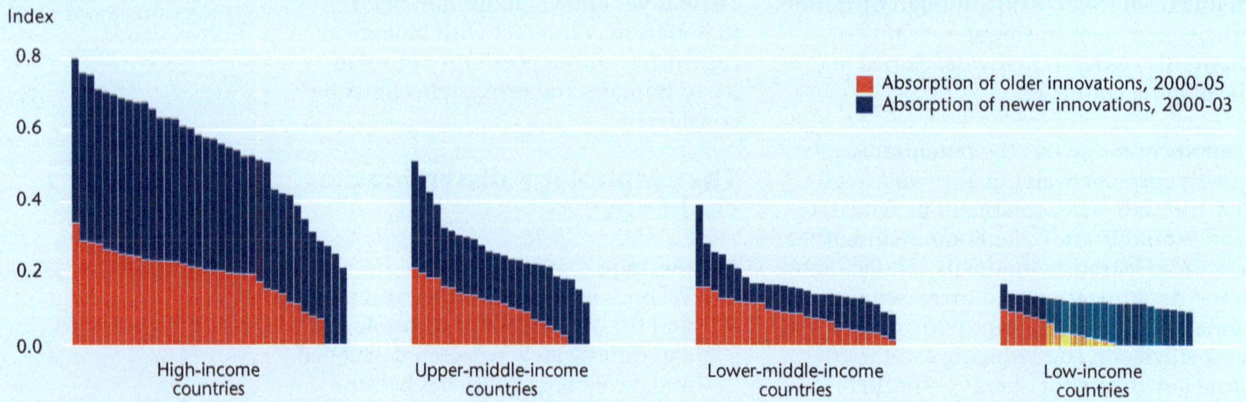

Note: Each bar represents a single country.
Source: Based on World Bank 2008.

which technologies diffuse across countries) and its ability to absorb, adapt and use the technologies to which it is exposed (the pace at which technologies diffuse within a country). Successful use depends on the technological absorptive capacity of the economy – the macroeconomic and governance environment, which influences the willingness of entrepreneurs to take risks on new and new-to-the-market technologies – and the level of technological literacy and advanced skills in the population.

Government policy also has a crucial role. Governments are often the primary delivery channel for technologies such as electricity, fixed-line telephones, transportation infrastructure and medical and educational services. And government policy can create a business environment that facilitates firm entry and exit and that is not hostile to exploiting new technologies. Too often, government regulations or features of the domestic market prevent firms from making money by exploiting a new technology, thus impeding the spread of technology within a country. Policy should also ensure that R&D and dissemination efforts give priority to creating and introducing products for which a market (domestic or foreign) exists and to helping firms exploit those opportunities.

Investing in research and development

Countries do well to invest in technology research and development. Research and extension programmes in agriculture, the sector that consumes the most water, have exceptionally high internal economic rates of return (table 3.1).

Many resource constraints can be overcome by technological capital and supporting institutions. Productivity gains, including genetic improvements that enable more production per unit of land, also enable more production per unit of water. For most developing countries gains in agricultural productivity arise from investments in adaptations of inventions produced in developed countries.

Challenges

A major technology challenge is how to balance the benefits and risks of new technologies. For the first time in human history, technology has provided humanity with the means to reshape the structure and functioning of the natural environment and thus to alter the possibilities for future development. The natural environment has an internal system of checks and balances for its own maintenance and that of the animals and plants that inhabit it. Humanity has acquired technologies that can radically affect these natural checks and balances.

Many positive impacts are associated with technological advances, such as a reduced burden of disease and loss of life due to medical advances, decreased malnutrition due to the green revolution and other agricultural advances, and increased economic livelihoods due to industrialization and urbanization and attendant technologies. But maintaining a sustainable relationship between people and the natural environment requires maintaining a balance between the technologies we develop to meet human needs and nature's ability to supply them. And there is ample evidence that this balance is not being achieved in many places around the world, as demonstrated by excessive water abstractions, degraded water quality, and damaged aquatic ecosystems and biological communities. Some of these impacts result from ignoring the environmental consequences of human development actions. Others result from ignorance of the many, often subtle, interactions between the natural environment and the human activities that fundamentally affect it.

Consider crop-based bioenergy production. The increased production and use of bioenergy to reduce greenhouse gas emissions associated with the burning of fossil fuels must be balanced against the rising need for water resources, associated pollution and sufficient agricultural land on which to grow the crops to supply crop-based bioenergy. An unintended impact has been rising prices for some foods, as cereal crops are currently used for the production of bioenergy rather than for food. Our choices of technology require appropriate consideration of their benefits and costs, including their negative environmental impacts.

Maintaining a sustainable relationship between people and the natural environment requires maintaining a balance between the technologies we develop to meet human needs and nature's ability to supply them

Table 3.1 **Return on investments in agricultural research and extension**

Investment	Median internal rate of return (percent)[a]
Agricultural extension programmes	41
Applied research	49
Pre-invention science	60

a. The internal rate of return is the rate of discount at which the present value of benefits is equal to the present value of costs.

Source: FAO 2000.

Notes

1. Bergkamp and Sadoff 2008.
2. IEA 2007.
3. Pimentel and Patzek 2005.
4. Mitchell 2008.
5. US Department of Agriculture 2008.
6. Mitchell 2008.
7. Mitchell 2008.
8. FAO 2008.
9. Hillie et al. 2005, p. 43; Berger 2008.
10. World Bank 2008.

References

Berger, M. 2008. Nanotechnology and Water Treatment. Nanowerk website. www.nanowerk.com/spotlight/spotid=4662.php.

Bergkamp, G., and C. W. Sadoff. 2008. Water in a Sustainable Economy. In *State of the World: Innovations for a Sustainable Economy*. Washington, DC: Worldwatch Institute.

FAO (Food and Agriculture Organization of the United Nations). 2000. Agricultural Production and Productivity in Developing Countries. In *The State of Food and Agriculture 2000*. Rome: Food and Agriculture Organization of the United Nations.

———. **2008.** Soaring Food Prices: Facts, Perspectives, Impacts and Actions Required. HLC/08/INF/1. Background paper for the High-Level Conference on World Food Security: The Challenges of Climate Change and Bioenergy, Rome, 3-5 June 2008.

Hillie, T., M. Munasinghe, M. Hlope, and Y. Deraniyagala. 2005. Nanotechnology, Water and Development. Global Dialogue on Nanotechnology and the Poor: Opportunities and Risks. Meridian Institute, Washington, DC.

———. **2007.** *World Energy Outlook 2007*. Paris: Organisation for Economic Co-operation and Development and International Energy Agency.

Mitchell, Donald. 2008. A Note on Rising Food Prices. Policy Research Working Paper 4682, Development Prospects Group, World Bank, Washington, DC.

OECD (Organisation for Economic Co-operation and Development). 2008. *OECD Environmental Outlook to 2030*. Paris: Organisation for Economic Co-operation and Development.

Pimentel, D., and T. W. Patzek. 2005. Ethanol Production Using Corn, Switchgrass, and Wood; Biodiesel Production Using Soybean and Sunflower. *Natural Resources Research* 14 (1, March): 65-76.

US Department of Agriculture. 2008. Grain and Oilseeds Outlook for 2008. Prepared for the Agricultural Outlook Forum, February 21-22, Crystal City, VA. www.usda.gov/oce/forum/2008_Speeches/Commodity/GrainsandOilseeds.pdf.

World Bank. 2008. *Global Economic Prospects 2008: Technology Diffusion in the Developing World*. Washington, DC: World Bank.

Technological innovation

Chapter 4

Policies, laws and finance

Authors: Gunilla Björklund, Stefano Burchi, Richard Connor, William Cosgrove, Sarah Hendry, Patrick Moriarty, Walter Rast, Léna Salamé and James Winpenny

Contributors: Jack Moss, Monica Scatasta, Jon Martin Trondalen and World Water Assessment Programme Expert Group on Legal Issues

Coordinator: Tim Kasten (UNEP)

Facilitator: Richard Connor

Key messages

- Effective policy and legal frameworks are necessary to develop, carry out and enforce the rules and regulations that govern water use and protect the resource.
- Water policy operates within a context of local, national, regional and global policy and legal frameworks that must all support sound water management goals.
- Legitimate, transparent and participatory processes can effectively mobilize input for designing and implementing water resources policy and create a strong deterrent to corruption.
- Although water is often described as a 'gift of nature', harnessing and managing it for the wide variety of human and ecological needs entail financial costs.
- While there may appear to be many financing options for water resources development, governments still have only three basic means of financing them: tariffs, taxes and transfers through external aid and philanthropy.
- Policy-makers need to make political decisions on socially and environmentally acceptable trade-offs among different objectives and on who bears the costs of such compromise.

Policies and laws

Authors: Gunilla Björklund, William Cosgrove, Patrick Moriarty, Walter Rast and Léna Salamé

Contributors: Richard Connor, Jon Martin Trondalen and World Water Assessment Programme Expert Group on Legal Issues

Effective policy and legal frameworks are necessary to develop, implement and enforce rules and regulations for controlling water uses. Although policy and law go hand in hand, they are fundamentally different. Policy serves mainly as a guide for decision-makers. Law provides a set of enforceable rules.

Water policy, developed at international and national levels, can lead to the establishment of international, national and local laws. Effective implementation and enforcement require an adequate institutional and governance framework – one that is legitimate, transparent and participatory and that has proper safeguards against corruption. The legal system within which water law operates can be a strong instrument of change – or a severe impediment to progress.

Water law sets the framework for stakeholders' use of water resources and responds to pressures from demographic, economic and social drivers. Policy-makers

Because the political negotiations involved in global and regional conventions or water-sharing agreements are meant to avoid conflicts between different uses or users of water, they serve as drivers for water management

use water law to establish the rules of the game for water users within a given community, country or region.

International and regional water policy

International goals and objectives for water resources, negotiated at UN meetings, conferences and summits or in ministerial-level sessions of the World Water Forum, can be viewed as political benchmarks. Because the political negotiations involved in global and regional conventions or water-sharing agreements are meant to avoid conflicts between different water uses or users, they serve as drivers for water management. The global policy framework for water began with the Stockholm Declaration of 1972, followed by other important international milestones over the years (see appendix 2).

Ratifying conventions means assenting to implement the actions and activities agreed to by the involved parties. Implementation requires that the proper institutions exist, that national laws are compatible with convention requirements and that political and financial measures are in place to ensure popular participation. It also requires a policy framework with operational goals, objectives and follow-up processes. As an example, the EU Water Framework Directive, negotiated by the EU member states, requires intranational, multilevel institutional structures, including legal systems, to ensure implementation of the directive for transboundary river basins and groundwater as well as national river basins (box 4.1).

International and regional legal frameworks

International water law is part of public international law. The rules of international law apply to sovereign states. But because there is generally no higher authority to enforce such rules, individual countries must generally ensure their own compliance. The first step in enforcement is identifying the applicable rules.

These rules are found in treaties, international custom, general principles of law and the writings of 'learned publicists'.[1] Treaties usually provide the most accessible source of law, but the other sources cannot be ignored. In the non-navigational uses of international watercourses, rules of customary law are often invoked by countries in the absence of codified law. A treaty applies only to parties to the treaty and only after the treaty has come into force and is thus legally binding. Finally, the normative content (requirements) of the treaty rules must be established and agreed to by all parties involved to determine whether a country's actions are in accordance with its treaty obligations.

Law may also be developed at a regional level. Such law typically supersedes national law. Treaties may operate regionally between two or more countries. Regional bodies such as the European Union may also create law for their members. EU law, unlike international law, can be directly binding on its members and has strong enforcement mechanisms.

In most cases the directly applicable law is national law, which ensures implementation of any international treaties that a country has signed. Within national law the specific law-making powers and hierarchies of laws are determined by the constitutional arrangements within a jurisdiction. National law also includes customary law as well as water laws directly relating to water resources (for example, pollution control and water abstraction permits). In addition to the formal legal framework and the customary laws that national law formally codifies and recognizes, there are also water rules and rights by which water user collectives and other actors abide.

Box 4.1 The EU Water Framework Directive – uneven implementation

The EU Water Framework Directive for water protection and management provides for the identification of European waters and their characteristics on the basis of individual river basin districts and the adoption of management plans and measures for each water body. Entered into force 22 December 2000, the directive seeks to prevent and reduce pollution, promote sustainable water use, protect the aquatic environment, improve the status of aquatic ecosystems and mitigate the effects of floods and droughts through the management of inland surface waters, groundwater, transitional waters and coastal waters.

Within four years of the directive's entry into force, member states were to complete an analysis of the characteristics of each river basin district, a review of the impacts of human activities on their water resources, an economic analysis of water use and a registry of areas requiring special protection. Within nine years they were to produce a management plan and programme for each river basin district.

In a 2007 report the European Commission noted that several EU member states may fail to meet the targets, particularly because of the physical deterioration of aquatic ecosystems as a result of overexploitation of water resources, and the high levels of pollution from diffuse sources. The report also cited problems in meeting the deadlines for incorporating the directive into national law. However, the establishment of river basin districts and the designation of competent national authorities appear to be well under way. The European Commission finished with recommendations for addressing the reported shortcomings, integrating sustainable management of water resources into national policies, maximizing public participation and giving advance notice of its plans for future European water management policy.

Source: European Parliament and Council 2000; CEC 2007.

These hybrid sets of water rules, common in most parts of the world, are often crucial in everyday water affairs and conflict resolution.

There are also many other areas of law not directly addressed to water issues that nevertheless affect management of the water environment. These include land use planning, environmental assessment, nature conservation and environmental law. Public health laws influence the supply of water and sanitation, as does land tenure reform. Individuals are reluctant to invest in sanitation where they have no security of tenure, nor will water companies lay pipes in such land. Legal provisions on freedom of information and access to justice, human rights and other constitutional measures are also important parts of a governance framework.

Conflicts and regional instability (or stability) can influence water demand and use, particularly in water-scarce regions. This is the case where competition arises between different water uses within a country or where water disputes exist between countries, as between Bangladesh and India over the Ganges River and among the riparian countries along the Danube River. (This subject is discussed further in chapter 9.) There are more than 400 registered agreements over shared watersheds,[2] most between two riparian countries. Although the UN Convention on the Law of the Non-navigational Uses of International Watercourses was adopted by the UN General Assembly in 1997, it has not yet been ratified by a sufficient number of countries to enter into force. One of the most successful conventions on water resources is the regional United Nations Economic Commission for Europe Convention on the Protection and Use of Transboundary Watercourses and International Lakes, convened in Helsinki in March 1992.[3] This convention, entered into force in 1996 and currently ratified by 35 countries, serves as a driver for water management in participating countries.

National legal framework: managing water resources and service delivery

Law and policy are interconnected, with particular legislation derived from water policy in many cases.[4] Making laws operational is often a painstaking process, because of the need to develop implementation regulations and manuals on interpreting the law. Often implementation is by trial and error, requiring feedback and the establishment of practices and cases on how to interpret aspects of water law.

For developing countries the long-term goal of such legislation is poverty reduction through a well managed and sustainable water sector. Associated goals include efficient service delivery, protection of consumer rights, financial sustainability and service coverage to the poor in both urban and rural areas.

Governance of the water sector is complex and involves actors beyond the water sector. The actors can be national legislatures and governments, other sector agencies, local governments, river basin authorities, representatives of indigenous peoples, consumer bodies, private companies and others. Who is involved may differ with the issues concerned – for example, surface waters, groundwater, coastal waters or wetlands. Effective action on such a complex group of interests requires open communication and strong coordination facilitated by an appropriate legislative and regulatory framework. The Government of Australia recognized this need when it adopted the Commonwealth Water Act in 2007 and subsequent regulations (box 4.2).[5]

There are fundamental differences between managing water resources and delivering water services. Managing water resources involves a wide range of institutions at local, state, national, regional and international levels. Delivering water services (including administration) usually falls under the authority of elected local officials and specific local institutions. It is misleading, therefore, to discuss resources management and services delivery in the same institutional context.

There are fundamental differences between managing water resources and delivering water services. It is misleading, therefore, to discuss resources management and services delivery in the same institutional context

Decentralization, for example, can affect how water resources and water distribution services are managed. It is a political process, however, not necessarily a water-specific solution to providing improved water services. It requires that water institutions integrate the physical watershed and administrative boundaries, nesting these within each other at different scales. Success with such integration for catchment bodies below the river basin scale has been limited, however, with evidence from countries like South Africa suggesting that such integration may often be too complex to implement.

Table 4.1 shows the range of measures required to address water rights and water management. Table 4.2 shows additional measures that may be required to address the provision of water services. The tables draw from a study across four jurisdictions

Box 4.2 Australian water law reform

Australian states have been reforming their water laws within a framework set out by the Commonwealth government, called the National Water Initiative. The initiative is intended to provide security of entitlements to water, including ecosystems use. It has a formula for sharing risk between government and users should water availability change in the future due to climate change or other factors.

The National Water Initiative and related policies require water trading, which enables water to be properly valued and allocated to higher-value uses. But this means that water rights have to be separated from land rights, which can in turn make it difficult for small-scale farms to survive. This has implications for equity and the potential need for structural adjustment funds. It also has consequences for supportive legal regimes. For example, there may be a need for a secure registry of water rights, similar to a registry of land rights.

There are also consequences for infrastructure. For example, Queensland separates the ownership and management of distribution facilities (irrigation networks) from the storage infrastructure (dams), and users of the irrigation networks cannot opt out of operating costs without the consent of the licence holder, to avoid leaving the system without an owner. If water were to be traded out of an irrigation area, the previous owners might no longer pay for the system, leaving the new owners with a liability but no income.

However, Australian states do not rely on water trading to manage water. In every state there will be a structure for river basin management and stakeholder engagement through water resources plans produced by the states. These plans will allocate water, and only when a plan is in place will it be possible to trade water, as for example under the Queensland Water Act of 2000.

The first requirement is thus to have a sound system to manage water and allocate it to users, which should be the focus of water law reforms, especially where human and financial resources are limited. Only a planned system can account for the public good elements of water. Markets alone cannot.

Source: www.nwc.gov.au/nwi/index.cfm; Roper, Sayers, and Smith 2006; Queensland Government 2000; Hendry 2006.

– Scotland, England, South Africa and Queensland, Australia.[6]

Key policy and regulatory issues

Although water allocation systems can be difficult to establish, managing competing water uses requires clear, widely accepted allocation rules, especially where water is scarce. Water allocation systems should balance equity and economic efficiency. Environmental concerns also require equal attention, though they are often neglected in the process. In Chile, for example, the environment is not granted any water licences, while in South Africa decision-makers are debating how to put water law on environmental protection into practice. Lawmakers must address public policy implications, including equity and water reallocations in times of drought or other emergencies. And permit systems should be sufficiently flexible to adapt to global changes and climate variability.

Figure 4.1 Formal and informal legal framework of water rights

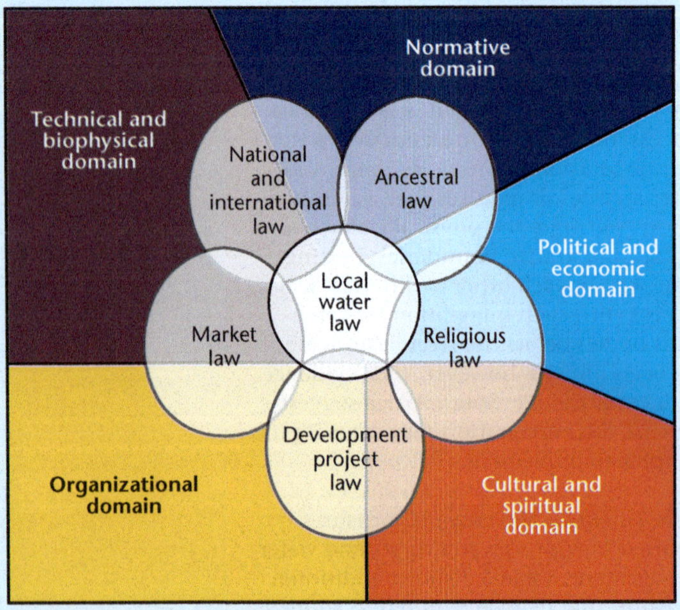

Source: Based on Boelens 2008.

Much water governance takes place outside formalized legal systems, particularly in developing countries (figure 4.1). Such 'traditional' rights systems form a dynamic mixture of rules, principles and organizational forms of different origins. They combine local, national and global rules and often mix indigenous, colonial and contemporary norms and rights. Important sources for these complex, local rights systems tend to be state laws, religious laws (whether formal or indigenous), ancestral laws, market laws and the rights frameworks of multiple water project interventions, which often set their own regulations.

Local water rights thus exist in conditions of legal pluralism, where rules and principles of different origins and legitimization coexist and interact.[7] In the eyes of water users in many parts of the world, legitimate water authority and water rights are not restricted to official law. Water users also clearly distinguish water rights as defined by lawyers (officially codified or recognized) from their own, living rights systems.

Table 4.1 **Laws addressing water rights and water management**

Legislative requirement	Options for provision		
	Integrated water resources management and river basin planning	**Water rights and abstraction licensing**	**Water quality and water pollution**
High-level principles, purposes and duties	• Primary law • Ownership or trusteeship of resource • Equity, water efficiency and integration • Priority uses in law or policy (e.g., basin plans)	• Ownership or trusteeship of resource if not in water resources management law • High-level duties on water users (e.g., sustainable and beneficial use) • Priority uses in law or policy (e.g., basin plans)	• High-level duties on water users (e.g., sustainable and beneficial use, no waste, efficiency)
Catchment planning	• Catchment based • Alignment with administrative boundaries • Coordination with other strategic planning processes (e.g., land use, biodiversity)	• Licence in accordance with catchment plans if they exist	• Licence in accordance with catchment plans if they exist
Define water environment	• Surface and groundwater • Coastal waters • Wetlands	Control all waters: • Surface and groundwater • Coastal waters • Wetlands	Control all waters: • Surface and groundwater • Coastal waters • Wetlands
Regulatory structure	Water authority: • Government department • Agency • Stakeholder-led	Water authority: • Government department • Agency • Stakeholder-led	Water authority or environmental authority: • Coordination mechanisms
Participation	Stakeholder engagement for planning in primary law	Stakeholder engagement through water resources management framework	Stakeholder engagement through water resources management framework
Licensing	Status of plan: • Regulatory (direct licensing) • Indirect (sets targets) • Managerial (sets targets, incentives)	Integrated water use licences for abstraction and discharge?	Integrated water use licences for abstraction and discharge (dependent on regulatory structure)?
Tiered system (proportionate)		• Tiered system (e.g., general rules and full licences) • Exemptions (e.g., domestic use, subsistence use, volume limits)	• Tiered system (e.g., general rules and full licences) • Emission, quality and ecological standards – progressive approach • Management of diffuse pollution
Licence conditions		Duration, review periods and tests for grant, review and reallocation	Duration, review periods and tests for grant, review and reallocation
Water trading		Prohibit, permit and encourage	

Source: Hendry 2008.

Understanding the nature of water rights in each system and water territory thus requires taking into account their multilayered bundles: their rights to use and withdraw, operate, supervise and manage, and control. Focusing only on the local level is clearly inadequate. Multistakeholder platforms or other arenas for achieving common goals and establishing patterns of governance – which include recognizing informal water rights, empowering marginalized social and ethnic groups and representing all interested parties in allocation and decision-making – have the potential to ensure fairer and smoother reallocation of water resources (box 4.3).

Participatory water management

Participants in the consensus-building World Water Vision exercise indicated that the Vision could be achieved only if empowered individuals and communities participated at all levels of decision-making on water resources management. Their concerns were pragmatic, driven more by considerations of governance systems than by equity. Thus the *World Water Vision* report concluded that

> Both public and private management of water will improve through greater accountability, transparency, and rule of law. Incentives

Table 4.2 Laws addressing provision of water services

Legislative requirement	Options for provision	
Regulators Economic, duties of supply and quality standards Environmental	• Ministry, sector agency (e.g., water industry commission, office of water services) or multiutility (e.g., competition authority) • Ministry or environment agency • Separate consumer body?	
Providers	• Local government • Water board or agency	• Private company
Vertical disintegration and integration Horizontal disaggregation and aggregation	• Abstraction, treatment, distribution and supply • Regional ('competition by comparison'?)	
Private sector involvement	• Forbidden? • Public sector preference? • Short-term contracts	• Build-operate-transfer • Leases and concessions • Divestiture
Constitutional and human rights	• High level • Additional enforcement mechanisms	
High-level duties (on regulators, providers and users)	• Universal service obligation • Conservation, efficient use (water efficiency) • Sustainable and secure supply	• Consumer protection • Competition • Economic efficiency and return on capital
Duties of supply	• Universal (progressive?) • In service areas	• Reasonable cost • Drinking water customer service standards
Tariffs, metering and disconnections	• Banded • Two-part • Free basic service • Participation in tariff-setting	• Presumption of metering • Powers to disconnect or limit supply for non-payment
Emergency powers	• Climate and drought • Pollution incident	• Infrastructure failure • Ministers, water providers and regulators
Storm water	• Incorporate storm water management into water services provision (and potentially into abstraction licensing and pricing)	
Conservation and demand management	• High-level duties on conservation and efficient use • Highest appropriate standards for built environment, grey water reuse	

Source: Hendry 2008.

must improve for all stakeholders. More community participation will provide a sense of ownership and empowerment to local stakeholders. The role of education in making this process possible cannot be overestimated. Public access to information will provide an incentive to elected officials and private operators, who will be held responsible for results, including maximizing social welfare. It will also reduce opportunities for corruption and for capture of the system by powerful elites.[8]

Role of non-governmental organizations. Non-governmental organizations (NGOs) can play a valuable role in a country or local community. Normally operating outside the formal government, NGOs may be community-based organizations, poverty-focused large external organizations such as Oxfam and the Bill and Melinda Gates Foundation, and charities funded by religious organizations. Issue-related organizations such as the Council of Canadians and IUCN–International Union for Conservation of Nature may also be important contributors at a national or regional scale. Where such organizations exist, they should be involved in the participatory processes. A thorough analysis of the contributions of the NGO sector to the Millennium Development Goals and to water management, showing the unique characteristics of different kinds of NGOs, their contributions, their limitations and a perspective on their future role would be a useful contribution to the related literature.

Participation in the irrigation sector. Along with market tools (such as privatization and removal of subsidies), water management policies have been shaped by calls for a more participatory development approach that advocates smaller government and local participation in governance, management and financing. At the same time, to comply with the structural adjustments required by international financial institutions in the last few decades, governments have decreased public spending in most sectors and disengaged from them. Such strategies have led to major changes in water management, particularly in irrigation, where governments have embarked on reform.[9]

One of the most important and far-reaching reforms is the irrigation management transfer that has been taking place in more than 57 countries on 5 continents (box 4.3). Overall, this transferring of responsibility and authority for managing irrigation systems from the public sector to the community has forced a new look at how services are provided to users and a move from supply-driven to demand-driven approaches. And the closer involvement of water user associations has resulted in increased accountability, transparency and responsibility, as has been reported in China and Mexico, for example.[10]

Participation reduces corruption

Global Corruption Report 2008: Corruption in the Water Sector estimates that corruption in the water sector can raise the investment costs of achieving the Millennium Development Goals target for water and sanitation by almost $50 billion.[11] Corruption in the sector includes falsified meter readings, distorted site selection of boreholes or abstraction points for irrigation, collusion and favouritism in public procurement and nepotism in the allocation of public positions.

Corruption remains a poorly addressed governance issue in the water domain. This domain is a high-risk sector for corruption because water service provision is a near natural monopoly. The resource is becoming increasingly scarce in many countries, and the water domain involves large and often complex construction contracts. Furthermore, water has multifunctional characteristics and is used and managed by a mix of private and public stakeholders.[12]

Corruption – on a petty or grand scale – occurs across the water spectrum and among all water sector actors. According to *Global Corruption Report 2008,* in some countries corruption siphons off as much as 30% of the budget.[13] By diverting funds from investment or operation and maintenance, corruption reduces access to water. And for many poor people, paying bribes is the only means of securing access to water supplies.

Corruption undermines poverty reduction efforts and impedes economic, social and sustainable development. Poor people generally suffer most from corruption, overpaying for water service delivery or bearing the health burdens arising from lack of sustainable access to safe drinking water. The indirect costs of corruption are also high. There is a strong correlation between access to adequate water and sanitation and infant mortality (see chapter 6).

Corruption can lead to uncontrolled pollution of water sources, overpumping and depletion of groundwater, lack of planning, uncontrolled degradation of ecosystems, weakened flood protection, urban expansion leading to heightened water tensions, and other harmful effects. In water-scarce southern Spain tens of thousands of properties have been developed illegally, particularly in seaside resorts. In the Andalucian city of Ronda this practice led to a severe governance crisis, and pollution now jeopardizes water supplies.[14] Where corruption is widespread, achieving the Millennium Development Goals will take much longer – not only with much higher cost in direct investment in services and management, but also in indirect costs, such as water-related diseases and loss of lives, ecosystems and productive capacities. It also threatens

Corruption remains a poorly addressed governance issue in the water domain

Box 4.3 Experience with irrigation management transfer

Irrigation management transfer is the transfer of responsibility and authority for managing irrigation systems to water user associations. It began in the 1960s in Taiwan, Province of China; Bangladesh; and the United States; in the 1970s in Mali, New Zealand and Colombia; and in the 1980s in the Philippines, Mexico, Tunisia and the Dominican Republic. It peaked in the 1990s in Turkey, Morocco, Australia, Peru, Albania and Zimbabwe, but still continues in countries such as Pakistan and Sudan (2000), India (2001) and China (2002), each with unique experiences and results.

The Food and Agriculture Organization of the United Nations and International Water Management Institute database on experience with irrigation management transfer provides information on the key factors that motivated adoption of the new policy. The most commonly cited is a shortage of government funds for operating and maintaining the system, followed by difficulties with fee recovery.

Overall, the results have been mixed. Financial sustainability and lack of clarity about the financial and technical assistance provided to water user associations by the government have been a concern. Monitoring and evaluation are essential. The concept of 'farmer participation' was often translated into a fixed set of principles, such as volumetric control, cost recovery, water pricing, economic, water use efficiencies and downstream control, that were not relevant in all contexts. For example, Andean farmer-managed irrigation systems typically have upstream control techniques and management structures, which provide transparency and ease of operation. For them, participating in water distribution usually means involving everyone in the system's management and decisions on water distribution to individual fields.

Source: Garces-Restrepo, Vermillion, and Muñoz 2007; Boelens 2008.

Corruption is an important driver of uncontrolled pollution of water sources, overpumping and depletion of groundwater, lack of planning and uncontrolled degradation of ecosystems, diminishing flood protection, urban expansion leading to heightened water tensions and other harmful effects

existing achievements by undermining institutions and the sustainability of infrastructure.

Legitimate, transparent and participatory processes can be effective in garnering support for the design and implementation of water resources policy and of deterring corruption. But participatory processes require adequate institutional, policy, legal and economic instruments. Political leadership is required to put these processes in place and demonstrate support for them. International assistance and interventions to reduce corruption will have only limited effect if political will is absent.

Implementing regulations

Water resources management is underpinned by a functioning legal system that includes:

- Water resources legislation, the province of the legislature and the executive.
- Implementation and administration of legislation, the province of the executive.
- Adjudication of civil disputes among water litigants, the province of the judiciary.
- Prosecution of criminal offenders by the executive and the judiciary.

Following adoption of a law by the legislature, the executive needs to address relevant details not included in the legislation by preparing implementation regulations. Neither legislation nor implementation regulations will make much difference, however, unless they are effectively administered by the water resources administration. Nor will they secure rights in the resource unless a judiciary can adjudicate disputes effectively, impartially, expeditiously and transparently. Finally, the legislation needs vigorous enforcement and systematic monitoring, using a set of indicators to gauge effectiveness and improve system performance.

Challenges

Water resources development and management in the interests of national development objectives require effective policy and legal frameworks that also respect deeply rooted customary practices. Participatory processes that take account of the social, economic and cultural characteristics of the country or community will make a significant contribution to meeting this challenge. But a greater challenge is to ensure that such laws and the regulations that support them make a difference on the ground by effectively administering and enforcing them. And the security of rights in the resource must be adjudicated by an effective, impartial, expeditious and transparent judicial system.

Financing – the missing link

Authors: Richard Connor, William Cosgrove, Walter Rast and James Winpenny

Contributors: Jack Moss and Monica Scatasta

Virtually all water-related activities, whether structural (infrastructure) or not (planning, data collection, regulation, public education and so on), require money to develop, implement and carry out. Even if all the necessary policies and laws are in place, lack of funding will bring necessary actions to a standstill. Adequate funding and the willingness to invest in water management and infrastructure are therefore major determinants of the availability of sufficient quantities of water of acceptable quality.

Although water is often described as a 'gift of nature', harnessing and managing water for human and ecological needs entail financial costs. These costs are often widely ignored, underestimated or underfunded, with the result that important functions and assets are neglected and underprovided, while existing assets and services deteriorate.

Three functions are involved in water management, each with associated costs:

- Water resources management and development, including watershed and river basin development, storage, flood-risk management, environmental protection and pollution abatement.
- Water services to municipalities and households, commerce and industry, agriculture, and other economic sectors, including the costs of wastewater treatment, rehabilitation, operations and maintenance and inadequate infrastructure.
- Integrative functions, such as water sector policy development, research, monitoring, administration, legislation (including compliance and enforcement) and public information.

The costs associated with these functions are either capital (investment) costs or annual recurrent costs, both variable and fixed. To function properly, the water sector must cover all costs – not just those of major physical infrastructure – in a sustainable way. That means ensuring reliable, predictable finance from government revenues (taxes), the sale of water services or long-term aid commitments.

Financing is often a limiting factor in effectively managing the water sector. The solution is to focus not only on increasing flows of funds to the sector but also on achieving a realistic balance between the demand for and supply of financing to ensure financial sustainability. Demand for funds needs to be rationalized by developing realistic investment plans, minimizing the recurrent costs of service delivery and ensuring the sustainability of water resources and the safe and reliable delivery of services to maintain users' willingness to pay.

The logic differs for the three sources of finance for the water sector. The rationale for local user financing is users' consumption of the resource and local authorities' responsibility in most cases for the main decisions about water services and tariffs. The rationale for national government finance is often the national or regional benefits to be gained from managing the resource. Capital investment costs tend to be covered largely by governments, except where assets are privately owned (for example, farmers who have their own infrastructure). The international community provides mainly 'catalytic' funding to jumpstart projects, which includes providing financial guarantees. Decisions made at the international and national government levels are most likely to be outside the 'water box', while local user concerns more directly address specific systems for water supply and sanitation.

Investment in water management capacity

The water sector has been plagued by lack of political support, poor governance, underresourcing and underinvestment. These ills are manifested in non-transparency, lack of accountability, unsustainable economics, high levels of unaccounted for water and low revenue collection. They have led to infrastructure deterioration, the breakdown of services and ultimately customer dissatisfaction. Figure 4.2 illustrates how this combination of factors creates a vicious cycle of low funding, weak political support and poor service provision. Breaking this vicious cycle will require more than investments in hardware.

Lack of political support, poor governance, underresourcing and underinvestment in the water sector have led to infrastructure deterioration, the breakdown of services and ultimately customer dissatisfaction

Figure 4.2 If the vicious cycle of low funding is reversed, the benefits to society will be enormous

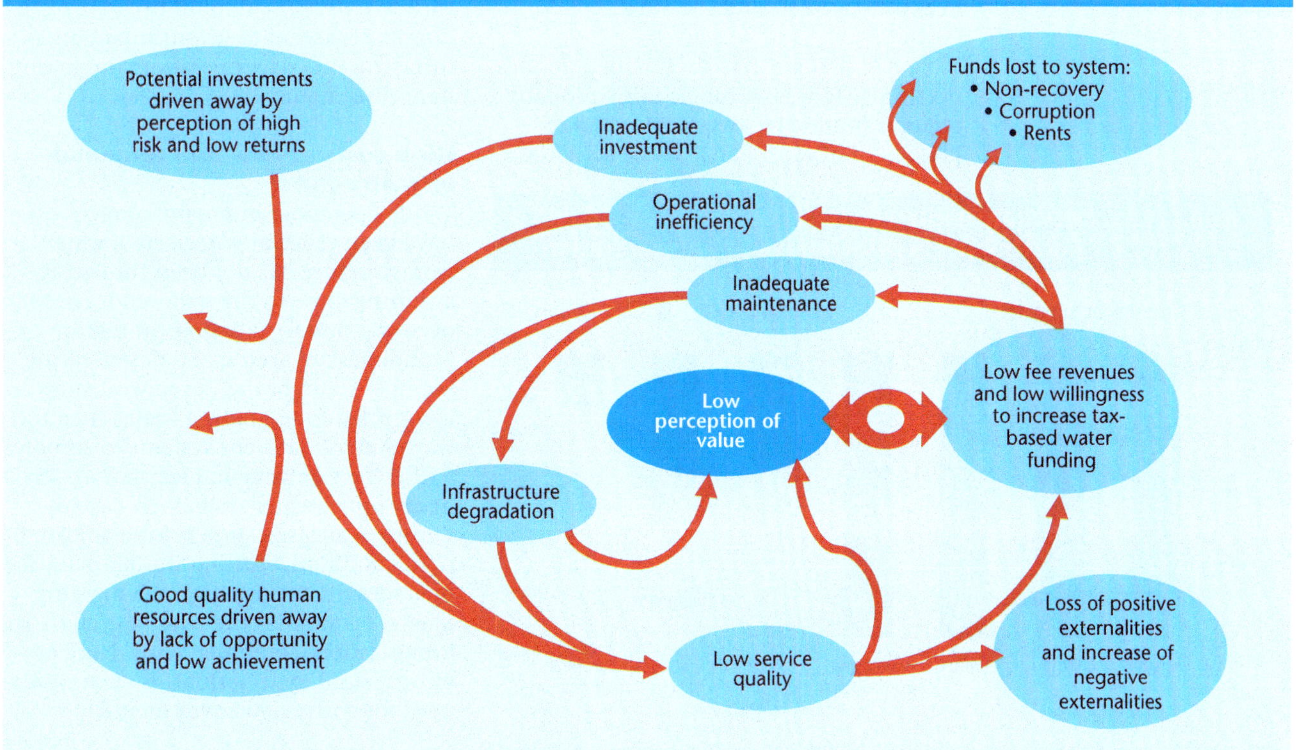

Source: J. Moss based on ideas from A. Mathys; Moss et al. 2003.

In most urban public water systems charges barely cover the recurrent costs of operation and maintenance. In rural areas neglect of operation and maintenance and cost recovery contribute to widespread non-functionality

Investment is also required in the operation and maintenance of physical infrastructure so that it meets appropriate standards and functions efficiently. Operations and maintenance are neglected nearly everywhere in favour of new infrastructure investments, regardless of the country's level of development. In the United States bringing water supply and sewerage infrastructure up to current standards will cost more than $1 trillion over the next 20 years, with hundreds of billions more required for dams, dikes and waterway maintenance.[15] The World Business Council for Sustainable Development estimates that the total costs of replacing ageing water supply and sanitation infrastructure in industrial countries may be as high as $200 billion a year.[16] Investment in physical infrastructure must be accompanied by the 'soft' infrastructure of policies and legal systems (as described earlier) and human capacity.[17] Yet much bilateral aid for sanitation and drinking water fails to achieve a balance between soft and hard infrastructure (figure 4.3).

In most urban public water systems charges often barely cover the recurrent costs of operation and maintenance, leaving little or no funds to recover the capital costs of modernization and expansion. A survey of such systems in 132 cities in high-, middle- and low-income countries found that 39% did not recover even their operation and maintenance costs (true of 100% of cities in South-East Asia and the Maghreb).[18]

Moreover, water infrastructure deteriorates over time. To keep it functioning properly requires routine repairs, service and replacement of worn parts. These activities, easy to postpone, are widely neglected. The result is infrastructure that deteriorates to a level that can no longer provide reliable access to safe drinking water to those who are nominally receiving the service. Leakage (loss) rates of 50% are not uncommon in urban distribution systems. Much of the apparatus for treating wastewater is also failing. According to a report by the Task Force for the Implementation of the Environmental Action Program for Eastern Europe, Caucasus and Central Asia, municipal water utilities have now become the main polluters of surface waters in many East European, Caucasus and Central Asian countries. The task force reports that up to 90% of nitrogen and phosphorus discharges into the Black and Caspian Seas originate from riverine inputs, which mostly transport municipal wastewaters.[19]

Figure 4.3 New infrastructure seems to dominate donor investments in drinking water and sanitation

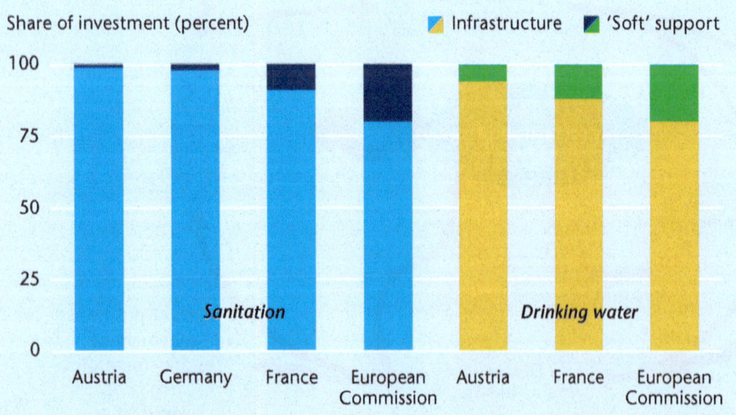

Note: Soft support includes support for policies, legal systems and human capacity building.
Source: Based on data from UN-Water 2008.

In rural areas neglect of operation and maintenance budgets and cost recovery contribute to widespread non-functionality. A recent survey of almost 7,000 rural water schemes in Ethiopia found that 30%–40% were non-functional.[20] A shortage of finance for wages, fuel, materials and spare parts was a common factor.

The deficit in financing, especially for operation and maintenance costs, is a substantial addition to the investment costs of achieving the Millennium Development Goals. Although governments often turn to external aid to fill financing gaps, donors also seem to favour financing new infrastructure over operation and maintenance (see figure 4.3).

High costs of new and remedial infrastructure

While operation and maintenance costs have been especially neglected, water infrastructure has not been funded at anything close to the required level. Many networks and installations in mature economies are ageing and deteriorating. Member states of the European Union are committed to upgrading their water and wastewater treatment systems to comply with EU environmental legislation. But many urban water systems in Eastern Europe, Caucasus and Central Asia are in poor condition, with no similar plans for upgrades. In developing and emerging market economies the pace of growth and urbanization, combined with rising environmental expectations, is creating the need for costly new investments.

Table 4.3 gives a sense of the magnitude of investment requirements over the next

20 years for water supply and wastewater services infrastructure and the gap between financing needs and projected revenues, by region. These calculations do not include other sizeable funding requirements, such as water resources development and management and water governance.[21] Raising and spending these huge sums will require quickening the pace of reform (including water pricing) across the water services industry, with implications for both regulators and consumers.

The cost of rehabilitating or decommissioning existing infrastructure is likely to be enormous. Repairing, strengthening or modifying older dams, for instance, will entail sizeable outlays. In extreme cases decommissioning a dam may be a rational decision (where it has outlived its purpose, where it is old and unsafe, where sedimentation is high or where river flows need to be maintained for fisheries and other ecosystems). Rehabilitating or decommissioning also depends on whether the costs of maintaining the dam exceed its expected future economic and financial benefits. Both rehabilitation and decommissioning costs are site specific.[22]

The cost of new water supply is rising. In developed countries and in many places elsewhere, the easiest investments for exploiting water resources have already been made. With available dam sites decreasing, water tables falling and the distances between the point of abstraction and water use increasing, the costs of exploitation and supply are rising. Costs are also pushed up by the growing need to treat water before use.

Sanitation has been severely neglected. Estimates of the cost of achieving the 2015 Millennium Development Goal target for sanitation vary widely, due to differences in approach and a weak information base. The World Health Organization estimates the total annual cost of meeting the target at just over $9.5 billion.[23] If estimates of current costs are correct, resources in the sanitation sector would have to be almost doubled to meet the 2015 target (although estimates of current spending probably underestimate the contributions by households to their own sanitation services). If the full cost of tertiary wastewater treatment for waste streams in urban areas is added, the total rises to $100 billion, the current value of total annual official development assistance. More cost-effective alternatives need to be explored – urgently – if the sanitation target is to be met.

Table 4.3 Annual capital requirements for water supply and wastewater services and water financing gaps, by region, 2006-25

(US$ billions)

Region	Capital needed	Low gap	Medium gap	High gap
Eastern Europe, Caucasus and Central Asia	28.1-40.5	13.4	20.0	26.1
North America	23.9-46.8	3.3	4.9	21.4
Latin America	4.3-6.5	2.9	4.0	5.1
Developed Asia and China	38.2-51.4	29.5	32.9	36.5
Rest of world	14.3-22.6	18.5	22.4	26.1
Total	92.4-148.0	67.5	84.2	115.2

Note: The gaps refer to the difference in projected investment needs for three different estimates of their size and existing sources of revenue from tariffs, official development assistance and government budgets and loans.
Source: Owen 2006.

UN-Water Global Annual Assessment of Sanitation and Drinking-Water: 2008 Pilot Report – Testing A New Reporting Approach (GLAAS report) looks at the constraints to progress towards the sanitation target from the human resources, institutional capacity and financial system capacity perspectives (figure 4.4).

Operation, maintenance and rehabilitation remain critical challenges. Respondents to the GLAAS survey indicated that flooding events and earthquakes were the main causes of damage to infrastructure.[24] Increased weather variability linked to

Figure 4.4 In the few countries surveyed financial system constraints weighed heavily on achieving the Millennium Development Goals sanitation target

Sanitation sector capacity in GLAAS pilot countries with data; latest available data, 2005-07 (number of countries)

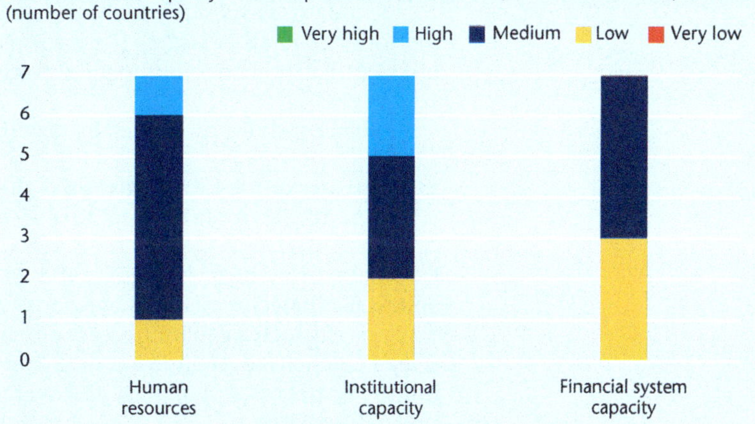

Note: GLAAS is Global Annual Assessment of Sanitation and Drinking-Water.
Source: Based on data from UN-Water 2008.

Figure 4.5 Household expenditure and private sector investments in drinking water supply are generally unknown

Funding sources for drinking water in GLAAS pilot countries with data, latest available data, 2006-07 (percent)

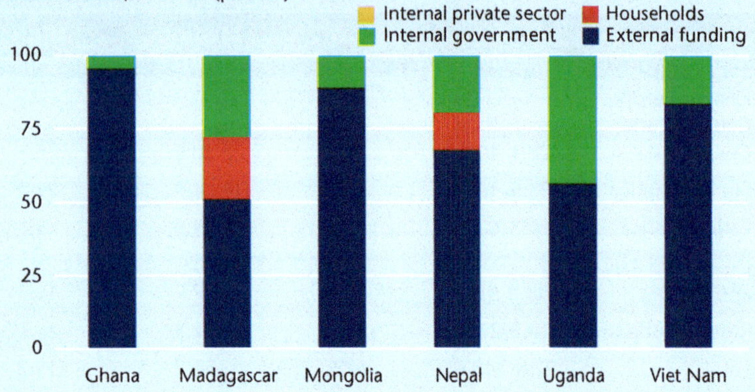

Note: GLAAS is Global Annual Assessment of Sanitation and Drinking-Water.
Source: Based on data from UN-Water 2008.

climate change and armed conflict bring added risk.

African countries, recognizing the urgency of the situation, signed the eThekwini Declaration in February 2008 in Durban, South Africa, committing them to prepare or update national sanitation and hygiene policies, allocate budget funds for sanitation, improve sanitation information and monitoring tools and increase capacity. The declaration also called on external support agencies to provide financial and technical assistance for sanitation and hygiene promotion and to improve aid coordination.

Sources of financing

There are three sources of revenue for financing water supply and sanitation services:

- User tariffs, including payment for environmental services, which can include cross-subsidies within the sector or from other sectors (for example, electricity or other municipal services).
- Public expenditures funded by taxation.
- Transfers in the form of external aid, from official or philanthropic sources.

External borrowing (debt, equity and bonds, facilitated by risk-management instruments such as guarantees) can help spread payments over time for large up-front investments and manage the overall cost of financing.

The 2008 GLAAS pilot report raises concerns about countries' limited resources to invest in operation, maintenance and capital rehabilitation for drinking water and sanitation systems.[25] These expenditures are difficult to assess because many are hidden in sector budgets or are not accounted for, as in the case of many private sector and household investments (figure 4.5). Data in the GLAAS report, available for three of the seven pilot countries, indicate that external funding – for many countries the main source of funds for drinking water and sanitation system investments – is directed mainly to infrastructure projects.

Reviewing and revising investment needs (the demand side of financing) by reducing costs are as important in closing the financing gap as finding new sources of funds. The full cycle of expenditures has to be considered, from operation and maintenance to technological choices about equipment and its eventual replacement or upgrading. For example, improving collection efficiency and reducing unaccounted for losses in distribution systems can make more water available for new consumers and help fund operations. Demand-side considerations also include such underlying determinants as coverage levels, services levels and environmental regulations.

Fund disbursements can also be accelerated, so that disbursement delays do not cause new funding to be postponed. Inefficient budgeting and budget allocation processes can lead to such disbursement delays. To ensure that funds are disbursed more efficiently during the budget period, funds can be allocated to regions of a country or to local authorities according to their relative capacity to implement projects. Finalizing the budget process before the budget year starts makes it possible to begin disbursements in the first quarter of the year.

A strategic financial plan, based on an in-depth examination of all demand- and supply-side aspects affecting the financing gap, will help ensure the financial sustainability of projects. It will direct investment choices towards the most financially and functionally appropriate processes and technologies, thus maximizing benefits. And it will make projects more attractive to external financiers by reducing the perception of risk.

A lot of funds move through the water system but is used inefficiently. Examples include high payments to informal providers outside the public networks, payments to corrupt operators to obtain water from networks and large public subsidies that end up in the wrong hands. Households spend large sums on coping strategies, such as time and money spent on alternative sources of water and on household water filters.

Tariffs – pricing water and willingness to pay. The obvious source of finance for the recurrent costs of water services is user charges, supplemented by government subsidies. The continuing underpricing of water to consumers encourages waste and use of water for low-value purposes in all sectors, depriving the sector of essential funds. This is a major contributor to underinvestment in water infrastructure, management and services and imposes heavy costs on society.

Maintaining the quality and reliability of services is essential (box 4.4), even if there is a parallel push for increased access, since these characteristics affect users' willingness to pay. Transparency, accountability and operational efficiency in service provision are also essential to user satisfaction. Affordability also needs to be determined. It is based on macroaffordability – of investment choices (driven by coverage, service levels, technology and other choices) and the cost efficiency of service provision – and household affordability, determined by current expenditure on water and sanitation services (including the hidden costs of securing access when people lack access to formal services and the consequences of access to unsafe services) and their willingness to pay for improved service levels.

Charging for water. Although prices can be strong drivers of positive change in a well functioning economic system, in practice, prices have had a relatively minor role in managing water demand. Many people are deeply ambivalent about using water prices to manage water resources or are strongly opposed to pricing water at the cost required to deliver it to consumers, especially in the politically sensitive segments of agriculture and urban households. As a result, water is often grossly underpriced.

One survey of municipal water utilities in low-income countries found that 89% had no cost recovery measures in place, 9% had partial cost recovery of operation and maintenance costs, and only 3% made any effort to recoup the costs of capital outlays.[26]

A common yardstick for assessing the affordability of water charges for households is that payments should not exceed 3% (in some cases 5%) of net household income. In practice, surveys show that in developed countries households connected to urban public systems pay on average 1% of incomes on water bills, including the cost of sewerage, which may be double that for water. Such an average is not a very reliable indicator, however, especially given the wide variability among income levels in a country. Generally speaking, poorer groups tend to pay a higher share of household income for water. In developing countries the picture is complicated by the widespread use of informal and small-scale private water distributors charging full market prices; in these cases the poorest households can pay 3%–11% of income on water.[27]

As recognition of this inequitable economic burden on the poor has spread, pressure on governments and service providers has increased to ensure delivery of a minimal supply of potable water to all households at a reasonable price. Achieving this objective would require tariff rates based on a household's ability to pay and subsidies that cover the excess cost of service delivery for those who can least afford to pay.

Where pricing is used to cover water supply costs (for example, cities committed to water demand management, private irrigation schemes, markets for irrigation water and penalties for water pollution), it is an important driver of reforms. Where

Box 4.4 Dalian water supply project in China – successful expansion of services

Dalian, a port city at the southern end of the Liaodong Peninsula, in northeastern China, was declared an 'open' coastal city in 1984 and given considerable autonomy in its economic planning. The Dalian Economic and Technology Development Zone, established in 1988, has become one of China's most successful economic zones. By the early 1990s, however, water shortages had become a serious constraint to economic growth. Many areas had water service for only a few hours a day. Frequent service disruptions had major public health implications. The Dalian Water Supply Project, begun in the mid-1990s, provided new infrastructure to address the water shortages and meet increasing water demands.

The project achieved its objectives. All constructed facilities were operating satisfactorily, and the 73,000 residential water connections in Dalian exceeded the predicted number. The project also increased water supply to commerce and industry, removing potential constraints to economic expansion and improving the investment environment. The project evaluation confirmed two important findings. Local government commitment was the most important contributor to the success of the project. And water consumers will accept the need for higher tariffs once they are convinced that services are adequate and reliable. Water tariffs were increased substantially from 1995 to 2001, at an average annual rate of 12.8%.

Source: ADB 2004.

Where prices cannot adjust to financial realities, stresses emerge as water shortages, water waste, inefficient water use, inadequate water infrastructure investments and poor water-related services

prices cannot adjust to financial realities, stresses emerge as water shortages, water waste, inefficient water use, inadequate water infrastructure investments and poor water-related services. Water quality may be inconsistent, and maintenance and rehabilitation of distribution systems may be neglected. Capital investment may also be inadequate, resulting in the failure to develop adequate water supply and sanitation services. However, even in situations where pricing is actively used to cover water supply costs, the long history of water as a public good means that water prices have been heavily subsidized by tax-funded distributions from individuals and corporations that may not be direct beneficiaries of the services provided.

In agriculture some farmers rely on public irrigation systems while others have private arrangements (for example, groundwater and water harvesting systems). In privately owned systems energy subsidies (for pumping water) are a key factor affecting efficiency. Farmers using public irrigation systems often pay little or nothing towards recurrent costs and usually nothing towards the capital costs of the irrigation infrastructure. This affects how farmers use water, as one survey in India discovered:

> Farmers have no incentive to use water efficiently as charges are too low and are based on the area irrigated. Inefficient water use has led to severe environmental problems – rising groundwater levels, water-logging and soil salinity. Administration is ineffective. Assessment and collection of fees is often carried out by different departments, or a department not related to irrigation. Farmers need to be involved in setting rates, because at present they simply oppose any suggestion of an increase in price.[28]

Though widely accepted, the 'polluter pays' principle has not had a major impact on polluters' behaviour or on raising funds that could be allocated for environmental purposes, with the exception of developed countries and a much smaller number of developing countries. Although not a financing source, the alternative method of water pollution quotas has similarly been limited to the industrial, urbanized economies, with almost negligible successful examples from developing countries (see chapter 8).[29]

Multipurpose water projects that cross-subsidize irrigation and household water use from hydropower revenues are another form of tariff-based financing. The hydropower components of dams and water storage schemes tend to perform better financially than the associated irrigation projects, which often fail to recover both operating and capital costs. Thus, the power element cross-subsidizes irrigation and other water users – and often navigation, flood control and other public goods as well. In the United States this kind of cross-subsidy was a planned part of the management of the Grand Coulee Dam in the Columbia River Basin and of the major river basin development works of the Tennessee Valley Authority.[30]

Role of the private sector. Several reports conclude that the private sector provides very few water supply and sanitation services in developing countries. The United Nations Development Programme's *Human Development Report 2006,* for example, estimates that although the number of people served by the private water sector grew from roughly 50 million in 1990 to 300 million in 2002, less than 3% of people in developing countries are covered by private or partially private companies.[31] These figures almost certainly understate the real scale of private sector service provision, since they consider only larger-scale private operations and investments. Private operators also include small and medium-size companies with fixed or mobile distribution systems as well as the much larger spread of informal operators that cover huge swathes of low-income urban areas.

The substantial role of small and medium-size entrepreneurs and operators is just beginning to be studied (figure 4.6). A World Bank report found 10,000 small service providers in a limited sample of 49 countries,[32] while an International Institute for Environment and Development study estimates that the global number may exceed 1 million.[33] In addition, the provision of infrastructure by property developers has not been examined but could be substantial.

The landscape for private water operators today is very different from that of a decade ago. Several major multinationals have withdrawn from international projects, leaving just two or three to pursue system concessions, build-operate-transfer and management contracts, especially in the Middle East, China and South-East and East Asia. The gap is being filled by new private water companies based mainly in China, South-East Asia, the Russian Federation and Latin America, servicing emerging local and regional markets. Small-scale and informal water providers have continued to enlarge their share of urban markets in developing countries.

A high proportion of the contracts won by the new market entrants are for desalination and wastewater treatment, which address the growing water scarcity in arid regions and the serious pollution caused by untreated municipal wastewater. The diversity of the new market entrants, their access to local sources of finance and their typically good political connections augur well for successful implementation.

External private investment in the water sector is significant, of the same order as that of official development assistance (figures 4.7 and 4.8). The domestic private sector is becoming a water funding source in some middle-income countries, where powerful local conglomerates are moving into water services, drawing on their own equity and that of other local commercial sources. Further down the financial scale small informal operators dominate large portions of the water market in urban and peri-urban communities. Although some of these operators invest in networks, most use mobile facilities, financed by their own equity or short-term credit. At the street level bottled water sellers have proliferated. A necessity in many areas across developing countries, where failing public supply systems are often contaminated by wastewater or storm water, the use of bottled water is a lifestyle choice in developed countries.

Government financing from public revenues. The public sector accounts for more than 70% of investment in the water sector.[34] There are marked differences in how – and how much – governments finance and subsidize the water sector. In many poor countries, where fiscal constraints are severe, water supply is only one of many priority sectors that governments are under domestic pressure or international commitment to finance.

Funding for infrastructure has varied with economic development and urbanization. At earlier stages the central government generally supports infrastructure provision through subsidies and administrative assistance (box 4.5). As countries develop, the portion of central government support declines, and the cost of environmental

Figure 4.6 Private water operators have a substantial role in developing and developed countries

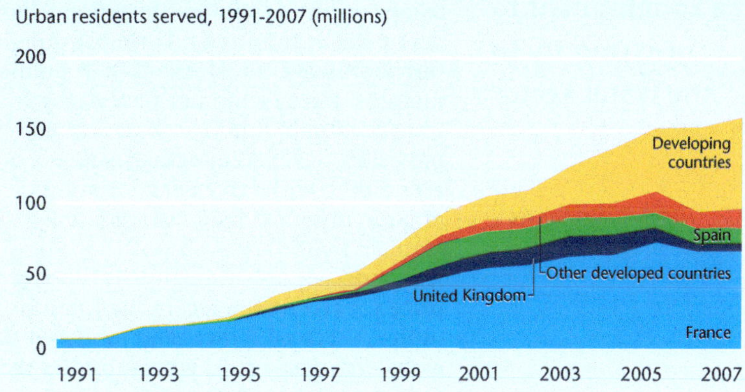

Source: Based on data from Marin 2009.

Figure 4.7 External private investment in the water sector, though variable, has been significant since the early 1990s

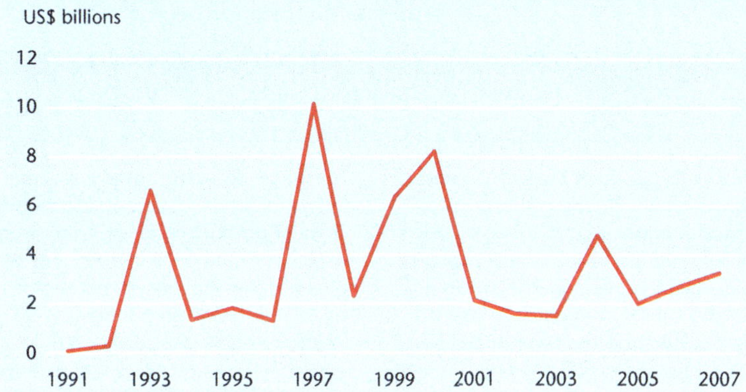

Note: Refers to management and lease contracts, concessions (or management and operation contracts with major private capital commitments), greenfield projects and divestitures for potable water generation and distribution and sewerage collection and treatment.
Source: World Bank Private Participation in Infrastructure Database (http://ppi.worldbank.org).

Figure 4.8 Official development assistance to the water supply and sanitation sector is rising again after a decline during the 1990s

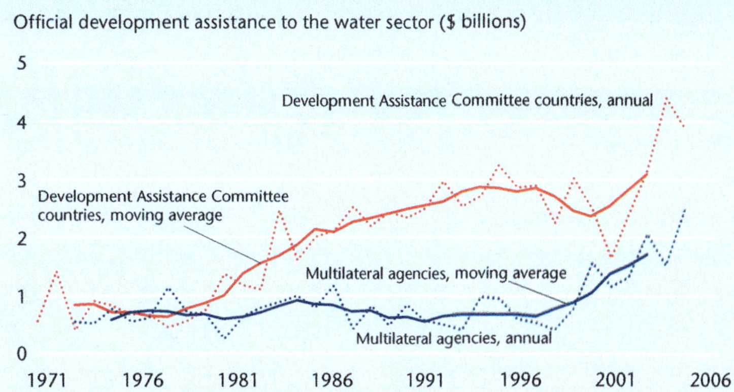

Source: Based on OECD-DAC 2008.

G-8 leaders in June 2002 made a commitment to give priority to the water sector

services is transferred to users, polluters and local governments.

Some countries that have benefited from debt relief or from the oil and commodities boom have transformed their public finances, but this has not necessarily translated into improved water service provision. Several emerging market economies, with large concentrations of poor, unserved populations, are in stronger budgetary positions than they were a decade ago, though this is being placed at risk by the recent fluctuations in the cost of oil, power and food and the global financial crisis, with subsidies rising accordingly. Improving budgetary circumstances provide opportunities for increasing investments in the development of the water sector.

Financing through external aid. Official development assistance from donor countries and multilateral donors to the water supply and sanitation sector increased during the 1970s and 1980s but decreased during the 1990s, with less aid for large infrastructure, before rising again in 2000 (see figure 4.8).

Support from multilateral agencies remained relatively stagnant from the 1970s – when it was about the same as bilateral assistance – until about 2000, when both sources of financial aid began to increase. But it still remained substantially less than official development assistance from bilateral sources.

Leaders at the meeting of the G-8 in Evian, France, in June 2002 made a commitment to give priority to the water sector. Official development assistance increased substantially in the years immediately thereafter. While the amount going to the water supply and sanitation sector increased, aid to the other water sectors remained relatively unchanged (table 4.4). However, overall lending for water remained at less than 6% of total official development assistance, and the share of total lending declined.

External assistance from philanthropic sources, such as foundations and religious groups, highlights an awareness of the importance of water and sanitation. Although these funds are generally much lower than those from multilateral and bilateral sources, a few of the largest foundations (for example, the Bill and Melinda Gates Foundation) can rival some bilateral sources.

Box 4.5 Subsidizing water supply and sanitation in the Republic of Korea

In the Republic of Korea the central government provides direct subsidies for water supply and sanitation infrastructure to local governments or service providers. The amount of the subsidy depends on the size of the city and the type of facility. Subsidies differ for construction and operation. Typically, subsidies are 50%-80% for water source development in rural areas and 50% for local waterworks improvements. Wastewater treatment is eligible for a 50% grant, and sludge treatment for loans of 30%-70%.

For municipal water supply, revenue from tariffs now covers an increasing share of production costs, rising from 69% in 1997 to 83% in 2005. For regional water supply systems supplied by the national water company, Korea Water Resources Corporation (K-water), full cost recovery was achieved by 2004. Tariffs still fall short of actual costs for sewage treatment. During 1997-2004 the central government paid 53% of the total investment costs for sewage treatment, using proceeds from the national liquor tax.

Source: OECD forthcoming.

Table 4.4 Commitments of official development assistance from bilateral and multilateral agencies, 2004-06

(US$ millions)

Sector	2004	2005	2006
Water transport	416	503	304
Hydropower plants	755	480	652
Agricultural water resources	608	830	790
Water supply and sanitation	3,127	4,405	3,879
Total water sector	4,951	6,218	5,625
Total all sectors	79,431	107,078	104,369
Water sector as share of all sectors (%)	6.2	5.8	5.4

Source: OECD, DCD/DAC 2007.

Recent financing initiatives – a new financing agenda

Over the last five years there have been several key initiatives on shaping the agenda of international water financing, notably the World Panel on Financing Water Infrastructure (chaired by Michel Camdessus), the Task Force on Financing Water for All (chaired by Angel Gurria) and the UN Secretary-General's Advisory Board on Water and Sanitation (UNSGAB). *Financing Water for All,* the report of the World Panel on Financing Water Infrastructure, addresses the financial architecture of the global water sector, including many proposals to improve its governance.[35] The Gurria task force report focuses on factors influencing the demand for finance and the scope for developing the financial capacity of subnational entities.[36] The UNSGAB stresses the importance of capacity building, especially in local authorities, and inspired creation of the Global Water Operators Partnership Alliance for peer group support.[37]

These initiatives occurred while domestic savings in emerging market economies were growing rapidly and local capital markets were developing. Sharp rises in the price of oil and other primary commodities had enriched producer countries and transformed their public finances, while causing budgetary problems in primary-commodity-importing countries. International commercial finance for water has become sharply polarized. Lenders and portfolio[38] investors have eagerly pursued opportunities in sound water companies, solvent municipalities and profitable projects (such as desalination), but many countries and municipalities have been relegated to financial backwaters.

Recent policy developments

A number of policies and financing tools have been developed to respond to this new agenda:[39]

- *Increasing commitments of official development assistance for water – and in more user-friendly forms.* International aid for water has bottomed out and commitments are starting to rise, led by a few donor agencies.

- *Using official development assistance to leverage other financial sources.* An approach that has made a promising beginning in Kenya and elsewhere is to use output-based aid to promote microfinance.

- *Establishing national water financing strategies.* Governments in Africa, Eastern Europe and the Caucasus and Central Asia and elsewhere are producing coherent financing strategies, supported by programmes of the Organisation for Economic Co-operation and Development, the EU Water Institute, the World Bank Water and Sanitation Program and other agencies and programmes.

- *Promoting finance to subsovereign entities.* In most countries responsibility for water services is devolved to subsovereign layers of administration. Donors have been adapting their products and procedures to facilitate the provision of finance to subsovereign agencies.

- *Establishing facilities to provide finance at decentralized levels.* Much of the development of household water and sanitation services arises from community initiatives, organization and finance. Among recently created finance facilities that operate at this level are the African Water Facility, the EU Water Facility and the Rural Water Supply and Sanitation Initiative of the African Development Bank.

- *Developing guarantees and risk-sharing instruments.* Guarantees and other forms of credit enhancement can lift local borrowers and bond issuers over the critical threshold of creditworthiness and mitigate specific risks. International financial institutions and other agencies have improved their capacity for risk sharing, and several new bodies have been formed specifically for this purpose (such as GuarantCo).[40]

- *Developing local capital markets and local-currency finance.* A number of countries (such as India and South Africa, some countries in Latin America and South-East and East Asia) have municipalities and utilities with sufficient financial standing to attract loan finance or to issue their own bonds. A significant proportion of the unserved populations (almost a half for water and more than a third for sanitation) live in countries classified as middle income, with the potential to raise subsovereign finance of this type.

- *Increasing role of small-scale local water providers.* It is estimated that small-scale providers serve 25% of the urban population in Latin America and East Asia and 50% in Africa and South East Asia.[41]

- *Instituting tariff reform and the principle of sustainable cost recovery.* In most cases tariffs will be the main source of revenue for covering the recurrent costs of water services, although full cost recovery through tariffs is rarely feasible in poor countries. Sustainable cost recovery focuses on securing all three of the basic sources of revenue for water and sanitation services (tariffs, taxes and external aid) as predictable sources of revenue for water operators, which can be used to leverage other sources of funding.

- *Paying for environmental services.* Environmental goods and services take many forms, including potable water supply, irrigation water, flood control benefits, water for transportation and aesthetic benefits. Payment

Lenders and investors have pursued opportunities in sound water companies, solvent municipalities and profitable projects, but many countries and municipalities have been relegated to financial backwaters

systems for such environmental services are easier to implement and administer for more visible and direct uses (such as admission costs for recreational uses).

Challenges

Developing and managing water resources to meet human needs and maintain essential ecosystems entail financial costs. The challenge is both to have more funds flow to the water sector and to ensure its financial sustainability. Sound, strategic financial planning is needed to balance funding requirements with cost-effective management that focuses on demand as well as supply. Full cost recovery has been advocated as a solution to the water financing crisis for many years. In the real world, however, water resources management and services delivery always receive some level of subsidy. Keeping in mind the obligation to meet the basic water services needs of all, the challenge for policy-makers is to make decisions about the acceptable trade-offs among different objectives and about who bears the costs.

Notes

1. United Nations 1945.
2. Transboundary Freshwater Dispute Database (www.transboundarywaters.orst.edu).
3. www.unece.org/env/water/.
4. This section draws on Boelens 2008.
5. Government of Australia 2008.
6. Hendry 2008.
7. See von Benda-Beckmann, von Benda-Beckmann, and Spiertz 1998.
8. Cosgrove and Rijsberman 2000, p. 64.
9. Garces-Restrepo, Vermillion, and Muñoz 2007.
10. Garces-Restrepo, Vermillion, and Muñoz 2007.
11. Transparency International 2008.
12. Stålgren 2006.
13. Transparency International 2008.
14. Transparency International 2008.
15. ASCE 2008.
16. WBCSD 2005.
17. United Nations 2008, p. vii.
18. Global Water Intelligence 2004.
19. EAP Task Force 2007.
20. Winpenny 2008.
21. Rees, Winpenny, and Hall 2008.
22. World Commission of Dams 2000.
23. Hutton and Haller 2004.
24. UN-Water 2008.
25. UN-Water 2008.
26. Olivier 2007.
27. UNDP 2006.
28. Bosworth et al. 2002.
29. Kraemer et al. 2003.
30. World Commission on Dams 2000.
31. UNDP 2006.
32. Kariuki and Schwartz 2005.
33. McGranahan and Owen 2006.
34. UNDP 2006.
35. Winpenny 2003.
36. van Hofwegan and Task Force on Financing Water for All 2006.
37. UNSGAB 2006.
38. Purchase of a fixed-interest security, such as a bond, or equity shares giving less than 10% ownership of a company.
39. For a full description of these and other policies and tools, see Winpenny 2003 and van Hofwegen and Task Force on Financing Water for All 2006.
40. Winpenny 2005.
41. Dardenne 2006; McIntosh 2003.

References

ADB (Asian Development Bank). 2004. *Evaluation Highlights of 2003*. Manila, Philippines: Asian Development Bank.

ASCE (American Society of Civil Engineers). 2008. 2005 Report Card on America's Infrastructure. 2008 Update. American Society of Civil Engineers. www.asce.org/reportcard/2005/index.cfm.

Boelens, R. 2008. The Rules of the Game and the Game of the Rules: Normalization and Resistance in Andean Water Control. PhD diss., Wageningen University, The Netherlands.

Bosworth, B., G. Cornish, C. Perry, and F. van Steenbergen. 2002. Water Charging in Irrigated Agriculture: Lessons from the Literature. Report OD 145, HR Wallingford, Ltd., Wallingford, UK. www.dfid-kar-water.net/w5outputs/electronic_outputs/od145.pdf.

CEC (Commission of the European Communities). 2007. Towards Sustainable Water Management in the European Union – First Stage in the Implementation of the Water Framework Directive 2000/60/EC. Communication from the Commission to the European Parliament and the Council [SEC(2007) 362] [SEC(2007) 363], Brussels.

Cosgrove, W., and F. Rijsberman. 2000. *World Water Vision: Making Water Everybody's Business*. London: Earthscan.

Dardenne, B. 2006. *The Role of the Private Sector in Peri-urban or Rural Water Services in Emerging Countries*. ENV/EPOC/GF/SD(2006)2. Organisation for Economic Co-operation and Development, Environment Directorate, Paris.

EAP Task Force for the Implementation of the Environmental Action Program for Eastern Europe, Caucasus and Central Asia. 2007. *Financing Water Supply and Sanitation in EECCA Countries and Progress in Achieving the Water-Related Millennium Development Goals*. Paris: Organisation for Economic Co-operation and Development.

European Parliament and Council. 2000. Directive 2000/60/EC of the European Parliament and of the Council of 23 October 2000: Establishing a Framework for Community Action in the Field of Water Policy. *Official Journal of the European Communities* 22 (12). http://eur-lex.europa.eu/LexUriServ/LexUriServ.do?uri=OJ:L:2000:327:0001:0072:EN:PDF.

Garces-Restrepo, C., D. Vermillion, and G. Muñoz. 2007 *Irrigation Management Transfer: Worldwide Efforts and Results*. FAO Water Reports No. 32. Rome: Food and Agriculture Organization of the United Nations.

Global Water Intelligence. 2004. *Tariffs: Half Way There*. Oxford, UK: Global Water Intelligence.

Government of Australia. 2008. Water Act 2007. Act 137. C2007A00137. www.comlaw.gov.au/ComLaw/Legislation/Act1.nsf/0/80C5168EF63926C2CA25741200026703/$file/1372007.pdf.

Hendry, S. 2008. Analytical Framework for National Water Law Reform/Analytical Framework for Reform of Water Services Law. PhD diss., University of Dundee, United Kingdom.

Hutton, Guy, and Laurence Haller. 2004. *Evaluation of the Costs and Benefits of Water and Sanitation Improvements at the Global Level*. Geneva: World Health Organization.

Kariuki, M., and J. Schwartz. 2005. Small-Scale Private Service Providers of Water Supply and Electricity. World Bank Working Paper 074, World Bank, Washington, DC.

Kraemer, R. A., Z. G. Castro, R. S. da Motta, and C. Russell. 2003. *Economic Instruments for Water Management: Experiences from Europe and Implications for Latin America and the Caribbean.* Washington, DC: Inter-American Development Bank.

Marin, Philippe. 2009. Public-Private Partnerships for Urban Water Utilities: A Review of Experiences in Developing Countries. Public-Private Infrastructure Advisory Facility and World Bank, Washington, DC.

McGranahan, G., and D. L. Owen. 2006. *Local Water and Sanitation Companies and the Urban Poor.* Human Settlements Discussion Paper (Water 04). London: International Institute for Environment and Development.

McIntosh. A. C. 2003. *Asian Water Supplies: Reaching the Urban Poor.* Manila: Asian Development Bank, and London: International Water Association.

Moss, J., G. Wolff, G. Gladden, and E. Guttieriez. 2003. Valuing Water for Better Governance: How to Promote Dialogue to Balance Social, Environmental, and Economic Values. White paper for the Business and Industry CEO Panel for Water. Presented at the CEO plenary session of the Third World Water Forum, 10 March, Osaka, Japan. www.wbcsd.org/DocRoot/8d4hpTIQ6FCa4jn7Y5CI/Valuing_water_report.pdf.

OECD (Organisation for Economic Co-operation and Development). Forthcoming. *Strategic Financial Planning for Water Supply and Sanitation* [provisional title]. Paris: Organisation for Economic Co-operation and Development.

OECD (Organisation for Economic Co-operation and Development)-DAC (Development Assistance Committee). 2008. Measuring Aid to Water Supply and Sanitation, Statistics on Aid to the Water Sector. Organisation for Economic Co-operation and Development, Paris. www.oecd.org/dataoecd/32/37/41750670.pdf.

OECD, DCD/DAC (Organisation for Economic Co-operation and Development, Development Co-operation Directorate, Development Assistance Committee). 2007. Development Database on Aid Activities: Creditor Reporting System: Aid Activities in Support of Water Supply and Sanitation. Paris. www.oecd.org/document/0/0,2340,en_2649_34447_37679488_1_1_1_1,00.html.

Olivier, A. 2007. Affordability: Principles and Practice. Presentation to Organisation for Economic Co-operation and Development Expert Meeting on Sustainable Financing for Affordable Water Services: From Theory to Practice, 14-15 November, Paris.

Owen, David Lloyd. 2006. *Financing Water and Wastewater to 2025: From Necessity to Sustainability.* New York: Thomson Financial.

Queensland Government. 2000. Water Act 2000. Act 34. Brisbane, Queensland. www.legislation.qld.gov.au/LEGISLTN/ACTS/2000/00AC034.pdf.

Rees, J., J.Winpenny, and A. Hall. 2008. Water Financing and Governance. GWP Technical Committee, Background Paper 12, Global Water Partnership.

Roper, H., C. Sayers, and A. Smith. 2006. Stranded Irrigation Assets. Staff Working Paper, Australian Government Productivity Commission, Melbourne. www.pc.gov.au/research/staffworkingpaper/strandedirrigation

Stålgren, P. 2006. Worlds of Water: Worlds Apart. How Targeted Domestic Actors Transform International Regimes. Göteborg University, Göteborg, Sweden.

Transparency International. 2008. *Global Corruption Report 2008: Corruption in the Water Sector.* Cambridge, UK: Cambridge University Press.

UNDP (United Nations Development Programme). 2006. *Human Development Report 2006: Beyond Scarcity: Power, Poverty and the Global Water Crisis.* New York: Palgrave Macmillan.

———. 2008. *Creating Value for All: Strategies for Doing Business with the Poor.* New York: United Nations Development Programme.

United Nations. 1945. Statute of the International Court of Justice. Charter of the United Nations. United Nations, New York. www.icj-cij.org/documents/index.php?p1=4&p2=2&p3=0.

———. 2008. *World Economic and Social Survey 2008: Overcoming Economic Insecurity.* New York: Department of Economic and Social Affairs, United Nations.

UNSGAB (United Nations Secretary General's Advisory Board on Water and Sanitation). 2006. Hashimoto Action Plan (formerly known as The Compendium of Actions). United Nations Secretary General's Advisory Board on Water and Sanitation, New York.

UN-Water. 2008. *UN-Water Global Annual Assessment of Sanitation and Drinking-Water (GLAAS): 2008 Pilot – Testing a New Reporting Approach.* Geneva: World Health Organization.

van Hofwegan, Paul, and Task Force on Financing Water for All. 2006. *Enhancing Access to Finance for Local Governments: Financing Water for Agriculture.* Chaired by Angel Gurria. Marseille, France: World Water Council.

von Benda-Beckmann, K., F. von Benda-Beckmann, and J. Spiertz. 1998. Equity and Legal Pluralism: Taking Customary Law into Account in Natural Resource Policies. In *Searching for Equity: Conceptions of Justice and Equity in Peasant Irrigation,* ed. R. Boelens and G. Dávila. Assen, The Netherlands: Van Gorcum.

WBCSD (World Business Council for Sustainable Development). 2005. Water Facts and Trends. World Business Council on Sustainable Development, Washington, DC.

WCD (World Commission of Dams). 2000. *Dams and Development: A New Framework for Decision-Making.* London: Earthscan.

Winpenny, James. 2003. *Financing Water For All.* Report of the World Panel on Financing Water Infrastructure, chaired by Michel Camdessus. Kyoto: World Water Council, 3rd World Water Forum, and Global Water Partnership.

———. 2005. *Guaranteeing Development? The Impact of Financial Guarantees.* Paris: Organisation for Economic Co-operation and Development, Development Centre Studies, Paris.

———. 2008. *Financing Strategies for Water Supply and Sanitation: A Report from the OECD Task Team on Sustainable Financing to Ensure Affordable Access to Water Supply and Sanitation.* COM/ENV/EPOC/DCD/DAC(2008)4. Paris: OECD Environment Directorate and OECD Development Co-operation Directorate.

Chapter 5

Climate change and possible futures

Authors: Richard Connor, Gilberto Gallopín, Molly Hellmuth and Walter Rast

Contributors: Joeseph Alcamo, BertJan Heij and World Water Assessment Programme Expert Group on Scenarios

Coordinator: Tim Kasten (UNEP)

Facilitator: Richard Connor

Key messages

- There is evidence that the global climate is changing. The main impacts of climate change on humans and the environment occur through water.
- Climate change is a fundamental driver of changes in water resources and an additional stressor through its effects on other external drivers.
- Policies and practices for mitigating climate change or adapting to it can have impacts on water resources, and the way we manage water can affect the climate.

Chapters 2-4 have described how external drivers exert pressure on water resources. These drivers of change are strongly interconnected, creating complex challenges and opportunities for water managers and decision-makers. Apart from extreme events (such as droughts and floods), climate change is seldom the main stressor on sustainable development, although the direct and indirect impacts of increasing climate variability can impede and even reverse development gains (see figure 5.1 for a depiction of climate change processes, characteristics and major threats). Climate change may not fundamentally alter most of the world's water challenges, but as an additional stressor it makes achieving solutions more pressing.

All of the potential impacts of climate-related disasters, including economic losses, health problems and environmental disruptions, will also affect – and be affected by – water.

Climate change differs from the other drivers. It is the only supply-side driver, ultimately determining how much water we have; the other drivers are demand-side drivers, influencing how much water we need. Climate change can directly affect the hydrologic cycle and, through it, the quantity and quality of water resources (see chapter 11). It can lower minimum flows in rivers, affecting water availability and quality for its flora and fauna and for drinking water intake, energy production (hydropower), thermal plant cooling and navigation. Anthropogenic climate change can also directly affect demand for water, when demand for crops increases in certain seasons, for instance (see chapter 7 for the implications of climate change on uncertainty in agriculture). The other drivers, by contrast, exert pressure on various water use sectors that, in turn, affect water resources.

Managing water has always been about managing naturally occurring variability. Climate change threatens to make this variability greater, shifting and intensifying the extremes, and introduces greater uncertainty in the quantity and quality of supply over the long term (see part 3). More subtly, climate change may alter the timing, magnitude and duration of precipitation events, which could pose problems

Figure 5.1 **Climate change: processes, characteristics and threats**

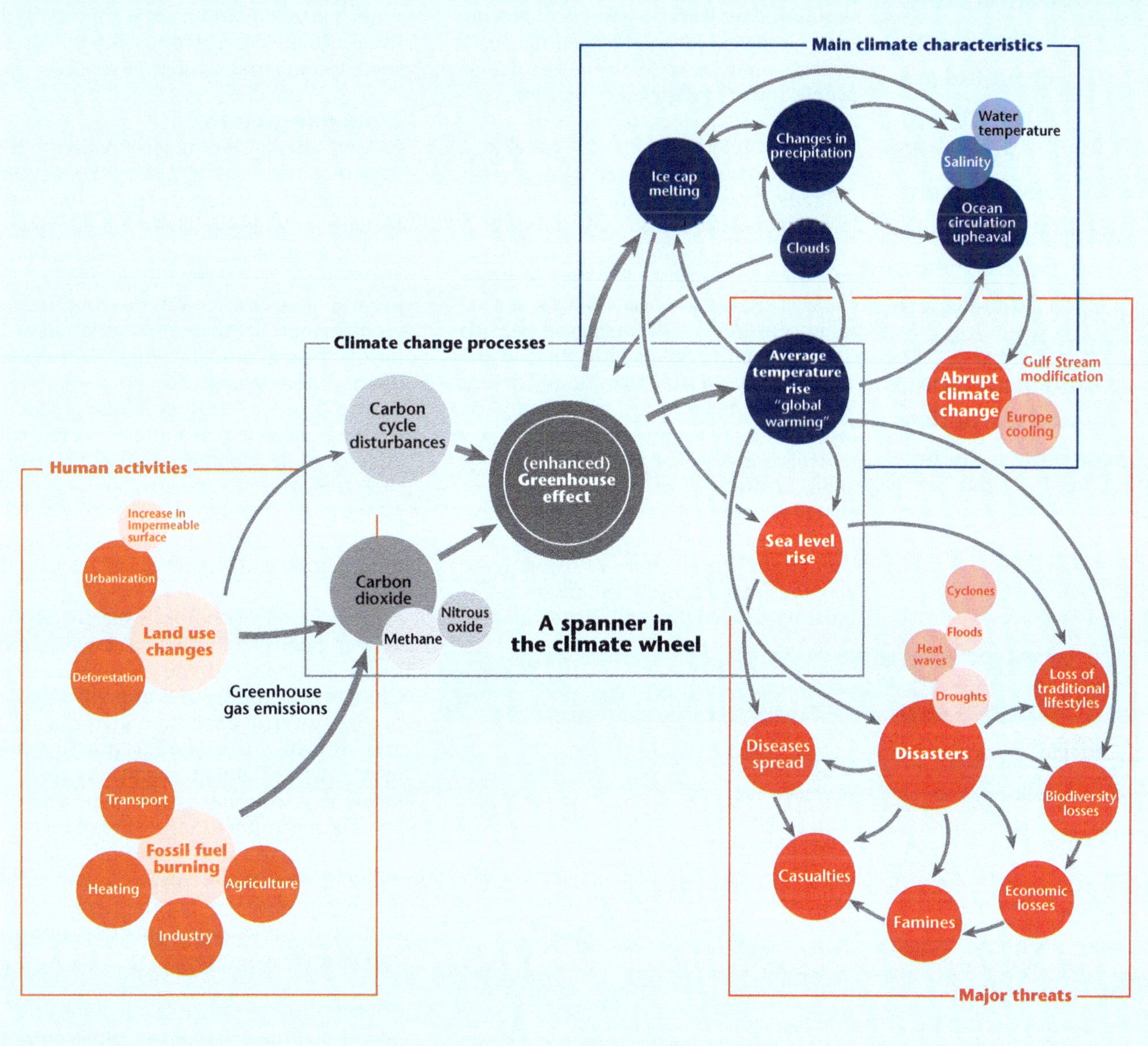

Source: Based on UNFCCC 2007a.

for the sustainability of water supplies and the continuity of treatment.

The decisions and policies put in place today for mitigation (such as reducing greenhouse gas emissions, applying clean technologies and protecting forests) and adaptation (such as expansion of rain-water storage and water conservation practices) can have profound consequences for water supply and demand both today and over the long term.[1] Climate change also adds to the uncertainty surrounding all the other drivers. Thus, examining climate change forces considerations of the interconnectedness of all the drivers. This chapter focuses on the pressures that climate change can exert on the other drivers and outlines a process for taking these interlinking pressures into account in identifying scenarios, or 'possible futures'.

The influence of climate change on the other drivers of change

The relationships between climate change and the other drivers are complex and interwoven. This section summarizes the influence of climate change on the other five major drivers: demographic processes, economic growth, social change, technological innovation and policies, laws and finance.

Demographic processes

The impacts of anthropogenic climate change, including increased water scarcity and flooding and accelerated glacial

Weather-related disasters such as floods and droughts are undermining economic development in many of the world's least developed countries, causing human suffering and disrupting economic activities

melting and sea level rise, have the potential to accelerate human migration. Drought, desertification and other forms of water scarcity are already estimated to affect as many as one-third of the world's people and are predicted to worsen.

The recent Intergovernmental Panel on Climate Change (IPCC) report notes that millions of people in densely populated low-lying coastal areas risk increasing exposure to flooding by storm surges over the 21st century.[2] The IPCC expects sea level rise to exacerbate floods, storm surges, erosion and other coastal hazards. Global warming can expand the endemic zones of water-related infectious diseases like dengue, malaria and schistosomiasis, making it increasingly difficult for people to remain in affected areas. Recurring floods or storm surges, if not managed effectively, could drive large numbers of people permanently from their homes. Current IPCC projections of rising temperatures and sea levels and increased intensity of droughts and storms suggest that substantial population displacements will take place within the next 30-50 years, particularly in coastal zones. All of these climate change refugees will require shelter, water and sanitation services.

Figure 5.2 **GDP growth tracks rainfall variability in Ethiopia (1983-2000) and Tanzania (1989-99)**

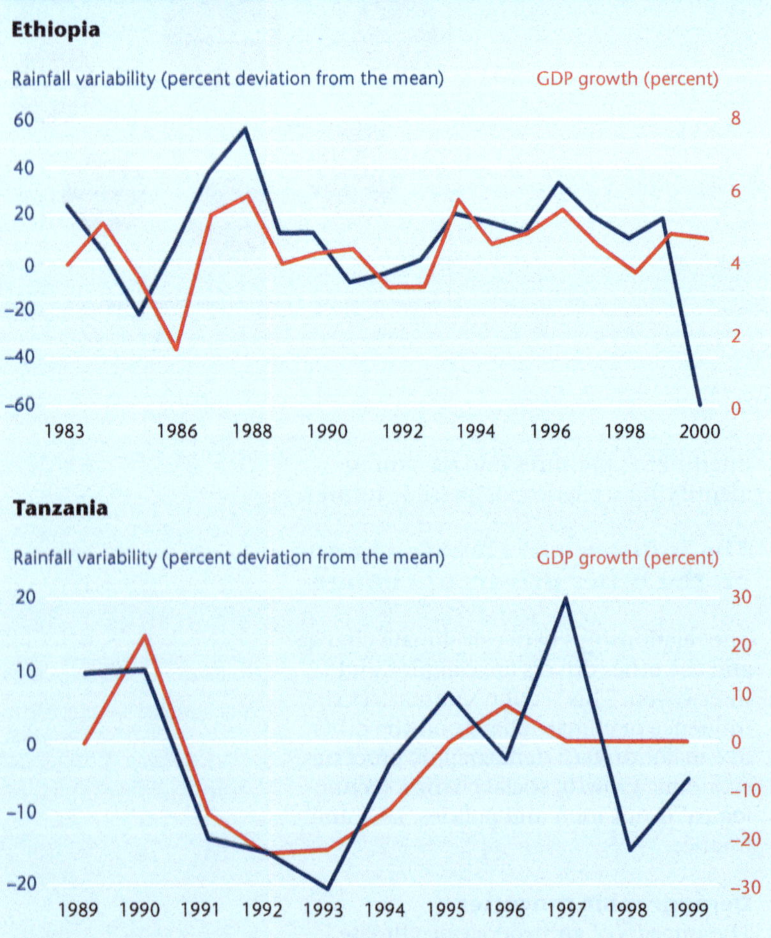

Source: Based on van Aalst, Hellmuth, and Ponzi 2007.

Economic growth

Climate change and its accompanying risks have direct and indirect effects on development and economic growth. Sea level rise, climate variability and weather extremes such as heat waves, floods and droughts are severe, direct threats to human life and property (see chapter 12). Tackling them requires mobilizing resources that may have to be reallocated from other investments. Their damage can substantially harm a country's gross domestic product (GDP). Economic performance is especially affected in developing countries because of their high and direct dependence on natural resources, notably rain-fed agriculture (see chapter 7), and their inadequate access to economic and technological resources.

Adverse climate conditions such as increased floods and droughts can also result in the underperformance of investments. Climate uncertainty and unpredictability can be powerful barriers to investments, and ultimately to economic growth, even in years when climate conditions are favourable. The changing climate also complicates infrastructure design and long-term investment planning. And internal and cross-border migration, driven by growing pressure on natural resources, can create tension among population groups and between countries.[3]

There is clear evidence of a relationship between climate variability and economic performance in countries in which agriculture is a large share of GDP, as in Ethiopia and Tanzania (figure 5.2). Evidence also suggests a strong relationship between economic development and vulnerability to disaster. Across developing countries losses associated with disasters are so large as to undermine development and poverty reduction goals. And yet climate risks are seldom adequately considered in infrastructure designs, agriculture investments and water management plans.

Weather-related disasters such as floods and droughts are undermining economic development in many of the world's least developed countries, causing human suffering (see table 12.1 in chapter 12) and disrupting economic activities (table 5.1). And substantial financial and other development resources are being diverted each year to post-disaster relief, emergency assistance, reconstruction and rehabilitation. Poorly managed climate risks also exact an indirect toll by discouraging private investment. With climate change and inadequate climate risk management, investors lack the reliable infrastructure, predictable human resources and stable markets needed to promote investment.

An estimated 40% of development investments are currently at risk, according to analyses by the Organisation for Economic Co-operation and Development (OECD).[4] These analyses indicate that while many development efforts contribute to reducing vulnerability to climate variability and change, climate risks are seldom explicitly factored into development projects and programmes. Similar issues affect sector and national development strategies.

The potential impacts of climate change on the global economy received international attention with the release of *The Stern Review* in 2006.[5] It concluded that by 2050 extreme weather could reduce global GDP by 1% and that, unabated, climate change could cost the world at least 5% in GDP each year. If even more dramatic predictions come to pass, the cost could rise to more than 20% of GDP. Such declines could in turn lead to an overall drop in official development assistance, exacerbating the struggle of poor people and countries to adapt and develop their water resources. Some other estimates of the costs of adapting to climate change are in box 5.1.

Social change

Unlike the more obvious effects of climate change on demographic processes or the global economy, the additional pressures that climate change is likely to exert on social change are often more subtle. Managing climate-related risk is a key enabler of development. Identifying and reducing the risks associated with climate-related hazards – including droughts, floods, cyclones, rising sea levels and extreme temperatures – can help to protect people, livelihoods and assets, thereby promoting the achievement of economic development goals.

Climate change and greater climate variability will increasingly affect the poorest and most marginalized groups, making them even more vulnerable to the impacts of climate change. Climate uncertainty – the inability to anticipate climate extremes – hurts investment and innovation and limits the success of other development interventions. In inhabited hazard-prone areas disasters and losses are inevitable unless societal exposure and vulnerability are reduced.

Table 5.1 **Economic impacts of flood and drought in Kenya, 1997-2000**

Impact area	Costs ($ millions)	Share of total (percent)
1997-98 El Niño flood impact		
Transport infrastructure	777	89
Health sector	56	6
Water supply infrastructure	45	5
Total flood impact	878	
Share of GDP 1997-98 (percent)		11
1998-2000 La Niña drought impacts		
Industrial production	1,400	58
Hydropower	640	26
Agricultural production	240	10
Livestock	137	6
Total drought impact	2,417	
Share of GDP 1998-2000 (percent)		16

Source: World Bank 2004.

Box 5.1 **The cost of adapting to climate change**

Estimates of the costs of climate change impacts vary because they depend on future greenhouse gas emissions, mitigation measures and assumptions about anthropogenic climate change itself and about how effectively countries will adapt to it. The following are some estimates of the costs of adaptation for developing countries:

- World Bank estimates of the additional costs to adapt or climate-proof new investments range from $9 to $41 billion a year. And a recent update by the United Nations Development Programme put the mid-range of the costs of adaptation at about $37 billion a year in 2015.
- The United Nations Framework Convention on Climate Change estimates additional investments for adaptation to climate change at $28-$67 billion and as high as $100 billion a year several decades from now. Estimates of the additional investments needed in water supply infrastructure in 2030 are $11 billion, 85% of it in developing countries.
- Oxfam estimates the current costs of adaptation to climate change for all developing countries at more than $50 billion a year.

While there is considerable debate about these estimates, they provide useful order-of-magnitude numbers for assessing resources available for adaptation. Current Global Environment Facility funds (about $160 million) are several orders of magnitude too little to meet these projected needs.

Source: World Bank 2006; UNDP 2007; UNFCCC 2007b; Oxfam 2007.

The most likely societal effects of climate change will come from changes in lifestyle and consumption patterns. Reflecting human needs and wants, changes in lifestyle and consumption patterns are among the most important drivers of change (see chapter 2). In emerging market economies rising standards of living are boosting demand for high-level goods and services, many with a large ecological and water

Some interventions in the water system might be counterproductive when evaluated in terms of mitigation of climate change

footprint. In the world's richest countries, meanwhile, growing awareness of climate change is slowly inducing people to alter their lifestyles and live in a more sustainable manner. Large cars are being replaced by smaller, more energy-efficient vehicles in some places, and governments are offering subsidies for purchasing energy-efficient appliances. But these changes alone are unlikely to substantially counteract the pressure from rising living standards in emerging market economies.

Technological innovation

Climate change will be a major driver of technological innovation and transfer.[6] Massive amounts of new investments will be required over the next 30 years to meet the growing energy needs of developing countries. Investments in adaptation will be necessary to safeguard vulnerable groups and infrastructure.

The relationship between climate change mitigation measures and water can be reciprocal. Mitigation measures can adversely influence the quantity and condition of water resources and their management, while some water management policies and measures can increase greenhouse gas emissions and affect other sectoral mitigation measures. Thus, interventions in the water system might be counterproductive when evaluated in terms of mitigation of climate change.

For example, many developed countries are shifting energy production from thermal energy plants that burn fossil fuels and emit large quantities of greenhouse gases to 'clean' energy sources. Thus, significant increases in the development of hydroelectric installations, a source of clean electricity, could be anticipated as the global community unites to combat climate change, although in many developed countries most of the 'best' sites for hydropower installations have already been developed (see map 7.6 in chapter 7). Some of the climate-related benefits of hydropower are illustrated in box 5.2.

However, there is evidence that hydroelectricity generation can also generate considerable amounts of greenhouse gases, which are released from sediment and decaying organic matter at the bottom waters of reservoirs.[7] Artificially flooded reservoirs of sufficient depth can experience anaerobic conditions as organic matter decomposes and, when the bottom waters are disturbed, emits large quantities of methane and other greenhouse gases. The problem arises most frequently in warmer climates, where reservoirs are prone to stratification and where there is year-round algal growth.

Biofuels, an alternative to fossil fuels in transport, are another means of reducing greenhouse gas emissions. Higher oil prices in recent years have made bioenergy more competitive. *World Energy Outlook 2006* projected an average rate of growth of bioenergy production of 7% a year.[8] By 2030 biofuels are expected to meet 4% of road-transport fuel demand worldwide, up from 1% today. But careful attention also must be given to minimizing negative externalities associated with producing bioenergy, such as upward pressure on food prices and the impact on food security.[9]

Developing countries will need to rely on technology development and transfer in mitigating and adapting to climate change. That will require removing obstacles to technology transfer and providing incentives for accelerating and scaling up transfers, along with cooperating on research and development (see chapter 3). According to the United Nations Framework Convention on Climate Change (UNFCCC), most technologies for adapting to climate change are already available in developing countries, and examples of successful implementation and operation abound, from coastal revetment to vaccination programmes.[10]

Policies, laws and finance

Climate change can stress political governance structures by increasing management and budget requirements for public services to mitigate climate change or to cope with its impacts, including public health care (box 5.3), disaster risk reduction and

Box 5.2 Micro-hydro plants in Nepal are expected to provide electricity access to 142,000 households and to reduce greenhouse gas emissions

Nepal has vast hydro resources. And while only about 27% of rural households are connected to a power grid (the urban share is 90%), off-grid power generated by micro-hydro plants provides many rural households with electricity for lighting, milling and other needs. The generating capacity of these plants varies from 5 to 500 kilowatts.

Through a project supported by the World Bank, the United Nations Development Programme and the governments of Denmark and Norway, micro-hydro plants are being installed for local communities by prequalified private companies that receive subsidies and technical assistance. Installation of micro-hydro plants will be phased in until 2011. The micro-hydro power plants, which qualify for emission reduction credits under the Clean Development Mechanism, will reduce greenhouse gas emissions by replacing diesel fuel used for lighting and milling.

Source: http://go.worldbank.org/9G19LTLEH0.

public security. As stress mounts, the resilience of already unstable social and political structures lessens, especially in countries with limited resources. At the international level pressures build on governance systems to combat climate change, mainly through the UNFCCC and growing public awareness.

Most efforts have focused on mitigation strategies, which are especially important for policies in energy (a major water use sector), international trade and transportation. In many countries climate change issues fall under the authority of the ministry for environment or natural resources. But as regional carbon trading markets emerge and as economies become ever more carbon-constrained, the ministries of finance and planning will need to become more directly involved.

Most governance structures today are too weak to tackle current water problems, much less prepare for emerging problems, including climate change. And there is still very little evidence about which types of governance responses work in which contexts and what their impacts are on water equity, efficiency and sustainability. Water reforms in most countries have not considered the implications of climate change or other major drivers of water use and the need for long-term planning.

Effective funding mechanisms are lacking for developing countries to support adaptation to climate change, which affects development at many levels. In Africa the impacts of climate change are expected to range from increased energy shortages, reduced agricultural production, worsening food security and malnutrition to the increasing spread of disease, more humanitarian emergencies, growing migratory pressures and increased risks of conflict over scarce land and water resources. Africa is least able to meet the costs of adapting to these impacts, yet it receives the least from current carbon finance mechanisms. Its governance structures and capacity are not ready for the intersectoral action that adaptation requires.[11]

Supporting developing country efforts to design adaptation strategies also requires better analysis. Information is needed at the local level, incorporating country-specific characteristics and sociocultural and economic conditions. At the macro-level information on both rich and poor countries is required to support international negotiations and to identify the overall costs of adapting to climate change.

Box 5.3 Health and climate change

Climate change can affect health through multiple pathways, such as greater frequency and intensity of heat waves, fewer cold-related illnesses, increased floods and droughts, changes in the distribution of vector-borne diseases and effects on the risk of disasters and malnutrition. The overall balance of impacts on health is likely to be negative, and populations in low-income countries are likely to be particularly vulnerable to the adverse effects. However, many of the projected impacts of climate change on health are avoidable. Climate change is expected to exacerbate some health problems rather than cause new diseases to emerge. Strengthening public health prevention strategies, including improving water supply and sanitation services and disease surveillance, would be an essential part of any effective response.

Source: Haines et al. 2006; Campbell-Lendrum, Corvelan, and Neira 2007.

Challenges for the impact of climate change on water resources and management

One of the most pressing challenges of climate change is addressing the vulnerability of human populations, particularly the poor, to the impacts of extreme hydrologic events, such as floods, storm surges and droughts. Over the longer term the effects of incremental climate change are likely to influence decisions about food security, energy security and land use, all with vital implications for water resources and management and environmental sustainability (see chapter 7). In this context climate change can intensify existing pressures, thereby increasing risk, vulnerability and uncertainty.

For water managers anthropogenic climate change poses a new set of challenges – because they can no longer plan, design and operate hydrologic systems based on historical statistics. Climate change means learning to manage under increased uncertainty. Climate change is a new risk to be taken into account in policy development, planning and operations at the global, basin, national, local and company levels. It calls for increasing use of 'climate knowledge' to better understand climate variability at different time scales, to assess the socioeconomic impacts observed in the past, to monitor current conditions of relevant environmental factors (climate, vegetation, water, diseases) and to provide the best possible information on future climate, from seasons to decades, for specific decisions and activities. Addressing the threats and opportunities of climate change and its impacts on water resources and supplies is vital for even the most remote rural areas as part of a broader developmental agenda.

> **The water drivers interact and can have even more of an impact on future water resources collectively than they can individually**

Although water is an important component in most energy-generating processes, its role in climate change mitigation policy is minor. Where water and climate change are most strongly linked is in adaptation policy, which functions in highly dynamic hydrologic, social, economic and demographic contexts. For water adaptation measures to be effective, however, there must be complementary climate change mitigation measures outside the water sector.

Because climate variability and change affect all the major water drivers, adaptation measures are needed in all sectors. Over the long term adaptation means applying a long-term, climate-focused approach to existing policies and programmes. But because the poor are the most vulnerable and the least able to cope with change, it is particularly important to strengthen the link between adaptation to climate change and economic development – a difficult challenge. Over the shorter term the best approach might be to manage climate variability by prioritizing risk-reduction strategies and reinforcing the capacity of hydrometeorological services to provide information for development needs.

Each country will face its own challenges and must determine how to respond in the short, medium and long run. With multiple challenges but limited financial and natural resources and capacities, countries will need to make hard choices about water use and allocation.

There tends to be a push and pull effect between identifying adaptation needs based on a climate change rationale and anchoring response options in baseline development activities. This separation between climate adaptation and development is artificial. Governments need to design climate-smart development policies and programmes, in part by strengthening sectoral capacities.

Identifying possible futures: the need for scenarios

Each of the water drivers is dynamic and continues to evolve, as do the direct and indirect pressures they exert on water resources. Thus, it is difficult to draw a comprehensive picture of the future by examining each driver independently. The drivers interact and can have even more of an impact on future water resources collectively than they can individually. Future scenarios that consider these interactions offer a more holistic picture.

Scenarios, which are sets of equally plausible futures, differ from forecasts, which are individual interpretations of a most probable future based on extrapolation of the best available information. Scenarios are not forecasts. Because the real world is so complex, forecasts are often wrong – especially those involving a time horizon of 20 years or more. Scenarios provide a means of looking beyond the water sector in search for an adequate causal understanding of different water issues.

Scenarios can contribute to several goals in the pursuit of sustainable water resources:

- *The need for a long-term view.* A long-term view of water for sustainable development requires taking into account the slow unfolding of some hydrologic, environmental and social processes and allowing time for water-works investments and water mitigation schemes to yield results.
- *The need to make decisions in a context of high uncertainty.* Decision-makers in the water sector must often address water management issues against a background of rapidly changing environmental conditions and increasing uncertainty. The uncertainty results from both a limited understanding of human and ecological processes and the intrinsic indeterminism of complex dynamic systems. Further, water resources futures depend on future human choices, which are unknown.
- *The need to include non-quantifiable factors.* The world's water system includes and is influenced by many factors that are difficult to quantify (such as cultural and political variables and processes), as well as factors that can be quantified and modelled mathematically (such as hydrologic and climatological dynamics and economic factors). Qualitative scenario analyses can provide insight into these factors that simulation models cannot.
- *The need for integration and breadth.* Water resources must be viewed holistically, considering both their natural state and the need to balance competing demands – domestic, agricultural, industrial and environmental – to ensure sustainability. Decisions on land use can affect the availability and condition of water resources, while decisions about water resources can also affect the environment and land

use. Decisions about economic and social futures can affect hydrology and ecosystems. And decisions at the international, national and local levels are connected. Sustainable management of water resources requires systemic, integrated decision-making that recognizes the interdependence of decisions; scenarios are particularly helpful for this purpose.

- *The need for perspective.* Qualitative scenarios provide guidance, perspective and context for computer models and sectoral studies, while models and studies provide consistency and feasibility checks for some elements of water scenarios, as well as numerical estimates of the modelled variables. Further, global scenarios provide a context for scenarios on a smaller geographic scale (local, watershed, national or regional). Many important changes in a river basin are determined by factors from outside the study area.
- *The need to organize understanding for decision-making.* Decision-makers may have difficulty identifying the elements from different studies that are most relevant for their decisions. Scenarios are developed with decision-making in mind. They are constructed to focus attention on causal processes and decision points, the unfolding of alternatives and the branching points at which human actions can significantly affect the future.
- *The need for an arena for conversation among water stakeholders.* Scenarios provide common frameworks for mapping and highlighting critical concerns of diverse stakeholders and identifying alternatives – setting the stage for discussions, debates and negotiation.

Over the past decade several global scenarios have been developed for the water sector. One of the most comprehensive was the scenario work for the World Water Vision in 2000.[12] The Vision generated three scenarios: a technology, economics and private sector scenario in which private sector initiatives lead research and development and globalization drives economic growth, but the poorest countries are left behind; a values and lifestyles scenario in which sustainable development is a global priority, with emphasis on research and development in the poorest countries; and a business-as-usual scenario.

In 2006 the World Business Council for Sustainable Development (WBCSD) produced three scenarios focusing on the role of business and water.[13] Its three storylines focus on efficiency (more value per drop), security (meeting the basic needs of all) and interconnectivity (a 'whole system' approach; table 5.2). In another example using driver categories similar to those in this Report, the *Global Environment Outlook* (GEO4) report of 2007 generated four different scenarios: markets first, policy first, security first and sustainability first.[14]

Despite these recent endeavours, experience indicates that new global water scenarios are needed. Existing global water scenarios do not fully incorporate each of the drivers described in this chapter. The scenarios are either outdated (those of the World Water Vision) or partial, incomplete or sectoral (WBCSD, GEO4). In addition, the evolution of the drivers and the logic behind their storylines need to be examined and possibly redefined in view of developments both inside and outside the water sector since 2000. Finally, important new policy initiatives have emerged since the last world water scenarios, such as the adoption of the Millennium Development Goals.

New global water scenarios are needed. Existing global water scenarios do not fully incorporate each of the drivers described in this chapter

Challenges for summarizing the pressures of external drivers on water resources

Multiple external drivers exert pressures on water resources through changes in water demands and uses. Some of these pressures are summarized in the table at

Table 5.2 The three water scenarios of the World Business Council for Sustainable Development, to 2025

Scenario	'Hydro'	'Rivers'	'Ocean'
Water challenge	Efficiency (more drops for less and more value per drop)	Security (quantity and quality for all)	Interconnectivity (taking the whole system into account)
Business challenge	Innovation	Social license to operate	Business role in water governance
The five key story themes	• Hard times in huge towns • Huge opportunities • High-stakes innovation • Hydro economy • Beyond legacy systems	• The security deficit • Two sides of the river • The trust deficit • Access and equity • Political reallocation – local solutions	• Unintended consequences • Global Fair Water Movement • The tipping point • Accountability tools • Networked global water governance

Source: WBCSD 2006.

the beginning of part 2. The challenge is to get decision-makers inside and outside the water sector to adopt appropriate measures to reduce the negative pressures on water and increase the positive pressures.

Making this challenge more difficult are the links between drivers (as illustrated by the pressures imposed by rising living standards), which involve demographic, social and economic factors but are also influenced by technology and governance. Generating a picture of this complex future would be greatly assisted by the development of a set of future scenarios.

Notes

1. IPCC 2008.
2. Nicholls et al. 2007.
3. van Aalst, Hellmuth, and Ponzi 2007.
4. OECD 2005.
5. Stern 2006.
6. IPCC 2008.
7. Giles 2006.
8. IEA 2006
9. FAO 2008.
10. UNFCCC 2006.
11. van Aalst, Hellmuth, and Ponzi 2007.
12. Cosgrove and Rijsberman 2000.
13. WBCSD 2006.
14. UNEP 2007.

References

Campbell-Lendrum, D., C. Corvalán, and M. Neira. 2007. Global Climate Change: Implications for International Public Health Policy. *Bulletin of the World Health Organization* 85: 235-37.

Cosgrove, W., and F. Rijsberman. 2000. *World Water Vision: Making Water Everybody's Business.* London: Earthscan.

FAO (Food and Agriculture Organization of the United Nations). 2008. *The State of Food and Agriculture 2008: Biofuels: Prospects, Risks, and Opportunities.* Rome: Food and Agriculture Organization.

Giles, J. 2006. Methane Quashes Green Credentials of Hydropower. *Nature* 444 (7119): 524-25.

Haines, A., R. S. Kovats, D. Campbell-Lendrum, and C. Corvalan. 2006. Climate Change and Human Health: Impacts, Vulnerability and Public Health. *Public Health* 120 (7): 585-96.

IEA (International Energy Agency). 2006. *World Energy Outlook 2006.* Paris: Organisation for Economic Co-operation and Development, and International Energy Agency.

IPCC (Intergovernmental Panel on Climate Change). 2008. Technical Paper on Climate Change and Water. IPCC-XXVIII/Doc.13, Intergovernmental Panel on Climate Change, Geneva. www.ipcc.ch/meetings/session28/doc13.pdf.

Nicholls, R. J., P. P. Wong, V. R. Burkett, J. O. Codignotto, J. E. Hay, R. F. McLean, S. Ragoonaden, and C. D. Woodroffe. 2007. Coastal Systems and Low-lying Areas. In *Climate Change 2007: Impacts, Adaptation, and Vulnerability*, eds. M. L. Parry, O. F. Canziani, J. P. Palutikof, P. J. van der Linden, and C. E. Hanson. Cambridge, UK: Cambridge University Press.

OECD (Organisation for Economic Co-operation and Development). 2005. *Bridge over Troubled Waters: Linking Climate Change and Development.* Paris: Organisation for Economic Co-operation and Development.

Oxfam. 2007. Adapting to Climate Change – What's Needed in Poor Countries, and Who Should Pay. Oxfam Briefing Paper 104, Oxfam International, Oxford, UK.

Stern, N. 2006. *The Stern Review: The Economics of Climate Change.* London: Cabinet Office, HM Treasury.

UNDP (United Nations Development Programme). 2007. *Human Development Report 2007/2008: Fighting Climate Change. Human Solidarity in a Divided World.* New York: Palgrave Macmillan.

UNEP (United Nations Environment Programme). 2007. *Global Environment Outlook 4 (GEO4): Environment for Development.* Nairobi: United Nations Environment Programme.

UNFCCC (United Nations Framework Convention on Climate Change). 2006. Application of Environmentally Sound Technologies. Technical Paper FCCC/TP/2006/2, United Nations Framework Convention on Climate Change, New York. http://unfccc.int/resource/docs/2006/tp/tp02.pdf.

———. **2007a.** *Climate Change: Impacts, Vulnerabilities and Adaptation in Developing Countries.* Bonn, Germany: United Nations Framework Convention on Climate Change.

———. **2007b.** Investment and Financial Flows to Address Climate Change. Background paper, United Nations Framework Convention on Climate Change, New York.

van Aalst, M., M. Hellmuth, and D. Ponzi. 2007. Come Rain or Shine: Integrating Climate Risk Management into African Development Bank Operations. African Development Bank Working Paper 89, African Development Bank, Tunis.

WBCSD (World Business Council for Sustainable Development). 2006. *Business in the World of Water: WBCSD Scenarios to 2025.* Washington, DC: World Business Council for Sustainable Development.

World Bank. 2004. Towards a Water-Secure Kenya. Water Resources Sector Memorandum, Report 28398-KE. World Bank, Washington, DC.

———. **2006.** Clean Energy and Development: Towards an Investment Framework. Paper DC2006-0002. Development Committee, World Bank, Washington, DC.

PART 2

Using water

Chapter

Chapters 6-9
Coordinator
Jean-Marc Faurès
(FAO)

Facilitator
Domitille Vallée

Author: Domitille Vallée

History shows a strong link between economic development and water resources development. Abundant examples can be drawn of how water has contributed to economic development and how development has demanded increased harnessing of water. Such benefits came at a cost and in some places led to increasing competition and conflicts between users and pressure on the environment.

While demand from all sectors is on the rise, in most places it is agriculture that accounts for the bulk of water use. Steadily rising demand for agricultural products to satisfy the diverse needs of a fast growing population (for food, fibre and now fuel) has been the main driver behind agricultural water use – and such demand is expected to continue to grow. In parallel, changing lifestyles and consumption patterns and rapidly growing cities and industries are claiming increasing amounts of water and are putting heavy pressure on local resources. The effects of water depleting and polluting activities on human and ecosystem health remain largely unreported or difficult to measure, and the need grows stronger for effective protection of ecosystems and the goods and services they produce – on which life and livelihoods depend. As competition among demands on water increases, society will need to respond with improved water management, more effective policies and transparent and efficient water allocation mechanisms. The drivers described in part 1 create pressures on society that lead to changes in water use (see table).

The 2003 and 2006 editions of The United Nations World Water Development Reports examined many aspects of water use. Some, such as the use of groundwater, are covered more extensively in this edition. Similarly, the availability of new information is reflected in the treatment here of water supply and sanitation.

Drivers create pressures that influence water use patterns

Users	Demographic growth	Economic growth	Social change	Technological innovation	Policies, laws and finance	Climate change
Agriculture	Rising demand for food and subsequent pressure on land and water resources	Rising demand for meat, fish and high-value agricultural products	Environmentally sensitive behavioural changes can lead to more vegetarian diets	Greater agricultural water productivity	Agriculture and trade policy (subsidies, import/export quotas, etc.) dictates crop yields and water requirements	Shifts in crop patterns, greater reliance on irrigation in places, generally greater crop evapotranspiration
Energy	Rising demand and pressure to develop more energy sources	Rising demand and pressure to develop more energy sources, sometimes 'dirty' resources (e.g., tar sands)	Awareness can lower demand Consumption lifestyles can increase demand	Greater efficiency (production and supply) Development of new or 'dirty' sources	Energy policy (and price speculation) dictates supply sources (hydro and renewables, fossil, nuclear)	Change in production patterns, with different water demands (quantity and quality implications)
Health	Urbanization and potential for increased disease transmission	Greater access to medical services, safe water and sanitation	Education increases good health possibilities	Increasing quality of health care Unexpected negative impacts (e.g., pesticides)	Health care and education policy (e.g., universal coverage, subsidies)	Shifting limits and timing of vector-borne diseases Greater vulnerability of the poor (floods, droughts, disease outbreaks)
Industry	Increased demand for basic goods and services	Positive feedback loop Greater resources needs and environmental degradation	Rising living standards change demands for consumer products	Can increase or decrease environmental impacts (both in some cases)	Can promote or impose standards	Increased uncertainty and risk Can prompt energy and water efficiency
Environment	Increased competition for land and resources	Can increase natural resource use and pollution	Awareness can lower impact Consumption lifestyles can increase impact	Can increase or decrease impacts – sometimes both	Can impose protection measures	Threatens ecological balances Leads to shifting habitats
Poverty focus	Growth of informal human settlements	Can aid in poverty reduction if services and opportunities are available Increased need for natural resources to fuel economic growth	Increasing expectations for poor communities	Low-cost technologies are increasingly accessible	Can impose equity rules on allocation and pricing policies May hinder efficient provision of needed services	Will affect the poor the most Impacts will affect developing countries (with limited resources) more than developed countries

Source: Compiled by Richard Connor.

Chapter 6

Water's many benefits

Authors: Gunilla Björklund, Andy Bullock, Molly Hellmuth, Walter Rast, Domitille Vallée and James Winpenny

Contributors: Moses Abukari, Cecile Brugère, Rudolf Cleveringa, Richard Connor, Jennifer Hauc, Walter Huppert, Jean Margat, Audrey Nepveu, Mary Renwick, Guido Santini, Eva Schiffer and World Water Assessment Programme Expert Group on Storage

Coordinator: Jean-Marc Faurès (FAO)

Facilitator: Domitille Vallée

Key messages

- Water has always played a key role in economic development, and economic development has always been accompanied by water development.
- Investment in water management has been repaid through livelihood security and reductions in health risks, vulnerability and ultimately poverty.
- Water contributes to poverty alleviation in many ways – through sanitation services, water supply, affordable food and enhanced resilience of poor communities faced with disease, climate shocks and environmental degradation.
- Water of the right quality can improve health through better sanitation and hygiene and, when applied at the right time, can enhance the productivity of land, labour and other productive inputs. In addition, healthy freshwater ecosystems provide multiple goods and services essential to life and livelihoods.

People have traditionally settled near water sources. An adequate and dependable source of water is needed to sustain humanity and to support future growth and development. Investment in water management has been repaid through increased livelihood security and reductions in health risks, vulnerability and ultimately poverty.[1] Poverty reduction is closely linked to enhanced access to water.[2]

Where economic growth has been strong and prosperity has been fairly equitably distributed, poor individuals and households have been able to reach the targets of the Millennium Development Goals. Conversely, where governments are unable or unwilling to deliver the basic services, water emerges among the most pressing issues (box 6.1).

Experience shows that access to water is fundamental for economic growth and livelihoods. In rural and agriculture-based economies water is often the most important factor for agricultural production and other livelihood activities.[3] In urban-based, labour-intensive manufacturing economies water is needed for nearly all productive activities.[4] Secure access to water with reliable storage and irrigation has boosted economic growth in many of the developed economies of the Americas and Europe, and through the green revolution in Asia has enabled the transformation of agriculture-based economies to industrial and emerging market economies.[5]

Past efforts of development and water use have often ignored the water needs of life on Earth and have placed at risk the resources on which life depends (see chapter 8). The links connecting water resources, the environment and economic sectors are complex. As a result, our understanding of all the ways that natural processes influence human well-being remains incomplete, impeding our ability to ensure sustainable economic and social development. This chapter explores our current understanding of the links between water and growth, poverty reduction, health and the environment.

Water for economic development

Water development is essential to growing economies. Over the centuries the world has witnessed an unprecedented expansion in urban water supply, irrigation, dam storage, drainage, water transport facilities and other water schemes as development has occurred at different rates in different regions.

Water development and growth

Water infrastructure supports growth and poverty reduction and should be planned by taking the possible impacts into account (box 6.2). The principal drivers of growth and change have often come from outside the domain of water managers. Water development has largely responded to and been affected by developments in the wider political economy, such as market-oriented reforms, openness to global trading systems and advances, supply chains and regional production networks. Storage, irrigation, urban water supply and wastewater have all been part of the enabling infrastructure. These have been led by public policies and microeconomic developments (productivity changes, capital and input accumulation, and technology). In some cases

Box 6.1 Water services are a crucial element of nation-building in fragile states

The importance of water services is especially apparent in societies where normal social life and political structures have broken down. Categorizing them as fragile states, the UK Department for International Development defines these as countries 'where the government cannot or will not deliver core functions to the majority of its people, including the poor'. Among the most important functions of the state for poverty reduction is 'the ability to protect and support the ways in which the poorest people sustain themselves'. While each fragile state is fragile in different ways and for different reasons – war, post-conflict recovery, major natural catastrophe, prolonged mismanagement and political repression – a striking commonality in reports from aid agencies is the prominence of water and sanitation in relief and reconstruction programmes. The rapid restoration of viable water services is often a crucial ingredient of nation-building in these fragile states.

Source: DFID 2005, p. 7; OECD 2008.

Box 6.2 Storing water for development

For millennia people have tried to control and store irregular water flows by creating reservoirs and storing water to regulate seasonal flows, limit floods and overcome dry spells.

Today, in parts of many countries demand exceeds available runoff. These countries depend on dams and water harvesting systems to control irregular storm runoff. The situation is particularly acute in arid and semi-arid areas where rainfall periods are short and floods can be especially destructive. Demands are often seasonal, relating not only to agriculture, but also to peak demands for tourism and hydropower production. Increasingly, it will be impossible to do without some form of water storage, either surface (reservoirs or water-harvesting systems) or underground (cisterns and aquifers). Global changes, in particular the impacts of climate change, elevate the need for water storage to a higher priority.

Food production has always been an important driver of water storage. In countries where the majority of the people live in rural areas, irrigation is increasingly indispensable to ensure reliable supplies of water during the growing season.

Satisfying demand for energy through hydropower has also led to the construction of dams. This becomes more imperative with the highly fluctuating cost of fossil fuels and the need to shift towards cleaner energy production. Emerging market economies with fast-growing industries and cities need to secure more energy. China, India and Thailand and many countries in Latin America are looking to invest in water infrastructure in neighbouring countries (as South Africa has done in Lesotho), if necessary to secure their water futures.

Water storage is a particularly important component of flood management. Its importance will likely increase in a changing global climate, especially in regions where the severity of storms is projected to intensify and where precipitation may be higher. The potential for increased storms and extreme rainfall events means that dams and other large-scale infrastructure will need to be built to higher engineering standards, to withstand future risks.

Small- and large-scale storage complement one another. Smaller decentralized and participatory water harvesting systems have increased water availability and, consequently, agricultural production, at household and community levels, especially among the poor. A diversity of storage types and capacities reduces vulnerability to catastrophic events.

Large storage projects may represent a more appropriate solution for multipurpose projects that provide hydropower, irrigation, flood control and drought management, and water supply for large urban areas. Their management is complex because storage can frequently compromise needs for other uses (for example, the need to lower reservoir water levels for flood control, maintain levels for energy production and replicate natural flows for protection of species). Integrated water management at the basin level using real-time hydrologic information from weather radar and computer models of individual reservoirs allows optimum management of storage and release to satisfy domestic, agricultural, industrial and environmental requirements.

Dams, especially large dams, are controversial, as they leave a heavy footprint on the natural environment and often displace large numbers of people, sometimes disrupting traditional societies. Nevertheless, many countries continue to plan for such large infrastructure projects to increase storage capacities and meet other needs considered vital to improve development and avoid crises. Such projects should strive to balance the desired objectives – economic growth and reduced vulnerability – with the likely associated environmental and social costs. Each storage project must evaluate the trade-offs involved. The World Commission on Dams has provided a basic framework for such an assessment.

Source: WWAP Expert Group on Storage 2008; WCD 2000.

Figure 6.1 **The shift of economies from agriculture-based to industrialized, 1965-2001**

Agriculture's contribution to growth in three types of economies, 1990-2005 (percent)

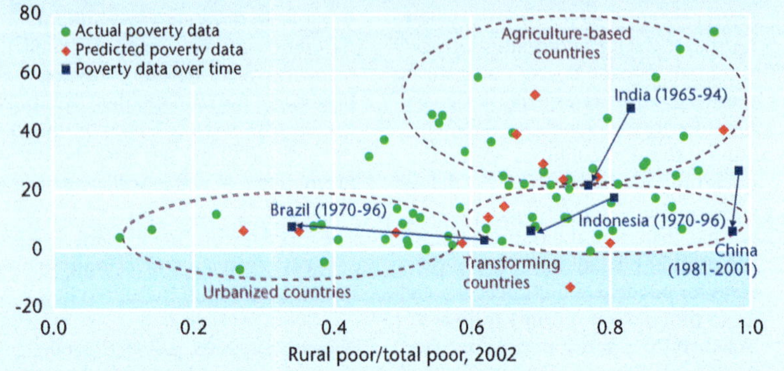

Note: Arrows show paths for Brazil, China, India and Indonesia.
Source: Based on World Bank 2007.

infrastructure development has been promoted by specific sectors in the economy that directly benefit from them, while the costs are usually borne by society at large.

Agriculture – especially food production – has historically been a first stage of national development (figure 6.1). For example, the Republic of Korea's industrial take-off in the 1960s was preceded by decades of rapid agricultural growth, with productivity driven by comprehensive land reform that saw smallholdings displace traditional tenant farming. In Thailand poverty fell from 57% in 1962 to 10% in 2002, with initial declines led by growth in agricultural production. Viet Nam laid the foundation for rapid post-war economic growth through liberalization of markets and macroeconomic stability, together with increased security of land tenure that permitted transfers of land-use rights. Between 1990 and 2003, as the economy grew at 7.5% a year and agriculture at 4.2%, the $1 a day poverty index dropped from 50.7 to 13.1. Agricultural productivity improvements largely drove gains in the early reform period as Viet Nam became the world's second largest exporter of rice, coffee and pepper. During this period of sustained economic growth, Asia witnessed a major expansion in irrigation infrastructure in water storage, urban water supply and wastewater treatment.

Development in a country or community is constrained by the abilities to gain access to water and – should the resource become scarce – to make the necessary economic, social and environmental trade-offs (see chapter 16). Changes in rainfall and more variable runoff as a result of climate change are likely to reduce water availability and represent a clear challenge for development (see chapter 11). Where runoff is extremely variable, the potential ability to store floodwater has the dual advantage of saving water for later use while protecting human settlements and development infrastructure.

Can we afford not to invest in water?

Evidence of the macroeconomic returns to investment in water is growing. The cost of a series of major typhoons and resulting flood damage in post-war Japan has been estimated at 5%-10% of GNP. Investment in soil conservation and flood control following legislation in the early 1960s reduced the impact of flood damage to less than 1% of GNP.[6] A rise in investment in domestic water use was accompanied by a sharp drop in reported illnesses and death from infectious water-borne diseases and a virtual end to related infant deaths.[7]

There are even more examples of the economic cost of lack of investment in water. In Kenya the combined impact of the winter floods of 1997/98 and drought between 1998 and 2008 has been estimated at $4.8 billion – effectively a 16% reduction in GDP (see table 5.1 in chapter 5).[8] Evidence suggests that floods and drought in Kenya translate into a direct annual loss of 22% of GDP over a 2.5 year period. The Mozambique floods of 2000 caused a 23% reduction in GDP and a 44% rise in inflation. Inability to tackle hydrologic variability in Ethiopia has been estimated to cause a 38% decline in GDP and a projected 25% increase in poverty for 2003-15.[9] Worldwide, more than 7,000 major disasters have been recorded since 1970, causing at least $2 trillion in damage and killing at least 2.5 million people.[10]

Improving water management would help countries reduce the damage of climate variability and the extreme events that can cripple economies. Year after year, the human costs of delayed investments are mounting.

GDP, water investments and water use

While the links between water development and GDP are strong, they are also complex. *Asian Water Development Outlook 2007* emphasizes the need for a multidisciplinary and multisector perspective on water in the Asia and Pacific region to face the challenges of sustaining growth.[11] The report highlights the need to address the links between water and other important development-related sectors, such as energy, food and the environment.

Actions that target rural economies will benefit the largest number of people. As of

2007, 3 billion people live in rural areas, most of them dependent on agriculture for their livelihood. Agricultural economies are especially vulnerable to lack of water during critical crop-growing seasons. Their performance is influenced by the ability to secure and control water through infrastructure, such as water harvesting storage, reservoirs and canals, and the ability to transport it to crops when required.

Investments in physical infrastructure must be accompanied by investments in 'soft' infrastructure, the dense network of institutions and human capacity needed to secure spaces in which individuals, households, firms and communities are able to pursue their day-to-day activities with a reasonable degree of predictability and stability and with due regard for the interests of others.[12] Investments are also required for the operation and maintenance of physical infrastructure (see chapter 9).

However, while there is a strong relation between water investment and growth, the relation between the quantity of water used and a country's level of development is inconclusive (figure 6.2). Many water-poor economies have developed, while the ratio of water use to GDP in many developed countries has been declining (figure 6.3).

Water and poverty reduction

The relationship between water and poverty is widely discussed in the development literature.[13] Water contributes to poverty alleviation in many ways – through sanitation services, water supply, affordable food and enhanced resilience of poor communities to disease, climate shocks and environmental degradation. Water of appropriate quantity and quality can improve health and, when applied at the right time, can enhance the productivity of land, labour and other inputs.

The daily water supply for multiple household uses is determined by the time, labour and financial costs required to access water. The economic and social returns from water access for different uses determine net livelihood benefits or losses.

Figure 6.2 **The relation between freshwater use and level of development is inconclusive**

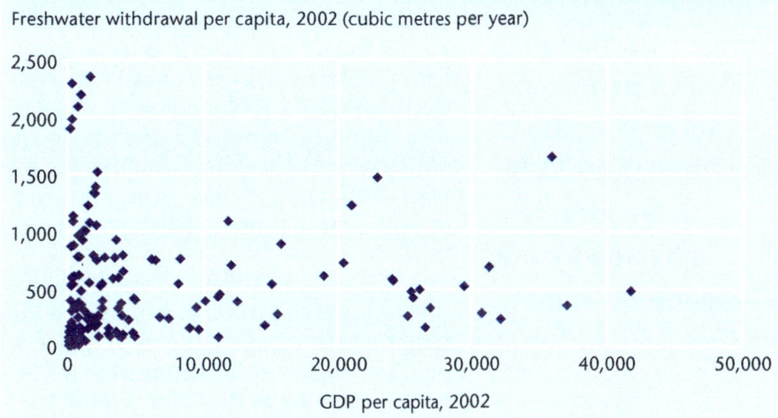

Source: Data on water withdrawal, AQUASTAT; data on GDP per capita, World Bank.

Figure 6.3 **The ratio of water use to GDP has been declining in many countries**

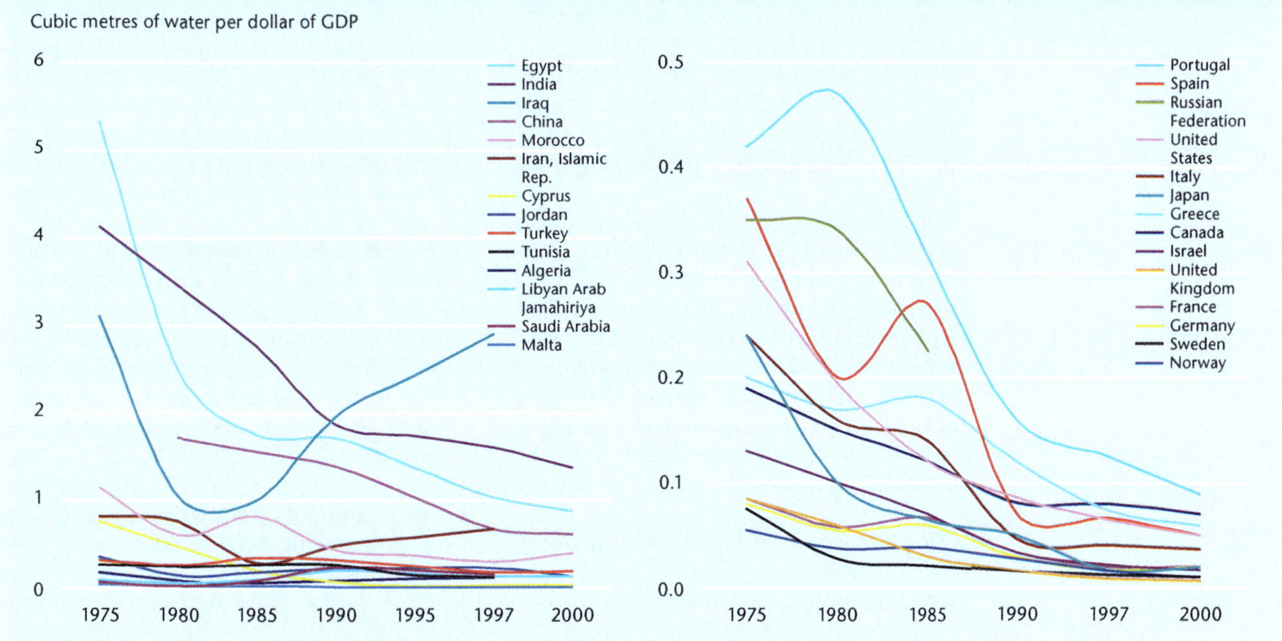

Source: Based on Margat and Andréassian 2008.

The poorest populations in the world have the lowest access to water supply and sanitation services and are the most dependent on water resources for sustainable livelihoods. They are at the losing end of the equity curve, most vulnerable to changing environmental and social conditions and most likely to be adversely affected by the vagaries of climate

Benefits or losses may take the form of reduced vulnerability to shocks, increased productive capacity, increased social benefits and increased capacity to maintain service levels. Studies in India show clear evidence of poverty alleviation and income gains for all rural groups, even the landless, through increased working days as a result of improved access to water.[14]

Distributing the benefits of growth

A major lesson of *Human Development Report 2006* on poverty and the global water crisis is that the distribution of economic growth affects the rate at which the growth is converted into poverty reduction.[15] Thus, every 1% increase in growth has reduced poverty by about 1.5% in Viet Nam – twice the 0.75% reduction in Mexico, with its larger income gap. The report records that some countries, such as Bangladesh and Thailand (for sanitation) and Sri Lanka and Viet Nam (for water), have performed far better than expected solely on the basis of income, as compared with others, such as India and Mexico (for sanitation). The lesson is that income matters but that policy shapes the conversion of income into human development.

The poorest populations in the world have the lowest access to water supply and sanitation services and are the most dependent on water resources for sustainable livelihoods. They are at the losing end of the equity curve, most vulnerable to changing environmental and social conditions and most likely to be adversely affected by the vagaries of climate. As noted in *Human Development Report 2006*, access to water in many developing countries mirrors the distribution of wealth. The Millennium Development Goals and other poverty reduction efforts such as Poverty Reduction Strategy Papers have been designed specifically to address these types of inequities.

One clear message from the past decade of poverty reduction initiatives is the diversity of approaches to development. Unlike health and education, which are firmly ensconced in the arena of social services, water management has often tended to fall between the economic and growth agenda or the human development and basic services agenda. Neither agenda has held sway across all countries.

Human Development Report 2006 unequivocally identifies the crises in drinking water and sanitation as a crisis for the poor, on the evidence that almost two in three people lacking access to safe drinking water survive on less than $2 a day and one in three on less than $1 a day.[16] More than 660 million people without adequate sanitation live on less than $2 a day, and more than 385 million on less than $1 a day. This evidence highlights clearly the financing difficulties of improving access through household investment. This is important because households, not public agencies, often make the largest investment in basic sanitation, with the ratio of household to government investment typically being 10 to 1.[17]

With household poverty widespread, the burden has shifted to governments. A strong social and economic case has been made over the past decade for investments in water supply and sanitation as essential prerequisites for economic growth. There is also evidence that more equitable economic growth in Asia has delivered improved water supply and sanitation – with rising wealth making possible household investments in basic services and higher government expenditure for basic services. WaterAid in its 2005 response to the Camdessus Report of 2003 points to growth as an enabler of government finance for the provision of basic services: 'For national governments in developing countries to double their allocations to water, their national incomes need to rise substantially. This requires, amongst other things, a healthy balance of trade and a growing national economy.'[18]

Rising levels of income inequality may make access to services more difficult for those who most need it. The UN Department of Economic and Social Affairs suggests a package of universal social policies and targeted economic policies tailored to individual country conditions. The package would be based on a strong 'social contract' to provide a 'global social floor' that provides a minimum level of security – including water security – in line with the Universal Declaration of Human Rights.[19]

The different situations and water needs of the urban and rural poor

Some 1.4 billion people are classified as poor:[20] 44% in South Asia, about 24% each in sub-Saharan Africa and East Asia, and 6.5% in Latin America and the Caribbean. The water-related needs of poor people differ in urban and rural contexts.

The urban poor often live in informal settlements following rapid urban growth: 77% of the population in Latin America is urban; 38% in Africa. Those figures are expected to rise over the next few decades with projected urban expansion. People in informal settlements live without many of life's basic necessities: safe drinking water, adequate sanitation services, access to health services, durable housing and secure tenure.[21] Affordable, safe, piped water is available to only a small share of low-income urban dwellers. The financing of water services is the key to expanding access, but the illegal status of the large majority of slum dwellers is often a barrier to access to finance or support (box 6.3). Many informal settlements are in flood-prone areas and are especially vulnerable to environmental hazards.[22] Thus, large populations of slum dwellers live at high risk of disease.

For the rural poor, who make up some 75% of the world's poorest people, access to water is essential both for basic needs and for productive purposes. Lack of access is often the main factor limiting their ability to secure their livelihoods. The Food and Agriculture Organization of the United Nations and the International Fund for Agricultural Development have been working on a response matrix for rural poverty in Africa to provide planners and policy-makers with a conceptual framework to identify appropriate context-specific interventions tailored to the needs of diverse groups of rural people (table 6.1 and figure 6.4).[23]

Although interventions are needed in several areas, water is a key factor because it plays a central role in agriculture, it is a frequent constraint on production and it provides a focal point around which other interventions can be organized (box 6.4). Strategies to reduce rural poverty need to focus on improving productivity in agriculture – for most, the main source of income. Gains require substantial interventions to improve farm-level access and control and management of water resources.

There are many links between rural and urban areas, and in most places there are no sharp boundaries between rural and urban spaces. Families often depend on both urban and rural locales to make a living. The share of rural household incomes from non-farm sources, including migrants' remittances, is 60% in South Asia, 40% in Latin America and 30-50% in sub-Saharan Africa (reaching as high as 80%-90% in Southern Africa).[24]

Questions remain about how best to realize the potential benefits of water management opportunities to assist the poor – particularly about how to most effectively engage the potential of limited water resources for all poor people. For example, evidence from the Zambezi Basin shows that even full development of the basin's irrigation potential could benefit no more than 18% of the basin's rural poor, even considering off-farm multipliers.[25] This argues for a dual approach, with interventions to reduce farmers' vulnerabilities to rain-fed agriculture as an essential accompaniment to the development of both smallholder irrigation and large-scale infrastructure to support macroeconomic growth.

Unless the growth and poverty-reducing contributions of water resources are made more explicit and specific at the country level, development-oriented finances are unlikely to follow. Those specifics will influence decisions about the sources, costs, viability, sustainability and instruments of development finance. But only national and local action plans can secure

Box 6.3 Land tenure and access to water and sanitation

Provision of safe water and adequate sanitation is often affected by systems of land tenure. Where sanitation is provided by individuals, as is common in rural areas, there is little incentive to invest without security of tenure. Where governments seek to improve the provision of public services through government provision, which may involve contracts with private sector providers, service providers usually require a land right (ownership or lease tenure) before concluding a contract to provide services.

Resolving these issues requires an integrated approach across government departments and related institutions to ensure that the agencies responsible for land and property rights recognize the public and personal health benefits of sanitation and consequent improvements in the social and economic standing of the poor. In Manila, a megacity with a large informal population, the utilities offered to extend water lines to the perimeter of slums. Metered connections to the households would be managed by resident associations or non-governmental organizations and paid for by residents. The residents and the utilities gained from this solution as both water costs in slums and illegal connections were cut by 25%. There are other examples where improved land tenure led to better service to the poorest and to better performance by public agencies, such as strategies to facilitate access to tenure title (India) or integrated land tenure security or improvement programmes (Morocco, South Africa, Thailand and Zambia).

Source: Ben Fawcett and Diana Mitlin, International Water Association, based on UNDP 2006; http://esa.un.org/iys; World Water Assessment Programme Expert Group on Legal Issues.

Table 6.1 Water and the characteristics of rural livelihoods

Characteristic of rural livelihood	Manifestation of characteristic among low-income populations	Manifestation of characteristic among high-income populations
Agricultural output (crops and livestock yield)	Low	High
Health and water access	Poor	Good
Direct natural resource dependency	High	Low
Susceptibility to flood and drought risk	High	Low
Knowledge and adaptive capacity	Traditional	Sophisticated

Source: Sullivan et al. forthcoming.

Figure 6.4 **Different categories of rural inhabitants in Africa**

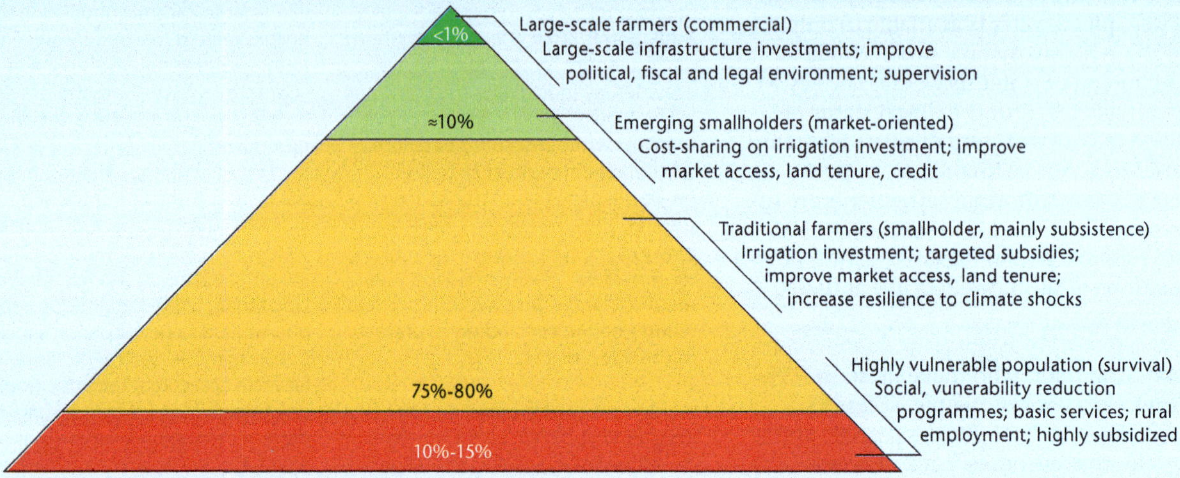

Source: Based on Faurès and Santini 2008.

Box 6.4 **Four water dimensions of rural livelihoods**

A livelihood approach puts people at the centre of development strategies, connecting their ability to move out of poverty with their capacities and assets. Water's importance as an asset is determined by the quantity available daily for household, agriculture and livestock consumption and by its ability to stimulate economic and social returns.

A 2008 joint pilot project of the Food and Agriculture Organization and the International Fund for Agricultural Development proposes a rural water livelihoods index and a draft framework for assessing the performance of water-related interventions for reducing rural poverty. The index considers four water-related components that influence rural livelihoods: access to basic water services, crop and livestock water security, clean and healthy water environment, and secure and equitable water entitlement.

The rural water livelihoods index is established on the model of both the human development index and the water poverty index (a composite index that attempts to capture the relationship between water and poverty). It is limited to eight subindicators for which data are available at the country level for most of the world. Though imperfect, a world mapping of the index shows that the highest returns to water investment are to be found where income and livelihoods are the lowest, in sub-Saharan Africa.

World map of the rural water livelihoods index, 2008

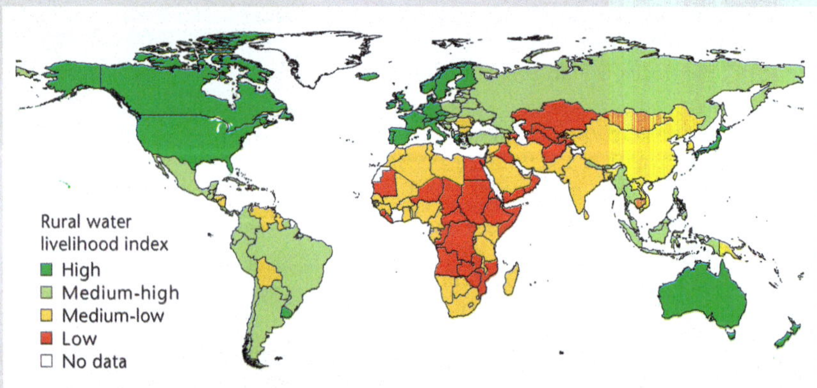

Note: Lower values reflect relatively worse conditions.
Source: Sullivan et al. forthcoming.

Source: Poverty Environment Partnership 2006; Faurès and Santini 2008; Sullivan 2002; Sullivan et al. 2003; Sullivan et al. forthcoming.

the necessary alignments among water resources, economic growth and poverty reduction. Formalizing those alignments within a new round of poverty reduction strategies, which are more growth-oriented, will help make the essential connections explicit.

Benefits of multiple-use approaches

A multiple-use approach to meeting the water needs of poor communities can bring multiple benefits. Poor households throughout the world depend on subsistence activities such as small-scale vegetable gardening, fish rearing, livestock watering, brick making, basket making, textile weaving, beer brewing and other handicrafts that require water. These activities often also provide a much-needed source of income. Better water access for domestic and agricultural use is likely to result in improved outcomes for poor households, by improving household productivity and health and releasing labour into the household production system, stimulating household income growth.[26]

Poor households use water from natural streams, from human-made structures (irrigation canals and wells) and from rainwater harvested for small-scale irrigation or domestic purposes. Although few water sources have a single use or user, a single-use perspective has dominated thinking on water development and services (box 6.5), particularly where communities provide water supply systems in rural and poor peri-urban areas.

Multiple-use water systems yield both financial and non-financial benefits (figure 6.5). Evidence from around the world indicates that if the multiple-use reality were acknowledged and investments were made to upgrade single-purpose systems to serve multiple functions, more than 1 billion poor people could benefit.[27] For example, upgrading domestic systems to provide 100 litres per person per day could generate income of $40-$80 per capita per year. In South Africa this approach resulted in improved income as the productive use of domestic water rose from 17% of average household income in villages with limited water to 31% in villages with adequate provision.[28] The income generated by multiple-use services can enable repayment of initial and ongoing costs for most service levels and technology options, making multiple-use services more likely to be sustained.

When poor people can access water for such household use and for small-scale productive water points, they are better able to avoid hunger and survive droughts. A multiple-use approach can improve health and reduce the incidence of waterborne diseases and lower child mortality. It can also contribute to gender equity by reducing the time women spend fetching water. In rural areas this approach could

Box 6.5 Defining water services: single or multiple uses?

Water services are defined as the provision of water of a given quality, quantity and reliability at a specified place. The definition emphasizes outputs – what people receive – rather than infrastructure.

Single-use approaches involve design, finance and arrangement of water services for a single intended use, such as for irrigation or domestic purposes, the most important single-use services in rural areas. Single-use approaches are the standard model of water service delivery.

Multiple-use approaches involve planning, finance and management of integrated water services for multiple domestic and productive uses based on consumer demand. 'Domestic +' approaches involve provision of water services for domestic as well as productive activities. 'Irrigation +' approaches involve provision of water services for irrigation as well as domestic and non-irrigation productive activities.

Source: Renwick et al. 2007.

Figure 6.5 Benefits of a multiple-use approach to water

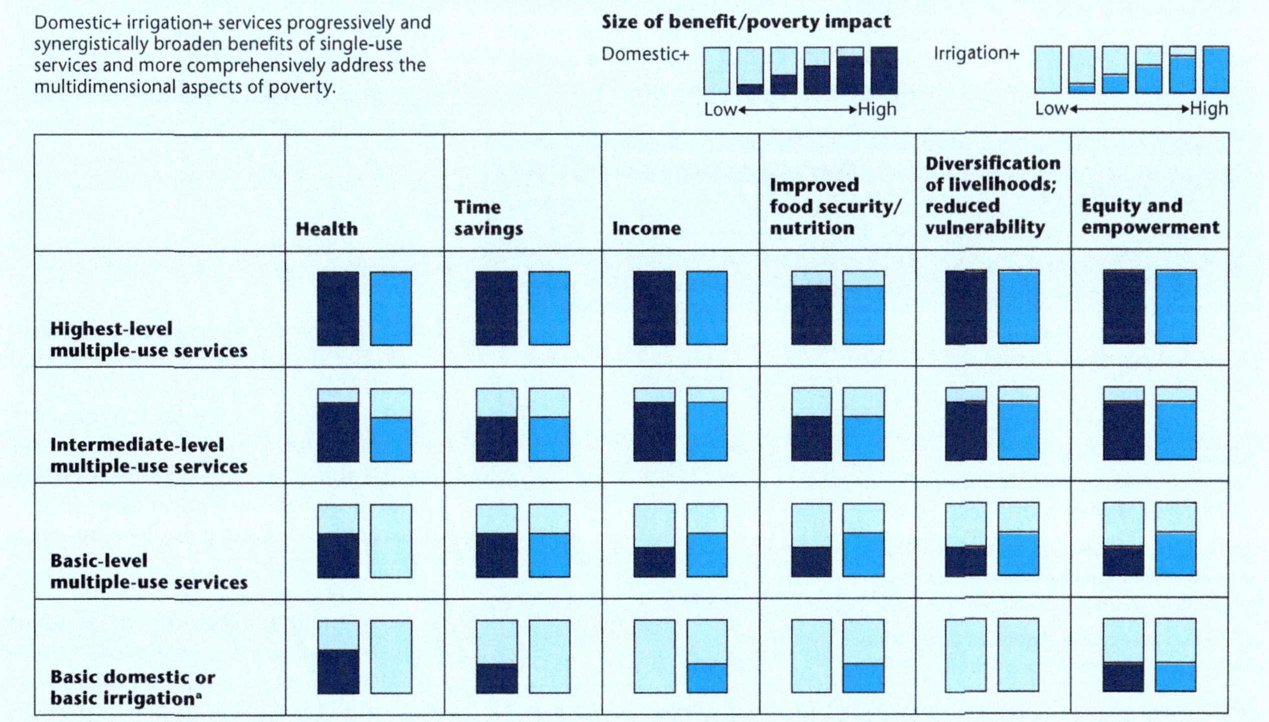

a. Assumes no unplanned uses as they cannot ensure sustainable generation of benefits.
Source: Renwick et al. 2007.

Access to safe water and adequate sanitation services has proved to be one of the most efficient ways of improving human health

provide water for garden plots to meet household needs.

Water and health

Access to safe water and adequate sanitation services has proved to be one of the most efficient ways of improving human health. The World Health Organization has estimated the economic costs that can be avoided through adequate sanitation and the economy-wide returns to various levels of investment in water supply and sanitation services (table 6.2). Every $1 invested in improved water supply and sanitation yields gains of $4-$12, depending on the type of intervention.

Almost one-tenth of the global disease burden could be prevented by improving water supply, sanitation, hygiene and management of water resources. Such improvements reduce child mortality and improve health and nutritional status in a sustainable way. They yield multiple social and economic benefits, enhancing well-being and, indirectly, people's access to health-related services. Upgrading water supply and sanitation services could also improve education outcomes by enabling more girls to attend school instead of fetching water, while providing sanitary facilities in schools encourages higher female enrolment in secondary schools and improves the working environment for female teachers. There is thus a strong conceptual case that improved coverage of drinking water and sanitation contributes to meeting the Millennium Development Goals and to accelerated growth.

Table 6.2 **Benefit-cost ratio by water and sanitation intervention in developing regions and Eurasia**

Intervention	Annual benefits ($ millions)	Benefit-cost ratio
Halving the proportion of people without access to improved water sources by 2015	18,143	9
Halving the proportion of people without access to improved water sources and improved sanitation by 2015	84,400	8
Universal access to improved water and sanitation services by 2015	262,879	10
Universal access to improved water and improved sanitation and water disinfected at the point of use by 2015	344,106	12
Universal access to a regulated piped water supply and sewerage connection by 2015	555,901	4

Note: Benefit-cost ratio is total benefits divided by total costs. The higher the ratio, the greater the benefits relative to the costs. Projects with a benefit-cost ratio greater than 1 have greater benefits than costs.
Source: Prüss-Üstün et al. 2008.

Saving children

The under-five mortality rate is an important social indicator of development. It is an indicator of the quality of life, including the income and education of parents, the efficacy of health services and access to safe drinking water and sanitation services. It is also easily measured and so is considered a good indicator of progress towards the Millennium Development Goals. In 2000 diarrhoea accounted for 17% of the 10.6 million deaths in children younger than five, and malaria for 8%.[29] Undernutrition is an underlying cause of 53% of all deaths in children younger than five.

Global under-five mortality has fallen from 93 per 1,000 live births in 1990 to 72 per 1,000 in 2005 – a decline of 22.5% – but the pace of progress has been uneven across regions and countries. The decline has been slowest in sub-Saharan Africa.

Benefits of improved access to water and sanitation for health

Improvements in drinking water, sanitation, hygiene and water resources management could have a particularly large impact on diarrhoea, malaria and malnutrition (table 6.3). But investments in improved water supply and sanitation would also make a difference for many neglected tropical diseases (such as intestinal nematode infections, lymphatic filariasis, trachoma and schistosomiasis) that have environmental transmission pathways. Another water health issue of increasing concern is naturally occurring chemical contaminants – notably arsenic and fluoride (see box 8.3 in chapter 8). Such contaminants underline the need for simple and reliable water quality monitoring systems and have important implications for the definition of 'safe' drinking water used to monitor progress towards international targets, such as the Millennium Development Goals.

Benefits go well beyond human health

Access to safe drinking water and adequate sanitation services is vital to human health but has other important benefits ranging from the easily identifiable and quantifiable (costs avoided, time saved) to the more intangible and difficult to measure (convenience, well-being, dignity, privacy and safety).

In cost-benefit analyses the major benefits of improving access to water and sanitation derive from the time savings associated with closer location of facilities. Easy access translates into increased production, higher school attendance and more leisure time. The case is exceptionally

Table 6.3 **Major diseases attributable to environmental factors**

Disease	Annual global burden attributable to water, sanitation and hygiene: Deaths (thousands)	Annual global burden attributable to water, sanitation and hygiene: DALY[a] (thousands)	Percent of total burden attributable to environmental factors	Environmental pathways
Diarrhoea	1,523	52,460	94	Water supply, sanitation, hygiene
Malnutrition	863	35,579	50	Water supply, sanitation, hygiene, water resources management
Malaria	526	19,241	42	Water resources management
Lymphatic filariasis	0	3,784	66	Water supply, sanitation
Intestinal nematodes	12	2,948	100	Sanitation
Trachoma	0	2,320	100	Water supply, hygiene, flies
Schistosomiasis	15	1,698	100	Water supply, sanitation, water resources management
Japanese encephalitis	13	671	95	Water resources management
Dengue	18	586	95	Water supply, sanitation

a. Disability adjusted life year, a summary measure of population health. One DALY represents one lost year of healthy life.

Source: Adapted from Prüss-Üstün and Corvalán 2006; Prüss-Üstün et al. 2008.

strong for sanitation, where the economic cost of inaction is enormous. Without improving sanitation, it will be difficult to fully achieve the Millennium Development Goals.

In addition, urban sanitation systems comprise a range of processes that represent new business opportunities. These may include small-scale service provision for construction of system components and the collection, transport, storage and processing of products from sanitation systems (into biogas, fertilizer, soil conditioners or irrigation water, for example).

Reducing diarrhoeal diseases

Some 1.4 million children die each year from preventable diarrhoeal diseases. Ordinary diarrhoea remains the major killer among water-, sanitation- and hygiene-related diseases, contributing to 43% of deaths.[30] Sub-Saharan Africa and South Asia are the most affected regions (map 6.1).

Map 6.1 **Diarrhoea deaths in 2004**

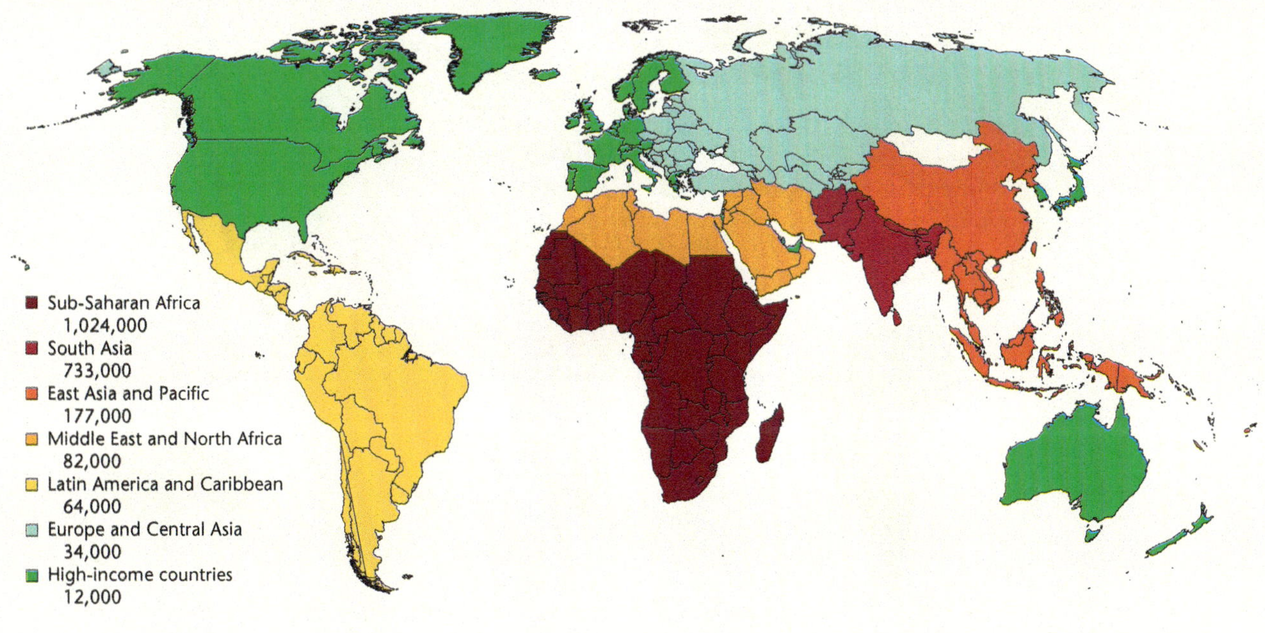

Source: WHO 2008.

Adequate sanitation and hand-washing after defecation helps break the chain of transmission of fæcal-oral disease

Adequate sanitation and hand-washing after defecation helps break the chain of transmission of fæcal-oral disease.[31] There is strong evidence that hand-washing with soap prevents not only diarrhoea but also acute respiratory infections, both major killers of children ages 1 month to 5 years. For example, in squatter settlements of Karachi, Pakistan, hand-washing with soap cut episodes of diarrhoea and acute respiratory infections in half.[32] Hand-washing even without water (in sand, for example), also significantly reduces the likelihood of diarrhoea, emphasizing the importance of hygiene as well as water provision in improving health.

Combating malnutrition

Malnutrition accounts for about a third of the disease burden in low- and middle-income countries.[33] Lack of access to adequate, safe food, partly related to water resources management, is one cause of malnutrition, but up to 50% of malnutrition is related to repeated diarrhoea or intestinal nematode infections as a result of unclean water, inadequate sanitation or poor hygiene (see table 6.3).

Low height-for-age (stunting), an indicator of chronic undernutrition, is useful for long-term monitoring of the cumulative effects of poverty, unclean water and inadequate sanitation and a high infectious disease burden (map 6.2).

Fighting malaria

World Malaria Report 2005 states that malaria continues to exact an unacceptable toll on the health and economic welfare of the world's poorest communities.[34] During the past decade malaria resurged or increased in intensity in Africa and South-East Asia after interruption of eradication efforts and re-emerged in several Central Asian and Transcaucasian countries. Of the estimated 350-500 million clinical disease episodes occurring annually, around 60% are in sub-Saharan Africa, as are 80% of the deaths. Most of the more than 1 million Africans who die from malaria each year are children under age five.

How much malaria could be eliminated by managing the environment – by eliminating stagnant water bodies, modifying reservoir contours, introducing drainage or improving irrigation management – differs across regions with variations in vector habitats, with a global average of 42% (see table 6.3). Malaria control programmes that emphasize environmental management are therefore highly effective in reducing malaria illness and malaria-related death.[35] Much of the attention in international malaria research and control has focused on medical solutions, such as drugs and vaccines, but developing new tools and approaches for malaria prevention and control, including innovation in vector control, is essential.

Map 6.2 **Geographical pattern of stunting in children under age five on a country basis**

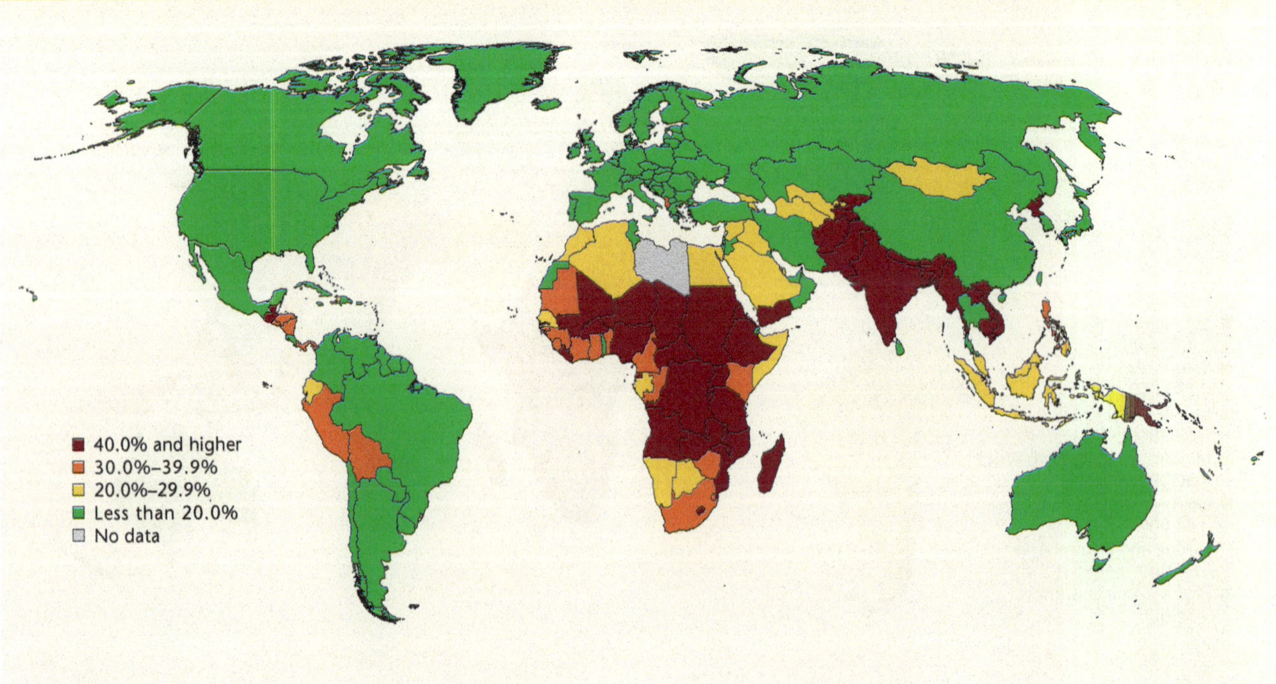

Source: WHO 2007.

Maintaining ecosystem services

How water is managed affects the health of ecosystems. Growing pressure on water resources affects ecosystems and threatens the ecosystem goods and services on which life and livelihoods depend.

The natural environment provides food, essential natural resources and other life-supporting goods, services and benefits for people, animals and plants. In addition to supporting the production of food and fibres, freshwater ecosystems regulate environmental flows, purify wastewater and detoxify wastes, regulate climate, provide protection from storms, mitigate erosion and offer cultural benefits, including significant aesthetic, educational and spiritual benefits.

The diversity of ecosystem services varies with the type of ecosystem and the ways in which this multifunctionality is managed (figure 6.6).[36] Aquatic ecosystems produce significant economic benefits, including flood control, groundwater recharge, shoreline stabilization and protection, nutrition cycling and retention, water purification, preservation of biodiversity, and recreation and tourism. The Nakivubo swamp, for example, provides wastewater treatment services worth some $363 million to the citizens of Kampala, Uganda.[37] In Uganda alone the use of inland water resources is worth almost $300 million a year in forest catchment protection, erosion control and water purification services. Almost 1 million urban dwellers rely on natural wetlands for wastewater retention and purification services.[38] Natural wetlands in the Zambezi Basin in Southern Africa have a net present value of more than $64 million – $16 million in groundwater recharge, $45 million in water purification and treatment services and $3 million in attenuation of flood-related damage costs.[39] Rice fields – human-made wetlands in the Ramsar Convention typology – offer a large range of ecosystem services that can be enhanced or diminished depending on management decisions about the use of water and its productive functions (see figure 6.6).

Growing pressure on water resources affects ecosystems and threatens the ecosystem goods and services on which life and livelihoods depend

Figure 6.6 **Agricultural systems can be managed to produce one ecosystem function or a range of ecosystem services**

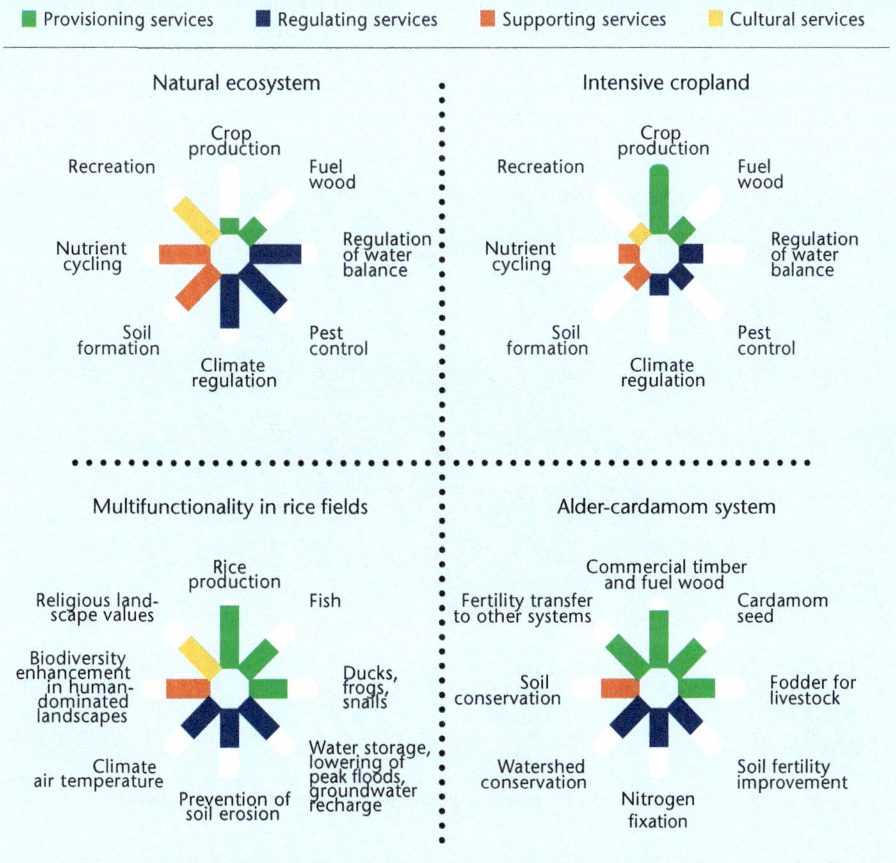

Source: Comprehensive Assessment of Water Management in Agriculture 2007, figure 6.4.

It is critical that allocations to the 'environment' or 'nature' not be considered 'wasted water'. Most such allocations can be considered in terms of benefits to people

Across developing countries 10% of undernourished people depend on direct access to natural resources, in particular freshwater ecosystems.[40] They are vulnerable to any degradation of these ecosystems or to changes in the water cycle that affect their functioning. This is the case for pastoralists moving with their herds from one water source and pasture area to another, for capture fishers vulnerable to water pollution and river water depletion and for forest-dependent people who are hurt when forests are cleared for agriculture or for construction of dams or other large infrastructure. These people are often as voiceless as ecosystems in the water allocation process.

Because of the interconnection between freshwater ecosystems and their services, developing one service (for example, food production through increased irrigation) automatically affects others. The management objective is to balance the delivery of all services collectively so that ecosystems are used optimally and development becomes sustainable.

Nature has to be recognized as a water stakeholder because it provides important services to society. Ways of valuing ecosystem services remain highly controversial, however, and implementation of environmental regulations is still limited (see chapter 9). In any case, defining an environmental water requirement – even if imperfect – provides a voice for nature in allocation decisions for water withdrawal at the basin level. It is critical that allocations to the 'environment' or 'nature' not be considered 'wasted water'. Most such allocations can be considered in terms of benefits to people, recognizing that these in-situ uses may constrain other uses, particularly during dry periods. Concerns for environmental services often happen too late, when water use has gone beyond the capacity of the environment to cope and when competition is critical. This is the result of decision processes that do not promote informed, impartial and balanced outcomes – and would not do so even if better valuations were at hand. Water still continues to often be allocated on a first-come, first-served sector basis.

Map 6.3 pinpoints areas where respecting environmental water requirements has become urgent because water use is reaching limits that threaten to undermine our life and development support base – particularly for people who are most vulnerable and dependent on the environment for their livelihood.

Our reliance on nature and its abundance must be matched by the care we take of the agro-ecosystems on which we depend. There are pastures in the Alps, oases in Morocco and irrigation systems in the Philippines that have been used for centuries with no diminution of their productive capacities or beauty. Rice terraces cascading down the Ifugao in the Philippines represent the

Map 6.3 **Water stress level of major river basins, around 2002**

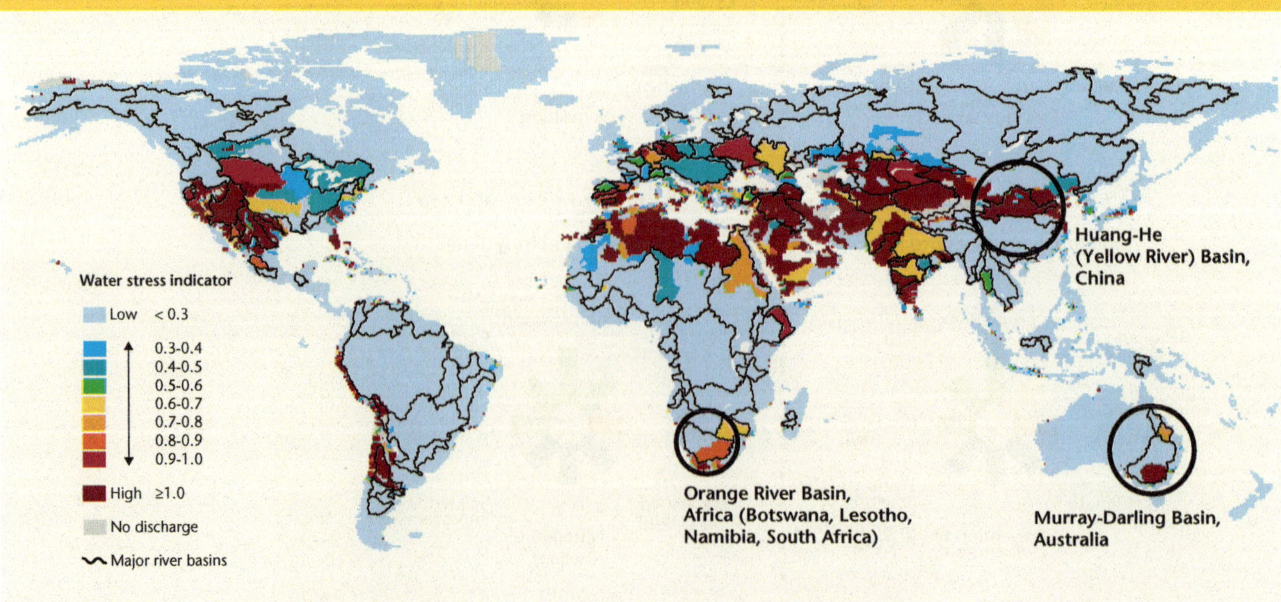

Source: Based on Smakhtin, Revenga, and Döll 2004.

Box 6.6 Agro-ecosystems and sustainability: an example from Peru

The central Andes are a primary centre of origin for potatoes. Up to 177 varieties have been domesticated by generations of Aymara and Quechua in the Cusco and Puno valleys. Many cultural and agricultural treasures from the Inca civilization have been carefully preserved and improved over centuries to guarantee living conditions in areas that are more than 4,000 metres above sea level.

One of these is the terracing system used to control land degradation and that allows cultivation of steep slopes at altitudes ranging from 2,800 metres to 4,500 metres. Maize is cultivated in the lower areas (2,500-3,500 metres above sea level), and potatoes mainly at medium altitudes (3,500-3,900 metres above sea level). Areas at and above 4,000 metres are used mostly as rangeland but are also cultivated with high-altitude crops. In the high plateau, around Lake Titicaca, farmers dig trenches around their fields and fill them with water. Warmed by sunlight during the day, the water gives off warm steam when temperatures drop at night. the steam serves as frost protection for several varieties of potatoes and other native crops, such as quinoa. This method is under consideration for use in irrigation areas in Peru as an adaptation to climate change.

Source: www.fao.org/sd/giahs.

collective efforts of countless generations of farmers who developed an ingenious irrigation system that allowed them to share water and develop rice varieties that survive at over 1,000 metres. In the combination rice-fish systems of Zhejiang Province in China, which date from the Han Dynasty 2,000 years ago, fish not only provide food, but they also eat larvae and weeds in the flooded fields, reducing the cost, labour and pollution risks involved in fertilization and insect control.[41] The sophisticated terracing system in the central Andes in Peru allows cultivation of steep slopes at different altitudes (box 6.6).

Notes

1. Poverty Environment Partnership 2006.
2. World Bank 2007.
3. Comprehensive Assessment of Water Management in Agriculture 2007.
4. UNIDO 2007.
5. World Bank 2007.
6. Japan Water Forum and World Bank 2005.
7. Prüss-Üstün et al. 2008.
8. Gichere, Davis, and Hirji 2006.
9. Biemans et al. 2006; Grey and Sadoff 2008.
10. United Nations 2008.
11. APWF 2007.
12. United Nations 2008.
13. GWP Technical Committee 2003; Hussain and Hanjra 2003; Lipton, Litchfield, and Faurès 2003; UNDP 2006.
14. World Bank 2005a.
15. UNDP 2006.
16. UNDP 2006.
17. DFID 2008. Research in India has shown that between 1985/86 and 1991/92 government investment in the construction of latrines increased coverage by 2.2%. At the same time, a national census found that access to latrines was much higher, suggesting that about 8% of rural households across the country had invested their own time and money in the construction of latrines (DFID 2008).
18. Narayanan 2005; Winpenny 2003.
19. United Nations 2008.
20. World Bank 2007.
21. UN-HABITAT 2006.
22. Worldwatch Institute 2007.
23. Faurès and Santini 2008.
24. Tacoli 2007.
25. World Bank 2005b.
26. Faurès and Santini 2008.
27. Renwick et al. 2007.
28. Renwick et al. 2007.
29. WHO 2007.
30. Prüss-Üstün et al. 2008.
31. Ejemot et al. 2008; Fewtrell et al. 2005.
32. Luby et al. 2005.
33. Laxminarayan, Chow, and Shahid-Salles 2006.
34. Roll Back Malaria, WHO, and UNICEF 2005.
35. Keiser et al. 2005.
36. MEA 2005.
37. Worldwatch Institute 2007.
38. UNEP 2007.
39. Turpie et al. 1999.
40. Comprehensive Assessment of Water Management in Agriculture 2007.
41. Lu and Li 2006.

References

APWF (Asia-Pacific Water Forum). 2007. *Asian Water Development Outlook 2007: Achieving Water Security for Asia.* Manila, Philippines: Asian Development Bank.

Biemans, Hester, Ton Bresser, Henk van Schaik, and Pavel Kabat. 2006. Water and Climate Risks: A Plea for Climate Proofing of Water Development Strategies and Measures. 4th World Water Forum, Cooperative Program on Water and Climate, Wageningen, The Netherlands.

Comprehensive Assessment of Water Management in Agriculture. 2007. *Water for Food, Water for Life: A Comprehensive Assessment of Water Management in Agriculture.* London: Earthscan, and Colombo: International Water Management Institute.

DFID (Department for International Development). 2005. *Why We Need to Work More Effectively in Fragile States.* London: Department for International Development.

DFID (Department for International Development) Sanitation Reference Group. 2008. Water Is Life, Sanitation Is Dignity, Final Draft1. DFID Sanitation Policy Background Paper, Department for International Development, London. www.dfid.gov.uk/consultations/past-consultations/water-sanitation-background.pdf.

Ejemot, R., J. Ehiri, M. Meremikwu, and J. Critchley. 2008. Hand Washing for Preventing Diarrhoea. *Cochrane Database of Systematic Reviews*, Issue 3. Art. No: CD004265. DOI: 10.1002/14651858.CD004265.pub2.

Faurès, J.-M., and G. Santini, ed. 2008. Water and the Rural Poor – Interventions for Improving Livelihoods in Sub-Saharan Africa. Rome: Food and Agriculture Organization of the United Nations and International Fund for Agricultural Development.

Fewtrell, L., R. B. Kaufmann, D. Kay, W. Enanoria, L. Haller, and J. M. Colford. 2005. Water, Sanitation, and Hygiene Interventions to Reduce Diarrhoea in Less Developed Countries: A Systematic Review and Meta-analysis. *Lancet Infectious Diseases* 5 (1): 42-52.

Gichere, Samuel, Richard Davis, and Rafik Hirji. 2006. Climate Variability and Water Resources Degradation in Kenya: Improving Water Resources Development and Management. World Bank Working Paper Series 69, World Bank, Washington, DC.

———. **2008.** Power point presentation at session on Infrastructure Platform for Achieving Water Security 2008. First African Water Week, Tunis.

GWP (Global Water Partnership) Technical Committee. 2003. *TEC Background Paper 8: Poverty Reduction and IWRM.* Sweden: Global Water Partnership.

Hussain, I., and M. A. Hanjra. 2003. Does Irrigation Water Matter for Rural Poverty Alleviation? Evidence from South and South-East Asia. *Water Policy* 5 (5-6): 429-42.

Japan Water Forum and World Bank. 2005. A Study on Water Infrastructure Investment and Its Contribution to Socioeconomic Development in Modern Japan. World Bank, Washington, DC.

Keiser, J., B. Singer, and J. Utzinger. 2005. Reducing the Burden of Malaria in Different Settings with Environmental Management: A Systematic Review. *The Lancet Infectious Diseases* 5 (11): 695-708.

Laxminarayan, R., J. Chow, and S. A. Shahid-Salles. 2006. Intervention Cost-Effectiveness: Overview and Main Messages. In *Disease Control Priorities in Developing Countries*, 2nd edition, ed. D. T. Jamison, J. G. Breman, A. R. Measham, G. Alleyne, M. Claeson, D. B. Evans, P. Jha, A. Mills, and P. Musgrove. Washington, DC: World Bank, and New York: Oxford University Press.

Lipton, M., J. Litchfield, and J.-M. Faurès. 2003. The Effects of Irrigation on Poverty: A Framework for Analysis. *Water Policy* 5 (5-6): 413-27.

Lu Jianbo and Xia Li. 2006. Review of Rice-Fish Farming Systems in China – One of the Globally Important Ingenious Agricultural Heritage Systems (GIAHS). *Aquaculture* 260 (2): 106-13.

Luby, S. P., M. Agboatwalla, D. R. Feikin, J. Painter, W. Billhimer, A. Altaf, and R. M. Hoekstra. 2005. Effect of Handwashing on Child Health: A Randomised Controlled Trial. *Lancet* 366 (9481): 225-33.

Margat, J., and V. Andréassian. 2008. *L'eau, les Idées Reçues.* Paris: Editions le Cavalier Bleu.

MEA (Millennium Ecosystem Assessment). 2005. *Ecosystems and Human Well-Being: Wetlands and Water Synthesis.* Washington, DC: World Resources Institute.

Narayanan, Ravi. 2005. Financial Water Report – Critical Issues to Be Addressed. Letter. Third World Water Forum and Water Aid. www.financingwaterforall.org/fileadmin/Financing_water_for_all/Stakeholders_responses/Narayanan_Response_to_Camdessus_Report.pdf.

OECD (Organisation for Economic Co-operation and Development). 2008. Service Delivery in Fragile Situations: Key Concepts, Findings and Lessons. OECD/DAC Discussion Paper. Off-print of the *Journal on Development* 9 (3). Organisation for Economic Co-operation and Development, Paris. www.oecd.org/dataoecd/11/8/40581496.pdf.

Poverty-Environment Partnership. 2006. *Linking Poverty Reduction and Water Management.* Poverty-Environment Partnership. http://esa.un.org/iys/docs/san_lib_docs/povety%20reduction%20and%20water.pdf.

Prüss-Üstün, A., and C. Corvalán. 2006. *Preventing Disease through Healthy Environments. Towards an Estimate of the Environmental Burden of Disease.* Geneva: World Health Organization. www.who.int/quantifying_ehimpacts/publications/preventingdisease/en/index.html.

Prüss-Üstün, A., R. Bos, F. Gore, and J. Bartram. 2008. *Safer Water, Better Health: Costs, Benefits and Sustainability of Interventions to Protect and Promote Health.* Geneva: World Health Organization.

Renwick, M., D. Joshi, M. Huang, S. Kong, S. Petrova, G. Bennett, R. Bingham, C. Fonseca, et al. 2007. Multiple Use Water Services for the Poor: Assessing the State of Knowledge. Final Report, Winrock International, Arlington, VA.

Roll Back Malaria, WHO (World Health Organization), and UNICEF (United Nations Children's Fund). 2005. *World Malaria Report 2005.* Geneva: World Health Organization.

Smakhtin, V. U., C. Revenga, and P. Döll. 2004. *Taking into Account Environmental Water Requirements in Global-Scale Water Resources Assessments Comprehensive Assessment Research Report 2.* Colombo, Sri Lanka: Comprehensive Assessment Secretariat, International Water Management Institute.

Sullivan, C. A. 2002. Calculating a Water Poverty Index. *World Development* 30 (7): 1195-210.

Sullivan C. A., A. Cohen, J.-M. Faurès, and G. Santini. Forthcoming. *The Rural Water Livelihoods Index: A Tool to Prioritize Water-related Interventions for Poverty Reduction.* Rome: Food and Agriculture Organization of the United Nations and International Fund for Agricultural Development.

Sullivan, C. A., J. R. Meigh, A. M. Giacomello, T. Fediw, P. Lawrence, M. Samad, S. Mlote, C. Hutton, J. A. Allan, R. E. Schulze, D. J. M. Dlamini, W. Cosgrove, J. Delli Priscoli, P. Gleick, I. Smout, J. Cobbing, R. Calow, C. Hunt, A. Hussain, M. C. Acreman, J. King, S. Malomo, E. L Tate, D. O'Regan, S. Milner, and I. Steyl. 2003. The Water Poverty Index: Development and Application at the Community Scale. *Natural Resources* 27 (3): 189-99.

Tacoli, Cecilia. 2007. Links between Rural and Urban Development in Africa and Asia. Paper presented at United Nations Expert Group Meeting on Population Distribution, Urbanization, Internal Migration and Development, United Nations Secretariat, New York, 21-23 January, 2008.

Turpie, J., B. Smith, L. Emerton, and J. Barnes. 1999. *Economic Valuation of the Zambezi Basin Wetlands.* Harare: IUCN–The World Conservation Union Regional Office for Southern Africa.

UNDP (United Nations Development Programme). 2006. *Human Development Report 2006. Beyond Scarcity: Power, Poverty and the Global Water Crisis.* New York: Palgrave MacMillan.

UNEP (United Nations Environment Programme). 2007. *Global Environment Outlook GEO 4: Environment for Development.* Nairobi: United Nations Environment Programme.

UN-HABITAT (United Nations Human Settlements Programme). 2006. *State of the World Cities 2006/7.* London: Earthscan.

UNIDO (United Nations Industrial Development Organization). 2007. Symposium Report on Water Productivity in the Industry of the Future for Technology Foresight Summit, 27-29 September 2007, Budapest.

United Nations. 2008. *World Economic and Social Survey 2008: Overcoming Economic Insecurity.* New York: Department of Economic and Social Affairs, United Nations. www.un.org/esa/policy/wess/wess2008files/wess08/overview_en.pdf.

WCD (World Commission on Dams). 2000. *Dams and Development. A New Framework for Decision-Making.* London: Earthscan.

WHO (World Health Organization). 2007. *World Health Statistics 2007.* Geneva: World Health Organization.

———. **2008.** *The Global Burden of Disease: 2004 Update.* Geneva: World Health Organization.

Winpenny, James. 2003. *Financing Water for All. Report of the World Panel on Financing Water Infrastructure.* Chaired by Michel Camdessus. Kyoto: World Water Council, 3rd World Water Forum, and Global Water Partnership.

World Bank. 2005a. India's Water Economy: Bracing for a Turbulent Future. World Bank, Washington, DC. http://go.worldbank.org/QPUTPV5530.

———. 2005b. Zambezi River Basin, Sustainable Water Resources Development for Irrigated Agriculture. Draft Report, TFESSD Africa Poverty and Environment Program, Environment, Rural and Social Development Department – AFTS1, Finance, Private Sector and Infrastructure Department – AFTU1, Regional Integration and Cooperation Department - AFC16, Africa Region, World Bank, Washington, DC.

———. 2007. *World Development Report 2008: Agriculture for Development.* Washington, DC: World Bank.

Worldwatch Institute. 2007. *State of the World: Our Urban Future.* London and New York: Norton Institute.

WWAP (World Water Assessment Programme) Expert Group on Storage. 2008. Summary Paper of Online Discussion on Storage. World Water Assessment Programme.

Chapter 7
Evolution of water use

Authors: Richard Connor, Jean-Marc Faurès, Johan Kuylenstierna, Jean Margat, Pasquale Steduto, Domitille Vallée and Wim van der Hoek

Contributors: José Aguilar-Manjarrez, Rizaldi Boer, Robert Bos, Cécile Brugère, Ann Cann, Olivier Dubois, Karen Frenken, Arjen Y. Hoekstra, Jippe Hoogeveen, John Jorgensen, Jon Lane, Darren Saywell and Stefania Vannuccini

Coordinator: Jean-Marc Faurès (FAO)

Facilitator: Domitille Vallée

Key messages

- While most of the old challenges of water supply, sanitation and environmental sustainability remain, new challenges such as adaptation to climate change, rising food and energy prices, and ageing infrastructure are increasing the complexity and financial burden of water management. Population growth and rapid economic development have led to accelerated freshwater withdrawals.

- Trends in access to domestic water supply indicate substantial improvement in the past decade, putting most countries on track to achieve the water supply target of the Millennium Development Goals. However, sanitation is lagging well behind, and most sub-Saharan African countries and many rural areas still show unsatisfactory records for both water supply and sanitation.

- Steadily increasing demand for agricultural products to satisfy the needs of a growing population continues to be the main driver behind water use. While world population growth has slowed since the 1970s and is expected to continue its downward trend, steady economic development, in particular in emerging market economies, has translated into demand for a more varied diet, including meat and dairy products, putting additional pressure on water resources.

- After agriculture, the two major users of water for development are industry and energy (20% of total water withdrawals), which are transforming the patterns of water use in emerging market economies. Water and energy share the same drivers: demographic, economic, social and technological processes put pressure on both energy and water. The recent acceleration in the production of biofuels and the impacts of climate change bring new challenges and add to the pressures on land and water resources.

- Freshwater ecosystems provide an extensive array of vital services to support human well-being. A variety of economic and recreational activities such as navigation, fisheries and pastoral activities depend on direct use of water in healthy ecosystems. Yet some environmental services receive inadequate policy attention and are endangered by the way development sectors use water.

The previous chapters have demonstrated the multiple benefits of water: its use as an economic backbone and an essential element for industrial and energy production systems, its use in human activities and its vital importance to the well-being of people and ecosystems. Water plays a strategic role for both on-stream uses (navigation, fisheries and freshwater ecosystems) and off-stream ones (productive sectors, human well-being and terrestrial ecosystems).

While most of the old challenges of water supply, sanitation and environmental sustainability remain unsolved, new challenges – including adaptation to climate change, volatile food and energy prices and ageing infrastructure – are adding to the complexity and financial burden of water management.

The 2003 and 2006 editions of the *World Water Development Report* detailed the multiple sectoral uses of water for human well-being and ecosystems. Over the last three years exhaustive reviews of selected issues have brought to light additional information on water supply and sanitation,[1] agriculture[2] and the environment.[3] This chapter summarizes some of the findings of these studies, focusing on the main challenges confronting the water community in the immediate and long-term future.

Water use in the world

Population growth and rapid economic development have accelerated freshwater withdrawals (see map 10.1 in chapter 10 on mismatch in distribution of runoff and population). Our somewhat patchy knowledge of water use shows high variability of use globally, both within sectors and across users. Growing uncertainty regarding water resources – particularly linked to climate change (described in chapter 5) – is expected to exacerbate water scarcity trends.

A challenge in managing water resources is our scattered knowledge of patterns of water use (box 7.1; see also chapter 13). Monitoring systems and modelling abilities require substantial improvement to measure progress in addressing challenges for water uses.

Total global freshwater use is estimated at about 4,000 cubic kilometres (km^3) a year.[4] Another 6,400 km^3 of rainwater is also used 'directly' in agriculture. Nature is the most important user of water. An estimated 70,000 km^3 of water a year are evaporated from forest, natural (uncultivated) vegetation and wetlands.[5] Evaporation from human-made reservoirs is difficult to estimate but is considerable in arid areas and is estimated to be about 200 km^3 a year. For example, an estimated 10 km^3 – about 12% of the total storage in Lake Nasser, upstream of the High Aswan Dam at the high storage level – are lost through evaporation each year.[6]

A challenge in managing water resources is our scattered knowledge of patterns of water use

Box 7.1 How much do we know about water uses?

Our knowledge of water use is as poor as our knowledge of water resources – perhaps poorer. Information is largely incomplete – particularly for agriculture, the largest user – and is lacking altogether for some countries. Only limited disaggregated information exists, and even this shows deficiencies of validity and homogeneity and provides extremely poor information on trends.

The quality of information systems varies with each country, but there are common difficulties:

- Statistics on the magnitude of demand and withdrawal are often estimated rather than based on data that are measured or collected from censuses. The level of uncertainty varies, but is particularly high for agriculture.
- Sectors of use are not defined homogeneously and are not well disaggregated.
- Adequate historical datasets are rare, and the dates of available statistics are not always explicit.
- Lack of agreed terminology leads to discrepancies in data compilation and analyses.

As an example, the table below shows the extent of metering of agricultural water use and self-supplied industries in the six major French river basins. Only half the water used in agriculture is effectively metered by a volumetric device. That means that water withdrawal figures are a mix of measurements and estimates (when no metering is available).

Uncertainty of statistics on water uses: importance of metering agriculture and industrial withdrawals in France

(Percentage of water withdrawals that is metered unless otherwise indicated)

	Use for agriculture or irrigation		Use for self-supplied industries		
Basin	From surface water	From groundwater	From surface water	From groundwater	Total use (cubic kilometres a year)
Adour-Garonne	72	62	82	66	2.30
Artois-Picardie	90	100	95	100	0.67
Loire-Brittany	80	95	40	69	3.62
Rhine-Meuse	0	0	90	81	5.05
Rhone-Mediterranean	30	57	87	86	17.13
Seine-Normandy	75	89	37	91	3.06
Total	43	74	73	84	31.81
Total volume (cubic kilometres a year)	3.39	1.38	2.72	1.48	

■ No metering □ Less than 45% of withdrawals metered □ Less than 75% of withdrawals metered

Source: IFEN 2006, based on 2001 data from the basin agencies.

But we know only part of what humans use: only the volume of water used off-stream (withdrawn) is generally measured (or estimated), and only a part of what is withdrawn is effectively consumed. Most of the flow is returned – usually at a lower quality – to the water systems, where it can be reused. Agriculture is by far the most significant consumer of water, particularly in dry areas where irrigation has been developed.

The consumptive uses of freshwater from agriculture, industry and domestic sectors place the greatest pressures on natural systems (see table 10.5 in chapter 10), both in quantity (withdrawals) and quality (returns of lower quality; for definitions of key terms relating to water use, see box 7.2). For this reason these are discussed in greater detail in the rest of chapter 7 (uses) and in chapter 8 (impacts).

The many realities of water use

Water use is uneven across countries. The 10 largest water users (in volume) are India, China, the United States, Pakistan, Japan, Thailand, Indonesia, Bangladesh, Mexico and the Russian Federation (map 7.1). Total water use at the country level ranges from 646 km^3 a year (India) to less than

Box 7.2 Water withdrawal, demand and consumption

Water use refers to water that is being put to beneficial use by humans. Detailed water accounting, however, requires more precise definition of terms.

Water withdrawal is the gross amount of water extracted from any source in the natural environment for human purposes. Differentiating withdrawals by type of source is useful to understand the pressure put on different parts of the system.

Water demand is the volume of water needed for a given activity. If supply is unconstrained, water demand is equal to water withdrawal.

Water consumption or *consumptive use* refers to that part of water withdrawn that is evaporated, transpired, incorporated into products or crops, consumed by humans or livestock, or otherwise removed from the immediate water environment.

Water consumption is high in agriculture, where water is consumed by evapotranspiration of crops, and typically low in other sectors (domestic industries; see table). The return flow may be restored to the water system either as groundwater infiltrated through permeable soils or directly drained into rivers or other freshwater bodies.

Consumption by sector

Uses	Consumption of water withdrawn (percent)
Domestic (urban)	10-20
Industry	5-10
Energy (cooling)	1-2
Agriculture (irrigation)	
Surface irrigation	50-60
Localized irrigation	90

Source: Margat and Andréassian 2008.

Map 7.1 Water withdrawals highlight discrepancies between regions and between the largest and smallest consumers, around 2001

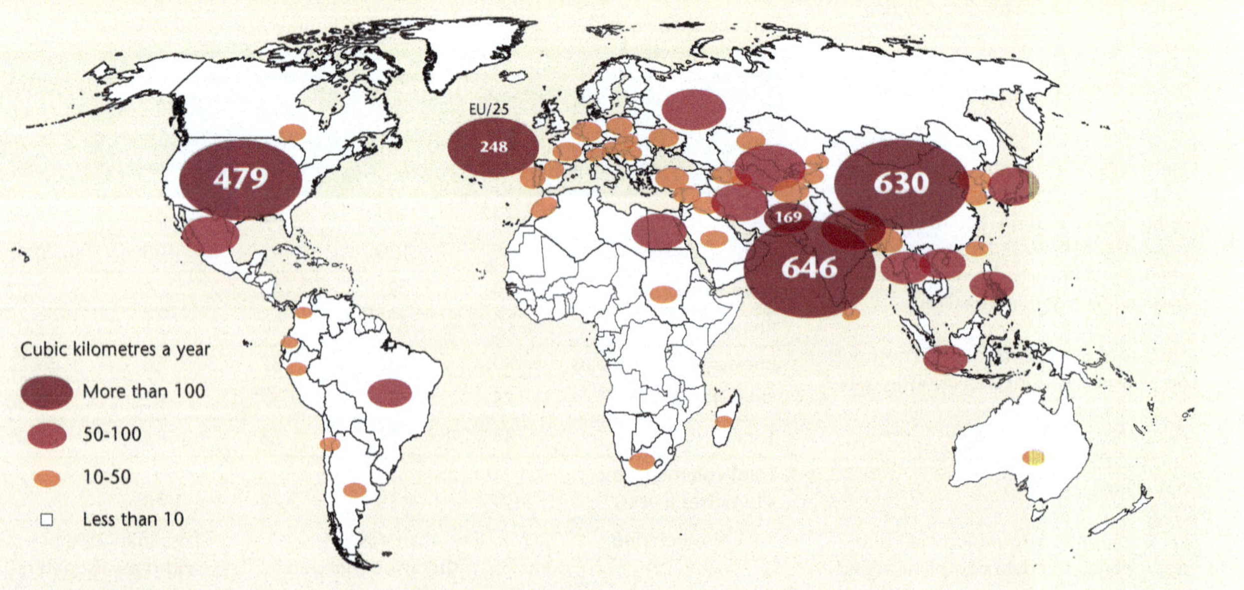

Source: FAO-AQUASTAT.

30 million cubic metres (m^3) a year in some tropical African countries (Cape Verde and the Central African Republic). National averages may hide large discrepancies in water resources and withdrawals between and within countries. In large countries such as China and the United States water demands are concentrated in limited parts of the country, in general where agriculture needs to be irrigated or where economic development is occurring. In addition, long-term annual or multiannual averages mask large temporal differences.

Water withdrawals per person is a better indicator of the impact of population on water. Water withdrawals per person range from 20 m^3 a year in Uganda to more than 5,000 m^3 in Turkmenistan, with a world average of 600 m^3 (map 7.2). Water withdrawals are highest in arid and semi-arid areas, where irrigation is most needed for agricultural production, and are lowest in tropical countries.

Water use is uneven across sectors. Agriculture is by far the main user of water. Irrigated agriculture accounts for 70% of water withdrawals, which can rise to more than 80% in some regions (table 7.1). Although increasing in urbanized economies, industrial (including energy) use accounts for only 20% of total water use and domestic use for about 10%.[7] Water withdrawals for energy generation – hydropower and thermo-cooling – are on the rise, but energy is one of the economic sectors that consumes the least water and

Map 7.2 **Annual water withdrawals per person by country, world view, 2000**

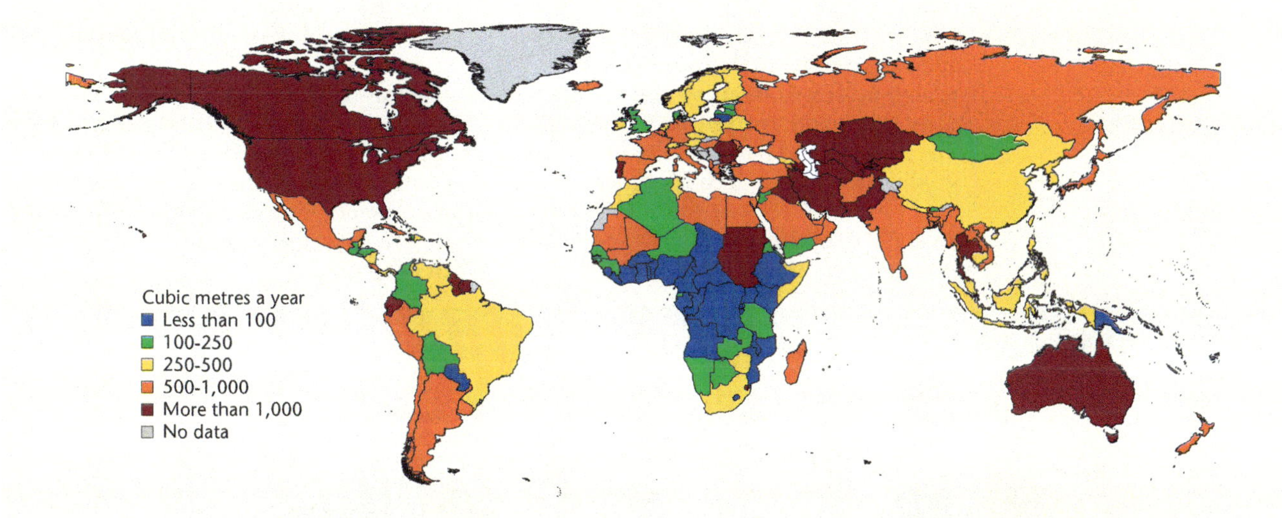

Source: Based on FAO-AQUASTAT global maps (www.fao.org/nr/water/aquastat/globalmaps/index.stm).

Table 7.1 **Water resources and withdrawals, 2000**

(Cubic kilometres per year unless otherwise indicated)

Region	Renewable water resources	Total water withdrawals	Water withdrawals: Agriculture Amount	Agriculture Percent	Industry Amount	Industry Percent	Domestic (urban) Amount	Domestic (urban) Percent	Withdrawals as percent of renewable resources
Africa	3,936	217	186	86	9	4	22	10	5.5
Asia	11,594	2,378	1,936	81	270	11	172	7	20.5
Latin America	13,477	252	178	71	26	10	47	19	1.9
Caribbean	93	13	9	69	1	8	3	23	14.0
North America	6,253	525	203	39	252	48	70	13	8.4
Oceania	1,703	26	18	73	3	12	5	19	1.5
Europe	6,603	418	132	32	223	53	63	15	6.3
World	43,659	3,829	2,663	70	784	20	382	10	8.8

Source: Based on Comprehensive Assessment of Water Management in Agriculture 2007.

Around 20% of total water used globally is from groundwater sources (renewable or not), and this share is rising rapidly, particularly in dry areas

it returns most of the water withdrawn back to the water system (about 95%). This is only a partial picture of sectoral usage as there are many unaccounted-for uses. Little is known about water use in informal urban settlements or informal irrigation systems, both of which are generally unaccounted for in official statistics.

Finally, there are numerous on-stream uses (such as fishing, navigation and ecosystems), which although generally non-consumptive, depend on a certain level of flows and water quality to function. Such uses cannot be measured in volume terms, and these uses are therefore not reflected in statistics on water use.

From a water use perspective the world can be divided into two groups. In one group of countries (in Africa, most of Asia, Oceania, Latin America and the Caribbean) agriculture is by far the main water user, while in the other group (in Europe and North America) withdrawals are related mostly to industry and energy. The domestic supply is essential to life (drinking, hygiene and bathing) but remains the smallest water user for both groups.

Most (99%) of the 4,000 km^3 a year in off-stream water uses – irrigation, domestic, industry and energy – are met by withdrawals from renewable sources, either surface water or groundwater. Less than 1% (currently estimated at 30 km^3 a year) comes from non-renewable (fossil) aquifers mainly in three countries – Algeria, Libyan Arab Jamahiriya and Saudi Arabia – which are the main source of water in these countries.

Around 20% of total water used globally is from groundwater sources (renewable or not), and this share is rising rapidly, particularly in dry areas.[8] This rise has been stimulated by the development of low-cost pumps and by individual investment for irrigation and urban uses. Private investment in self-supply of groundwater – essentially uncontrolled and unmonitored – has mushroomed in response to inadequate public services. As a result, groundwater withdrawals rose fivefold during the 20th century, leading to a rapid drawdown of aquifers in some areas, putting at risk the sustainability of the uses that rely on it (see chapter 8).

In areas of scarce freshwater resources, brackish water and wastewater are often used to meet water demand. While accounting for less than 5% of global water use, the potential is substantially greater (figure 7.1; see also figure 9.3 and box 9.5 in chapter 9).

Figure 7.1 Sources of water use globally and for major sectors, 2000

Withdrawals by supply source

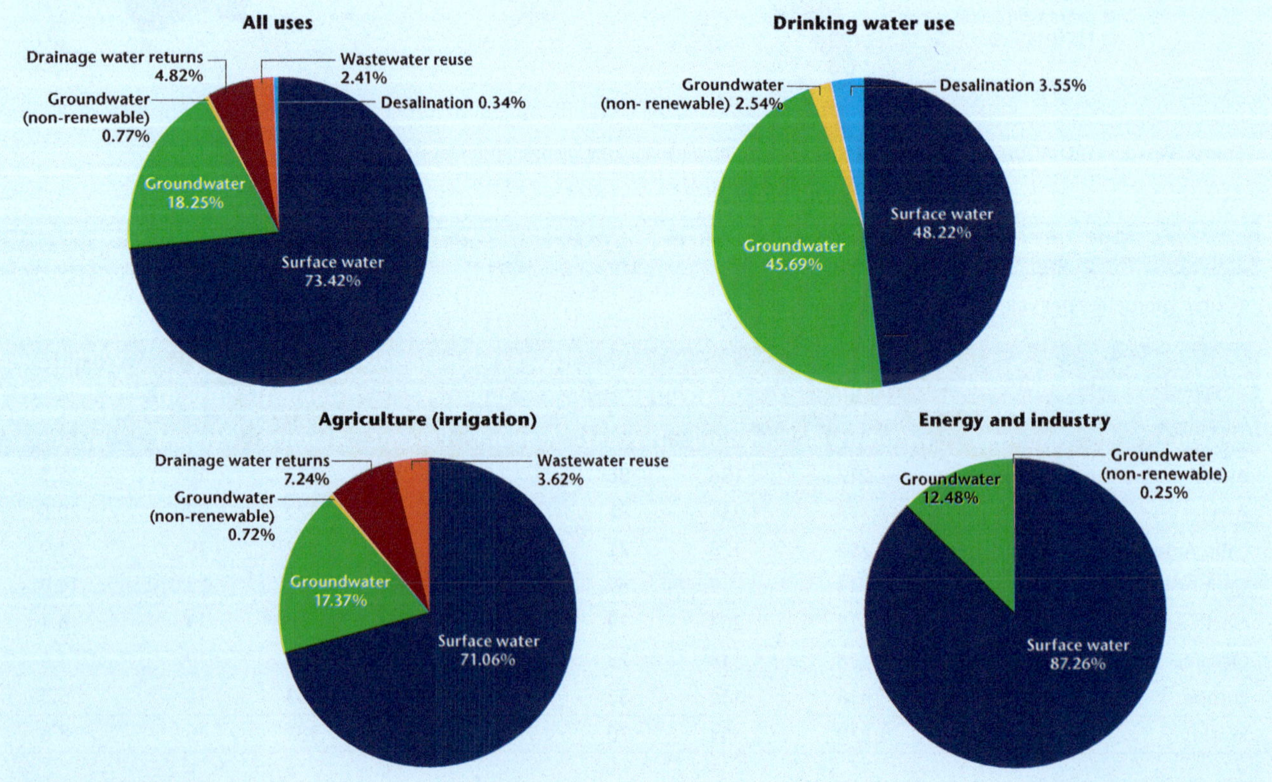

Source: FAO-AQUASTAT.

Trends in water use

Recent trends. With rapid population growth water withdrawals have tripled over the last 50 years. This trend is explained largely by the rapid increase in irrigation development stimulated by food demand in the 1970s and by the continued growth of agriculture-based economies.[9]

Emerging market economies (such as China, India and Turkey) still have an important rural population dependent on water supply for food production. They are also experiencing rapid growth in domestic and industrial demands linked to urbanization and related changes in lifestyle. There are hot spots in these countries where rural and urban demands are in competition. Urbanized and industrial economies (such as the European Union and the United States) import increasing amounts of food and manufactured products, while water use in industrial processes and urban environments has been declining, thanks to both technological changes in production processes and pollution mitigation efforts.

Expected trends over the next 50 years. There is general agreement that population growth, economic growth, urbanization, technological change and changing consumption patterns are the main factors influencing water use (see chapter 2). There is still substantial uncertainty, however, on the scale of future demands. Between 2000 and 2050 the world's population is projected to grow from 6 billion to 9 billion, and demand for food and other goods will increase significantly. Will water resources be adequate? How will the level of development affect demand? How will urbanization influence changes in diet and lifestyle? Where will demands be highest? How will societies address the competition among growing demands? These are difficult questions for water managers, who are planning water development works for future decades, particularly because of the rapid rate of global change and the many uncertainties that lie ahead.

One of the biggest uncertainties is the effect of climate change on water resources, uses and users (see chapter 5), which calls for a complete revisiting of past scenarios (such as the World Water Vision scenarios of 2000, the Millennium Ecosystem scenarios of 2005, and the Comprehensive Assessment of Water Management in Agriculture of 2007[10]) to explore hypotheses and options for action.[11] At the national level countries are already revising their long-term plans. The Mediterranean Action Plan is exploring possible futures for agriculture-based economies that are most vulnerable to anticipated climate change effects (figure 7.2).[12]

Water footprints. The concept of a water footprint helps show the extent and locations of water use in relation to consumption patterns (see chapter 2). The water footprint is defined as the total volume of water used in the production of the goods and services consumed by an individual or community or produced by a business. A country's water footprint is the volume of water used in the production of all the goods and services consumed by inhabitants of the country. The United States has a water footprint of 2,480 m^3 per capita a year; China has a footprint of 700 m^3 per capita a year. The global water footprint is 1,240 m^3 per capita a year. The four major factors determining a country's water footprint are volume of consumption, consumption pattern (for example, high or low meat consumption), climate (growing conditions) and agricultural practices (water use efficiency).

A country's internal water footprint is the volume of water used from domestic water resources; its external footprint is the water used in other countries to produce the goods it imports ('virtual water', discussed in box 2.1 in chapter 2). Whether water consumed by rainfed agriculture should be accounted for in calculating water footprints is a subject of debate. The ratio of internal to external water

With rapid population growth water withdrawals have tripled over the last 50 years

Figure 7.2 **Future water demands in Mediterranean region countries for 2025 in Blue Plan business as usual scenario (trend scenario)**

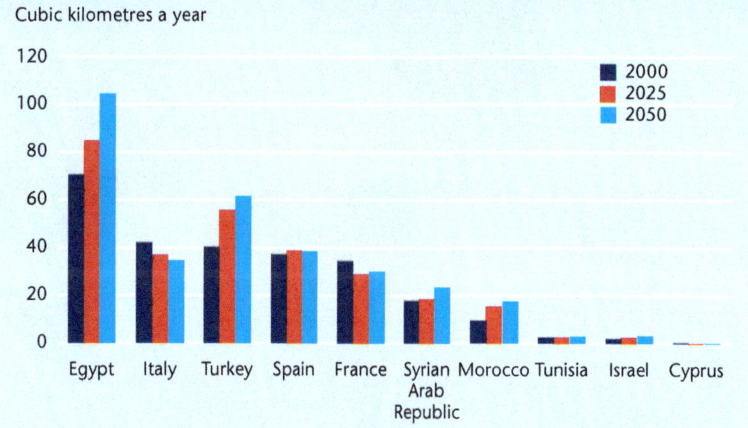

Note: Projections for 2025 are based on Blue Plan 2005 scenarios established on the basis of national planning documents. Projections for 2050 are based on the optimistic assumption that 2050 water demand per inhabitant is similar to that for 2025.
Source: Based on Margat 2008.

Externalizing the water footprint means increasing dependence on foreign water resources but also passing on environmental impacts

footprints is relevant, because externalizing the water footprint means increasing dependence on foreign water resources but also passing on environmental impacts.

With increasing globalization, it is no longer sufficient to examine water issues only in a national context. Local decisions on water use in agriculture and industry are increasingly driven by decisions outside the local water domain. For example, the water footprint of inhabitants of Europe and North America has been externalized to other parts of the world (map 7.3). Europe is a large importer of cotton, which is produced in many water-scarce areas (and elsewhere) though it is one of the thirstiest crops. Through the global market European and U.S. consumption relies on water resources available outside Europe's boundaries, and thus European and North American consumers influence agricultural and industrial strategies elsewhere. About 80% of virtual water flows relate to trade in agricultural products (see box 2.1 in chapter 2).[13] Water-scarce countries such as Greece and Spain use large volumes of water to produce fruit and oil crops for export. The rationale for such uses will become increasingly questionable where climate change leads to reductions in water availability.

Domestic water supply and sanitation

While rapid progress has recently been made in water supply in all regions except sub-Saharan Africa, sanitation coverage still lags. To highlight the problem, the UN General Assembly declared 2008 the International Year of Sanitation in response to the recommendations of the UN Secretary-General's Advisory Board on Water and Sanitation.[14] The goal is to raise awareness and accelerate progress towards the target set for the Millennium Development Goal of reducing by half the proportion of people without access to basic sanitation between 1990 and 2015.

Current state of water supply and sanitation

In 2006, 54% of the world's population had a piped connection to their dwelling, plot or yard, and 33% used other improved drinking water sources. The remaining 13% (884 million people) relied on unimproved sources. Progress has been greatest in East Asia, with an increase in coverage of improved drinking water sources from 68% in 1990 to 88% in 2006.[15]

Except for sub-Saharan Africa and Oceania, all regions are on track to meet the Millennium Development Goal drinking water target. But if current trends continue, 2.4 billion people will still be without access to basic sanitation.[16] Coverage is much higher in urban than in rural areas for both water supply (figure 7.3) and sanitation (figure 7.4). Global and regional aggregates for water and sanitation coverage do not show the large differences between countries.

Map 7.3 **Average national water footprint per capita, 1997-2001**

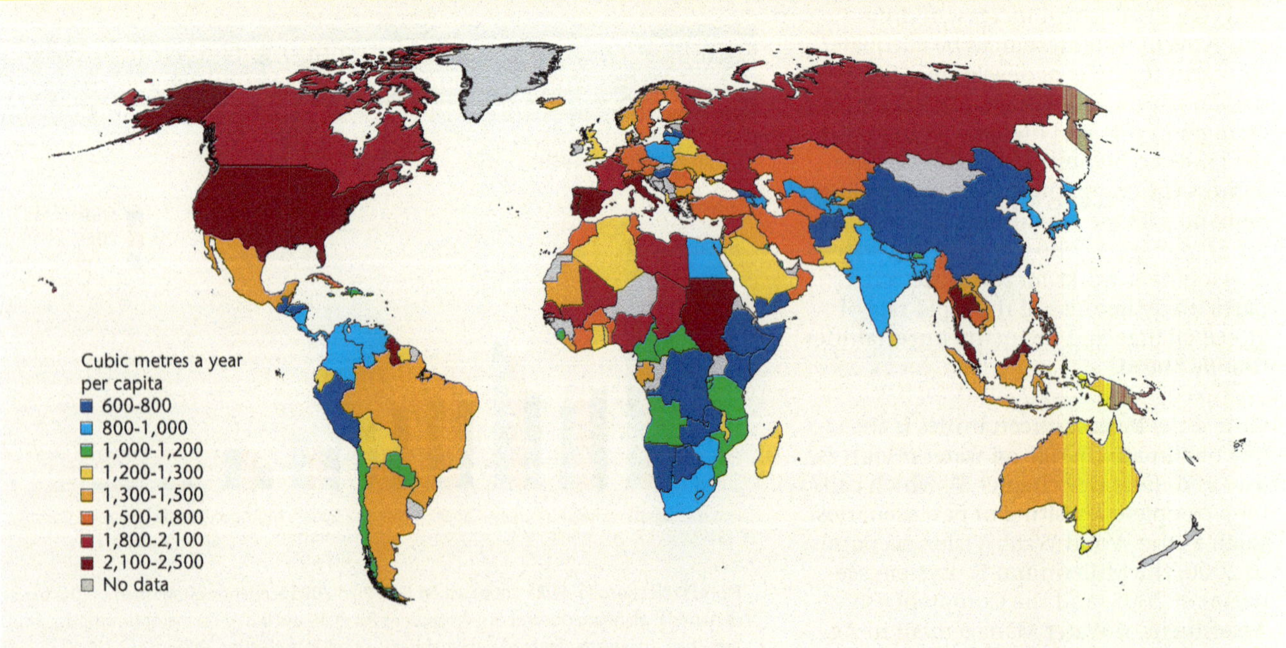

Source: Hoekstra and Chapagain 2008.

The forgotten challenge of small towns. Efforts to increase the number of people benefiting from improved water and sanitation services have focused on two areas: rural areas, where the favoured approach is community-based management of appropriate systems, and urban centres, where utilities (public or private) operate a range of services tailored to customer groups. The grey area between the two is small towns with populations of 2,000-50,000, where neither community management nor utility-based solutions are fully suitable.[17] Populations in these towns account for a large share of the overall population. Improving services will therefore require greater attention to management models, engineering designs, financing arrangements and professional support options that offer the demanded services and that can be expanded incrementally and sustained over the long term. Other aspects of success include financial and management autonomy, transparency and accountability, demand responsiveness, cost-effective design and operation, and professional capacity, as well as elements of competition and the ability to expand.

Monitoring water supply and sanitation. Before 2000 coverage data came primarily from government water and sanitation agencies. There were no common definitions for access to safe water and basic sanitation, so data were not comparable. Since 2000 the Joint Monitoring Programme[18] has reported on water supply and sanitation using population-based data gathered through household surveys and national censuses (box 7.3). Its assessments categorize water supply and sanitation technologies as 'improved' or 'unimproved'.

The Joint Monitoring Programme defines 'access to drinking water' as the availability of at least 20 litres of drinking water per person per day within 1 km of the dwelling (a 30 minute round-trip journey). In urban areas the distance to a source is usually not a problem, and in such densely populated areas a water-hauling trip of 30 minutes or less, including queuing time, is a more appropriate indicator of access. 'Safe' drinking water is water that meets accepted quality standards and poses no significant health risks. But determining the microbiological safety of drinking water in each household is impractical, and it is therefore assumed that water is safe if it comes from an 'improved' source such as piped water, a protected well or spring, or rainwater.

Figure 7.3 Regional and global water supply coverage, by urban and rural areas, 1990 and 2006

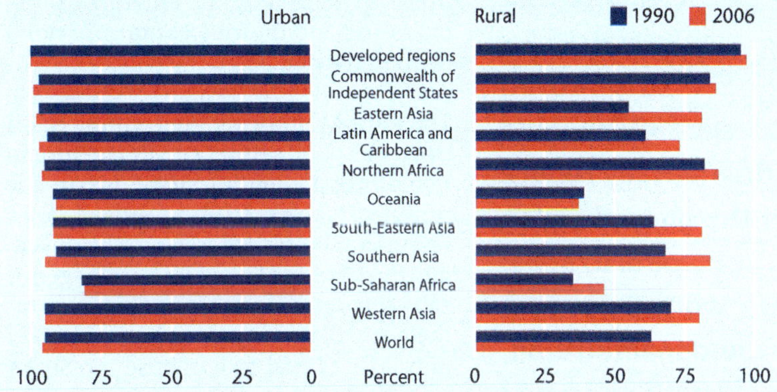

Note: For charting progress towards the Millennium Development Goals, the United Nations has classified countries into three regions: developed regions, countries in the Commonwealth of Independent States, and developing regions. The developing regions are further divided into subregions.
Source: WHO and UNICEF 2008b.

Figure 7.4 Regional and global sanitation coverage, by urban and rural areas, 1990 and 2006

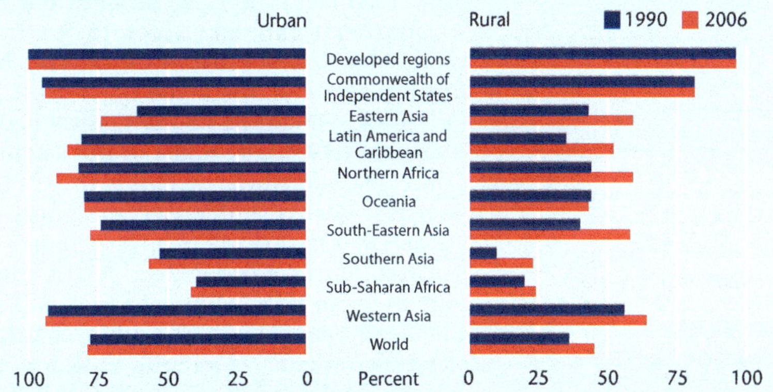

Note: For charting progress towards the Millennium Development Goals, the United Nations has classified countries into three regions: developed regions, countries in the Commonwealth of Independent States, and developing regions. The developing regions are further divided into subregions.
Source: WHO and UNICEF 2008b.

Box 7.3 Rapid assessment of drinking water quality

The World Health Organization (WHO) and the United Nations Children's Fund (UNICEF) have developed the Rapid Assessment of Drinking Water Quality (RADWQ) survey method, which they pilot-tested in China, Ethiopia, Jordan, Nicaragua, Nigeria and Tajikistan. It uses cluster sampling across a country to select individual drinking water sources for testing. The parameters to be tested depend on the extent of the survey and on local health hazards. The output is a snapshot of drinking water quality for each improved source tested.

The study in Tajikistan involved 1,620 samples in 53 clusters over a six-month period during 2004/05. The samples were taken from utility piped supplies and protected springs. Overall, 87% of the sites complied with the WHO guideline value for microbiological quality and arsenic, fluoride and nitrate concentrations.

Source: Aliev et al. 2006.

Despite strong epidemiological evidence that universal access to improved sanitation would dramatically reduce the global burden of diarrhoea, worm infections and malnutrition, sanitation receives substantially less priority and funding than water supply in virtually every country

While the assumption that improved sources are also safe generally holds with respect to fæcal contamination, it has been challenged by the emerging problem of arsenic and fluoride contamination of drinking water.

The Joint Monitoring Programme defines 'access to basic sanitation' as the proportion of the population (total, urban and rural) with access to an improved sanitation facility for defecation. Improved means that human excreta are hygienically separated from human contact or the immediate environment, essential for preventing fæcal-oral disease transmission.

Although the formulation of common definitions was a major step, controversies remain regarding the availability and quality of data sources. The global coverage of improved drinking water in urban areas is estimated at 95%,[19] but that figure seems difficult to reconcile with the reality of millions of people in slums in developing countries. These people might be connected to a piped water supply system and therefore benefit from an 'improved' technology according to the Joint Monitoring Programme criteria. In reality, however, the piped water supply system might be poorly maintained, be inoperable many hours of the day or provide polluted water. This is the controversy of 'served' and 'unserved'. Another problem is that many people remain unaccounted for in the data. Again, these are the poor people living in informal settlements that are not officially recognized by governments. As a result, millions of people are probably missing from the national statistics.[20]

Three key issues need to be addressed in monitoring programmes for water and sanitation services: access, quality and sustainability. A challenge is developing appropriate indicators that can be used in household surveys to collect information about disparities in access, affordability, per capita use and the sustainability and reliability of services. Also needed are simple and inexpensive water quality tests as a cross-check on the safety of improved drinking water sources and on safety at point of use.

Monitoring has traditionally focused on the household, overlooking the fact that household members are outside the house for most of the day. Across developing countries there are 600 million children of primary school age, and roughly half attend schools that lack safe drinking water and sanitation. In two of the pilot countries for the *UN-Water Global Annual Assessment of Sanitation and Drinking-Water (GLAAS): 2008 Pilot Report,* average sanitation coverage is 26% in schools and 75% in hospitals.[21] Sanitation in schools has indirect benefits as well as health benefits. In particular, separate facilities for boys and girls have been shown to stimulate girls' school attendance and to lead to greater attention in class by all pupils.

Putting sanitation in the spotlight. Despite strong epidemiological evidence that universal access to improved sanitation would dramatically reduce the global burden of diarrhoea, worm infections and malnutrition, sanitation receives substantially less priority and funding than water supply in virtually every country.[22]

Human Development Report 2006 analysed the sanitation problems facing the world and the reasons why sanitation lags behind water supply in attention and resources allocated.[23] The main barrier has been political reluctance. Sanitation is usually a low-priority item in national policy-making, planning, budgeting and implementation. As a result, it is often delegated to the lowest level of governance – struggling municipalities. There are, however, encouraging signs of change. In response to findings that in 2006, 62% of Africans lacked access to an improved sanitation,[24] 32 African ministers signed the eThekwini Declaration in February 2008 pledging to create separate budget lines for sanitation and hygiene and to commit at least 0.5% of GDP.[25]

Significantly, five countries along with private sector and voluntary sources have made contributions totalling $60 million to the Global Sanitation Fund launched in March 2008 by the Water Supply and Sanitation Collaborative Council and its partners – with an annual target of $100 million – to help meet the Millennium Development Goal sanitation target.

Trends in water and sanitation provision

Before the 1990s water industries were national monopolies in many countries. Since then, major water service reforms have taken place, mostly from centralized to decentralized public provision.[26] Asian countries, including Indonesia, Pakistan and the Philippines, undertook radical decentralization programmes. In many Latin American countries (for example, Argentina, Chile, Colombia, Panama and

Peru) national monopolies were broken into hundreds of municipal providers as part of a wider process of devolution across all areas of government. Rapid decentralization after the political turnaround in Eastern Europe and Central Asia devolved responsibilities to lower tiers of government, but financial means and capacity remained mainly at the central level. In Africa the picture is mixed, with many countries remaining centralized while some (such as Ethiopia and Tanzania) are decentralizing rapidly.

Decentralization was not a studied response to the specific problems of the water sector, but rather a by-product of wider state reforms. As a result, local governments often found themselves in charge of service delivery while lacking the capacity to fulfil this function. Private sector participation has proved difficult to implement. In larger urban centres this has been primarily for political reasons, while in smaller cities and rural areas economic viability is an additional problem. Thus, the real transition for most water consumers has not been from public to private, but rather from unregulated centralized public provision to regulated decentralized public provision. Today, most urban and peri-urban areas in the world are served by publicly owned and managed utilities, a model that is likely to continue.[27]

In many developing countries public utilities do not perform well because of low motivation, poor management, inadequate cost recovery and political interference. Public sector reform is one of the most important avenues for sustaining and increasing coverage and service. In addition, encouraging public sector utilities to extend services to informal urban settlements (in partnership with citizens groups or informal private sector operators) is a priority in cities where slum populations account for a large share of the urban population.

Lessons from the sanitation sector call for a shift in thinking:

- Technological solutions are available to satisfy almost any requirement within the sector, but technology interventions are very unlikely to succeed on an upscalable and sustainable basis without assessment of their relevance and full local social, political and economic ownership.
- To succeed, technology interventions must incorporate carefully considered approaches ('software') as well as technologies ('hardware'). Implicit is the realization that technologies should be demand-driven and that demand for certain technologies can be stimulated more effectively by using appropriate approaches – for example, through social marketing or the development of entrepreneurial local capacities.
- Providing the technology is not enough. The health gains of universal access to basic sanitation accrue only if people use the sanitary facilities properly and practice proper hygiene. This means that hygiene promotion and social marketing are always needed along with technology provision.[28]
- Household decision-making is crucial. Because behaviour change is central to achieving health gains from sanitation, service providers need to focus on this level. This means that hygiene promotion is central to any sanitation strategy and that the technology should be appropriate. In simple terms, it is no good promoting a type of toilet that people will not use.
- Change is needed across the entire community. While household behaviour is critical, individual households alone may not be able to influence health outcomes. For the majority of households in poor rural settings or overcrowded urban settlements, the actions of the community as a whole have great influence. Excreta need to be removed from the environment in which children play and adults labour. For this reason, interventions at the household level need to be coordinated across the community.
- Public and private (personal, household) benefits of sanitation need to be in balance. The public nature of sanitation remains important (primarily environmental protection and public health). Despite calls to scale down all public provision, it is not feasible for households or local communities to take responsibility for wider societal concerns. Governments have to find pragmatic ways of balancing local and household needs with wider societal ones. Linking household service provision with community-level planning can be vital in creating local mechanisms to achieve this balance.[29]

We need to transform the way we look at wastewater, recognizing it as a resource

In many developing countries public utilities do not perform well because of low motivation, poor management, inadequate cost recovery and political interference

Steadily increasing demand for agricultural products to satisfy the needs of a growing population is the main driver behind agricultural water use

rather than a problem, and manage it accordingly. Its various possible uses, such as farming, aquaculture, gardening and forest planting, need to be planned, and risk mitigation measures put in place to avoid the health costs (see chapter 8).

While technically and financially appropriate sanitation services are available, little is known about community perceptions, demand and acceptability of different sanitation solutions. There are many taboos surrounding defecation behaviour, making it difficult to study. Sanitation services, far more than water supply, must be adapted to the local situation to ensure that they are used by men, women and children. Sociocultural factors are fundamental to sustainability. Women need to be consulted when toilets are built and must be allowed to manage their sanitation facilities.

An example of the shift from supply-led to demand-driven approaches is the community-led total sanitation campaign in South Asia, which seeks to end open defecation by highlighting the problems caused to everyone in the community. It also ensures that every household either builds and uses its own low-cost toilet or has access to a shared toilet. In Bangladesh the total sanitation campaign was begun by local non-governmental organizations and has since been scaled up into a national programme, and in India the Sulabh Sanitation Movement has successfully scaled up a non-governmental organization model,[30] but elsewhere it has proved difficult for governments to scale up similar successes. The 2008 pilot survey of the UN-Water Global Annual Assessment of Sanitation and Drinking-Water (GLAAS) suggests that there is a need to reinforce education programmes with actions that provide better hygiene and sanitation in schools and hospitals.[31]

Water use in agriculture

Steadily increasing demand for agricultural products to satisfy the needs of a growing population is the main driver behind agricultural water use. Although population growth has slowed since the 1970s, economic development, in particular in emerging market economies, is translating into demand for a more varied, water-intensive diet, including meat and dairy products (see chapter 2 and box 7.4). To meet these future food needs, pressure to develop new supply sources or increase water allocation to agriculture will continue. This challenge was highlighted in the first two *United Nations World Water Development Reports*, and its implication for an emphasis on water for agriculture was clearly drawn by the International Water Management Institute[32] and the World Bank.[33] Climate change and the recent acceleration in biofuel production bring new challenges to agriculture and put further pressure on land and water resources. In a tighter global food market, where an increasing number of major agricultural systems are reaching the limits of their productive capacity, climate events increasingly influence food prices, with devastating social and humanitarian consequences.

Why is so much water needed for food production?

Agriculture accounts for 70% of freshwater withdrawals from rivers, lakes and aquifers – up to more than 90% in some developing countries. Furthermore, unlike in industrial and domestic uses, where most of the water returns to rivers after use, in agriculture a large part of water is consumed by evapotranspiration. Many irrigation systems, however, return a large amount of water to the system after use.

Biomass cannot be produced without water. The source of all food is photosynthesis, a process by which plants transform energy captured from the sun, carbon dioxide from the air and minerals from the ground into biomass. Water, stored in the soil, is pumped by the roots and transpired into the atmosphere through the leaves. Transpiration cools the leaves and enables mass flows of mineral nutrients and water from roots.

Biomass is processed through the food chain, which describes the flow of energy and feeding relationship between species: from primary producers (plants) to herbivores to carnivores. Despite substantial progress in agricultural research, energy flow efficiency in the food chain remains extremely low: about 10% for herbivores and 20% for carnivores. About 10 kilocalories (kcal) of grass are needed to produce 1 kcal of beef (box 7.4).

Rainfed agriculture covers 80% of the world's cultivated land, and is responsible for about 60% of crop production. In rainfed agriculture the soil stores the rain and releases it slowly to the plants. Rainwater used in agriculture, part of what is called 'green water', is a characteristic of the land on which it falls and is not usually subject to competition from other sectors.

Box 7.4 How much water is needed to produce food for a single day?

We can estimate how much water is needed to sustain our diets by calculating the water lost in evapotranspiration based on crop physiology. Depending on local climate, varieties and agronomical practices, it takes 400-2,000 litres of evapotranspiration daily to produce 1 kilogram (kg) of wheat, and 1,000-20,000 litres per kilogram of meat, depending on the type of animal, feed and management practices. Based on these values, researchers have estimated daily water requirements to support diets, ranging from 2,000 to 5,000 litres of water per person per day. The Food and Agriculture Organization of the United Nations uses 2,800 kilocalories (kcal) per person at the national level as a threshold for food security. As a rule of thumb, it can therefore be estimated that 1 litre of water is needed to produce 1 kcal of food. Because of the low energy efficiency of the food chain, protein-rich diets require substantially more water than vegetarian diets.

Value produced from a unit of water for selected commodities

Product	Water productivity: Kilograms per cubic metre	Dollars per cubic metre	Protein grams per cubic metre	Kilocalories per cubic metre
Cereal				
Wheat ($0.2 per kilogram)	0.2-1.2	0.04-0.24	50-150	660-4,000
Rice ($0.31 per kilogram)	0.15-1.6	0.05-0.18	12-50	500-2,000
Maize ($0.11 per kilogram)	0.30-2.00	0.03-0.22	30-200	1,000-7,000
Legumes				
Lentils ($0.3 per kilogram)	0.3-1.0	0.09-0.30	90-150	1,060-3,500
Fava beans ($0.3 per kilogram)	0.3-0.8	0.09-0.24	100-150	1,260-3,360
Groundnut ($0.8 per kilogram)	0.1-0.4	0.08-0.32	30-120	800-3,200
Vegetables				
Potatoes ($0.1 per kilogram)	3-7	0.3-0.7	50-120	3,000-7,000
Tomatoes ($0.15 per kilogram)	5-20	0.75-3.0	50-200	1,000-4,000
Onions ($0.1 per kilogram)	3-10	0.3-1.0	20-67	1,200-4,000
Fruits				
Apples ($0.8 per kilogram)	1.0-5.0	0.8-4.0	Negligible	520-2,600
Olives ($1.0 per kilogram)	1.0-3.0	1.0-3.0	10-30	1,150-3,450
Dates ($2.0 per kilogram)	0.4-0.8	0.8-1.6	8-16	1,120-2,240
Others				
Beef ($3.0 per kilogram)	0.03-0.1	0.09-0.3	10-30	60-210
Fish (aquaculture[a])	0.05-1.0	0.07-1.35	17-340	85-1,750

a. Includes extensive systems without additional nutritional inputs to superintensive systems.
Source: Based on Comprehensive Assessment of Water Management in Agriculture 2007.

In irrigation, by contrast, water is extracted from rivers, lakes and aquifers (also called 'blue water') and applied on land, where most of it is consumed by evapotranspiration. Irrigation thus competes with other sectors (including the environment) for blue water. The importance of irrigated and rainfed agriculture varies across regions (map 7.4), for the most part following climate patterns. While the contribution of irrigation to total crop production is modest (about 10%-20%), it allows crop growth in many permanently water-scarce or temporarily water-stressed environments.

More efficient use of water – higher socioeconomic returns and more crops per drop – can be obtained primarily through intensification (improved crop varieties plus better agronomic practices). Over the last 40 years major food yields have increased progressively and crop water productivity has doubled. However, yields in rainfed agriculture are still far from their potential (figure 7.5). Opportunities therefore exist to contain future increases in water use in agriculture by reducing the yield gap. In 2005 cereal yields were about 1-1.5 tonnes per hectare in sub-Saharan Africa, compared with 5 tonnes per hectare in Europe.[34] However, where land or water constrain future development, the yield gap is closing rapidly, leaving little prospect for easy improvement. China and Egypt, for instance, are close to realizing their maximum potential for major food crops.

Trends and current situation of water use in agriculture

The last 50 years have seen rapid acceleration in water resources development for agriculture.[35] Development in hydraulic infrastructure (dams and large-scale public surface irrigation), as well as private and

Map 7.4 **Relative importance of rainfed and irrigated agriculture**

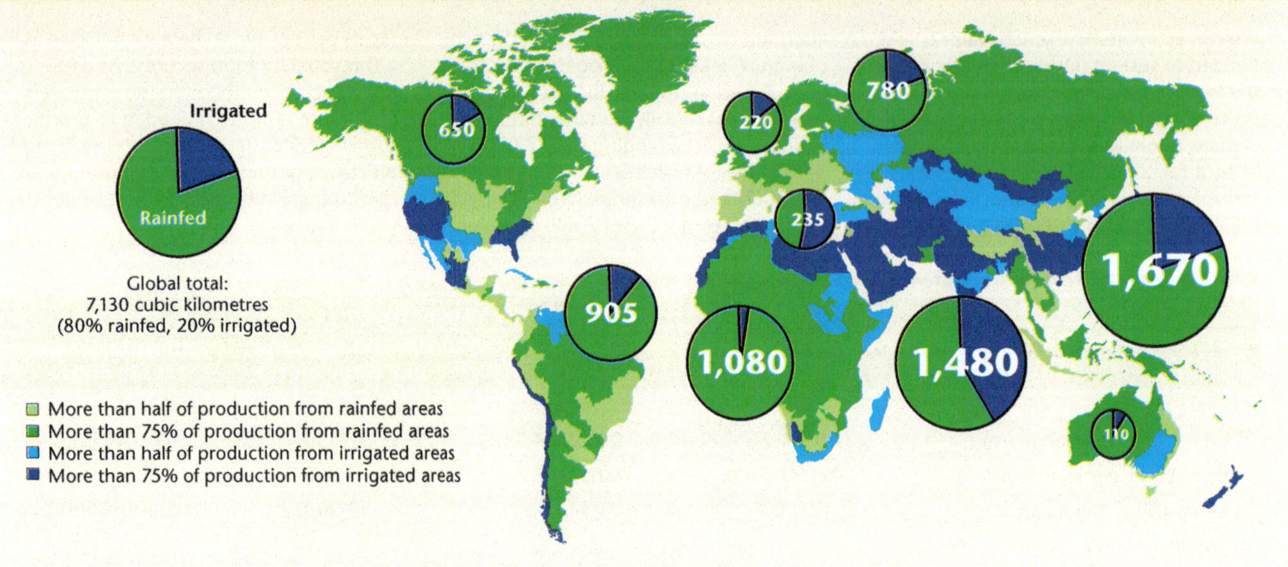

Source: Based on Comprehensive Assessment of Water Management in Agriculture 2007.

Figure 7.5 **Gaps are large between farmer's actual yields and achievable yields for major rainfed cereal crops**

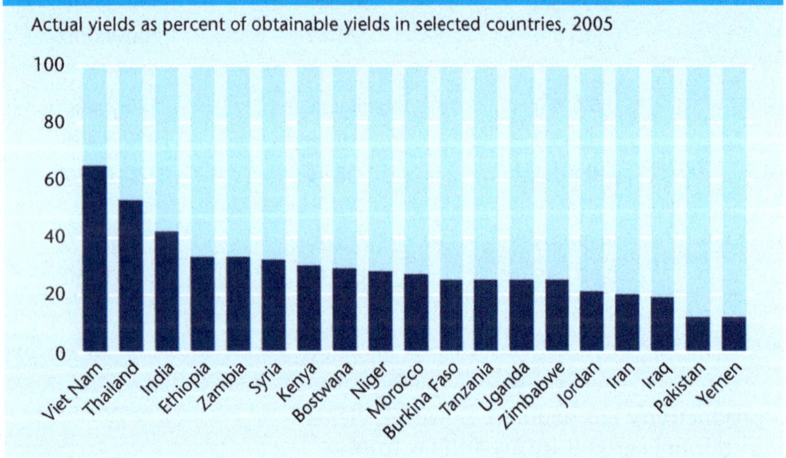

Source: Based on Comprehensive Assessment of Water Management in Agriculture 2007.

community schemes (particularly groundwater pumping; see chapter 8), have put water at the service of populations as part of the global effort to rapidly increase staple food production, ensure food self-sufficiency and avoid famines. As the global population grew from 2.5 billion in 1950 to 6.5 billion at the beginning of the 21st century, food production growth outstripped population growth,[36] irrigated area doubled (particularly in Asia), and water withdrawals tripled. Today, irrigated agriculture covers 275 million hectares – about 20% of cultivated land – and accounts for 40% of global food production (map 7.5).

This success in agricultural production led to a 30-year decline in food prices in most countries (figure 7.6), a trend that lasted until very recently. In real terms, food prices declined, until recently, to their lowest levels in history, so that consumers in many countries could eat better while spending less of their budget on food. Today, food supply accounts for a very small part of household income in rich countries, but it can constitute as much as 80% of income of poor people in developing countries.

Declining food prices, high agricultural productivity, improved trade and markets and progressive reduction in the risk of food shortage and famines also led to reduced investment in agriculture, particularly in irrigation, resulting in neglect of maintenance of public irrigation schemes and a sharp slowdown in the growth of irrigated agriculture.

Future water demand in agriculture

As population continues to grow and demand for food intensifies, pressure on water from agricultural activities are expected to increase, although at a slower pace. Agriculture is also facing competition for water from other sectors, and its allocation is decreasing in water-scarce areas, especially around urban centres.

Growth in world demand for food will mirror population growth, progressively declining from 2.2% a year in the last decades of the 20th century, to 1.6% in 2015, 1.4% in 2015-30, and 0.9% in 2030-50.[37] However, these global figures hide extremely large variations, with developing countries growing faster than

Map 7.5 **Percentage of cultivated areas equipped for irrigation, around 2003**

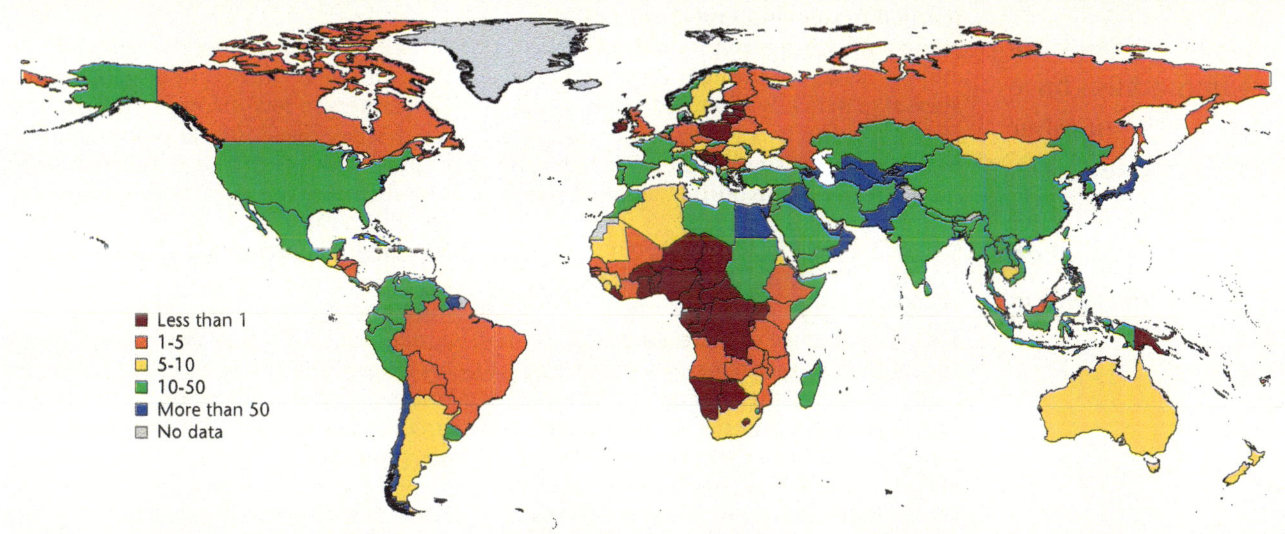

Source: FAO-AQUASTAT.

developed countries. Niger, the country with the highest population growth rate, is expected to grow from 10.7 million in 2000 to 53 million in 2050. Countries with high population growth rates and limited agricultural resources will likely see their food deficit increase, with serious implications for economic and food security.

The increasing number of areas where water has become a limiting factor for irrigated agriculture, associated with rising claims for releasing water to guarantee or restore environmental services, has tightened food production in some regions. The Middle East, for example, can no longer satisfy its food requirements and relies increasingly on food imports.

Part of the current pressure on water resources comes from increasing demands for animal feed (figure 7.7). Meat production requires 8-10 times more water than cereal production. With the increase in consumption of dairy products and meat, the production of feed grains has expanded rapidly, at the expense of other crops. With rising living standards and urbanization, consumption of meat and dairy products will continue to rise (it has more than tripled in China over past decades).

The latest projections available show an average increase of 0.6% a year in irrigated land from 1998 until 2030, compared with 1.5% over the 1950s-1990s. In the same period (1998-2030), because of continued increases in agricultural productivity, 36% more food will be produced

Figure 7.6 **As irrigation area expanded, food price fell for 30 years before starting to rise again**

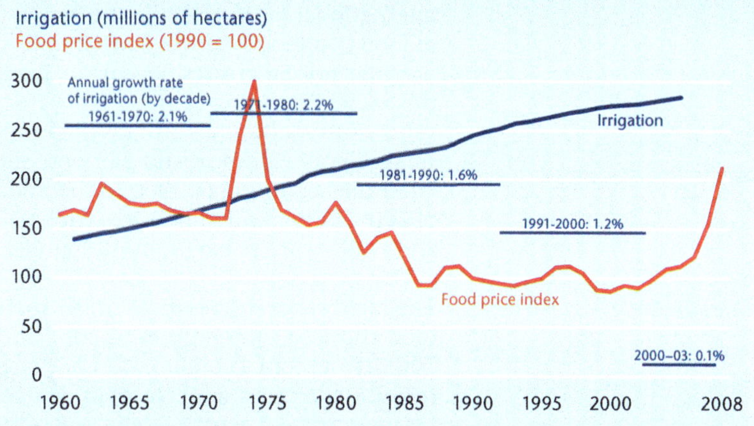

Source: Based on Comprehensive Assessment of Water Management in Agriculture 2007; FAO FAOSTAT.

Figure 7.7 **Feed demand drives future demand for grains**

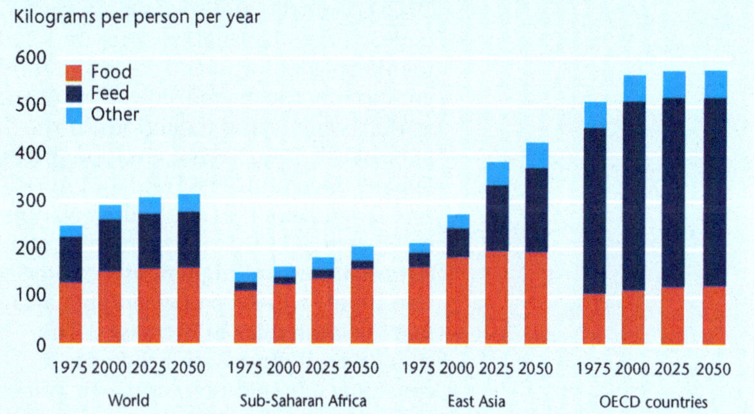

Source: Based on Comprehensive Assessment of Water Management in Agriculture 2007.

Recent increases in the prices of the main agricultural commodities have caused the number of people suffering from hunger to rise from 850 million to 963 million

with 13% more water.[38] Crop and animal breeding and biotechnology have already resulted in tremendous gains in yields, along with savings in production costs and pesticide use through improved resistance of genetically modified crops. Common grains such as wheat, maize and rice, which achieved significant gains from the 1960s to 1980s, are unlikely to see further gains.

The implications of food prices for food security

The old challenge of increasing and securing food supply remains a priority in many countries as the number of people suffering from hunger remains desperately high, most of them in rural areas of South Asia and sub-Saharan Africa.[39] Recent increases in the prices of the main agricultural commodities have caused the number of people suffering from hunger to rise from 850 million to 963 million. Between September 2007 and March 2008 the price of wheat, corn, rice and other cereals rose an average of 41% on the international market. While the increase in food commodity prices started in 2000, previous price increases have never been this rapid. The rising demand for high-value commodities has also resulted in surging prices for meat and dairy products. From the beginning of 2000 to the middle of 2008, butter and milk prices tripled, and poultry prices have almost doubled.

The rapid surge in the prices of the main food and feed commodities in 2007 and 2008 arose from a combination of causes, including long-term increases in demand for meat and dairy products in emerging market economies, a progressive reduction in the stocks of the main commodities and unfavourable climate in some of the largest exporting countries. The effects of these factors have likely been amplified by incentives for bioenergy production in OECD countries and speculation on food trade.[40] Prices have fallen since mid-2008 thanks to good prospects for world food production, the overall slowdown of the world economy and reductions in the price of oil. Future trends in food prices remain uncertain. The Food and Agriculture Organization of the United Nations (FAO) estimates that in the medium term tight markets and higher costs of production inputs will keep prices higher and more volatile than in the past.[41] This situation will hurt both producers and consumers. In addition, domestic prices of food in developing countries have not followed downward trends from the international market, and the prices of major staple commodities remain high in many places.

The effects of price increases on consumption vary by country and consumer group. Consumers in low-income countries are much more responsive and vulnerable to price changes than are consumers in high-income countries, because food expenditure can represent 50%-75% of their income.[42] Surges in food prices thus hurt the poorest populations the most. Should food prices remain high, investment in agriculture, including water development for irrigation, is likely to grow. Higher food prices may represent an opportunity for smallholder farmers if the right policies are adopted.[43]

How will bioenergy affect agricultural water use?

Bioenergy is energy derived from biological origins, such as grains, sugar crops, oil crops, starch, cellulose (grasses and trees) and organic waste. Liquid biofuel (bioethanol and biodiesel), while representing only a small percentage of all bioenergy products, currently dominates the debate because of its capacity to substitute for fossil fuel and because most of its source feedstock can also be used to produce food (see chapter 3).[44]

Current and projected trends in bioenergy. Around 10% of the total energy supply comes from biomass, and most of that (80%) comes from the 'traditional' biomass sources of wood, dung and crop residues. These represent a significant part of the energy used in many developing countries. Of commercial or 'modern' bioenergy, two-thirds is produced from fresh vegetable material and organic residue used to produce electricity and heat. About 5% of biomass is used to produce liquid biofuel for transport, which currently accounts for less than 2% of transport energy worldwide.

The quest for greater energy autonomy, the rise in oil prices until the second half of 2008 and concerns about the impacts of greenhouse gas emissions in OECD countries are behind a recent surge in transport bioenergy.[45] The production of bioethanol, from sugarcane, corn, sugar beet, wheat and sorghum, tripled between 2000 and 2007 to an estimated 77 billion litres in 2008.[46] Brazil (using sugarcane) and the United States (using mostly maize) are the main producers, accounting for 77% of global supply. Biodiesel production, derived from oil- or tree-seeds such as rapeseed, sunflower, soybean, palm oil, coconut or jatropha, increased 11-fold between 2000 and 2007, with 67% produced in the European Union.

In 2007 approximately 23% of maize production in the United States was used to produce ethanol, as was about 54% of Brazil's sugarcane crop. In the European Union about 47% of vegetable oil produced was used in the production of biodiesel, necessitating higher imports of vegetable oil to meet domestic consumption needs. In energy equivalence the 2008 ethanol share of the gasoline transport fuel market in these economies was estimated at 4.5% for the United States, 40.0% for Brazil and 2.2% for the European Union. The biodiesel share of the diesel transport fuel market was estimated at 0.5% for the United States, 1.1% for Brazil and 3.0% for the European Union.[47]

The international policy environment, national policy support and oil prices will strongly influence future demand for biofuel. Global ethanol production is projected to increase rapidly to 127 billion litres in 2017, with production concentrated in the United States, Brazil and, to a lesser extent, the European Union and China. Global biodiesel production is expected to reach 24 billion litres in 2017.[48]

The global potential of conventional biofuel is limited by the availability of suitable land and water for crops and the high cost of most conventional technologies. Technically, up to 20 exajoules from conventional ethanol and biodiesel, meeting 11% of total demand for liquid fuels in the transport sector, could be possible by 2050.[49] Energy yield is highest for feedstock grown in tropical conditions, in particular, sugarcane and palm oil.

Implication of increased crop demands for land, water and the environment. The potential impact of biofuel production on land and water resources varies with local agroclimatic conditions and policies. The potential impact on freshwater resources is greatest where agricultural production depends on irrigation and is practically negligible where rainfed production is practiced. Where agriculture requires irrigation, increased production of biofuel could result in reduced water allocation to other crop commodities.

Globally, irrigation water allocated to biofuel production is estimated at 44 km^3, or 2% of all irrigation water.[50] Under current production conditions it takes an average of roughly 2,500 litres of water (about 820 litres of it irrigation water) to produce 1 litre of liquid biofuel (the same amount needed on average to produce food for one person for one day). But regional variations can be substantial, depending primarily on the relative percentage of irrigation in biofuel crop production. The share of irrigation water used for biofuel production is negligible in Brazil and the European Union and is estimated to be 2% in China and 3% in the United States.[51] In India, where sugarcane is fully irrigated, nearly 3,500 litres of water are withdrawn for each litre of ethanol produced. The markets for biofuel and agricultural products are strongly meshed. Because of crop substitutability, all crops tend to compete for the same inputs, land, fertilizers and irrigation water, and farmers select crops that offer the best return on their investment.[52]

Implementing all current national biofuel policies and plans would take 30 million hectares of cropland and 180 km^3 of additional irrigation water. Although globally less than a few percentage points of total area and water use, the impacts could be large for some countries, including China and India, and for some regions of large countries, such as the United States. There could also be significant implications for water resources, with possible feedback into global grain markets. The volume of water and area of land used for biofuel production depend on the crop and the agricultural system (table 7.2). Private investors are showing increasing interest in land and irrigated schemes in Africa for agricultural products for biofuel production.

According to the OECD, growth of the bioenergy industry is likely to place additional pressure on the environment and biodiversity.[53] The potential of bioenergy to mitigate climate change is complex and varies by type of crop and farming system. Among current technologies only ethanol produced from sugarcane in Brazil, ethanol produced as a by-product of cellulose production (as in Sweden and Switzerland) and biodiesel produced from animal fats and used cooking oil can substantially reduce greenhouse gas emissions compared with gasoline and mineral diesel. The study concludes that all other conventional bioenergy technologies typically deliver greenhouse gas emissions reductions of less than 40% compared with their fossil fuel alternatives. When impacts such as soil acidification, fertilizer use, biodiversity loss and the toxicity of agricultural pesticides are taken into account, the adverse environmental impacts

The international policy environment, national policy support and oil prices will strongly influence future demand for biofuel

Table 7.2 Different types of biofuel and quantity of water needed to produce them in rainfed or irrigated conditions

Crop	Fuel product (energy density: biodiesel 35 megajoules per litre; ethanol 20 megajoules per litre)	Annual obtainable yield (litres per hectare)	Rainfed or irrigated	Evapotranspiration (litres per litre of fuel)	Irrigation water withdrawn (litres per litre of fuel)
Sugarcane	Ethanol (from sugar)	6,000	Irrigated	2,000	1,000
Sugar beet	Ethanol (from sugar)	7,000	Irrigated	786	571
Cassava	Ethanol (from starch)	4,000	Rainfed	2,250	na
Maize	Ethanol (from starch)	3,500	Irrigated	1,360	857
Oil palm	Biodiesel	5,500	Rainfed	2,360	na
Rapeseed/ mustardseed	Biodiesel	1,200	Rainfed	3,330	na
Soybean	Biodiesel	400	Rainfed	10,000	na

na is not applicable.
Note: Values are indicative only.
Source: Hoogeveen, Faurès, and van de Giesse forthcoming, adapted from Müller et al. 2008.

of ethanol and biodiesel can exceed those of petrol and mineral diesel. An exception was biofuel produced from woody biomass, which rated better than gasoline. A key question is how to ensure that production will be sustainable. One answer being explored is certification of conformity to a set of environmental and social standards on a life-cycle basis.[54]

More uncertainty for agriculture under climate change

The issues of agricultural production are complicated by increasing climate uncertainty. The relationship between agriculture and climate change is complex. Agriculture contributes to global warming through emissions of methane and nitrous oxide. Changes in land use practices (management of cropland and grazing land) are considered to be the best mitigation options.[55] Agriculture is also extremely sensitive to climate change, and it is anticipated that large areas of croplands, in particular in semi-arid zones, will need to adapt to new conditions with lower precipitation.

Climate change is expected to alter hydrologic regimes and patterns of freshwater resource availability (see chapter 5), with impacts on both rainfed and irrigated agriculture.[56] Projections converge in indicating a reduction in precipitation in semi-arid areas, greater variability in rainfall distribution, greater frequency of extreme events and rising temperature, particularly affecting agriculture in low latitudes. Severe reductions in river runoff and aquifer recharge (see chapter 11) are expected in the Mediterranean basin and in the semi-arid areas of Southern Africa, Australia and the Americas, affecting water availability for all uses (box 7.5).

The projected increase in the frequency of droughts and floods will hurt crop yields and livestock, with greater impacts coming earlier than previously predicted.[57] While climate change does not seem to threaten global food production, it will alter the distribution of agricultural potential. Most of the increase in cereal production will be concentrated in the Northern Hemisphere, while more frequent and severe droughts and floods will hurt local production, especially in subsistence sectors at low

Box 7.5 Coping with water scarcity and climate change in agriculture in the Near East

Arid and semi-arid conditions and widespread water scarcity prevail in the Near East Region. Agricultural production is projected to grow more than 60% between 2003/05 and 2030 and to more than double by 2050 as a result of increased food demand. Most of the increase will come from yield increases and higher cropping intensities.

Irrigation is crucial, with about 80% of production originating from irrigated agriculture and with irrigated land accounting for about a third of arable land. Irrigation water withdrawals could increase 29% by 2050. Under pressures of water scarcity water use efficiency is expected to improve from 52% in 2003/05 to 66% by 2050, as irrigation water requirements grow from 64% to 83% of renewable water resources – all very high values compared with global averages.

The situation may become critical if the expected impacts of climate change are considered, including the combined effect of changes in precipitation and evapotranspiration. These changes will alter current patterns of soil moisture deficits, groundwater recharge and runoff. Second-order impacts on streamflow, groundwater and lake and dam storage levels will translate into reduced availability of water for irrigation and other purposes. Under the Intergovernmental Panel and Climate Change B2 scenario the overall availability of renewable water resources may fall from 416 km^3 in the base situation to 397 km^3 in 2050, while irrigation water withdrawals would need to raise an additional 20 km^3. Total water withdrawals would represent the equivalent of 92% of the region's renewable water resources – even higher if the leaching requirements of agricultural areas affected by saltwater intrusion and leakages from brackish aquifers are considered.

Source: FAO 2008b; IPCC 2000.

latitudes.[58] Several densely populated farming systems in developing countries are at risk from the impacts of climate change. A combination of reduced river base flows, increased flooding and rising sea levels is expected to impact highly productive irrigated systems that help maintain the stability of cereal production.

The production risks will be amplified in alluvial plains dependent on glacier melt (Colorado, Punjab) and, in particular, in lowland deltas (the Ganges and Nile). Table 7.3 shows the expected impact of climate change on some major agricultural systems and analyses their vulnerability and adaptive capacities.

Table 7.3 Typology of climate change impacts on major agricultural systems

System	Current status	Climate change drivers	Vulnerability	Adaptability
Snow melt systems				
Indus	Highly developed, water scarcity emerging. Sediment and salinity constraints	20-year increasing flows followed by substantial reductions in surface water and groundwater recharge. Changed seasonality of runoff and peak flows. More rainfall in place of snow. Increased peak flows and flooding. Increased salinity. Declining productivity in places.	Very high (run of river), medium high (dams)	Limited possibility for adaptation (all infrastructure already built)
Ganges-Brahmaputra	High potential for groundwater, established water quality problems. Low productivity		High (falling groundwater tables)	Medium (still possibilities for groundwater development)
Northern China	Extreme water scarcity, high productivity		High (global implications, high food demand with great influence on prices)	Medium (adaptability is increasing due to increasing wealth)
Red and Mekong Rivers	High productivity, high flood risk, poor water quality		Medium	Medium
Colorado River	Water scarcity, salinity		Low	Medium, excessive pressure on resources
Deltas				
Ganges-Brahmaputra	Densely populated. Shallow groundwater, extensively used. Flood adaptation possible; low productivity	Rising sea level. Storm surges, and infrastructure damage. Higher frequency of cyclones (East and South-East Asia). Saline intrusion in groundwater and rivers. Increased flood frequency. Potential increase in groundwater recharge.	Very high (flood, cyclones)	Poor except salinity
Nile River	Highly dependent on runoff and Aswan Storage – possibly sensitive to upstream development		High (population pressure)	Medium
Yellow River	Severe weather scarcity		High	Low
Red River	Currently adapted but expensive pumped irrigation and drainage		Medium	High, except salinity
Mekong River	Adapted groundwater use in delta; sensitive to upstream development		High	Medium
Semi-arid and arid tropics: limited snow melt and limited groundwater				
Monsoonal: Indian subcontinent	Low productivity. Overdeveloped basin (surface water and groundwater)	Increased rainfall. Increased rainfall variability. Increased drought and flooding. Higher temperature.	High	Low (surface irrigation); medium (groundwater irrigation)
Non-monsoonal: sub-Saharan Africa	Poor soils. Flashy water systems. Overallocation of water and population pressure in places. Widespread food insecurity	Increased rainfall variability. Increased frequency of droughts and flooding.	Very high. Declining yields in rainfed systems. Increased volatility of production	Low
Non-monsoonal: Southern and Western Australia	Flashy water systems. Over-allocation of water. Competition from other sectors	Lower rainfall, higher temperature. Decreasing runoff.	High	Low

(continued)

Table 7.3 Typology of climate change impacts on major agricultural systems (continued)

System	Current status	Climate change drivers	Vulnerability	Adaptability
Humid tropics				
Rice: Southeastern Asia	Surface irrigation. High productivity but stagnating.	Increased rainfall. Marginally increased temperatures. Increased rainfall variability and occurrence of droughts and floods.	High	Medium
Rice: Southern China	Conjunctive use of surface water and groundwater. Low output compared with Northern China		High	Medium
Mediterranean				
Southern Europe	Increasing pressure on water	Significantly lower rainfall and higher temperatures. Increased water stress. Decreased runoff. Loss of groundwater reserves.	Medium	Low
Northern Africa	High water scarcity		High	Low
West Asia	Heavy pressure on water		Low	Low
Small islands				
Small islands	Fragile ecosystems. Groundwater depletion	Seawater rise. Saltwater intrusion. Increased frequency of cyclones and hurricanes.	High	Variable

Source: FAO 2008b.

In key food-insecure areas, dominated by rainfed agriculture (sub-Saharan Africa and peninsular India, in particular), anticipated reductions in production may have multiple impacts, including loss of livelihoods and displacement of rural populations. This will accentuate demand in global markets and put further pressure on irrigated production.

Changes in runoff affect water availability in rivers and aquifers, placing an additional burden on areas where human pressure on water resources is already high. In addition, rising temperatures and lower precipitation associated with diminishing runoff will increase crop water demand in irrigated areas. The impacts of climate change on irrigation water requirements may therefore be substantial.[59]

In large irrigation systems that rely on high mountain glaciers for water (Andes, Himalayas and Rocky Mountains), temperature changes will cause high runoff periods to shift to earlier in the spring, when irrigation water demand is still low (see chapter 12).[60] Such changes could incite demand for new water control infrastructure to compensate for changes in river runoff. Indonesia shows how climate change can influence weather variability and make current farming and cropping systems unsustainable (box 7.6).

Box 7.6 Impacts of water shortage on rice production in Indonesia

Many of the extreme climate events in Indonesia, particularly droughts, are associated with the el Niño Southern Oscillation. The end of the dry season occurs later than normal during el Niño and earlier during la Niña years, the onset of the wet season is delayed during el Niño and advanced during la Niña years, a substantial reduction of dry season rainfalls occur during el Niño and a substantial increase during la Niña years, and long dry spells occur during the monsoon period, particularly in Eastern Indonesia.

Historical data indicate that the national rice production system is vulnerable to extreme climate events. During el Niño years rice production declined due to drought, with losses during 1991-2000 averaging three times greater than during 1980-90.

It is very likely that in Bali and Java the rainy season may shorten as a result of climate change, though the amount of rainfall may be higher. This suggests exposure to higher flood and drought risks in the future. For regions north of the Equator, however, the pattern of change will be the opposite.

Changes in rainfall pattern and length of the rainy season will have serious implications for the agriculture sector and current cropping patterns. In most rice-growing areas of Indonesia two rice crops are planted each year. The second planting depends heavily on irrigation water. In years of extreme drought irrigation water becomes very limited, causing severe production loss. Under a changing climate, drought will occur more frequently, and so retaining this cropping pattern may expose farmers to more frequent crop failures.

Source: BertJan Heij, co-chair, WWAP Expert Group on Climate and Water, based on Government of Indonesia 2007.

Options for water management in agriculture

It is possible to produce enough food and other agricultural products at a global level to meet demand while reducing the negative impacts of water use in agriculture.[61] But doing so will require a change from today's food production and environmental trends, which, if continued, will lead to crises in many parts of the world. A combination of supply- and demand-side measures is needed to address the acute water challenges in the coming 50 years. The difficult task at hand is to manage the additional water supply in a way that minimizes the adverse impacts and – where possible – enhances ecosystem services and aquatic food production, while achieving the necessary gains in food production and poverty alleviation.

The Comprehensive Assessment of Water Management in Agriculture scenario analysis shows opportunities and options – in rainfed, irrigated, livestock and fisheries systems – for preserving, and even restoring, healthy ecosystems.[62] But gains require major changes in the way water is managed, especially by farmers. The behaviour of different categories of farmers is shaped not only by agricultural policies but also by the capacity to allocate water according to wider financial restrictions and by local capacity to overcome pollution and environmental damage in emerging market economies. China has succeeded over the last 10 years in improving its water use efficiency by around 10% without increasing its water allocation to agriculture (see box 14.20 in chapter 14).

Improved water management in agriculture includes reduced water wastage in irrigation. Irrigated agriculture is often seen as inefficient, in both water use and added value.[63] While on average only an estimated 37% of the water withdrawn for agriculture is effectively consumed by plants, a substantial share of the unused water returns to rivers and aquifers and is available for downstream uses. The net loss of water due to irrigation is therefore substantially less than may be apparent, and the potential gains from programmes aimed at increasing water use efficiency are often overestimated (see chapter 8).

Programmes aimed only at reducing losses in irrigation are unlikely to have a substantial impact on water use. Most large irrigation schemes also serve other functions, such as providing water for drinking, bathing, swimming, fishing and livestock, and water savings may take water away from these uses. Management thus needs to focus instead on multiple-use strategies.

Technological improvements can occur at all levels and affect all types of irrigation systems (see chapter 3). Better technologies are not necessarily new, expensive or sophisticated options, but rather ones that are appropriate to agricultural needs and demands, the managerial capacity of system managers and farmers, and the financial and economic capacity needed to ensure proper operation and maintenance. Better design and better matching of technologies, management and institutional arrangements are needed. Technological innovation will occur in broadly three categories:

- At the irrigation system level: water level, flow control and storage management within surface irrigation systems at all scales.
- On the farm: storage, reuse, water lifting (manual and mechanical) and precision application technologies such as overhead sprinklers and localized irrigation.
- Across sectors: multiple-use systems in rural areas and urban agriculture with wastewater.

Greater impact on irrigation efficiency can be expected from external drivers (see chapter 2) on the evolution of irrigation than from demand management programmes. The trend and forecast are for irrigation to serve an increasingly market-oriented agriculture, with progressive increases in the value of production and growing use of precision irrigation. This will lead to the progressive adoption of pressurized irrigation, thus reducing losses.[64]

Industry puts pressure on water resources more by the impacts of wastewater discharges and their pollution potential than by the quantity used in production

Water for industry and energy

Water use for industry and energy is growing coincident with rapid development, transforming the patterns of water use in emerging market economies. Industry and energy together account for 20% of water demand. The actual figure may be even larger, as many industries self-supply (these volumes are only partially metered and reported) or get their water directly from the urban distribution system (use that is difficult to separate from domestic use).

Trends in industrial water use

Water is used by industry in multiple ways: for cleaning, heating and cooling; for generating steam; for transporting dissolved substances or particulates; as a raw material; as a solvent; and as a constituent part of products (as in the beverage industry). The volume of water used by industry is

Energy and water are inextricably linked

low, less than 10% of total water withdrawals, but there are large differences in efficiency of use. Industry creates more pressure on water resources from the impacts of wastewater discharges and their pollution potential than by the quantity used in production.

There is no simple relation between a country's production index (volume, value and jobs) and its total industrial demand for water. Demand depends first on the composition of the industrial sector, the processes in use and the degree of recycling that is in place in each sector. Different industries demand different water quality (the high-technology industry requires water of a higher quality than drinking water) and quantities (table 7.4).

The diminishing quality of water supplies, increasing costs of water purchases and strict environmental effluent standards are forcing industries to target greater water efficiency and report on their progress (as in the Global Reporting Initiative, for example[65]). Industrial water productivity (ratio of value of water withdrawn to value of industrial output using the water) is a general indicator of performance in water use. The intensity of water use in industry, in overall terms, is believed to be increasing, as is the value added by industry per unit of water use. Industrial water use is only partially linked to a country's level of industrialization, as exemplified by the large difference in water productivity between two high-income countries: more than $138 per cubic metre in Denmark and less than $10 per cubic metre in the United States (figure 7.8).

After rising between 1960 and 1980, water withdrawal for industrial use in developed countries has stabilized and has even started to decline in some countries, as industrial output continues to expand while falling in absolute terms (because of efficiency gains and the energy transition).[66] In Eastern Europe demand for water in the industrial sector fell following advances in production technology and structural change.[67] In emerging market economies industrial demand for water is expected to rise with the region's rapid growth in manufacturing output.

Some industries, such as tourism, show large seasonal variations in water use that can lead (on coastlines, islands and mountain areas) to supply difficulties in peak seasons. Around the Mediterranean Sea seasonal water demands from the tourism industry increase annual water demand by an estimated 5%-20% (box 7.7).

Energy and water

Energy and water are inextricably linked. Water is an integral part of energy resource development and use; it is needed for cooling and energy production (figure 7.9) but is also consumed passively as reservoirs built for energy production and other purposes evaporate substantial amounts of water. Total evaporation from reservoirs in the 22 countries of the Mediterranean Action Plan is estimated at around 24 km^3 a year – nearly the water use of Argentina – almost half of it in Egypt.[68] For hydroelectric, wave or tidal energy production water offers an 'active' medium for transferring kinetic energy into electricity. For cooling thermal and nuclear plants or producing bioenergy, water plays a more passive, though equally important, role. The demand for energy is therefore a major

Table 7.4 Water use per tonne of product produced, selected industries

(cubic metres per tonne)

Product	Water use[a]
Paper	80-2,000
Sugar	3-400
Steel	2-350
Petrol	0.1-40
Soap	1-35
Beer	8-25

a. Amount varies with process used.

Source: Margat and Andréassian 2008.

Figure 7.8 Industrial water productivity varies greatly across countries

1995 US$ per cubic metre per year, latest year available

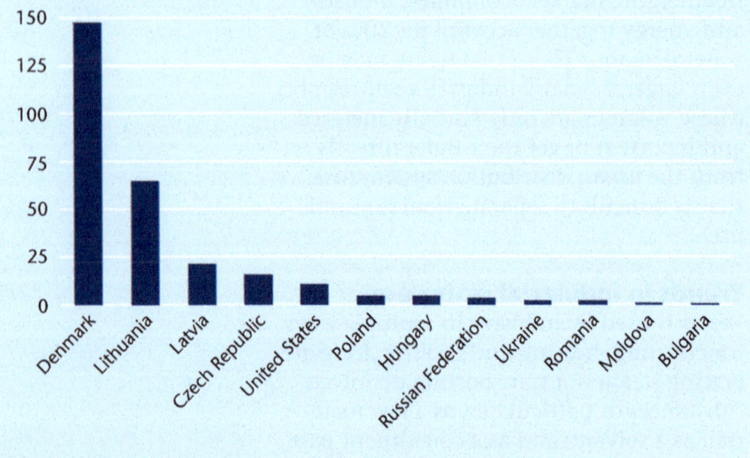

Source: Based on UNIDO 2007.

driver of water and agriculture development, creating pressures that strongly affect the quantity and quality of freshwater resources.

Energy is important for pumping water, transporting it, processing it and using it. Desalination is also an energy-intensive process. Energy demand is affected by many of the same drivers that are putting direct pressure on water resources: demographic, economic, social and technological processes, including changes in consumption patterns. Energy consumption is also the main driver behind climate change (see chapter 5), which threatens the sustainability of water resources. Growing pressure and efforts to curb greenhouse gas emissions are leading to increasing demand for 'cleaner' sources of renewable energy. Hydropower has been earmarked as one of the most important of these sources.

Energy requirements of the water sector. Energy can account for 60%-80% of water transportation and treatment costs and 14% of total water utility costs.[69] In 2005-06 water and wastewater companies in England and Wales spent $632 million on electric power (7,700 gigawatts), making it the largest non-staff operating cost item.[70] Efficiency and conservation are therefore not only good for water resources; they are also a means of conserving energy.

Much of water resources development during the 20th century took place in a context of low water and energy prices. Substantial energy subsidies are still provided in many parts of the world. The delivery of subsidized rural electricity services has boosted agricultural production in existing irrigated areas and introduced irrigation in areas beyond those with surface water. The energy provided through electricity and fuel, together with low-cost technology for pumping, has generated tremendous changes in agricultural water management. One result of energy subsidies has been overabstraction of groundwater in many irrigated agricultural areas. When properly used, subsidies can fulfil important socioeconomic goals, but the challenge is to find a balance: to encourage efficiency and to ensure that subsidies reach intended beneficiaries.

Energy prices increasingly tend to be volatile, and it will therefore be important to consider how future changes in energy prices and markets will affect water use and production patterns and costs. Apart from the effects on production costs, rising prices are likely to influence investment in areas where the price of energy is an important factor (from farms to businesses, impacts are likely to be very heterogeneous). Rising prices may increase the willingness to make efficiency-related investments while reducing overall sector investments, with possible impacts on food production. Indirect effects can also be considerable, such as boosting transport costs.

Box 7.7 Tourism water demand in the Mediterranean coastal area

With 364 million tourists in 2000 the Mediterranean region is the world's top tourism destination, and by 2025 the number of tourists could reach 637 million.

Knowledge of the water demands of tourism is limited because national statistics rarely distinguish between domestic water use and water for tourism. The annual additional demands from tourism are relatively modest, at 20% of domestic supply in Cyprus in 2006, 5% of total water demand in Malta during 1995-2000 and 5% of domestic demand in Tunisia in 2003.

Daily demands are generally more important for planners and managers than annual demands because they show the timing and volume of peak demands. The daily demands of a tourist are generally higher than those of a local resident. Tourism also creates a demand for seasonal services and leisure activities that demand a lot of water, such as golf courses.

Tourist demands often occur at the same time as peak demand for agricul ture, which is also seasonal, when resources are at their lowest. Satisfying such peak demands requires oversizing the drinking water production and distribution system as well as the wastewater collection and treatment infrastructure. In many places water supply for tourism relies on desalination of seawater, a promising option in places such as Cyprus, the Balearic Islands, Malta, Tunisia and certain Greek islands.

Source: Blue Plan, MAP, and UNEP 2005, 2007.

Figure 7.9 Interlinkages between energy and water

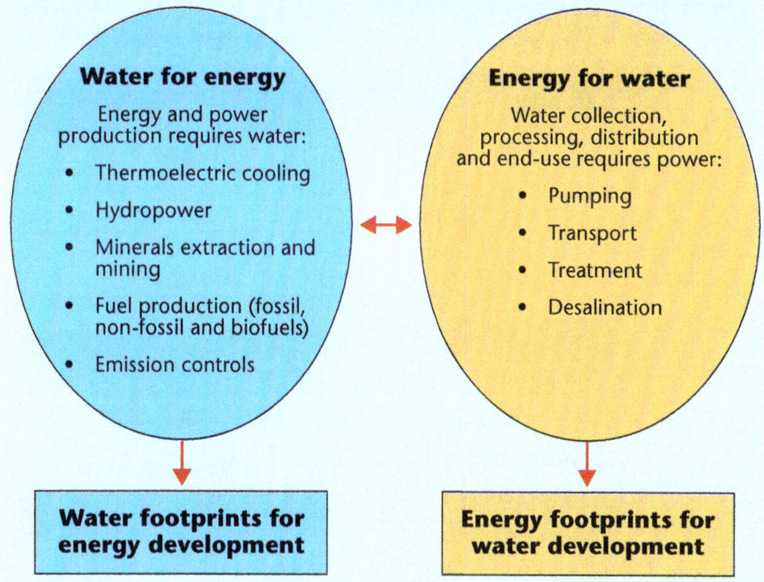

Source: Based on DHI 2008.

Hydropower supplies about 20% of the world's electricity

Water use for energy production. Cooling in the energy sector is one of the main industrial water uses, with final consumption (evaporation) estimated at around 5% of withdrawals. Outflows of water used in cooling nuclear power plants demand sufficient river flow to reduce the temperature in order to mitigate adverse ecological impacts. Thus non-directly productive but substantial flows are required.

Hydropower generation requires large quantities of water, but unlike in other major water use sectors (agriculture and domestic), the nature of the use is non-consumptive: water is returned to the river after passing through turbines. However, substantial losses occur through evaporation from reservoirs, and thus this use is not entirely non-consumptive.

Water use efficiency in the energy sector differs with the power generation technology used, as illustrated by an assessment of 19 power generation systems in the United States (figure 7.10). This suggests considerable potential for improving efficiency.

Figure 7.10 **Water consumption for various power generation technologies in the United States, 2006**

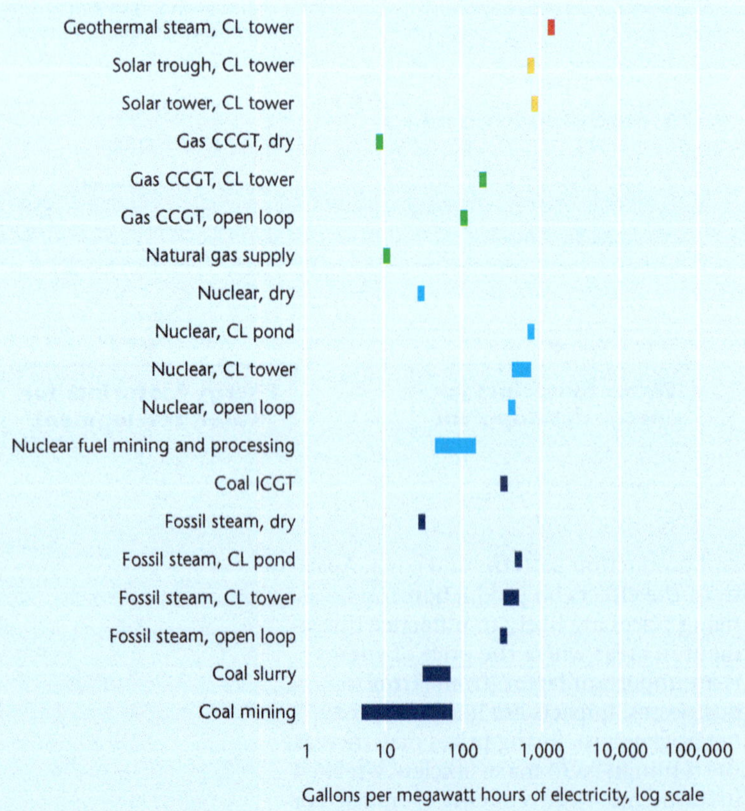

Note: CL is closed loop cooling, CCGT is combined cycle gas turbine, and ICGT is integrated gasification combined-cycle.
Source: US Department of Energy 2006.

Current and projected trends in hydropower. Hydropower supplies about 20% of the world's electricity,[71] a share that has remained stable since the 1990s. Hydropower stations are spread across the globe (map 7.6) and have shaped water infrastructure in many parts of the world. The first large hydropower stations were developed in Norway, Sweden, Switzerland, Canada, the United States, Australia and New Zealand. The largest hydropower station in operation is the Itaipu Dam on the Rio Parana River between Brazil and Paraguay, with an installed capacity of more than 14,000 megawatts. Brazil produces more than 90% of its electricity from hydropower.

Hydropower development was stimulated by the oil crisis of the 1970s, fell off for a few decades and then returned to the agendas of many countries in response to the energy demands created by development. Hydropower continues to be the most important and economic source of commercial renewable energy worldwide, and its popularity is increasing with the surge of interest in clean energy prompted by concerns about climate change. Hydropower plants, when managed for multiple uses, should also allow for flow regulation and flood management, water for irrigation and drinking water supply during dry seasons and rapid response to grid demand fluctuations due to peak demands.

According to the International Energy Agency, electricity generation from hydropower and other renewable energy sources is projected to increase at an average annual rate of 1.7% from 2004 to 2030, for an overall increase of 60% through 2030 (figure 7.11).

Development of hydropower will be limited by two main factors. The first is the modest spatial and geophysical potential for new hydropower installations. In many developed countries – including Australia, the United States and most of Western Europe – most of the suitable sites for hydropower installations have already been developed. Other limiting factors are investment capacity (including the availability of funds) and the social and environmental impacts of large dams and the controversy surrounding them, which collectively explain why so little hydroelectric potential has been tapped in developing countries.

Oil prices and energy choices. Volatile prices of oil and natural gas, which are expected

Map 7.6 **World potential and current hydropower production, 2004**

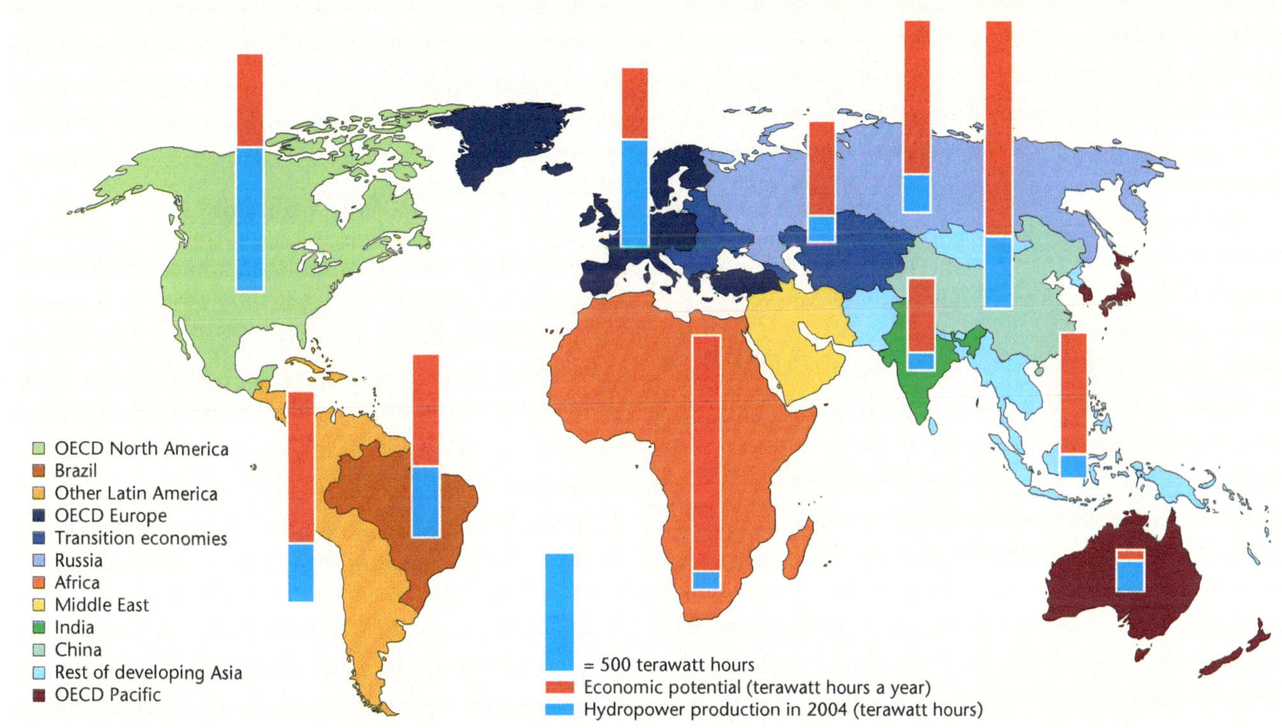

Source: IEA 2006.

to persist over the medium term, encourage the use of renewable energy, which is also attractive for environmental reasons. Government policies and incentives can also increase the use of renewable energy sources even when renewable energy cannot compete economically with fossil fuels. Nonetheless, the renewable energy share of world electricity production is projected to fall slightly, from 19% in 2004 to 16% in 2030, as growth in the consumption of coal and natural gas for electricity generation worldwide exceeds that in renewable energy sources. The capital costs of new power plants using renewable energy remain high compared with those for coal- or natural gas-fired plants.

There are complex, and partly competing, challenges associated with energy production and water resources management. While average commercial energy use in high-income countries is about 5,500 kilograms of oil equivalent per capita, it is still well below 500 kilograms in low-income countries.[72] Thus, energy consumption in developing countries will need to increase as part of any development strategy. Yet major challenges remain for energy development. The pressure to extend hydropower on the basis of its comparative sustainability will continue to stir discussions of its environmental and social impacts, including the regulation of hydropower dam releases to optimize downstream uses and minimize negative impacts on aquatic ecosystems. Likewise, expansion of thermal power-producing facilities will require cooling water, with the need to discharge heated water.

Figure 7.11 **Renewable energy sources are expected to meet only a small part of total energy demand to 2030**

Global production of electricity by energy source, 1971-2030 (thousands of terawatt hours)

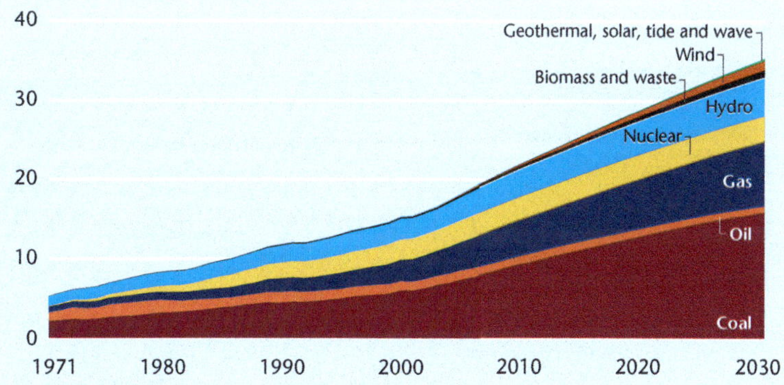

Source: IEA 2008.

Emerging challenges will affect both the energy sector and water resources. The most obvious is climate change. Political

Freshwater ecosystems provide an extensive array of services to support human well-being

pressures are mounting, and there are likely to be increased calls for action to deal with greenhouse gas emissions in coming decades in ways that can alter the energy production landscape. Yet the International Energy Agency's *World Energy Outlook 2007* forecasts that fossil fuels will continue to provide the major part of the increased energy demand.[73] Nonetheless, the pressure and prospects for hydropower development may also increase as part of efforts to mitigate climate change.

In-stream water uses

Freshwater ecosystems provide an extensive array of services to support human well-being, many of them extremely valuable. Yet some environmental services continue to receive inadequate policy attention. Shifting attitudes away from considering the environment as a victim of human uses of water and towards viewing environmental sustainability as central to sustainable development remains a significant challenge. (For further discussion, see chapters 6 and 9.)

Navigation has been critical to development and communication around the world. A variety of other economic activities such as fisheries and pastoral activities also depend on in-stream use of water. Recreational activities such as tubing and kayaking do not deplete water, but demand releases of water and maintenance of minimum flows at times that may not be compatible with demands by other users.

Water for transport

The 25 largest cities, the 25 largest production locations, the 25 most prosperous areas and the 25 most densely populated areas in the world are all located near waterfronts, almost all of them by the sea.[74] This has been the case for at least 2,000 years. River navigation was part of the Indus Valley Civilization in Northwest India around 3300 BCE. Many major rivers of the world are used for navigation today. The development of waterways for transport lies behind many large-scale river transformations and dam constructions. Of 230 major world rivers some 60% are considered to be seriously or moderately fragmented by dams, dikes and dredging, with improved river transport often being an objective.[75]

However, inland shipping remains an underdeveloped sector on most waterways (map 7.7). China alone has 110,000

Map 7.7 **Important waterways in the world, 2007**

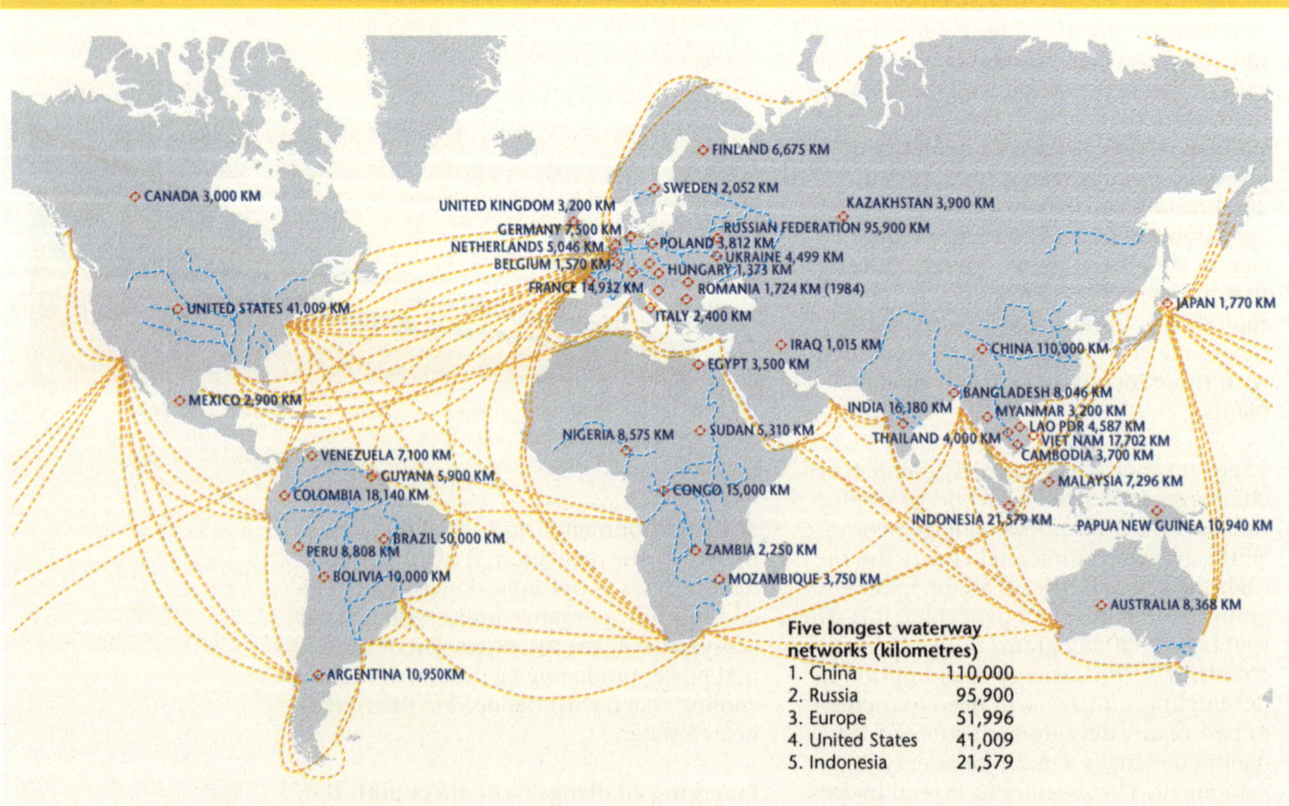

Source: Based on BVB 2008.

km of navigable waterways, the longest in the world.[76] Inland navigation is often the most cost-effective and least polluting means of transport and, with improved trade and exchange, has contributed to the development of mature economies.[77] Inland waterways can efficiently convey large volumes of bulk commodities over long distances. On the Mississippi River convoys of 30 barges (equivalent to 30 jumbo rail cars or 110 normal trucks) are common, while on rivers such as the Danube and the Rhine, container fleets 23 metres wide by 135 metres long can push convoys consisting of six barges, each about 64 metres by 270 metres.

However, developing rivers for navigation often results in irreversible transformation of river courses, with negative impacts on vulnerable groups and ecosystems (such as fish mortality from propeller impact and larvae stranding due to drawdown). Western Europe's Rhine River is perhaps the best-known example of a river altered by navigation schemes. To make an 880 km stretch navigable, some 450 dams and thousands of kilometres of banks were built, meanders were removed, and the Rhine became 25% shorter as a result.[78]

Fisheries and aquaculture

Inland fisheries are an important activity for many poor people in rural areas[79] and contribute significantly, at least locally, to economic development, poverty alleviation, increased protein in diets, and food security[80] – even though estimation is difficult, as many fishing activities do not fall within the economic domain.[81]

For small-scale fisheries, in particular, data remain patchy or not adequately disaggregated to allow detailed analysis, but case study information is available.[82] Some general indicators of the importance of the sector to national economies have been compiled (table 7.5).

At national levels inland fisheries contribute to GDP through multiplier effects and generate tax revenues and foreign exchange. In Bangladesh benefits from inland fisheries represented 80% of export earnings and about 50% of daily animal protein intake. Most fish exporters have well established aquaculture industries.[83] In Tanzania mainland fisheries generated $6.9 million in taxes, 97% from export taxes, and in Uganda the Nile perch from Lake Victoria accounted for 17% of the value of exports in 2002.[84]

Table 7.5 Contribution of inland and marine fisheries to exports, daily animal protein supply and employment in major fish-harvesting economies, latest year available

Economy	Percent of agricultural exports	Percent of average per capita daily animal proteins	Percent of economically active population
Argentina	7	4	0.08
Bangladesh	76	51	1.90
Brazil	2	4	0.37
Chile	39	9	0.82
Ecuador	31	8	3.29
India	22	13	1.35
Malaysia	3	37	1.07
Mexico	9	7	0.64
Morocco	58	16	0.90
Pakistan	12	3	0.52
Peru	62	20	0.68
Philippines	23	41	3.16
Thailand	38	40	0.95
Viet Nam	40	34	2.45
Average	30	20	1.30

Source: Thorpe et al. 2005.

At the household level inland fisheries are central to livelihood strategies, providing not only direct and indirect employment for some 100 million people, most in developing countries, but also a safety net activity for the poor through catch and trade. Not included in these estimates are the hundreds of millions of people engaged in temporary fishing activities, mostly in inland areas. In Africa men and boys engage in seasonal fishing along rivers or reservoirs when agricultural activities are slow.[85] In the Tonle Sap Lake area of Cambodia thousands of households split their time between fishing and cultivating crops.[86] In the floodplain areas of India occasional fishing by children, elders or women in male-headed households on the margins of water bodies or in waterways such as irrigation canals can involve up to 70%-80% of households during the flood seasons.[87]

Fish[88] consumption has undergone major changes over the past four decades. World fish capture and aquaculture production have grown fairly steadily (figure 7.12). Global per capita fish consumption has risen from an estimated average of 9.9 kg in the 1960s to about 16.7 kg in 2006.[89] However, not all regions have experienced the same rise. Over the last three decades

Figure 7.12 **Global freshwater fish production has grown rapidly in recent decades**

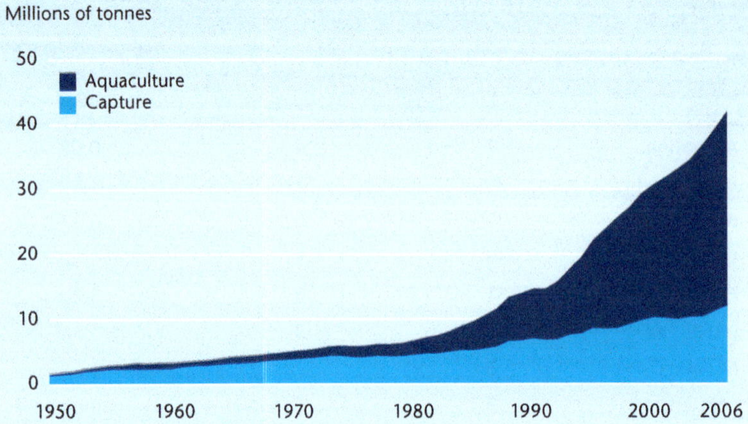

Source: FAO forthcoming.

Figure 7.13 **Inland capture fisheries vary greatly by region, 2006**

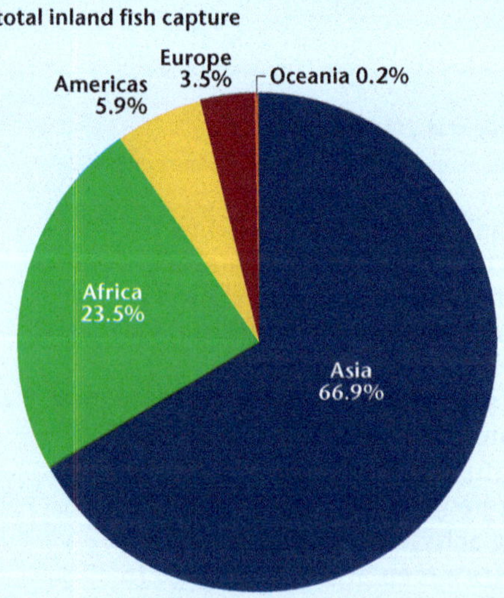

Source: FAO forthcoming.

the per capita fish supply has remained almost static in sub-Saharan Africa, while rising dramatically in East Asia (mainly in China) and in the Near East and North Africa.

The share of fish in total protein intake was 7.8% in 2005, about the same as in the mid-1980s. The contribution of inland fish to total protein intake grew 6.5%-8.5% during 1961-89, before declining gradually as consumption of other animal proteins rose. Although consumption in low-income, food-deficit countries has increased in the last four decades, especially since the mid-1990s (1.3% a year since 1993), per capita fish intake remains half that of industrial countries. Despite the relatively low level of fish consumption, the contribution of fish to total animal protein intake in 2005 was about 20%, and may be higher than indicated by official statistics because of the unrecorded contribution of subsistence fisheries. Fish contribute to diets in many regions of the world, offering a valuable supplement for a diversified and nutritious diet, including high-value protein and a wide range of essential micronutrients, minerals and fatty acids.

Landing more than 10 million tonnes in 2006, inland capture fisheries contributed 11% of global capture fisheries production. Although much lower than marine fisheries, fish and other aquatic animals from inland waters remain irreplaceable to the diets of both rural and urban inhabitants in many parts of the world, especially in developing countries. For demographic and cultural reasons, however, levels of exploitation differ considerably across major geographic regions (figure 7.13). And although statistics are improving in some countries, collecting accurate information on inland fisheries can be difficult and costly, and many governments still do not gather such information or assess the status of inland fishery resources.

The average contribution of aquaculture to per capita fish available for human consumption grew from 15% in 1996 to 47% in 2006 and is expected to reach 50% within the next decade. Aquaculture production has boosted consumption for several freshwater species, such as tilapia and catfish, as well as for high-value species such as shrimp, salmon and bivalves. Since the mid-1980s these species have shifted from being primarily wild-caught to being primarily aquaculture-produced, with a corresponding decline in price and a strong increase in commercialization. Aquaculture has also improved food security in several developing countries, particularly in Asia, through the production of certain low-value freshwater species destined mainly for domestic consumption. As the demand for fish is price-elastic, and with stable or declining fish prices, rising incomes and diversification of diets, there is a shift towards higher fish consumption in developing countries. These trends in fish consumption are expected to continue for the foreseeable future, driven by population and income growth, urbanization and dietary diversification. However, aquaculture has also contributed to serious water pollution when not well managed, a problem that is likely to intensify with increased aquaculture activities.[90]

Capture fisheries and aquaculture are sustained by aquatic ecosystems (including rice paddies). If the ecosystem is damaged, the quantity of fish decreases. Lack of water at a certain flow for critical periods of fish life can be detrimental. The health of capture fisheries is therefore a good indicator of the health of the ecosystem.

Notes

1. UNDP 2006; WHO and UNICEF 2008a.
2. Comprehensive Assessment of Water Management in Agriculture 2007; World Bank 2007.
3. MEA 2005; UNEP 2007.
4. Margat and Andréassian 2008.
5. Comprehensive Assessment of Water Management in Agriculture 2007.
6. Shiklomanov and Rodda 2003.
7. FAO-AQUASTAT database (www. fao.org/nr/water/aquastat/main/index.stm).
8. Comprehensive Assessment of Water Management in Agriculture 2007.
9. World Bank 2007.
10. Cosgrove and Rijsberman 2000; MEA 2005; Comprehensive Assessment of Water Management in Agriculture 2007.
11. This is one of the tasks set for the scenarios process that is under way by the World Water Assessment Programme, to be reported in the next World Water Development Report.
12. Blue Plan, MAP, and UNEP 2005.
13. Hoekstra and Chapagain 2007.
14. UNSGAB 2006.
15. WHO and UNICEF 2008b.
16. WHO and UNICEF 2008b.
17. Pilgrim et al. 2008.
18. The Joint Monitoring Programme for Water Supply and Sanitation, managed by the World Health Organization (WHO) and the United Nations Children's Fund (UNICEF), is the UN mechanism for monitoring progress towards the Millennium Development Goal target on access to water supply and sanitation.
19. WHO and UNICEF 2006.
20. UNDP 2006.
21. UN-Water 2008.
22. UN Millennium Project 2005.
23. UNDP 2006.
24. WHO and UNICEF 2008a.
25. AfricanSan +5 Conference on Sanitation and Hygiene 2008.
26. van Ginneken and Kingdom 2008.
27. A comprehensive review of approaches and technologies for water, sanitation and hygiene was carried out on behalf of the Bill and Melinda Gates Foundation by a consortium of Cranfield University, Aguaconsult and IRC International Water and Sanitation Centre. Various documents can be downloaded from www.irc.nl/page/35947.
28. EHP 2003.
29. Wright 1997.
30. UNDP 2006.
31. UN-Water 2008.
32. Comprehensive Assessment of Water Management in Agriculture 2007.
33. World Bank 2007.
34. FAOSTAT database (http://faostat.fao.org/).
35. See Comprehensive Assessment of Water Management in Agriculture 2007, chapter 9.
36. WWAP 2006.
37. FAO 2006b.
38. FAO 2006a.
39. FAO 2006b.
40. FAO 2008c.
41. FAO 2008c.
42. Worldwatch Institute 2008.
43. FAO 2008c.
44. Bioethanol and biodiesel are typically mixed with gasoline and diesel, respectively, as so-called flex-fuel. Blends vary between a few percent of biofuel to nearly 25% in Brazil.
45. Müller et al. 2008; De Fraiture, Giodano, and Yongsong 2007; OECD and FAO 2008.
46. OECD and FAO 2008.
47. FAO 2008a.
48. OECD and FAO 2008.
49. Dornbosch and Steenblik 2007.
50. De Fraiture, Giodano, and Yongsong 2007.
51. De Fraiture, Giodano, and Yongsong 2007.
52. Dornbosch and Steenblik 2007.
53. Dornbosch and Steenblik 2007.
54. Zah et al. 2007; Dornbosch and Steenblik 2007.
55. IPCC 2007b.
56. FAO 2008b
57. IPCC 2007a.
58. IPCC 2007c.
59. IPCC 2007a.
60. Bennett, Haberle, and Lumley 2000.
61. Comprehensive Assessment of Water Management in Agriculture 2007.
62. Comprehensive Assessment of Water Management in Agriculture 2007.
63. Water can accumulate in the soil profile through runoff and groundwater recharge and from irrigation if the rate of input exceeds the rate of crop consumption (see chapter 8). This accumulation can lead to waterlogging (when soil pores are filled with water and oxygen is lacking) and salinization (when the rising water in the soil profile brings diluted salts to the surface). Worldwide, about 10% of irrigated land suffers from waterlogging.
64. Comprehensive Assessment of Water Management in Agriculture 2007.
65. www.globalreporting.org/.
66. WWAP 2006.
67. Somlyódy and Varis 2006.
68. Blue Plan, MAP, and UNEP 2007.
69. Global Water Intelligence 2007.
70. Dornbosch and Steenblik 2007.
71. ICOLD 2007.
72. World Development Indicators database (2005 data; http://ddp-ext.worldbank.org/ext/ddpreports/).
73. IEA 2007.
74. BVB 2008.
75. WWF 2008.
76. BVB 2008.
77. PIANC 2008.
78. WWF website (www.panda.org/about_wwf/what_we_do/freshwater/problems/infrastructure/river_navigation/).
79. Comprehensive Assessment of Water Management in Agriculture 2007.
80. Béné, Macfayden, and Allison 2007.
81. This section is based on FAO forthcoming.
82. For example, Southern Africa, in Béné, Macfadyen, and Allison (2007).
83. Thorpe et al. 2005.
84. Wilson 2004.
85. Sana 2000.
86. Ahmed et al. 1998.
87. Thompson and Hossain 1998; Hoggarth et al. 1999.

88. Including fish, crustaceans and mollusks, and excluding aquatic mammals and aquatic plants.
89. FAO forthcoming.
90. Gowing 2006.

References

AfricaSan + 5 Conference on Sanitation and Hygiene. 2008. *The eThekwini Declaration and AfricaSan Action Plan.* WSP (Water and Sanitation Program) – Africa.

Ahmed, M., H. Navy, L. Vuthy, and M. Tiongco. 1998. *Socio-Economic Assessment of Freshwater Capture Fisheries in Cambodia.* Report 185. Phnom Penh: Mekong River Commission, Department of Fisheries and Danida.

Aliev, S., P. Shodmonov, N. Babakhanova, and O. Schmoll. 2006. Rapid Assessment of Drinking-Water Quality in the Republic of Tajikistan. Dushanbe, Ministry of Health of the Republic of Tajikistan, United Nations Children's Fund, World Health Organization.

Béné, C., G. Macfadyen, and E. Allison. 2007. Increasing the Contribution of Small-Scale Fisheries to Poverty Alleviation and Food Security. FAO Fisheries Technical Paper 481, Food and Agriculture Organization of the United Nations, Rome.

Bennett K. D., S. G. Haberle, and S. H. Lumley. 2000. The Last Glacial-Holocene Transition in Southern Chile. *Science* 290 (5490): 325-28. DOI:10.1126/science.290.5490.325.

Blue Plan, MAP (Mediterranean Action Plan), and UNEP (United Nations Environment Programme). 2005. *The Blue Plan's Sustainable Development Outlook for the Mediterranean.* Sophia Antipolis, France: Blue Plan. www.planbleu.org/publications/UPM_EN.pdf.

———. **2007.** *Water Demand Management, Progress and Policies: Proceedings of the 3rd Regional Workshop on Water and Sustainable Development in the Mediterranean Zaragoza, Spain, 19-21 March.* MAP Technical Reports Series 168. Athens: United Nations Environment Programme.

BVB (Bureau Voorlichtin Binnenvaart). 2008. *The Power of Inland Navigation: The Future of Inland Navigation on European Scale.* Rotterdam: Bureau Voorlichtin Binnenvaart. www.bureauvoorlichtingbinnenvaart.nl/images/download/waardevoltransport/The_power_of_inland_navigation.pdf. www.inlandshipping.com/

Comprehensive Assessment of Water Management in Agriculture. 2007. *Water for Food, Water for Life: A Comprehensive Assessment of Water Management in Agriculture.* London: Earthscan, and Colombo: International Water Management Institute.

Cosgrove, W., and F. Rijsberman. 2000. *World Water Vision: Making Water Everybody's Business.* London: Earthscan.

De Fraiture, C., M. Giodano, and Yongsong L. 2007. Biofuels: Implications for Agricultural Water Use: Blue Impact of Green Energy. Paper presented at the International Conference Linkages between Energy and Water Management for Agriculture in Developing Countries, 28-31 January 2007, Hyderabad, India.

DHI. 2008. Linking Water, Energy and Climate Change. A Proposed Water and Energy Policy Initiative for the UN Climate Change Conference, COP15, in Copenhagen 2009, Draft Concept Note, Danish Hydrological Institute, Horsholm, Denmark.

Dornbosch, Richard, and Ronald Steenblik. 2007. Biofuels: Is the Cure Worse than the Disease? Round Table on Sustainable Development, 11-12 September 2007, Paris. SG/SD/RT(2007)3, Organisation for Economic Co-operation and Development. www.foeeurope.org/publications/2007/OECD_Biofuels_Cure_Worse_Than_Disease_Sept07.pdf.

EHP (Environmental Health Project). 2003. *The Hygiene Improvement Framework: A Comprehensive Approach for Preventing Childhood Diarrhoea.* Washington, DC: Environmental Health Project.

FAO (Food and Agriculture Organization of the United Nations). 2006a. *World Agriculture towards 2030/2050. Prospects for Food, Nutrition, Agriculture, and Major Commodity Groups. Interim Report.* Rome: Food and Agriculture Organization of the United Nations.

———. **2006b.** *The State of Food Insecurity in the World 2006. Eradicating World Hunger – Taking Stock Ten Years after the World Food Summit.* Rome: Food and Agriculture Organization of the United Nations.

———. **2008a.** Report of the Co-Chairs of the Informal Open-ended Contact Group. Contribution to the High-Level Conference on World Food Security: The Challenges of Climate Change and Bioenergy, 3-5 June 2008, Rome. ftp://ftp.fao.org/docrep/fao/meeting/013/k2359e.pdf.

———. **2008b.** Climate Change, Water and Food Security. Technical background document from the expert consultation held in Rome, 26-28 February 2008. Contribution to the High Level Conference on World Food Security: The Challenges of Climate Change and Bioenergy, 3-5 June 2008, Rome.

———. **2008c.** Soaring Food Prices: Facts, Perspectives, Impacts and Actions Required. Contribution to the High Level Conference on World Food Security: The Challenges of Climate Change and Bioenergy, 3-5 June 2008, Rome.

———. **Forthcoming.** *The State of World Fisheries and Aquaculture 2008.* Rome: Food and Agriculture Organization of the United Nations.

Global Water Intelligence. 2007. *Global Water Market 2008: Opportunities in Scarcity and Environmental Regulation.* Oxford: Global Water Intelligence.

Government of Indonesia. 2007. *Indonesia Country Report. Climate Variability and Climate Changes and Their Implications.* Jakarta: Ministry of Environment, Republic of Indonesia. www.undp.or.id/pubs/docs/Final%20Country%20Report%20-%20Climate%20Change.pdf.

Gowing, J. 2006. A Review of Experience with Aquaculture Integration in Large-Scale Irrigation Systems. In Integrated Irrigation and Aquaculture in West Africa: Concepts, Practices and Potential, eds., M. Halwart and A. A. van Dam. Rome: Food and Agriculture Organization of the United Nations.

Hoekstra, A. Y., and A. K. Chapagain. 2007. Water Footprints of Nations: Water Use by People as a Function of Their Consumption Pattern. *Water Resources Management* 21 (1): 35-48.

———. **2008.** *Globalization of Water: Sharing the Planet's Freshwater Resources.* Oxford: Blackwell Publishing.

Hoggarth, D. D., V. J. Cowan, A. S. Halls, M. Aeron-Thomas, J. A. McGregor, C. A. Garaway, A. I. Payne, and R. Welcomme. 1999. *Management Guidelines for Asian Floodplain River Fisheries. Part 1: A Spatial, Hierarchical and Integrated Strategy for Adaptive Co-management. FAO Fisheries Technical Paper, 384/1.* Rome: Food and Agriculture Organization of the United Nations.

Hoogeveen, J., J.-M. Faurès, and N. van de Giesse. Forthcoming. Increased Biofuel Production in the Coming Decade: to What Extent Will It Affect Global Freshwater Resources? *Irrigation and Drainage Journal.*

ICOLD (International Commission on Large Dams). 2007. *Dams and the World's Water. An Educational Book that Explains How Dams Help to Manage the World's Water.* Paris: International Commission on Large Dams.

IEA (International Energy Agency). 2006. *World Energy Outlook 2006.* Paris: Organisation for Economic Co-operation and Development and International Energy Agency.

———. **2007.** *World Energy Outlook 2007.* Paris: Organisation for Economic Co-operation and Development and International Energy Agency.

———. **2008.** *World Energy Outlook 2008.* Paris: Organisation for Economic Co-operation and Development and International Energy Agency.

IFEN (Institut Français de L'environnement). 2006. *L'environnement en France.* Les Syntheses. Octobre 2006 ed. Orléans, France: Institut Français de L'environnement.

IPCC (Intergovernmental Panel on Climate Change). 2000. *Emissions Scenarios.* Cambridge, UK: Cambridge University Press.

———. **2007a.** *Climate Change 2007: Impacts, Adaptation and Vulnerability. Contribution of Working Group II to the Fourth Assessment Report of the Intergovernmental Panel on Climate Change,* eds., M. L. Parry, O. F. Canziani, J. P. Palutikof, P. J. van der Linden, and C. E. Hanson. Cambridge, UK: Cambridge University Press.

———. **2007b.** *Climate Change 2007: Mitigation. Contribution of Working Group III to the Fourth Assessment Report of the Intergovernmental Panel on Climate Change.* Cambridge, UK: Cambridge University Press.

———. **2007c.** *Climate Change 2007: Synthesis Report. Contribution of Working Groups I, II and III to the Fourth Assessment Report of the Intergovernmental Panel on Climate Change.* Geneva: Intergovernmental Panel on Climate Change.

Margat, J. 2008. Preparatory Documents to the 5th World Water Forum 2009, 16-22 March, Istanbul. Internal Documents for Blue Plan/MAP/UNEP.

Margat, J., and V. Andréassian. 2008. *L'Eau, les Idées Reçues.* Paris: Editions le Cavalier Bleu.

MEA (Millennium Ecosystem Assessment). 2005. *Ecosystems and Human Well-Being: Wetlands and Water Synthesis.* Washington, DC: World Resources Institute.

Müller, A., J. Schmidhuber, J. Hoogeveen, and P. Steduto. 2008. Some Insights in the Effect of Growing Bio-energy Demand on Global Food Security and Natural Resources. *Water Policy* 10 (Supp.1): 83-94.

OECD (Organisation for Economic Co-operation and Development) and FAO (Food and Agriculture Organization of the United Nations). 2008. *OECD-FAO Agricultural Outlook 2008-2017.* Paris: Organisation for Economic Co-operation and Development/Food and Agriculture Organisation of the United Nations. www.fao.org/es/ESC/common/ecg/550/en/AgOut2017E.pdf.

PIANC (The World Association for Waterborne Transport Infrastructure). 2008. Guidelines for Environmental Impacts of Vessels. InCom working group 27. PIANC.

Pilgrim, N., B. Roche, J. Kalbermatten, C. Revels, and M. Kariuki. 2008. *Principles of Town Water Supply and Sanitation. Water Working Note No. 13.* Washington, DC: World Bank.

Sana, B. 2000. Systèmes d'Aménagement Traditionnels des Pêches et Leur Impact dans le Cadre de la Lutte contre la Pauvreté sur les Petites Pêcheries Lacustres du Sud-Ouest du Burkina Faso. Report of the Seminar on Livelihoods and Inland Fisheries Management in the Sahelian Zone, Ouagadougou, Burkina Faso.

Shiklomanov, I. A., and John C. Rodda, eds. 2003. *World Water Resources at the Beginning of the 21st Century.* Cambridge, UK: Cambridge University Press.

Somlyódy, László, and Olli Varis. 2006. Freshwater under Pressure in Groundwater Management and Policy: Its Future Alternatives. *International Review for Environmental Strategies* 6 (2): 181-204.

Thompson, P. M., and M. M. Hossain. 1998. Social and Distributional Issues in Open Water Fisheries Management in Bangladesh. In *Inland fishery enhancements: Papers Presented at the FAO/DFID Expert Consultation on Inland Fishery Enhancement, Dhaka, Bangladesh, 7-11 April 1997,* ed., T. Petr. Rome: Food and Agriculture Organization of the United Nations.

Thorpe, A., C. Reid, R. van Anrooy, and C. Brugere. 2005. When Fisheries Influence National Policy-Making: An Analysis of the National Development Strategies of Major Fish-Producing Nations in the Developing World. *Marine Policy* 29 (3): 211-22.

UN Millennium Project. 2005. *Health, Dignity, and Development: What Will It Take?* Task Force on Water and Sanitation. London: Earthscan.

UNDP (United Nations Development Programme). 2006. *Human Development Report 2006: Beyond Scarcity: Power, Poverty and the Global Water Crisis.* New York: Palgrave Macmillan.

UNEP (United Nations Environment Programme). 2007. *Global Environment Outlook GEO4: Environment for Development.* Nairobi: United Nations Environment Programme.

UNIDO (United Nations Industrial Development Organization). 2007. Symposium Report on Water Productivity in the Industry of the Future for the Technology Foresight Summit, 27-29 September 2007, Budapest.

UNSGAB (United Nations Secretary General's Advisory Board on Water and Sanitation). 2006. Hashimoto Action Plan (formerly known as The Compendium of Actions). United Nations Secretary General's Advisory Board on Water and Sanitation, New York.

UN-Water. 2008. *UN-Water Global Annual Assessment of Sanitation and Drinking-Water (GLAAS): 2008 Pilot Report – Testing a New Reporting Approach.* Geneva: World Health Organization.

US Department of Energy. 2006. Energy Demands on Water Resources. Report to Congress on the Independency of Energy and Water, U.S. National Research Council, Washington, DC.

van Ginneken, M., and B. Kingdom. 2008. *Key Topics in Public Water Utility Reform. Water Working Note 17.* Washington, DC: World Bank.

WHO (World Health Organization) and UNICEF (United Nations Children's Fund) Joint Monitoring Programme. 2006. *Meeting the MDG Drinking Water and Sanitation Target: The Urban and Rural Challenge of the Decade.* New York: United Nations Children's Fund, and Geneva: World Health Organization.

———. **2008a.** *Progress on Drinking Water and Sanitation: Special Focus on Sanitation.* New York: United Nations Children's Fund, and Geneva: World Health Organization.

———. **2008b.** *A Snapshot of Sanitation in Africa.* New York: United Nations Children's Fund, and Geneva: World Health Organization.

Wilson, J. D. K. 2004. *Fiscal Arrangements in the Tanzanian Fisheries Sector. FAO Fisheries Circular 1000.* Rome: Food and Agriculture Organization.

World Bank. 2007. *World Development Report 2008: Agriculture for Development.* Washington, DC: World Bank.

Worldwatch Institute. 2008. *State of the World.* New York: Norton Institute.

Wright, A. M. 1997. *Toward a Strategic Sanitation Approach: Improving the Sustainability of Urban Sanitation in Developing Countries.* Washington, DC: World Bank.

WWAP (World Water Assessment Programme). 2006. *The United Nations World Water Development Report 2. Water: A Shared Responsibility.* Paris: United Nations Educational, Scientific and Cultural Organization, and New York: Berghahn Books.

WWF. 2008. Infrastructure Problems: River Navigation Schemes. WWF. www.panda.org/about_wwf/what_we_do/freshwater/problems/infrastructure/river_navigation/index.cfm.

Zah, Rainer, Heinz Böni, Marcel Gauch, Roland Hischier, Martin Lehmann, and Patrick Wäger. 2007. Life Cycle Assessment of Energy Products: Environmental Assessment of Biofuels – Executive Summary. Empa, Swiss Federal Institute for Materials Science and Technology, Technology and Society Lab, Gallen, Switzerland.

Evolution of water use

Chapter 8

Impacts of water use on water systems and the environment

Authors: Gunilla Björklund, Jake Burke, Stephen Foster, Walter Rast, Domitille Vallée and Wim van der Hoek

Contributions: Francesca Bernardini, Rudolf Cleveringa, Alasdair Cohen, Jean-Marc Faurès, Sasha Koo-Oshima, Christopher Kuonqui, Robson Mutandi, Laurent Stracasto, Ti Le-Huu and James Winpenny

Coordinator: Jean-Marc Faurès (FAO)

Facilitator: Domitille Vallée

Key messages

- The pattern and intensity of human activity have disrupted – through impacts on quantity and quality – the role of water as the prime environmental agent. In some areas depletion and pollution of economically important water resources have gone beyond the point of no-return, and coping with a future without reliable water resources systems is now a real prospect in parts of the world.
- While the intensity of groundwater use, partly encouraged by subsidized rural electrification, has led to the emergence of many groundwater-dependent economies, their future is now threatened by aquifer depletion and pollution. Prospects for relaxing use of these key aquifers, remediating water quality and restoring groundwater services to ecosystems look remote unless alternative management approaches are developed.
- Our ability to maintain the environmental services we depend on has improved but remains constrained by an incomplete understanding of the magnitude and impact of pollution, the resilience of affected ecosystems and the uses and management of water resources systems. A failure to monitor the negative impacts of water use on the environment and institutional weaknesses in many developing countries prevent effective enforcement of regulatory provisions.
- Relevant information about pollution loads and changes in water quality is lacking precisely where water use is most intense – in densely populated developing countries. As a result, the often serious impacts of polluting activities on the health of people and ecosystems remain largely unreported. Still, there are signs of progress in how pollution and the risks of pollution can be mitigated, and trends in environmental degradation reversed.

Progress in mitigating the negative effects of water development has been slow, while accelerated economic growth has placed additional burdens on resources. There is clear evidence of the degradation of water quantity and quality: drying rivers, aquifers and groundwater basins; bioaccumulation of agrochemicals and heavy metals in fish; algal blooms from high nutrient loads; and the silting of dams and nutrient loss because of the fragmentation of rivers. Many of these impacts are caused by cities, industry and agriculture that lack incentives or obligations to act and report on their performance on water use and pollution mitigation. Overexploitation and pollution are mainly externalities of the activities of users and polluters: users and polluters

An increasing number of river basins lack sufficient water to meet all the demands placed on them, and competition among users can be intense

seldom directly suffer the consequences of their actions. Internalizing these effects thus seems a good way to reduce misuse and pollution. Examples include providing incentives through payments for ecosystem services (such as to local communities for maintaining the integrity of forested watersheds) and for sustaining benefits provided to others (see box 14.23 in chapter 14) and applying the 'polluter pays' principle.

How water use affects water resources

Humans have settled near water bodies for millennia, and human alteration of coastlines, rivers, lakes and wetlands has gone hand in hand with social and economic development. While this has increased the demand to produce more food, urban growth and industrial development have pushed cities to look increasingly farther for the water they need, often taking water from, and perhaps hurting, other users – agriculture and nature.

Disturbing impacts: ecosystems under stress

Increasing water scarcity. Water scarcity occurs when so much water is withdrawn from lakes, rivers or groundwater that supplies can no longer adequately satisfy all human and ecosystem requirements, resulting in more competition among potential users (map 8.1). An increasing number of river basins lack sufficient water to meet all the demands placed on them, and competition among users can be intense (see chapter 9).

Available information hides the full reality of scarcity at local or basin level. This is particularly problematic in large countries such as the United States, where average water use nationally accounts for only 25% of available resources, but can reach 80% on a subnational scale.[1]

Degradation of ecosystems. The Millennium Ecosystem Assessment has demonstrated how modifying landscapes to increase food production and allow development has resulted in adverse ecological changes to many ecosystems, with accompanying loss and degradation of ecosystem services.[2] Synergistic and cumulative effects can make it difficult to attribute changes to a single cause. Losses have adverse effects on livelihoods and economic production,[3] and some ecosystems have passed thresholds into regime shifts, with a collapse in ecosystem services, making the cost of restoration (if possible) very high.

There are many instances where consumptive use and water diversion have severely degraded downstream wetlands or closed basins. Examples include the drastic shrinking Aral Sea in Central Asia and Lake Chapala in Mexico, the world's largest shallow lake. With some of the largest rivers becoming small streams close

Map 8.1 **Increasing water scarcity**

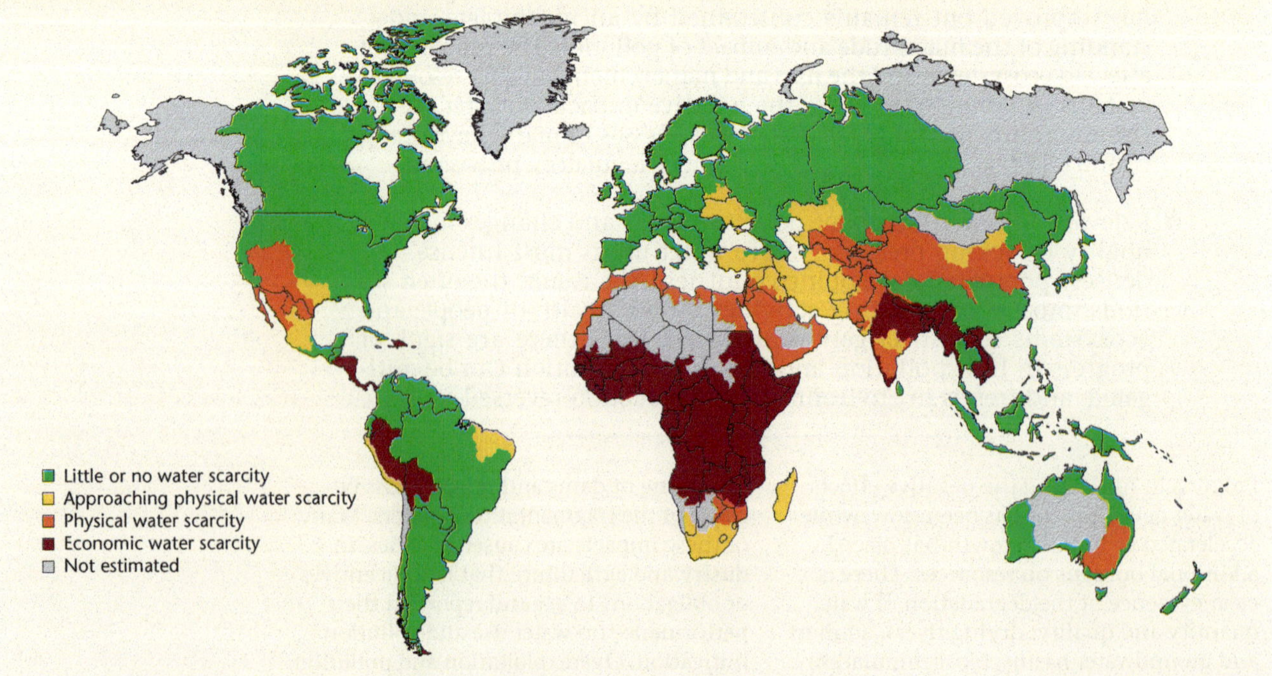

Source: Based on Comprehensive Assessment of Water Management in Agriculture 2007.

to their mouth (such as the Colorado, Murray-Darling, Nile and Yellow), flows are no longer sufficient to maintain the health of aquatic ecosystems.

Water regulation and drainage for agricultural development are the main causes of wetland habitat loss and degradation.[4] Although loss and degradation should be avoided, where they have occurred recovery can sometimes be quick if the right mechanisms are put in place. An example is the Mesopotamian Marshlands, which were deliberately drained but are now being reclaimed (map 8.2). Following over a decade of decline, more than 20% of the original marshland area was reflooded in less than a year between May 2003 and March 2004.[5]

Double-edged impact: more control, less resiliency

The Living Planet Index, developed by the United Nations Environment Programme's World Conservation Monitoring Centre and WWF, is based on trends in populations of vertebrate species. The index shows that on average freshwater species populations were reduced by half between 1970 and 2005, a sharper decline than for other biomes (figure 8.1).

Dams and large interbasin transfers fuelled the prosperity of many ancient civilizations. Construction of such structures has continued throughout history. During the last century the number of such river modifications increased massively. As of 2000 there were more than 50,000 large dams in operation. Some 589 large dams were built in Asia from 1999 to 2001. As of 2005, 270 dams of 60 metres or larger were planned or under construction.[6] The demand for reservoirs of all sizes is expected to continue to grow, particularly in regions with high water demands and a need to cope with the increased variability accompanying climate change (see chapter 11).

Modifications made for water-related development (dams, irrigation schemes, urban extension, aquaculture and the like) have major consequences for the key ecological components or processes of rivers, lakes, floodplains and groundwater-fed wetlands. Dams play a major role in altering water regimes. By substantially reducing discharges into oceans, they modify aquatic habitats and transform flowing systems (rivers) into still or semi-still

Figure 8.1 Biodiversity in freshwater species has declined by half since 1970

Freshwater, temperate and tropical Living Planet Indexes, 1970-2005

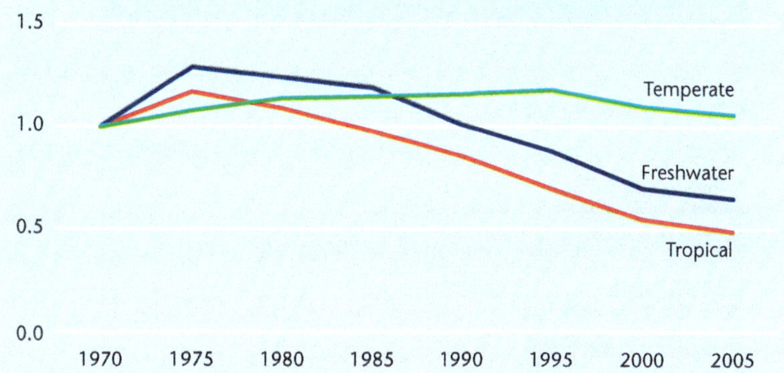

Note: The Living Planet Index tracks trends in populations of 1,313 vertebrate species around the world.
Source: WWF 2008.

Map 8.2 Restoration of the Mesopotamian Marshlands in Iraq, March 2003-December 2005

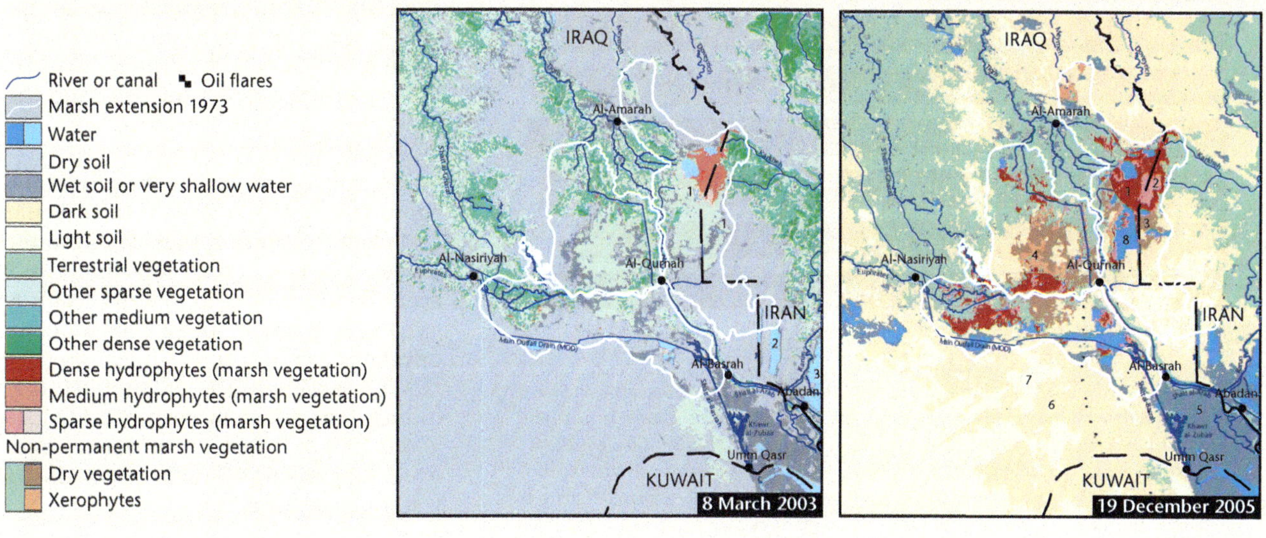

Source: Based on UNEP 2006.

systems (wetlands). Some ecosystems disappear when rivers are regulated or impounded because of the altered flow and new barriers to the movement of migratory species. Humans can also suffer from forced migration and population displacement, two well documented social impacts of dams.[7]

Of the world's 292 largest river systems in 2005[8] (accounting for 60% of the world's runoff), more than a third (105) were considered to be strongly affected by fragmentation, and 68 moderately affected.[9] Wetland ecosystems feel the greatest impact, but terrestrial ecosystems such as forests and grasslands are also affected (figure 8.2).

Ecosystem change has accelerated in many areas, and there is concern that large-scale changes will make some ecosystems more vulnerable to water-related agricultural activities. The non-linear dynamics of ecosystems may lead to abrupt changes that can affect their resilience and capacity to absorb disturbances.[10]

Variability and flexibility are needed to maintain ecosystem resilience. Attempts to keep systems in some perceived optimal state, whether for conservation or production, have often reduced long-term resilience, making the system more vulnerable to change.[11] Modifications of landscapes and reductions of other ecosystem services have diminished the capacity of ecosystems to cope with larger-scale and more complex dynamics through reduced ecosystem resilience locally and across scales.[12]

Water resources development and withdrawals have led to increased turbidity and salinity of water and soil, making land and water management more difficult. Siltation and heavier sediment loads in rivers, due partially to river regulation and the resulting erosion of river banks and the sides of reservoirs, have turned reservoirs into sediment traps and reduced the quantity of sediment being carried into delta areas. The Aswan Dam on the Nile River is such a trap; it stops sediment from reaching the delta front in the Mediterranean to compensate for sand lost to coastal erosion created by waves and currents. It has thus led to faster erosion of the delta and the coast from Egypt to Lebanon.[13]

Water can accumulate in the soil profile through runoff and groundwater recharge and from irrigation if the rate of input exceeds the rate of crop consumption. This accumulation can lead to water-logging (when soil pores are filled with water and oxygen is lacking) and salinization (when the rising water in the soil profile brings diluted salts to the surface). Worldwide, about 10% of irrigated land suffers from water-logging, resulting in a 20% drop in water-logged areas.[14]

Although a problem worldwide, salinization is particularly acute in semi-arid areas

Figure 8.2 Effects of river fragmentation and flow regulation vary by region and biome type, 2005

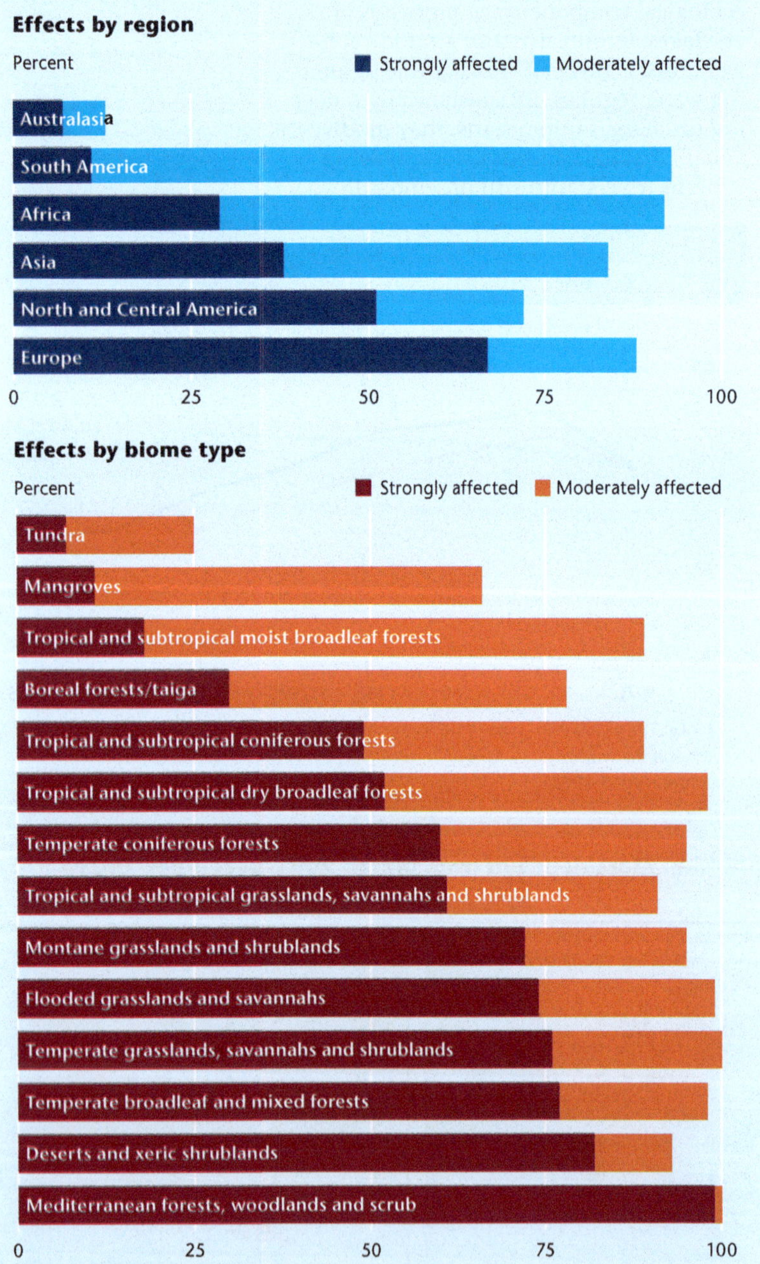

Note: The fragmentation and flow indicator was developed by Umea University in Sweden, in collaboration with the World Resources Institute, for assessing the state of large river systems. Unaffected river systems have no dams in their catchments but can have dams in tributaries if flow regulation is less than 2%. A river system is considered 'affected' if there are dams in the main channel but is never classified as 'strongly affected' if there are no dams in the main channel. All river systems with no more than a quarter of their main channel length left without dams are considered strongly affected.
Source: Based on WWF 2006.

that are heavily irrigated and are poorly drained and where salt is never completely flushed from the land. Such conditions are found in parts of the Middle East, in China's North Plain, in Central Asia and in the Colorado River basin in the United States, among others.

Seeking sustainable management of groundwater

Throughout history groundwater has been integral to human life and livelihoods and to stable agricultural production in the face of hydrologic variability. But groundwater is not evenly distributed around the world. Of total annual precipitation of 577,000 cubic kilometres (km^3) per year (based on long-term averages), 79% falls on the oceans, 2% on lakes and 19% on land.[15] Most of this evaporates or runs off into streams and rivers. Only 2,200 km^3, or 2%, is infiltrated into groundwater.

In many public debates declining groundwater levels or quality are cited as the main reason for the need for management action, but resource depletion and degradation are only part of the problem. Precise data on the status of groundwater resources are still not available in sufficient detail to make a global assessment (see chapters 10 and 13), and long-term continuous depletion has to be distinguished from medium-term system disequilibrium.[16] A more sobering conclusion drawn from detailed local aquifer studies is that where groundwater services are in heavy demand, much of the good quality groundwater has already been used. Contemporary recharge to shallow aquifers has become seriously (perhaps irrevocably) polluted, and relaxing water abstraction and pollution pressure on aquifers will take considerable time.[17]

The development of the power-driven pump in the mid-20th century led to the emergence of many groundwater-dependent economies (figure 8.3)[18] and recently to warnings of the potential adverse impacts of excessive abstraction and of aquifer pollution.[19]

Groundwater development and the response of groundwater systems have not been uniformly documented. Even in industrial countries political realization of the economic and social impact of access to groundwater has tended to come late (generally after some damage to groundwater systems has been done), requiring water resources management agencies to play 'catch-up'. Public investment in groundwater development and protection has been smaller and more dispersed than for surface water resources. Needed most are changes in human behaviour, which requires a much less technocratic approach.[20]

The Intergovernmental Panel on Climate Change has highlighted the implications of accelerated climate change for groundwater,[21] and changes in excess rainfall (recharge and runoff patterns) are expected to add to the resource management burden for both groundwater depletion and rising water tables, depending on the region. But these impacts are likely to be small (and possibly negligible) compared with the stresses placed on groundwater systems by current socioeconomic drivers.

Groundwater-use levels, demand patterns and economic benefits

It is not possible to objectively estimate the current state of groundwater use by country, economic sector or aquifer, though it is evident that some countries have made strong efforts to manage their aquifers sustainably and have achieved good results. The constraints of doing so have been highlighted for the growing importance of groundwater in agriculture through AQUASTAT[22] and for domestic water by the World Bank Groundwater Management Advisory Team (GW-MATE) programme.[23] There has been little

Where groundwater services are in heavy demand, much of the good quality groundwater has already been used and recharge to shallow aquifers has become seriously polluted

Figure 8.3 **Groundwater use has grown rapidly in some countries**

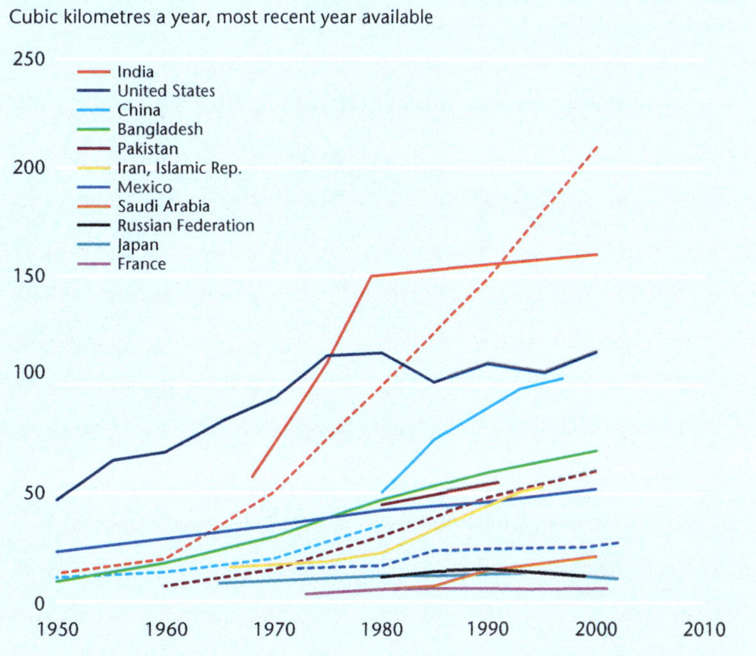

Note: Countries with multiple lines have different datasets that do not reconcile.
Source: Based on Margat 2008.

systematic updating and collection of data on national groundwater use and resource status inventories on a global scale.[24] The situation is improving in Europe, driven by the monitoring requirements of the EU Water Framework Directive and a joint information system for sharing data, the European Environment Information and Observation Network (EIONET). Map 8.3 presents the state of knowledge on renewable and fossil aquifer withdrawals.

Irrigated agriculture is the principal user of the major sedimentary aquifers of the Middle East, North Africa, North America and the Asian alluvial plains of the Punjab and Terai (map 8.4). Less evident is the conjunctive use associated with the concentration of irrigated agriculture and urban development in many alluvial fan and delta environments (such as those of the Chao Praya, Ganges-Brahmaputra, Godavari, Indus, Krishna, Mekong, Narmada, Nile, Mississippi, Po, Yangtze and Yellow Rivers). Reducing stress on these groundwater systems involves more than groundwater resources management. It requires reducing land-based pollution, rehabilitating degraded habitats and conserving water resources.

Map 8.3 **Annual withdrawals of renewable groundwater sources and non-renewable fossil aquifers, most recent year available, 1995-2004**

Renewable groundwater sources on a national basis

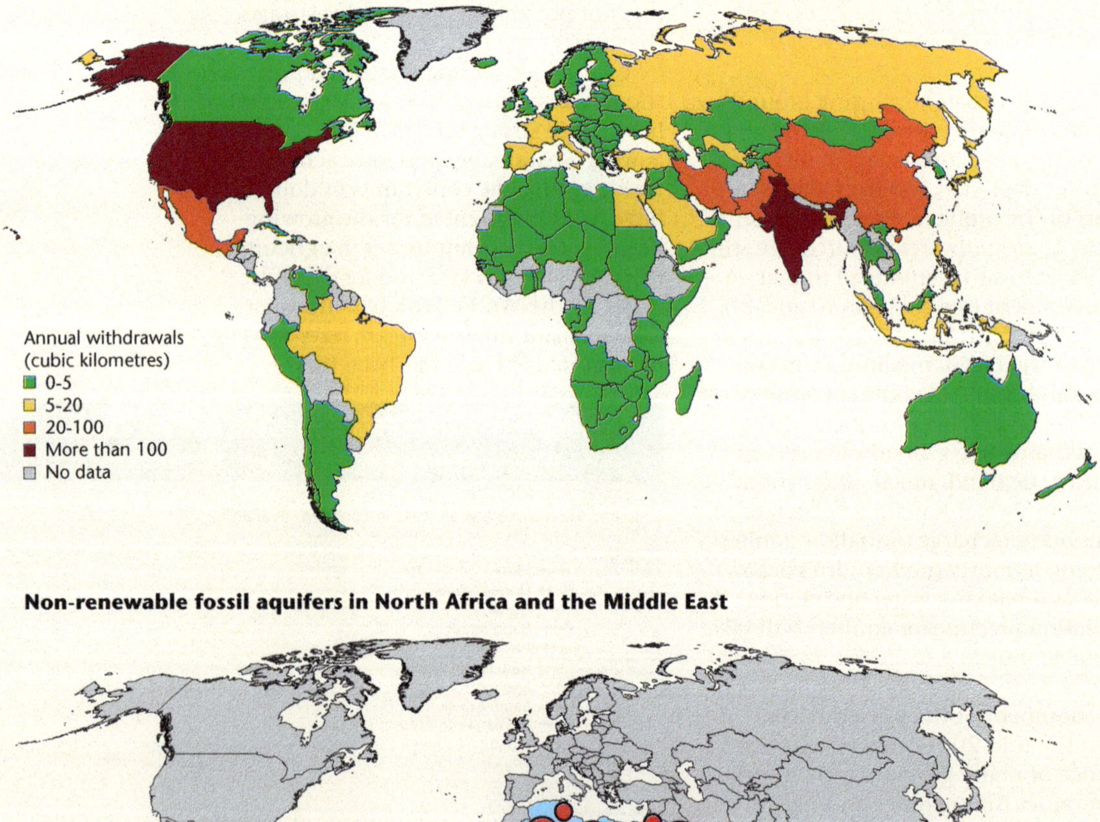

Non-renewable fossil aquifers in North Africa and the Middle East

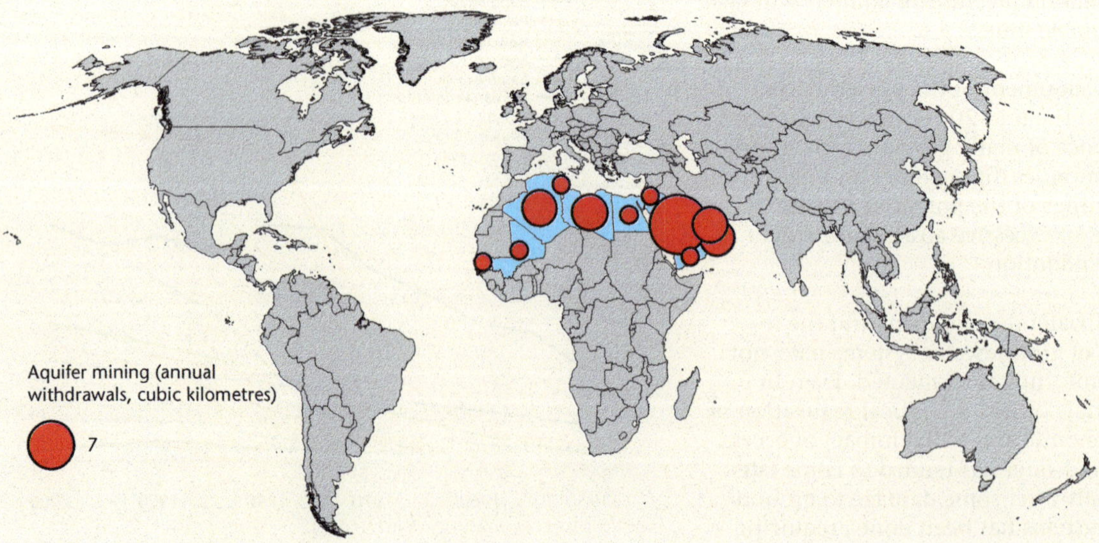

Note: National averages can mask the true situation, which can vary dramatically on a local scale.
Source: Based on Margat 2008.

Map 8.4 **Groundwater use for irrigation on a country basis, most recent year available, 1995-2005**

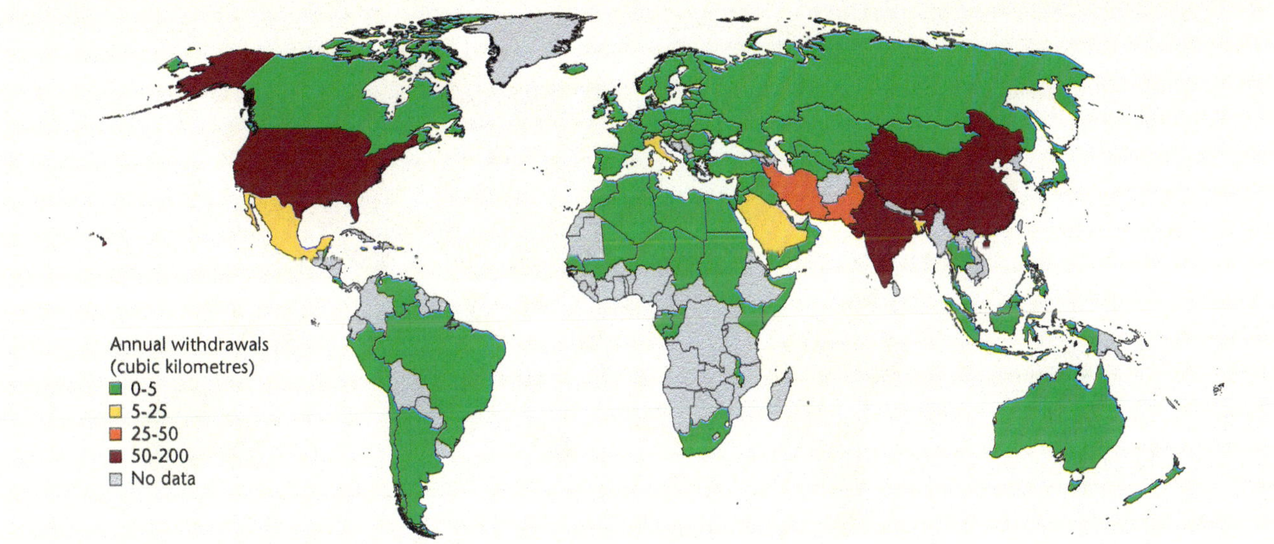

Source: Based on Margat 2008.

Socioeconomic drivers of groundwater development show substantial geographic differences unrelated to resource availability. Agricultural demand for groundwater has often been spurred by both explicit and hidden subsidies for rural electrification, irrigation equipment and occasionally water well construction. In South Asia, for instance, subsidized rural electrification to meet irrigation demands has been a key driver of groundwater use, especially in dryland areas with no surface water services. The concentration of drilling, pumping and water well maintenance services has progressively reduced the cost of groundwater exploitation. The flat-rate electrical energy policy in parts of South Asia (and subsidized rural electricity elsewhere) is not the major cause of groundwater resource overexploitation, but it has allowed grossly inefficient use of energy in pumping groundwater from shallow, low-storage aquifers in hard-rock terrains, effectively bankrupting state electricity providers.[25]

While the rate of agricultural growth has generally slowed over the past 25 years, the progressive adoption of precision agriculture (requiring on-demand, just-in-time irrigation) has considerably intensified groundwater use and boosted its productivity. For the most part, pumping by farmers has been determined less by groundwater management and more by the prices of basic commodities and cash crops compared with the costs of production, including energy for pumping.

Groundwater is a major source of urban water supply around the world (map 8.5), not just in megacities but also in thousands of medium-size towns. Some cities (for example, Beijing, Dhaka, Lima, Lusaka and Mexico City) are located on or near major aquifers, and urban water utilities have drawn heavily on groundwater for their supply. In other cities (for example, Bangkok, Buenos Aires and Jakarta), the share of water derived from groundwater has fallen considerably as a result of aquifer depletion, saline intrusion or groundwater pollution.

These trends have tended to obscure mushrooming growth over the past 10-15 years in private self-supply from groundwater by residential, commercial and industrial users in Latin America and South and South-East Asia. The scale of exploitation is generally determined by the cost of access (existence of shallow aquifers rather than their yield). The initial impetus for private urban groundwater use is the inadequacy of the main utility service level, but once investments in private wells have been made, water use tends to continue because, for those who can afford them, wells usually offer a less expensive supply of water than most water utilities, with larger users paying more because of cross-subsidized tariffs. As a result, groundwater accounts for more than 30% of urban water supply (and a higher proportion by number of consumers), even in many cities and towns far from major known aquifers. Heavy private use of groundwater in urban areas complicates the finance and operation of water utility services – both water supply and sewerage. Increasingly,

Map 8.5 **Groundwater use for drinking water on a country basis, most recent year available, 1995-2005**

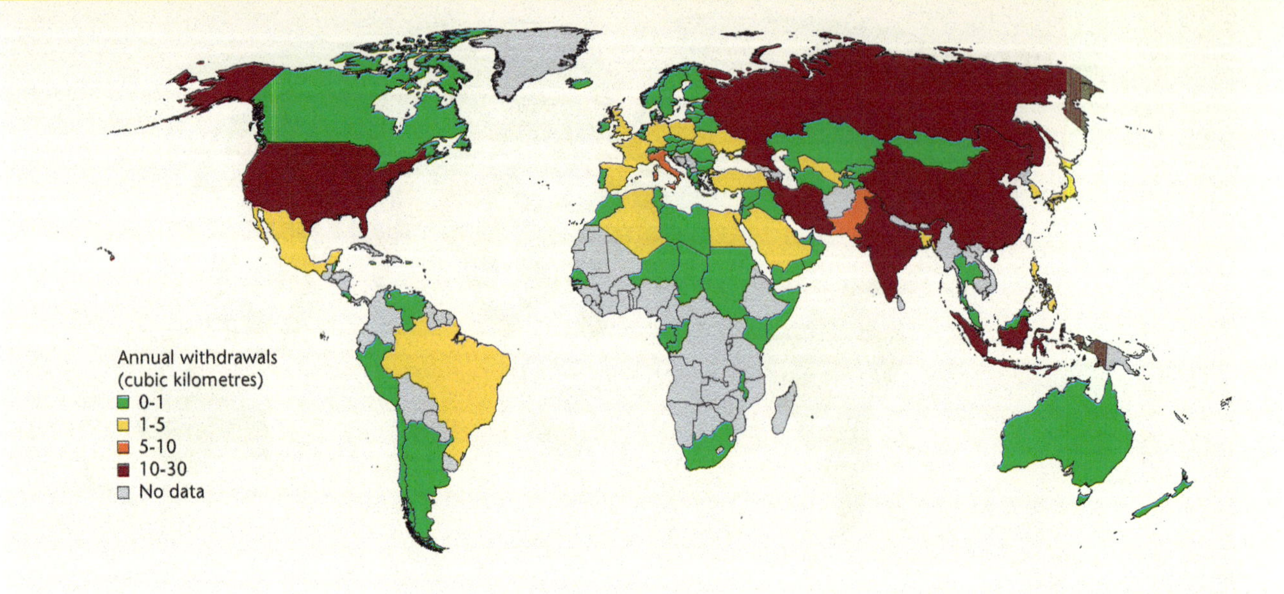

Source: Based on Margat 2008.

African cities are using boreholes to improve water security, with the aim of easing pressure on water facilities in densely populated suburbs.

Social, economic and environmental risks

Three aquifer characteristics determine whether groundwater services will ultimately prove sustainable:

- Vulnerability to pollution under contaminant pressure from the land surface.
- Susceptibility to irreversible degradation from excessive exploitation.
- Renewability of storage reserves under current and future climate regimes.

These characteristics vary widely by aquifer type and hydrogeologic setting. Vulnerability to pollution is generally linked to an aquifer's accessibility. Aquifers that are shallow and 'open' to regular and dependable recharge are more likely to suffer pollution from agrochemicals and urbanization (in particular, from low-cost wastewater disposal and careless disposal of industrial chemicals). Aquifer development and effluent disposal for urban water supply have far-reaching implications for public health, municipal planning and resource sustainability. The presence of groundwater initially relieves financial pressure on urban water service utilities, but often at the cost of degrading shallow aquifers and complicating public health emergencies. In Europe land use zoning is now used to protect vulnerable key aquifers that provide municipal water supply, or deeper confined groundwater sources are developed that are naturally protected from urban pollution.

A recent study of the water economics of the Middle East and North Africa region estimates that groundwater resource depletion has substantially reduced GDP in some countries – Jordan by 2.1%, Yemen 1.5%, Egypt 1.3% and Tunisia 1.2%.[26] Reductions in groundwater stocks (whether renewable or non-renewable) appear to have been translated into reduced economic productivity of water. More difficult to assess are the positive economic impacts of groundwater abstraction and where access to and use of groundwater results in economic liberation. Oman, Saudi Arabia and the United Arab Emirates, for example, rely almost exclusively on non-renewable groundwater and desalination for water supply. In all such calculations, however, it should be kept in mind that values for groundwater 'reserves' or 'stocks' and 'sustainable yield' are not precise, while the partition between 'renewable' and 'non-renewable' is a further complication.[27]

Sharp points of competition over groundwater resources between urban and rural users are now becoming more apparent. Expanding municipalities and expanding light industrial and commercial activities in peri-urban and linked rural areas are competing with agriculture over groundwater quantity and quality. Even if agriculture is, within certain limits, indifferent to groundwater quality, precision agriculture is likely to be located near urban areas, and it uses high quantities of chemical fertilizers, pesticides and fungicides, which contaminate shallow groundwater. While relatively benign organic alternatives to persistent chemical compounds exist, the impacts of large-scale commercial farming near cities on what may be key strategic groundwater reserves should not be underestimated. Municipalities, water and environmental regulators and agricultural agencies are all implicated, with the evidence pointing to an enormous disconnect between water and land use regulations. Effective measures for groundwater quality protection require that the two be considered together.

Complex management

Groundwater recharge processes are extremely complex, and there is still considerable uncertainty about their relationship with natural vegetation, land management and groundwater use. While many specific local-scale recharge studies are available, knowledge of the range of recharge modes across large river basins and their linked aquifers rarely comes together to form a systemic overview. For example, in the complex Ganges-Brahmaputra and Indus aquifers, snow and glacier melt determine indirect recharge, but shallow and deep groundwater circulation emerges in the piedmont fans and deep alluvial basins, with re-emergent flows, water-logging and salinization as a result of surface water and groundwater interactions perturbed by irrigation. These systems are at risk but have not had the same detailed hydrogeological evaluation as has the similarly complex Chalk Aquifer of Northwest Europe[28] or the large regional groundflow systems of the United States.[29]

For many heavily exploited aquifers groundwater abstraction and use are still poorly quantified, and dedicated groundwater monitoring networks have not been established (see chapter 13). Instead, periodic head observations are made of pumped wells, which give only an approximate measure and are completely inadequate for detecting response to recharge events. Many cities are working 'blind' when they distribute water supplied from groundwater sources. There are reported cases of drops in drinking water production, subsidence due to excess pumping (see box 12.4 in chapter 12), leakages from drinking water supply systems into shallow aquifers and even contamination (leaks from sanitation networks or leachates from wastes).

Much is expected of water harvesting and managed aquifer recharge – and many assumptions are made about the potential for recharge. In some hydrogeologic settings it is difficult to improve on the efficiency of natural recharge processes, while in others the economically feasible proportion of recharge enhancement over natural recharge is very limited, although the techniques can help solve local problems and improve groundwater quality. The highest management priority, though, will always be to protect the main recharge zones.

Resource prospects and future management needs

Population and income growth projections and associated increases in water demand will place unprecedented demands on aquifer systems. Further depletion and degradation of aquifers should be anticipated unless there is much more investment in effective governance and management practices. In addition, climate change will place some key aquifers under additional pressure.

In heavily populated areas with large numbers of irrigation systems using groundwater, communities will need to self-regulate resource use. Approaches pioneered in Tunisia are now being put forward in South Asia. Resource management institutions need to better understand the socioeconomic drivers to which irrigated agricultural users respond. Social marketing of basic groundwater information to farmers can have a positive impact, as in the case of Andhra Pradesh farmer-managed groundwater systems in India.[30] It is important to identify the kind of information that will catalyse self-regulation. Demand-side approaches that focus on obtaining consensus on aquifer use may have more success in the long run than technical supply-side or hardware-led approaches.

In and around large urban areas economic competition for water is forcing agriculture to adapt – by boosting its productivity and minimizing its environmental impact – or to disappear. Demand for precision agriculture will be unrelenting, and further concentration of agricultural activity can be expected as market chains are reinforced (cold storage, niche global markets and the like), along with further adoption of agronomy advances such as deficit irrigation and subsurface drip.

Municipalities, water and environmental regulators and agricultural agencies are all implicated in the failure to manage groundwater resources, with the evidence pointing to an enormous disconnect between water and land use regulations

Information about pollution loads and water quality changes is lacking in many countries because of inadequate monitoring systems

While the trend towards more precision agriculture will boost overall groundwater productivity, it will not necessarily relieve abstraction pressure on aquifers because it will potentially improve the feasibility of abstraction by larger pumping lifts.

The tension between private and public services derived from aquifers remains. More convergent and sustainable resource use will be achieved only through substantial investment in management operations on the ground, working primarily through community consultation and cross-sectoral policy dialogue. Such dialogue is supported by shared knowledge and common understanding of the current situation and future options, as illustrated in the country policy support programmes of the International Commission on Irrigation and Drainage that have been tested in river basins in China (Jiaodong Peninsula and Qiantang River) and India (Brahmani and Sabarmati). Dialogue on these basins was supported by a user-friendly model for investigating risks and adaptation strategies using a scenario approach for looking at supply and demand and quantity and quality for different sectors of use.

Growing risks: pollution and degradation of water quality

Despite improvements in some regions, water pollution is on the rise globally. And unless substantial progress is made in regulation and enforcement, pollution is expected to increase as a result of economic development driven by urbanization, industries and intensive agriculture systems. Many industries – some of them known to be heavily polluting (such as leather and chemicals) – are moving from high-income countries to emerging market economies (box 8.1), where they benefit from various incentives, a lower-cost workforce and, in some cases, less stringent environmental regulations.

Information about pollution loads and water quality changes is lacking in many countries because of inadequate monitoring systems. As a result, the often-serious impacts of polluting activities on human and ecosystem health remain unreported or underreported.

An overview of water quality issues: the increasing threat from pollution

Pollution's increasing threat to water quality. Human-generated water pollution is a serious threat to human and ecosystem health, but its impact is hard to quantify. Despite monitoring inadequacies, there are local signs that the declining quality of domestic water supply sources is becoming a major concern in many countries.

Pollution typically refers to chemicals or other substances in concentrations greater than would occur under natural conditions. Major water pollutants include microbes, nutrients, heavy metals, organic chemicals, oil and sediments; heat, which raises the temperature of the receiving water, can also be a pollutant. Pollutants are typically the cause of major water quality degradation around the world. Virtually all goods-producing activities generate pollutants as unwanted by-products (see chapter 10).

Water pollutants are categorized as point or non-point according to their primary sources (table 8.1). Point sources are pollutants from pipelines and other readily identifiable sources. Non-point sources are pollutants mobilized by precipitation as it flows over the land and infiltrates the soil. Non-point source pollutant loads in a drainage basin are a function of precipitation patterns and the range of human activities in the basin (especially agriculture). The ecological footprint of consumption[31] shows where human activities to meet a population's needs are well above the respective region's biocapacity (map 8.6), implying that such regions are running an ecological deficit and depend increasingly on the natural wealth and resources of other countries.

The most important water contaminants created by human activities are microbial pathogens, nutrients, oxygen-consuming

Box 8.1 Asian 'Tigers' and the hidden tip of the pollution iceberg

Industrial and economic development in Asia, especially among rapidly growing 'Asian Tiger' economies, has in many cases come at the expense of water resources. Rapid urbanization across Asia and the Pacific will continue to shape the parameters of water use trends, which affect prospects for water scarcity. Although rural populations in Asia are projected to remain stable over the next 20 years, urban populations are likely to increase by 60% before 2025. While attention focuses on the multiple challenges of megacities, smaller urban areas, with few financial and technical resources due to their weak political clout, are set to continue current trends of poor wastewater management, posing graver threats to water resources than physical scarcity does.

Current strategies for economic development have propelled river degradation to the top of the water use agenda. Malaysia, for example, recently witnessed an increase in the number of rivers deemed slightly polluted and a decline in the number of rivers considered clean. As recognition of the problem grows, more efforts are being directed towards river rehabilitation.

Source: Le-Huu Ti, Chief, Water Security Section, United Nations Economic and Social Commission for Asia and the Pacific, based on information from the Asian River Restoration Network and the fresh and coastal waters session at the 3rd Southeast Asia Water Forum Regional Workshop, 23 October 2007, Kuala Lumpur.

Table 8.1	Comparison of point and non-point sources of water pollution
Point sources	**Non-point sources**
Fairly steady volume and quality	Highly dynamic; occurs at random intervals closely related to hydrologic cycle
Variability of values typically less than one order of magnitude	Variability of values can range across several orders of magnitude
Most severe water quality impacts typically occur during low-flow summer periods	Most severe water quality impacts occur during or after storm events
Enters receiving waters at identifiable points, usually through pipelines or channel sources	Entry point to receiving waters usually cannot be identified; typically arises from extensive land areas
Can be quantified with traditional hydraulic techniques	Difficult to quantify with traditional techniques
Primary water quality parameters are organic water pollutants (biological oxygen demand), dissolved oxygen, nutrients, suspended solids and sometimes heavy metals and synthetic organic chemicals	Primary water quality parameters are sediments, nutrients, heavy metals, synthetic organics, acidity and dissolved oxygen
Control programmes typically applied by government agencies	Control programmes involve individuals not normally considered in pollution control programmes (such as farmers and urban homeowners)

Source: Thornton et al. 1999.

Map 8.6 Ecological footprints and biocapacity, 2000

Ecological footprints on a country basis

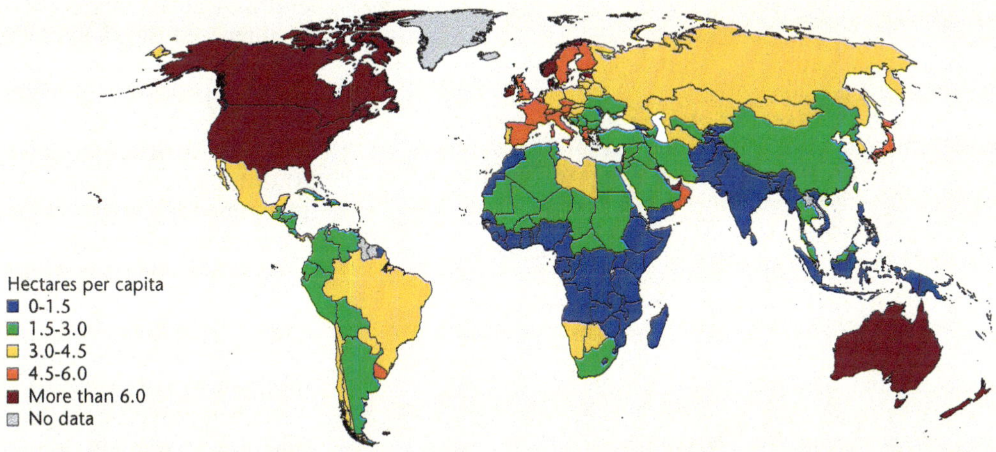

Biocapacity on a country basis

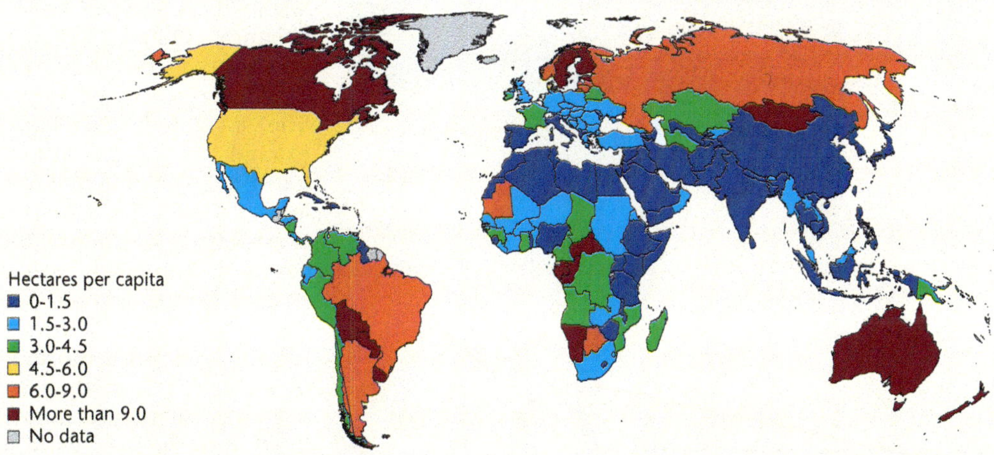

Note: Ecological footprint is a measure of the area needed to support a population's lifestyle. This includes the consumption of food, fuel, wood and fibres. Pollution, such as carbon dioxide emissions, is also counted as part of the footprint. Biocapacity measures how biologically productive land is. It is measured in 'global hectares' – a hectare with the world average biocapacity. Biologically productive land includes cropland, pasture, forests and fisheries.
Source: Ewing et al. 2008.

Globally, the most prevalent water quality problem is eutrophication

materials, heavy metals and persistent organic matter, as well as suspended sediments, nutrients, pesticides and oxygen-consuming substances, much of it from non-point sources. The most important pollutant affecting human health is microbial contamination. Inadequate sanitation facilities, improper wastewater disposal and animal wastes are the major sources of microbial pollution. In at least 8 of the United Nations Environment Programme's 13 Regional Seas Programme regions, over 50% of the wastewater discharged into freshwater and coastal areas is untreated, rising to over 80% in 5 regions.[32]

Pollution-causing activities

Globally, the most prevalent water quality problem is eutrophication, a result of high-nutrient loads (mainly phosphorus and nitrogen), which substantially impairs beneficial uses of water (see the example of the Baltic Sea in box 8.2). Major nutrient sources include agricultural runoff, domestic sewage (also a source of microbial pollution), industrial effluents and atmospheric inputs from fossil fuel burning and bush fires. Lakes and reservoirs are particularly susceptible to the negative impacts of eutrophication because of their complex dynamics, relatively longer water residence times and their role as an integrating sink for pollutants from their drainage basins.[33] Nitrogen concentrations exceeding 5 milligrams per litre of water often indicate pollution from human and animal waste or fertilizer runoff from agricultural areas.

Excessive nutrient inputs can also cause harmful algal blooms. Cyanobacteria, also known as blue-green algae, have increased in freshwater and coastal systems such as the East China Sea in recent decades (figure 8.4). The toxins produced by the excessive algal blooms are concentrated by filter-feeding bivalves, fish and other marine organisms and can cause fish and shellfish poisoning. In people they can cause acute poisoning, skin irritation and gastrointestinal illnesses. There are global warming implications associated with this phenomenon, as cyanobacteria have a competitive advantage over other types of algae at higher temperatures.

Organic materials, particularly from domestic wastewater treatment plants, food-processing discharges and algal blooms, are decomposed by oxygen-consuming microbes in water bodies, as measured by biochemical oxygen demand (BOD). Thermal stratification in nutrient-enriched lakes with high BOD levels can produce chemical conditions allowing nutrients and heavy metals in lake bottom sediments to re-enter the water column. Lake Erie's oxygen-depleted bottom zone, for example, has expanded since 1998, with harmful environmental impacts on the lake's fisheries. The eastern and southern coasts of North America, the southern coasts of China and Japan and large areas around Europe have also undergone oxygen depletion.[34] One of the world's largest 'dead zones' has appeared off the mouth of the Mississippi River in the Gulf of Mexico, attributed to excessive nitrogen loads from the river, with harmful impacts on biodiversity and fisheries.[35] Projected food production needs and increasing wastewater effluents associated with an increasing population over the next three decades suggest a 10%-15% increase in the

Box 8.2 Addressing eutrophication and its effects in the Baltic Sea

In 1998 approximately 90% of the coastal and marine biotopes in the Baltic Sea were threatened by loss of area or reduction in quality from eutrophication, contamination, fisheries and settlements. Agriculture, urbanization and atmospheric deposition were considered the root causes of eutrophication. For agriculture the primary causes were mainly inadequate adoption of modern agricultural technology and inadequate integration of environmental and agricultural practices. The causes of pollution from urban sources were lack of investment in wastewater facilities and high growth rates. Atmospheric pollution was caused by energy production and transport in response to population growth and urbanization, increased sea and road traffic, ineffective laws and regulations to control emissions and inadequate transport policy.

Environmental protection legislation and several new measures in the region have resulted in some improvement, as reported in the Helsinki Commission 2003 assessment. Phosphorus inputs have decreased considerably following measures taken by the Baltic Sea riparian countries, but eutrophication still remains an urgent problem in most coastal areas.

Source: Helsinki Commission 2007; Lääne, Kraav, and Titova 2005.

Figure 8.4 Increasing frequency of harmful algal blooms in East China Sea associated with increasing fertilizer use

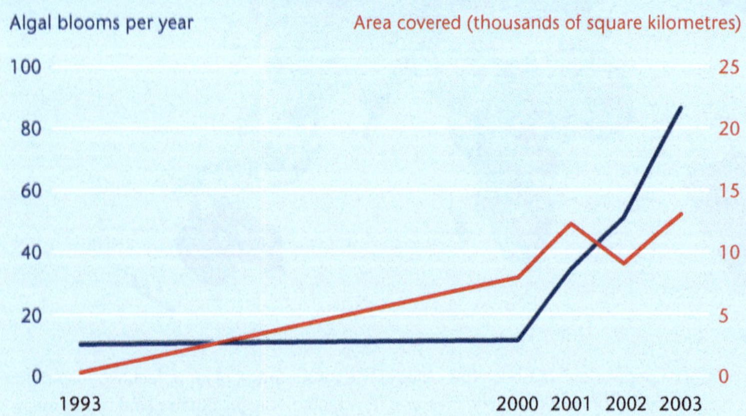

Source: UNEP/GIWA 2006.

river input of nitrogen loads into coastal ecosystems, continuing the trend observed during 1970-95.[36]

Heavy metals can accumulate in the tissues of humans and other organisms. An example is the serious human health impacts of the high natural concentrations of arsenic in groundwater in parts of Bangladesh and adjacent parts of India (box 8.3). Mercury and lead from industrial activities, commercial and artisanal mining and landfill leachates also threaten human and ecosystem health in some areas, with emissions from coal-fired power plants being a major source of the mercury accumulating in the tissues of fish at the top of fish trophic levels.

An emerging water quality concern is the impact of personal care products and pharmaceuticals, such as birth control pills, painkillers and antibiotics, on aquatic ecosystems. Little is known about their long-term human or ecosystem impacts, although some are believed to mimic natural hormones in humans and other species.

The level of pollution is a function of the structure of a country's economy and its institutional and legal capacity to address it. Groundwater systems are very vulnerable freshwater resources, since once contaminated, they are difficult and costly to clean – when cleaning is technically feasible at all. Pollutants from non-point sources, such as leaching of excess nitrates or pesticides used on agricultural lands or heavy metals in mines, can take decades to reach the aquifers and once they do it may be too late or too expensive to act. With an increasing load of chemical substances being discharged into water systems and onto agricultural lands, uncertainties persist about the long-term effects on human and ecosystem health. A recent study on drinking water in France estimated that more than 3 million people (5.8% of the population) were exposed to water quality that does not conform to World Health Organization (WHO) standards (for nitrates, non-conformity was found in 97% of groundwater samples).[37]

Storm-generated runoff from agricultural and urban areas is the most important non-point pollutant source (such as leached nitrates in runoff and its accumulation in rivers) in most countries. Further, as point sources (such as urban sewage discharge) are becoming more controlled, non-point pollutant loads are becoming an increasing concern. The US Environmental Protection Agency, for example, notes that agricultural activities contribute the largest quantity of pollutants to water bodies in the United States, and the situation is probably similar in many other countries.[38]

Box 8.3 The arsenic crisis: as yet no solution

Some 10 years have passed since the extent of the arsenic poisoning disaster in Bangladesh became widely known. High levels of arsenic were found in tubewells constructed for drinking water. Today, up to 70 million people in Bangladesh are exposed to water that contains more than the World Health Organization threshold value of 10 micrograms of arsenic per litre. Up to half the estimated 10 million tubewells in Bangladesh might be contaminated with arsenic. Large amounts of arsenic-contaminated groundwater have also been used for irrigation, leading to the appearance of arsenic in the food chain. Natural arsenic pollution of drinking water, although originally linked to Bangladesh and the state of West Bengal in India, is now considered a global threat with as many as 140 million people affected in 70 countries on all continents.

Source: Bagchi 2007; Fry et al. 2007.

Localized pollution occurs frequently with mining activities. If no mitigation measures are in place, the pollution can lead to serious environmental degradation and water contamination. Negative impacts include lowering of the water table – with negative consequences for vegetation, ecosystems and farming – and groundwater contamination with heavy metals in water drained from mines and waste (mine tailings) affecting downstream ecosystems and drinking water.

Problems with heavy metals go beyond the drinking water supply, also affecting food quality (as with the preferential accumulation of cadmium in rice grain when effluent from zinc mines is used for irrigation). The United Nations Economic Commission for Europe finds that mining activities have had severe impacts on water and the environment in Eastern Europe, South-East Europe, the Caucasus and Central Asia.[39] In some river basins (for example, in Belarus, Kyrgyzstan and Tajikistan) the mining industry is a major past or current pollution source, with numerous storage facilities (such as tailing dams for mining wastes) presenting substantial risks (box 8.4).

Progress in mitigating pollution

There are signs of progress in addressing pollution and pollution risks in different sectors. There is well documented evidence that the costs of inaction are high and that some impacts may be irreversible or nearly so (contamination of groundwater drinking water, ecosystem losses).[40] Polluted water has a high human health

Most countries have legislation to protect their water resources, but implementation often lags because responsibilities are dispersed and the costs are high

cost. One-tenth of the global burden of disease (measured in disability-adjusted life years) can be attributed to water, sanitation and hygiene, and water and environmental factors.[41] Other pollution costs include clean-up, additional treatment and damage to fisheries, ecosystems and recreation. Most countries have introduced legislation to protect their water resources, but implementation often lags because responsibilities are dispersed across multiple institutions and the costs of control and monitoring are high.

There are examples, mostly in develped countries, where the 'polluter pays' principle has stimulated changes in attitudes towards pollution and led to recycling and reuse, clean processes for industry, organic farming and so on or collect and treat effluents with an 'end of the pipe' approach that collects, controls, treats and monitors performance. However, investments for such approaches are lacking at all levels (from household sanitation and industrial processes to city waste treatment plants). Experience from high-income countries shows that a combination of incentives, including more stringent regulation, enforcement and targeted subsidies (where justified), can boost investments in water pollution prevention and treatment. Sanitation systems and sewage treatment (preferably with industrial discharges separated from domestic sewage) can reduce the risk of discharge into natural systems. Such practices increase production costs for water services, and effective cost-recovery systems are needed to ensure that pollution control remains sustainable.

Urban settlements are the main source of point-source pollution. In contrast,

Box 8.4 Long-term impacts of mining on ecosystems in transition economies

While pollution-abatement technologies exist for hazardous substances, their use in economies in transition is limited to a small number of economically viable industrial plants.

In January 2000 a dam holding tailings from gold extraction overflowed in Baia Mare in Romania, releasing about 100,000 cubic metres (m^3) of waste, including about 70 tonnes of cyanide and other heavy metals. The spill affected local rivers, interrupting the drinking water supply for about 24 communities and killing several thousand tonnes of fish. Following this accident, several EU directives and national laws were enacted to improve the safety of mining facilities, and most mining companies are now using improved technology.

On 29 August 2003 the Sasa lead and zinc mine in northeastern FYR Macedonia released some 486,000 tonnes of mine tailings into the Kamenicka River. Deposits of large quantities of toxic heavy metal (including arsenic, cadmium, copper, lead, manganese, nickel and zinc) and the release of acidic drainage to surface and groundwater led to acidification, degradation of water resources, accumulation of metals in sediments and bioaccumulation, with severe ecological impacts. This industrial pollution accident contaminated the water source that supplies irrigation water for 25,000 hectares of farmland, reducing economic, environmental and agricultural viability.

The map shows the multiple pollution hot spots around mining sites in the Balkans – the result of a long history of mining and industries with limited pollution prevention.

Hazardous industrial sites, water pollution and mining hotspots in the Balkans, 2007

Source: Based on UNEP/GRID Arendal 2007.

Source: UNEP/OCHA 2000; UNEP/GRID Arendal 2007; Koo 2003.

economic recession can reduce pollution (box 8.5).

Urban sewage treatment still limited mainly to high-income countries

To achieve pollution mitigation objectives for the environment and human health, improved sanitation must be accompanied by sewage treatment. Sewage treatment is the removal of physical, chemical and biological contaminants from wastewater, both surface drainage and domestic, using physical, chemical and biological processes. The objective is to produce a treated waste stream (or treated effluent) and solid waste or sludge suitable for discharge or reuse back into the environment.[42] Data on the rates and levels of collection and treatment of sewage are limited and often difficult to compare.

Sewage: a problem to manage? More than 80% of sewage in developing countries is discharged untreated, polluting rivers, lakes and coastal areas.[43] Even in some developed countries treatment of urban wastewater is far from satisfactory. The OECD online environmental compendium finds a broad range of applications of tertiary waste treatment, from 3.6% in Turkey to 90% in Germany.[44]

In most low- and middle-income countries wastewater is discharged directly into the sea or rivers without treatment. Urban wastewater constitutes a significant pollution load and is particularly hazardous when mixed with untreated industrial waste – a common practice. Many large cities still have no treatment plants or plants quickly become undersized as urban population growth outpaces investments. A nationwide survey in Pakistan found that only 2% of cities with a population of more than 10,000 had wastewater treatment facilities and that less than 30% of wastewater receives treatment in these cities.[45] Some 36% of wastewater is used in agriculture (2.4 million m^3 a day directly for irrigation and 400,000 m^3 a day is disposed of in irrigation canals), and 64% is disposed of in rivers or the Arabian Sea. In many developing countries waterborne sanitation systems and pollution mitigation facilities may not be the most sustainable option; other improved facilities may be more suitable (for example, using lagoons for collective units and ecosanitation units for rural households; see box 8.6).

In developed countries wastewater is progressively coming under control. Over the last 20 years Europe's Urban Wastewater Treatment Directive has resulted in significant improvements in treatment capacity, with more advanced wastewater treatment becoming increasingly common (box 8.7 and figure 8.5). Continuous progress is being made. Belgium, for example, put in operation a mega-treatment plant, which has improved its situation since 2006, which figure 8.5 reflects.

Box 8.5 Impact of economic recession on pollution in Eastern Europe

With the economic recession of the 1990s and a decline in highly polluting industries, the amount of wastewater and pollutants discharged in Eastern European countries fell considerably. Although many of these economies have since recovered and their industrial activity has increased, the structure of their industrial sectors has shifted towards less-polluting industries, especially in the new EU member states. In compliance with strict EU legislation, these countries have also constructed new wastewater treatment plants, further reducing pollution.

There has also been a marked reduction in pesticide and fertilizer use in agriculture, and river quality has consequently improved in many places. Water abstractions also declined, falling 70% in industry and agriculture in the new EU member states and 74% in agriculture and 50% in industry in Eastern Europe, Caucasus and Central Asia compared with 1990.

However, the economic recession also resulted in a breakdown of essential systems of water supply and wastewater treatment in the region. Thus, many rivers and drinking water supplies, especially downstream from cities and industrial and mining regions, are heavily polluted.

Source: EEA 2003.

Box 8.6 A wastewater river

The Musi River runs through Hyderabad, one of India's fastest growing cities, with a population of 6.8 million in 2005 and expected to exceed 10 million in 2015. The mostly untreated domestic and industrial wastewater from the city fills the dry riverbed, converting it into a perennial wastewater river. The wastewater provides livelihoods to low-income groups of urban dwellers and migrants from rural areas in a hidden economy, neither recognized nor supported by the local government. Downstream of the city, Musi water is retained in large and small reservoirs with the help of weirs and then diverted into irrigation canals and village tanks to be used by farmers for crop production, mainly rice.

A clear improvement in river water appearance and smell can be observed with increased distance from the city, and surveys have confirmed quality improvements. Infection rates of farmers with intestinal nematodes are significantly higher close to Hyderabad than further downstream.

Source: Ensink, Mahmood, and Dalsgaard 2007; Buechler and Devi 2003; van Rooijen, Turral, and Bigg 2005.

Water from the Musi River sampled at 0, 5, 10, 14, 18, 20, 30 and 40 km downstream of Hyderabad

Photo credit: Jeroen Ensink

Box 8.7 Setting target for pollution mitigation and limits for reuse – the example of Europe

The Urban Wastewater Treatment Directive of the European Union prescribes the level of treatment required before discharge. Collecting systems must be provided for agglomerations of more than 2,000 people, and secondary treatment (biological treatment) must be provided for agglomerations of more than 2,000 people discharging into freshwater and estuaries and agglomerations of more than 10,000 people discharging into coastal waters. Special requirements are placed on five determinants of treatment performance.

For smaller agglomerations and those equipped with a collecting system, treatment must be appropriate, meaning that the discharge allows the receiving waters to meet relevant quality objectives. The European AQUAREC project proposes seven categories of water quality for different types of reuse and compiles microbial and chemical limits for each category. The limits are based on recently published guidelines and risk estimates, including the most important microbial parameters.

Source: www.eea.europa.eu; Salgot et al. 2006.

Figure 8.5 Levels and types of wastewater treatment in OECD countries and selected European countries, 2006

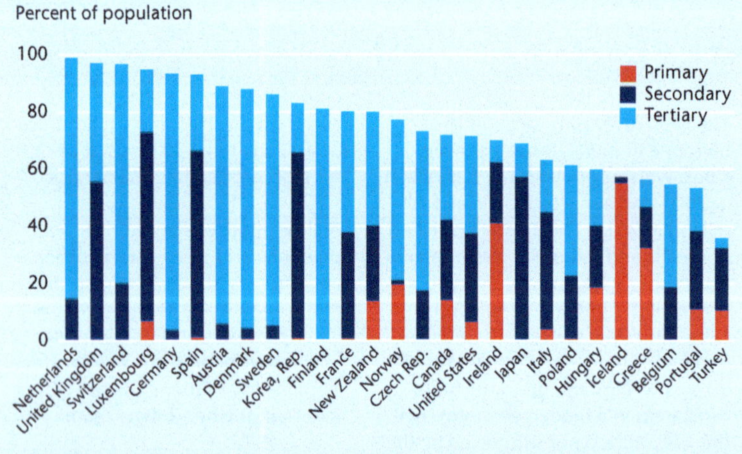

Source: Based on OECD 2008a.

Figure 8.6 High level of wastewater reuse in water-stressed countries, various years

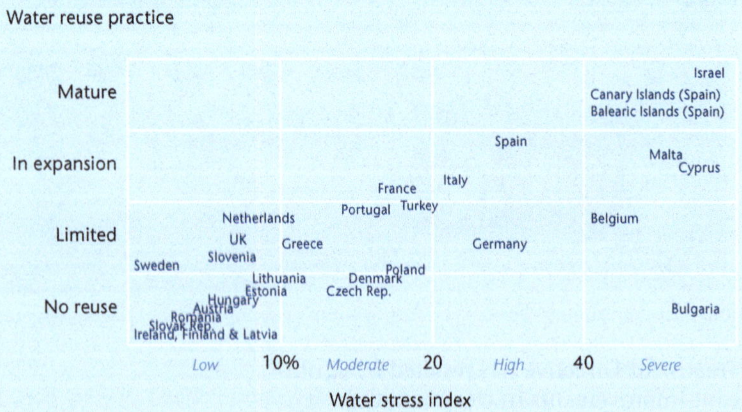

Source: Based on Wintgens and Hochstrat 2006.

As more wastewater is treated, cities must also deal with increasing volumes of sludge. Dumping and incineration can transfer pollution from water to soil or air. Some 60% of the sludge produced by treatment plants in France is used as fertilizer in agricultural areas after undergoing additional treatment, though convincing people to use the sludge remains a struggle.[46]

Sewage: a resource to use? Increasingly, sewage is being seen as a resource. A European Commission–funded project, AQUAREC (Integrated Concepts for Reuse of Upgraded Wastewater), identifies several uses for treated sewage: agricultural irrigation, urban landscaping and recreational uses, industrial cooling and processing and indirect potable water production (through groundwater recharge, for example).[47]

Wastewater is already being reused in water-stressed countries (figure 8.6). Farmers in peri-urban areas use streams for agriculture and aquaculture as in the past, but now increasingly also use wastewater and the nutrients in it. Wastewater flows are typically more reliable than freshwater sources and are rich in nutrients for the cultivation of high-value crops. Authorities are concerned about the practice because of the potential human health risks and because it presents obstacles to installing wastewater treatment plants. Enforcement of water quality standards is often complicated by ambiguous lines of authority; for example, should standards be enforced by health, agricultural or water supply and sanitation agencies?

There are no reliable figures on the extent of wastewater use in agriculture at the global level, but it is estimated that some 20 million hectares of agricultural land is irrigated by untreated, partially treated or wastewater-polluted river water.[48]

Restricting the crops that can be grown with wastewater is difficult because farmers grow crops that have high demand in the local market and that are thus the most profitable to cultivate. Municipal officers in Pakistan, where untreated wastewater is used in 80% of cities, indicated that enforcement was futile as farmers would reopen wastewater inlets within hours of their closure.[49] Courts in Pakistan have found in favour of farmers, ruling that access to irrigation water is a fundamental right and that the loss of livelihood overrides potential health risks.

WHO recently revised its *Guidelines for the Safe Use of Wastewater, Excreta and Greywater in Agriculture and Aquaculture.*[50] The third edition of the WHO guidelines allows for risk assessment and management

Improving hygiene is a cost-effective way to protect public health even before investing in wastewater treatment

along the entire chain, from wastewater generation to food consumption; wastewater treatment is considered one possible component in a cumulative and integrated risk management approach. The guidelines are based on actual risk (using the cost-effective quantitative microbial risk assessment approach) and therefore avoid needlessly strict and expensive treatment technologies to achieve standards. The high investment costs for wastewater treatment are hard to justify in many developing countries where the national burden of disease from wastewater irrigation is just a fraction of that resulting from continuing poor access to safe drinking water and adequate sanitation, and poor hygiene standards. Also, the cumulative risk management approach ensures maximum conservation of nutrients in the wastewater, reducing the need for poor peri-urban farmers to purchase commercial fertilizer.

The long-term goal of integrated wastewater management will always be to move from the unregulated use of untreated wastewater to the regulated use of wastewater that has been treated to some extent. Depending on local possibilities, the level of treatment can vary if a complementary health risk reduction strategy is in place, as explained in the WHO guidelines. This flexibility will be necessary in low-income countries so long as the provision of sanitation infrastructure lags behind urbanization rates. Options to develop Wastewater Safety Plans, following the principles of Water Safety Plans, are currently being studied.

But even where no wastewater treatment is available, health risks can still be significantly reduced. Studies in Ghana show how cessation of irrigation before harvest and inexpensive and easily adoptable wastewater irrigation methods can reduce fæcal contamination of crops.[51] And studies in Africa and Asia suggest that a major source of cross-contamination of food is unhygienic handling of produce at the market and in the kitchen. Thus, improving hygiene is a cost-effective way to protect public health even before investing in wastewater treatment.[52]

In Asia, in particular, wastewater is used in aquaculture. Producing fish and aquatic plants in wastewater-fed ponds provides income, employment and food for poor households living in peri-urban areas and also provides an important supply of affordable and nutritious food for the urban population.[53]

The risk of parasitic diseases associated with such wastewater-fed aquaculture has been well described.[54] Less information is available on the health effects of chemical pollutants. The chemistry of toxic chemicals such as heavy metals in aquatic environments is complex. However, concentrations of heavy metals reported in fish raised in aquaculture do not usually exceed levels recommended by the Codex Alimentarius Commission, even in fish harvested from highly polluted water with high metal concentrations.[55] A study on toxic element accumulation in fish from wastewater-fed fish ponds in Hanoi, Viet Nam, found generally low levels of arsenic, cadmium and lead concentrations – for many samples, below the detection limit.[56]

In Hanoi, Ho Chi Minh City and Phnom Phenh the most important health problem farmers associated with wastewater exposure was skin disease.[57] Epidemiological studies confirmed that contact with wastewater was an important risk factor for dermatitis among farmers engaged in peri-urban wastewater-fed aquatic food production.[58] Water spinach cultivated in wetlands that receive wastewater from Phnom Penh was highly contaminated with fæcal bacteria. However, natural biological and physical processes in the lake reduced bacterial numbers almost to WHO guideline levels for irrigation water, as shown by differences in bacterial counts at wastewater entry and exit points.[59]

Industrial pollution control is improving

Industries based on organic raw materials are the largest contributors of organic pollution, while oil, steel and mining industries represent the major risk for heavy metal release.[60] Measures of BOD per year in industrial wastewater have stabilized over the past 20 years in industrial countries, or even decreased slightly, as in Eastern Europe (see box 8.6).

Industrial pollution is expected to increase in emerging market economies with economic and industrial development. More than half of China's 21,000 chemical companies have factories along the country's two major river basins – the Yangtze and the Yellow – which supply drinking water for tens of million of people.[61] This means that industrial accidents could have disastrous consequences in these regions.

In the early years following the widespread introduction of environmental regulations, firms tended to invest in end-

Non-point pollution from agriculture and urban areas often constitutes a greater total pollutant load than industrial point-source pollution

of-pipe technologies, such as membrane technologies, which reduce wastewater effluent following production to restore water quality to appropriate standards. But much industrial wastewater is discharged without treatment to open watercourses, reducing the quality of larger volumes of water and sometimes infiltrating aquifers and contaminating groundwater resources.

The OECD reports evidence of changing production processes through investments in clean production technologies for industries.[62] There has been a steady growth in companies seeking certification through ISO 14001, the international standard for environmental management administered by the International Organization for Standardization. By the end of 2002 nearly 50,000 companies in 118 countries had received ISO 14001 certification; Japan and China have the largest number of certified companies (figure 8.7).[63] Many multinational enterprises apply high environmental standards to their activities worldwide, introducing

Figure 8.7 **The number of companies with a certified system of environmental management varied considerably by country in 2006**

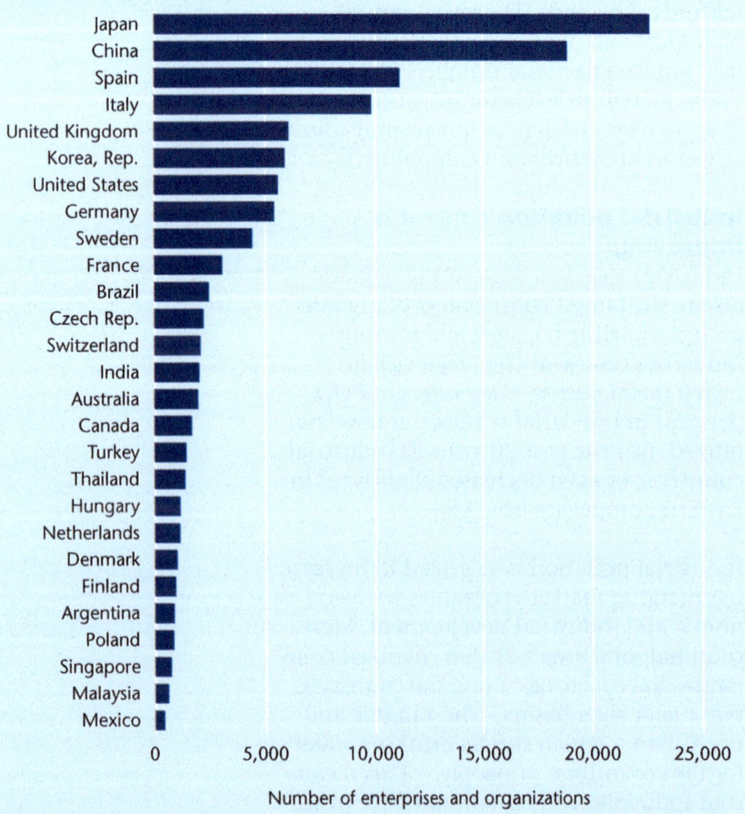

Note: Data are based on the ISO 14001 certification of the International Organization for Standardization and registration under the Eco-Management and Audit Scheme (EMAS) of the European Commission.
Source: ISO 2007.

environmental management systems to increase environmental performance, thus contributing to the globalization of better corporate practices.[64] But recent accidents involving large multinationals from OECD countries and the weak environmental performance of enterprises from emerging market economies underline the need for continued vigilance.

Mitigation of non-point source pollution needs to focus on the source

Rivers that drain major catchment areas in intensive agriculture areas (such as the Danube, Ebro, Mississippi, Nile, Po and Yellow) are major vectors for transmitting nutrients to the sea, where they often cause severe eutrophication. It is estimated that the nitrate load in the 80 main rivers flowing into the Mediterranean Sea doubled between 1975 and 1995.[65] Non-point pollution from agriculture and urban areas often constitutes a greater total pollutant load than industrial point-source pollution and is more difficult to control when leached into aquifers. Certain types of agricultural monoculture and in-situ sanitation may be incompatible with maintaining groundwater quality. More balanced agricultural land use and alternative sanitation measures will be required in these priority areas.

Pesticides tend to migrate throughout the environment, ending up in fatty tissues (fish) and sediments, creating adverse environmental effects. Pesticide contamination has increased rapidly since the 1970s, particularly in freshwater in developing countries, despite increased regulation of the use of these bioaccumulating and highly persistent substances.

Commercial fertilizer consumption in some high-income countries has stabilized, or even declined, since the 1990s, after a period of high growth during 1960-90.[66] In other countries fertilizer use is still growing rapidly (4% a year in Syria and 2% in Turkey) and could increase as much as 50%-70% by 2025 in Turkey.[67] There is less information on pesticide consumption, although data for France (the world's second biggest user) indicate that consumption has stabilized. Data on consumption provide only indirect information on the discharge of pollutants into the environment.

To improve environmental performance, intervention is needed for non-point source activities such as agriculture to limit the use of nutrients and change production processes. Interest in sustainable farming is growing, including conservation agriculture, integrated plant protection and plant nutrition management. The contribution of phosphorus to algae growth stabilized after its use in washing detergents was restricted in many locations.

Financing pollution mitigation and risk management

While the concept of charging polluters for the damage they impose on the environment is now widely accepted, this 'polluter pays' principle has had only a slight impact on behaviour, because the charges are rarely set high enough and because the authorities have difficulty setting the level of charges. To ensure effective pollution control, pollution charges – a source of revenue that can be earmarked for environmental funds and programmes – need to be combined with regulatory measures.

An alternative form of pollution control is a trading system for water pollution quotas. Authorities can begin by setting overall limits on the emissions of pollutants into specific water bodies, progressively reducing the limits over time to improve water quality. Emissions quotas would be allocated to polluters based on current or recent practices or auctioned to polluters, thereby raising revenue for public authorities. Polluters who are able to reduce emissions would sell their quotas to those less able to do so. All polluters would have an incentive to reduce emissions, and most abatement would be carried out by those who could do it most efficiently and affordably. Trading schemes must be set up to respect land zoning and must be monitored and controlled to protect 'invisible' water sources, in particular.

In practice, trading schemes for water pollution quotas are rare and difficult to apply successfully. They are especially difficult to apply to farmers – agriculture being an important source of non-point water contamination.[68]

Progress in achieving environmental sustainability

Our capacity to achieve environmental sustainability has improved but remains constrained by an incomplete understanding of the impact of pollution and the resilience of ecosystems, inadequate monitoring of the negative impacts of water use on the environment and institutional weaknesses that prevent effective implementation of legal instruments in many developing countries. There are some promising developments, however, and they are discussed briefly below.

Instruments implemented at a national level. The importance of the services provided by nature is now widely acknowledged, although the economic valuation of such environmental services and estimates of environmental flows[69] and benefits remain problematic (box 8.8). Implementation relies on multistakeholder dialogue and negotiations based on recognition of the value of environmental services (see chapter 14).

But evidence of effective implementation of environmental flows is still limited. Lessons from places where environmental flows have gained ground indicate that political support may be the most crucial element. Strong community interest, pressures from a river basin critically degraded because of overallocation, and donor-driven or -instigated overdevelopment all motivated implementation of the concept. Adoption and implementation of environmental flows have been particularly strong where national legislation and policies assign it a priority within an integrated water resources management framework and integrate it into natural resource management plans at the basin scale.

Examples abound, however, of cases where interest in environmental flows fails to be converted into legislative action. The chief reason for failure is lack of understanding of the socioeconomic benefits of environmental flows, of political will to support implementation and of appropriate legal,

The importance of the services provided by nature is now widely acknowledged

Box 8.8 Environmental flow assessment in Asia – from concept to reality

Environmental flows refers to a water regime within a river, wetland or coastal zone necessary for maintaining ecosystems and their services where there are competing water uses. The concept emphasizes how water management strategies may influence water and is intended as a tool for ending the disconnect of ecosystems from livelihoods.

Research in Asia and the Pacific shows that 23 of 48 countries are undertaking some activity on environmental flows (funded by Australia, Japan and New Zealand). These countries are moving towards implementing and integrating an environmental flows approach into local, regional and state planning processes and national legislation and policies. Cambodia, China, India, the Republic of Korea, Lao PDR, Malaysia, Nepal, Pakistan, Thailand and Viet Nam have all adopted the approach, and some of them have even incorporated it in national legislation. Explicit consideration of environmental flows in the national water accounts is a further step towards recognition of environmental water demands. Interest in adopting an environmental flows approach is also spreading in Bangladesh, Indonesia, Iran, the Philippines and Sri Lanka, as well as in a number of Central Asian countries.

Source: Adapted by UNESCAP from China Ministry of Water Resources 2005; IWMI 2005; Illaszewicz et al. 2005.

institutional and monitoring arrangements. In Asia and the Pacific large transboundary river basins are a special challenge. With 21 of 38 major watersheds shared between two or more countries, cross-border collaboration is needed to address environmental flows in river restoration and other projects.

International instruments for ecosystem protection and pollution mitigation. At the international level multiple frameworks (often adopted after a major crisis) support the protection of water systems and the mitigation of impacts (see appendix 2). The OECD monitors the level of commitment to these instruments by its member countries[70] as does the European Commission for its directives.[71]

There are also specific regulations on pollution control and water rights and allocation at regional and national levels. Such instruments for the regulation of use and systems are discussed in chapter 14. In the European Union member states have to comply with European Commission directives focused on water issues, including the EU Water Framework Directive (2000) and the EU Urban Waste Treatment Directive (1991), and deadlines for implementation. Few countries are yet in full conformity with those directives, and less than 50% of the urban waste load is treated.[72]

Notes

1. US Department of Energy 2006.
2. MEA 2005.
3. Comprehensive Assessment of Water Management in Agriculture 2007.
4. Comprehensive Assessment of Water Management in Agriculture 2007.
5. www.grid.unep.ch/activities/sustainable/tigris/index.php.
6. WWAP 2006.
7. WWAP 2006.
8. A large river system is one with a river channel section with a virgin mean annual discharge (discharge before any significant direct human manipulations) of at least 350 cubic metres per second anywhere in the catchment (Dynesius and Nilsson 1994, as cited in WWAP 2006, p. 176).
9. WWAP 2006.
10. Comprehensive Assessment of Water Management in Agriculture 2007.
11. Holling and Meffe 1996.
12. Gunderson and Holling 2002.
13. WWAP 2006.
14. Muir 2007.
15. Shiklomanov 2002.
16. Custodio and Llamas 2003.
17. Margat 2008.
18. Burke and Moench 2000; Comprehensive Assessment of Water Management in Agriculture 2007.
19. Foster and Chilton 2003. And contemporary groundwater issues and management options are discussed on the Websites and publications of the World Bank-Groundwater Management Advisory Team GW-MATE (www.worldbank.org/gwmate) and UNESCO's Groundwater Resources Assessment under the Pressures of Humanity and Climate Change (GRAPHIC) programme (www.unesco.org/water/ihp/graphic). A partnership led by UNESCO and the International Association of Hydrogeologists implements the Internationally Shared Aquifer Resources Management initiative, a multiagency effort to improve understanding of scientific, socioeconomic, legal, institutional and environmental issues related to the management of transboundary aquifers (www.isarm.net/).
20. Darnault 2008.
21. IPCC 2007.
22. Burke 2003.
23. www.worldbank.org/gwmate.
24. Some work has been done by Robert Dijon for the United Nations (1976-90) and by Jean Margat for UNESCO (2008).
25. Shah, Singh, and Mukherji 2006.
26. World Bank 2007.
27. Foster and Loucks 2006.
28. Downing, Price, and Jones 1993.
29. Alley 1993.
30. www.apfamgs.org.
31. WWF 2006.
32. UNEP/GPA 2006.
33. ILEC 2005; Lääne, Kraav, and Titova 2005.
34. WWAP 2006.
35. MEA 2005.
36. MEA 2005.
37. France, Ministry of Health 2007.
38. US EPA 2007.
39. UNECE 2007.
40. OECD 2008b.
41. Fewtrell et al. 2007.
42. There are three main types of wastewater treatment systems. *Primary treatment* (mechanical treatment technology) removes part of the suspended solids, but no ammonium. *Secondary treatment* (biological treatment) uses aerobic or anaerobic micro-organisms to decompose most of the organic matter and retain some of the nutrients (around 20%-30%) and removes around 75% of the ammonium. *Tertiary treatment* (or advanced treatment technology) removes the organic matter even more efficiently than secondary treatment. It generally includes phosphorus retention and in some cases nitrogen removal.
43. Scott, Faruqui, and Raschid-Sally 2004.
44. OECD 2008a.
45. IWMI 2003.
46. IFEN 2006.
47. Wintgens and Hochstrat 2006.
48. Scott, Faruqui, and Raschild-Sally 2004; Keraita et al. 2007a.
49. Ensink et al. 2004.
50. WHO 2006.
51. Keraita et al. 2007b.
52. Ensink, Mahmood, and Dalsgaurd 2007.
53. Leschen, Little, and Bunting 2005.
54. WHO 2006.
55. WHO 2006.
56. Marcussen et al. 2008.
57. Anh, van der Hoek, Cam et al. 2007.
58. Anh, van der Hoek, Ersbøll et al. 2007.
59. Anh, Tram et al. 2007.
60. WWAP 2003.
61. www.minesandcommunities.org/Action/press973.htm.
62. OECD 2008c.
63. IFEN 2006.
64. WBCSD 2005.
65. Benoit and Comeau 2005.
66. FAO-AQUASTAT database (www.fao.org/nr/water/aquastat/main/index.stm).

67. Benoit and Comeau 2005.
68. Kraemer et al. 2003.
69. See, for example, Dyson, Bergkamp, and Scanlon 2003 and the International Water Management Institute's Global Environmental Flow Calculator (GEFC) available at www.iwmi.cgiar.org/Tools_And_Resources/ Models_and_Software/ GEFC/index.aspx.
70. OECD 2008a.
71. http://ec.europa.eu/environment/water/water-framework/.
72. Wintgens and Hochstrat 2006.

References

Alley, W. M., ed. 1993. *Regional Groundwater Quality.* New York: Van Nostrand Reinhold.

Anh, V. T., N. T. Tram, L. T. Klank, P. D. Cam, and A. Dalsgaard. 2007. Faecal and Protozoan Parasite Contamination of Water Spinach (Ipomoea Aquatica) Cultivated in Urban Wastewater in Phnom Penh, Cambodia. *Tropical Medicine and International Health* 12 (S2): 73-81.

Anh, V. T., W. van der Hoek, P. D. Cam, and A. Dalsgaard. 2007. Perceived Health Problems among Users of Wastewater for Aquaculture in Southeast Asian Cities. *Urban Water Journal* 4: 269-74.

Anh, V. T., W. van der Hoek, A. K. Ersbøll, N. V. Thuong, N. D. Tuan, P. D. Cam, and A. Dalsgaard. 2007. Dermatitis among Farmers Engaged in Peri-Urban Aquatic Food Production in Hanoi, Vietnam. *Tropical Medicine and International Health* 12 (S2): 59-65.

Bagchi, S. 2007. Arsenic Threat Reaching Global Dimensions. *Canadian Medical Association Journal* 177 (11): 1344-45.

Benoit, G., and A. Comeau, eds. 2005. *A Sustainable Future for the Mediterranean: The Blue Plan's Environment and Development Outlook.* London: Earthscan.

Bixio, D., C. Thoeye, T. Wintgens, R. Hochstrat, T. Melin, H. Chikurel, A. Aharoni, and B. Durham. 2006. Wastewater Reclamation and Reuse in the European Union and Israel: Status Quo and Future Prospects. *International Review for Environmental Strategies* 6 (2).

Buechler, S., and G. Devi. 2003. Household Food Security and Wastewater Dependent Livelihood Activities along the Musi River in Andhra Pradesh, India. International Water Management Institute, Colombo. www.who.int/water_sanitation_health/wastewater/gwwufoodsecurity.pdf.

Burke, J. J. 2003. Groundwater for Irrigation: Productivity Gains and the Need to Manage Hydro-Environmental Risk. In *Intensive Use of Groundwater: Challenges and Opportunities*, ed., E. Custodio and M. R. Llamas. Lisse, The Netherlands: A. A. Balkema Publishers.

Burke, J. J., and M. Moench. 2000. *Groundwater and Society – Resources, Tensions and Opportunities.* New York: United Nations, Department of Economic and Social Affairs.

China, Ministry of Water Resources. 2005. Strategic Planning and Management of Water Resources in the Haihe and Huaihe River Basins of China. Beijing: Government of China.

Comprehensive Assessment of Water Management in Agriculture. 2007. *Water for Food, Water for Life: A Comprehensive Assessment of Water Management in Agriculture.* London: Earthscan, and Colombo: International Water Management Institute.

Custodio, E., and M. R. Llamas. 2003. Intensive Use of Groundwater: Introductory Considerations. In *Intensive Use of Groundwater: Challenges and Opportunities,* ed., E. Custodio and M. R. Llamas. Lisse, The Netherlands: A. A. Balkema Publishers.

Darnault, Christophe, ed. 2008. *Overexploitation and Contamination of Shared Groundwater Resources Management, (Bio)Technological, and Political Approaches to Avoid Conflicts.* NATO Science for Peace and Security Series C: Environmental Security. Dordrecht, Netherlands: Springer.

Downing, R. A., M. Price, and G. P. Jones. 1993. *The Hydrogeology of the Chalk of North-West Europe.* Oxford: Clarendon Press.

Dyson, M., G. Bergkamp, and J. Scanlon, ed. 2003. *Flow: The Essentials of Environmental Flows.* Gland, Switzerland: International Union for Conservation of Nature and Natural Resources.

EEA (European Environment Agency). 2003. *Europe's Environment – The Third Assessment.* State of the Environment Report 1/2003. Luxembourg: Office for Official Publications of the European Communities.

Ensink, J. H. J., T. Mahmood, and A. Dalsgaard. 2007. Wastewater-Irrigated Vegetables: Market Handling versus Irrigation Water Quality. *Tropical Medicine & International Health* 12 (2): 2-7.

Ensink, J. H. J., T. Mahmood, W. van der Hoek, and L. Raschid-Salty. 2004. A Nationwide Assessment of Wastewater Use in Pakistan: An Obscure Activity or a Vitally Important One? *Water Policy* 6 (2004): 197-206.

Ewing B., S. Goldfinger, M. Wackernagel, M. Stechbart, S. Rizk, A. Reed, and J. Kitzes. 2008. *The Ecological Footprint Atlas 2008.* Oakland, CA: Global Footprint Network.

Fewtrell, L., A. Prüss-Üstün, R. Bos, F. Gore, and J. Bartram. 2007. *Water, Sanitation and Hygiene: Quantifying the Health Impact at National and Local Levels in Countries with Incomplete Water Supply and Sanitation Coverage.* Geneva: World Health Organization.

Foster, S. S. D., and P. J. Chilton. 2003. Groundwater – the Process and Global Significance of Aquifer Degradation. *Philosophical Transactions of the Royal Society London B* 358 (1440): 1957-72.

Foster, S. S. D., and D. P. Loucks. 2006. *Non-Renewable Groundwater Resources. A Guidebook on Socially-Sustainable Management for Water Policy Makers.* Paris: United Nations Educational, Scientific, and Cultural Organization.

France, Ministry of Health. 2007. *L'eau potable en France, 2002-2004. Eau et sante, Guide technique.* Paris: Ministere de la Santé et des Solidarités.

Fry, R. C., P. Navasumrit, C. Valiathan, J. P. Svensson, B. J. Hogan, M. Luo, S. Bhattacharya, K. Kandjanapa, S. Soontararuks, S. Nookabkaew, C. Mahidol, M. Ruchirawat, and L. D. Samson. 2007. Activation of Inflammation/NF-κB Signaling in Infants Born to Arsenic-Exposed Mothers. *PLoS Genetics* 3 (11): e207.

Gunderson, L. H., and C. S. Holling, ed. 2002. *Panarchy: Understanding Transformations in Human and Natural Systems.* Washington, DC: Island Press.

Helsinki Commission. 2007. HELCOM Releases Annual Report on 2007 Activities. Press release, 24 June, Helsiniki Commission. www.helcom.fi/press_office/news_helcom/en_GB/2007AnnualReport/.

Holling, C. S., and G. K. Meffe. 1996. Command and Control and the Pathology of Natural Resource Management. *Conservation Biology* 10 (2): 328-37.

IFEN (Institut Français de l'Environnement). 2006. *L'environnement en France. Les Syntheses.* October 2006 ed. Orléans, France: Institut Français de l'Environnement.

ILEC (International Lake Environment Committee Foundation). 2005. *Managing Lakes and Their Basins for Sustainable Use: A Report for Lake Basin Managers and Stakeholders.* Kusatsu, Japan: International Lake Environment Committee Foundation.

Illaszewicz, J., R. Tharme, V. Smakhtin, and J. Dore, eds. 2005. *Environmental Flows: Rapid Environmental Flow Assessment for the Huong River Basin, Central Vietnam.* Gland Switzerland: International Union for Conservation of Nature.

IPCC (Intergovernmental Panel on Climate Change). 2007. *Climate Change 2007: Impacts, Adaptation and Vulnerability. Contribution of Working Group II to the Fourth Assessment Report of the Intergovernmental Panel on Climate Change*, ed.,

M. L. Parry, O. F. Canziani, J. P. Palutikof, P. J. van der Linden, and C. E. Hanson. Cambridge, UK: Cambridge University Press.

ISO (International Organization for Standardization). 2007. *The ISO Survey of Certifications 2006.* Geneva: International Organization for Standardization.

IWMI (International Water Management Institute). 2003. Confronting the Realities of Wastewater Use in Agriculture. Water Policy Briefing Issue 9. International Water Management Institute, Colombo. www.iwmi.cgiar.org/Publications/Water_Policy_Briefs/PDF/wpb09.pdf.

———. **2005.** *Environmental Flows: Environment Perspectives on River Basin Management in Asia* 2 (1). www.eflownet.org/download_documents/EnvFlowsNL_Vol_2_Issue_1.pdf.

Keraita, B., F. Konradsen, P. Drechsel, and R. C. Abaidoo. 2007a. Effect of Low-Cost Irrigation Methods on Microbial Contamination of Lettuce Irrigated with Untreated Wastewater. *Tropical Medicine and International Health* 12 (S2): 15-22.

———. **2007b.** Reducing Microbial Contamination on Wastewater-Irrigated Lettuce by Cessation of Irrigation before Harvesting. *Tropical Medicine and International Health* 12 (S2): 8-14.

Koo, Sasha. 2003. Preliminary Remedial Investigation of the Potential Impact of an Environmental Accident on Agriculture and Irrigation in the Affected Region. Macedonia Mission Report, 27-31 October, Food and Agriculture Organization of the United Nations, Rome.

Kraemer, R. A., Z. G. Castro, R. Seroa da Motta, and C. Russell. 2003. *Economic Instruments for Water Management: Experiences from Europe and Implications for Latin America and the Caribbean.* Regional Policy Dialogue Study Series. Washington, DC: Inter-American Development Bank.

Lääne, A., Kraav, E. and G. Titova. 2005. *Global International Waters Assessment.* Baltic Sea, GIWA Regional Assessment 17. Sweden: University of Kalmar, Global International Water Assessments.

Leschen, W., D. Little, and S. Bunting. 2005. Urban Aquatic Production. *Urban Agriculture Magazine* 14: 1-7.

Marcussen, H., K. Joergensen, P. E. Holm, D. Brocca, R. W. Simmons, and A. Dalsgaard. 2008. Element Contents and Food Safety of Water Spinach (Ipomoea aquatica Forssk) Cultivated with Wastewater in Hanoi, Vietnam. *Environmental Monitoring and Assessment* 139 (1-3): 77-91.

Margat, Jean. 2008. *Les eaux souterraines dans le monde.* Orléans, France: Editions BRGM.

MEA (Millennium Ecosystem Assessment). 2005. *Ecosystems and Human Well-being: Wetlands and Water Synthesis.* Washington, DC: World Resources Institute.

Molle, F., and J. Berkoff. 2006. *Cities versus Agriculture: Revisiting Intersectoral Water Transfers, Potential Gains and Conflicts.* Comprehensive Assessment of Water Management in Agriculture Research Report 10. Colombo: Comprehensive Assessment Secretariat.

Muir, P. S. 2007. Human Impacts on Ecosystems. Online Study Book for Students, Oregon State University, Corvallis, Oregon.

OECD (Organisation for Economic Co-operation and Development). 2008a. *OECD Environmental Data Compendium.* Paris: Organisation for Economic Co-operation and Development. www.oecd.org/document/49/0,3343,en_2649_34283_39011377_1_1_1_1,00.html.

———. **2008b.** *Service Delivery in Fragile Situations: Key Concepts, Findings and Lessons.* OECD/DAC Discussion Paper. Offprint of the *Journal on Development* 9 (3). Organisation for Economic Co-operation and Development, Paris.

———. **2008c.** *OECD Environment Outlook to 2030.* Paris: Organisation for Economic Co-operation and Development.

Salgot, M., E. Huertas, S. Weber, W. Dott, and J. Hollender. 2006. Waste Water Reuse and Risks Definition of Key Objectives. *Desalination* 187 (1-3): 29-40.

Scott, C. A., N. I. Faruqui, and L. Raschid-Sally, ed. 2004. *Wastewater Use in Irrigated Agriculture: Confronting the Livelihood and Environmental Realities.* Wallingford, UK: Cabi Publishing.

Shah, T., O. P. Singh, and A. Mukherji. 2006. Some Aspects of South Asia's Groundwater Irrigation Economy: Analyses from a Survey in India, Pakistan, Nepal and Bangladesh. *Hydrogeology Journal* 14 (3): 286-309.

Shiklomanov, I. A., ed. 2002. *Comprehensive Assessment of Freshwater Resources of the World: An Assessment of Water Resources and Water Availability in the World.* Stockholm: Stockholm Environment Institute.

Thornton, J. A., W. Rast, M. M. Holland, G. Jolankai, and S.-O. Ryding, ed. 1999. *Assessment and Control of Nonpoint Source Pollution of Aquatic Ecosystems.* Man and the Biosphere Series, Volume 23. Paris: United Nations Educational, Scientific, and Cultural Organization.

UNECE (United Nations Economic Commission for Europe). 2007. *Our Waters: Joining Hands across Orders. First Assessment of Transboundary Rivers, Lakes and Groundwaters.* Geneva and New York: United Nations. www.unece.org/env/water/publications/assessment/assessmentweb_full.pdf.

UNEP (United Nations Environment Programme). 2006. Iraq Marshlands Observation System. United Nations Environment Programme, Division of Early Warning and Assessment/GRID-Europe, Geneva. www.grid.unep.ch/activities/sustainable/tigris/mmos.php.

UNEP (United Nations Environment Programme)/GIWA (Global International Waters Assessment). 2006. *Challenges to International Waters: Regional Assessments in a Global Perspective.* Global International Waters Assessment Final Report. Nairobi: United Nations Environment Programme. www.giwa.net/publications/finalreport/.

UNEP (United Nations Environment Programme)/GPA (Global Programme of Action for the Protection of the Marine Environment from Land-Based Activities). 2006. *The State of the Marine Environment: Trends and Processes.* The Hague: United Nations Environment Programme/Global Programme of Action for the Protection of the Marine Environment from Land-Based Activities.

UNEP (United Nations Environment Programme)/Global Resource Information Database Arendal. 2007. *Balkan Vital Graphics: Environment without Borders.* United Nations Environment Programme/Global Resource Information Database Arendal, Arendal, Norway.

UNEP (United Nations Environment Programme)/OCHA (Office for the Co-ordination of Humanitarian Affairs). 2000. Cyanide Spill at Baia Mare. Spill of Liquid and Suspended Waste at the Aurul S.A. Retreatment Plant in Baia Mare. Assessment Mission Romania, Hungary, Federal Republic of Yugoslavia, United Nations Environment Programme, Geneva.

United Nations. 1976. *Groundwater in the Western Hemisphere.* Natural Resources/Water Series No. 4. New York: United Nations Department of Technical Cooperation for Development.

———. **1982.** *Groundwater in the Eastern Mediterranean and Western Asia.* Natural Resources/Water Series No. 9. New York: United Nations Department of Technical Cooperation for Development.

———. **1983.** *Groundwater in the Pacific Region.* Natural Resources/Water Series No.12. New York: United Nations Department of Technical Cooperation for Development.

———. **1986.** *Groundwater in Continental Asia.* Natural Resources/Water Series No.15. New York: United Nations Department of Technical Cooperation for Development.

———. **1988a.** *Groundwater in North and West Africa.* Natural Resources/Water Series No.18. New York: United Nations.

———. **1988b.** *Groundwater in Eastern, Central, and Southern Africa.* Natural

Resources/Water Series No.19. New York: United Nations Department of Technical Cooperation for Development.

———. **1990.** *Groundwater in Eastern and Northern Europe.* Natural Resources/ Water Series No.24. New York: United Nations Department of Technical Cooperation for Development.

US Department of Energy. 2006. Energy Demands on Water Resources: Report to Congress on the Interdependency of Energy and Water. US Department of Energy, Washington, DC.

US EPA (Environmental Protection Agency). 2007. *National Water Quality Inventory Report to Congress: 2002 Reporting Cycle.* Office of Water. Washington, DC: US Environmental Protection Agency. www.epa.gov/305b/2002report/ report2002305b.pdf.

van Rooijen, D. J., H. Turral, and T. W. Bigg. 2005. Sponge City: Water Balance of Mega-City Water Use and Wastewater Use in Hyderabad, India. *Irrigation and Drainage* 54 (S1): S81-S91.

WBCSD (World Business Council for Sustainable Development). 2005. Water Facts and Trends. World Business Council on Sustainable Development, Washington, DC.

WHO (World Health Organization). 2006. *WHO Guidelines for the Safe Use of Wastewater, Excreta and Greywater.* Geneva: World Health Organization.

Wintgens T., and R. Hochstrat, eds. 2006. AQUAREC Integrated Concepts for Reuse of Upgraded Wastewater: Report on Integrated Water Reuse Concepts. EVK1-CT-2002-00130, Deliverable D19, RWTH Aachen University, Aachen, Germany.

World Bank. 2007. *Making the Most of Scarcity: Accountability for Better Water Management Results in the Middle East and North Africa.* MENA Development Report. Washington, DC: World Bank.

WWAP (World Water Assessment Programme). 2003. *Water for People, Water for Life. The World Water Development Report 1.* Paris: United Nations Educational, Scientific, and Cultural Organization, and New York: Berghahn Books.

———. **2006.** *The United Nations World Water Development Report 2. Water: A Shared Responsibility.* Paris: United Nations Educational, Scientific, and Cultural Organization, and New York: Berghahn Books.

WWF. 2006. *Living Planet Report 2006.* Gland, Switzerland: WWF.

———. **2008.** *Living Planet Report 2008.* Gland, Switzerland.

Chapter 9

Managing competition for water and the pressure on ecosystems

Author: François Molle and Domitille Vallée

Contributors: Cécile Brugère, Thierry Facon, Jean Margat, Léna Salamé, Ti Le-Huu and Jon Martin Trondalen

Coordinator: Jean-Marc Faurès (FAO)

Facilitator: Domitille Vallée

Key messages

- Competition for water and shortcomings in managing it to meet the needs of society and the environment call for enhanced societal responses through improved management, better legislation and more effective and transparent allocation mechanisms.
- Challenges include wise planning for water resources, evaluation of availability and needs in watersheds, possible reallocation or storage expansion in existing reservoirs, more emphasis on water demand management, better balance between equity and efficiency in water use, inadequate legislative and institutional frameworks and the rising financial burden of ageing infrastructure.
- Water management choices should emerge from informed consultation and negotiation on the costs and benefits of all options after considering basin interconnectedness, relationships between land and water resources, and the consistency and coherence of decisions with other government policies.

Competition for water exists at all levels and is forecast to increase with demands for water in almost all countries. In 2030, 47% of world population will be living in areas of high water stress.[1] Water management around the world is deficient in performance, efficiency and equity. Water use efficiency, pollution mitigation and implementation of environmental measures fall short in most sectors. Access to basic water services – for drinking, sanitation and food production – remains insufficient across developing regions, and more than 5 billion people – 67% of the world population – may still be without access to adequate sanitation in 2030.[2]

Increased competition for water and shortcomings in its management to meet the needs of society and the environment call for enhanced societal responses through improved water management. Challenges include wise planning for water resources; evaluation of availability and needs in watersheds; possible needs for reallocation or additional storage; the need to balance equity, efficiency and ecosystem services in water use; the inadequacy of legislative and institutional frameworks and the increasing financial burden of ageing infrastructure. Substantial efforts are needed in regulation, mitigation and management, primarily through community consultation and cross-sectoral policy involving the private sector.

Type, extent and effect of competition for water

Competition among uses and users is increasing in almost all countries, as are the links connecting them, calling for more effective negotiation and allocation mechanisms.

Basin closure and interconnectedness

Abstraction of water has approached and in some cases exceeded the threshold of renewability of water resources in many river basins, leading to widespread damage to ecosystems. Demand for water is often highest when availability is lowest, and water shortages and conflicts have increased accordingly. This trend has been paralleled by degradation in the quality of surface water and groundwater from the combined effluents of cities, industries and agricultural activities. This has exacerbated economic water scarcity by rendering water unfit for certain uses and has harmed human and ecosystem health.

Hydrology, ecology and society are all connected. Water resources are increasingly diverted, controlled and used as countries develop. Water flowing out of sub-basins is often committed to other downstream uses, including several often overlooked functions: flushing-out sediments, diluting polluted water, controlling salinity intrusion and sustaining estuarine and coastal ecosystems. As water in a basin is increasingly allocated and river discharges fall short of meeting such commitments some or all of the time, basins (or sub-basins) are said to be closing or closed.[3] Water no longer flows out from the basin – as is happening in the Jordan River (box 9.1).

Perturbations of the hydrologic cycle in one location may affect another. This is most clearly illustrated by the common upstream-downstream effect, but can take diverse, often less visible, forms. Figure 9.1 provides examples of the water quantity, quality, timing and sediment load of upstream-downstream impacts.

Water also connects aquatic ecosystems. Relationships among land, water and biota are complex, and cross-impacts may not be evident immediately. Groundwater abstraction generally reduces flows from underground aquifers back to the surface, drying up springs and wetlands. In Azraq, Jordan, for example, groundwater use for cities and agriculture has resulted in the desiccation of a Ramsar-designated wetland associated with high biodiversity and migratory birds. Dams, through their impacts on flood-pulse regimes, have altered complex ecosystems that were providing valuable services and supporting livelihoods (such as fisheries, receding agriculture, pastures, reeds and medicinal plants). Examples include the Senegal Valley and the Hadejia-Jama'are plains in northern Nigeria.[4]

Competition and conflict for water

Conflicts about water can occur at all scales. Local-level conflicts are commonplace in irrigation systems, where farmers vie for limited resources. In Northern Thailand, for example, low flows in the dry season are diverted by upland farmers to irrigate their orchards, where use of pesticides sometimes leads to the pollution of streams. Conflicts also occur at the scale of large national river basins (multistate Indian rivers such as the Cauvery and the Krishna) or transnational river basins (the Jordan and the Nile). While conflict resolution mechanisms and adequate modes of governance will differ with scale, the

Conflicts around water can occur at all scales

Box 9.1 The closure of the lower Jordan River basin

The lower Jordan River, downstream of Lake Tiberias, flows through the Jordan rift valley before emptying into the Dead Sea. Because of Israel's redirection of the upper course, the river now receives water mostly from the Yarmouk River, a tributary originating in Syria, and from a few lateral wadis that incise the two mountain ranges that run parallel to the valley on each side. Most of the population and cities, together with the bulk of the country's rain-fed agriculture and increasing groundwater-based irrigation, are concentrated in these highlands. In the east bank of the valley some 23,000 hectares of irrigated land have been developed as a result of diversion of the Yarmouk and side wadis.

The lower Jordan River basin has undergone a drastic squeeze, with 83% of its flow consumed before it reaches the Dead Sea because of diversions in Israel and Syria, 45,000 hectares of irrigated land, mushrooming cities swollen by waves of refugees from Palestine and Iraq and immigrants from the Gulf countries, and the new Wehdah Dam reservoir on the Yarmouk River.

The consequences of this squeeze are broad, and some are dire:

- Limited (though still desirable) scope for efficiency improvement.
- Increased recycling and use of treated wastewater for irrigation.
- Reallocation of water from the valley (irrigation) to the highlands (cities).
- Environmental degradation (overdraft of aquifers in the Azraq oasis and a declining Dead Sea that now receives less than 250 million cubic metres of water).
- A surge in costly supply augmentation projects aimed at tapping distant aquifers, transferring water from the Red Sea to the Dead Sea or desalinating saline water.
- Increased irregularity and uncertainty in water supply for irrigation in the valley, the residual user.
- A more politicized and contested water policy, with costs and benefits apportioned across social and ethnic groups and subregions, yielding different levels of power.

Source: Courcier, Venot, and Molle 2005.

Figure 9.1 Examples of hydrologic interactions in river basins – upstream-downstream impacts

Upstream / Downstream / Variable	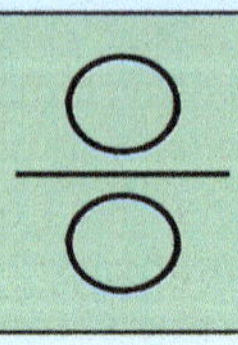	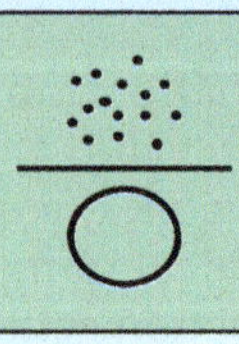	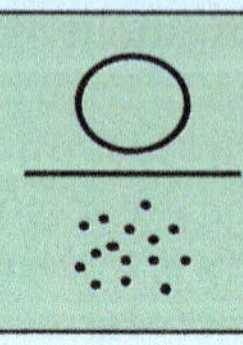	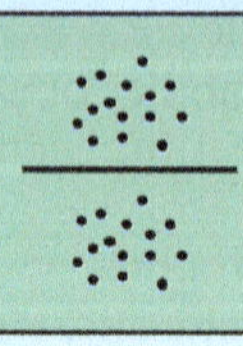
Quantity	Upstream diversion scheme on downstream irrigation area	Water harvesting (or small tanks) on a downstream dam	Cities on irrigation wells (out-pumping)	Wells on qanats[a]; deep wells on shallow wells
Quality	Cities or industries on irrigated agriculture	Diffuse pollution of agriculture on city supplies	Cities on groundwater used in pumping irrigation (contaminating)	Diffuse agricultural pollution on village groundwater-based water supply
Timing	Hydropower generation on large irrigation schemes or fisheries	Small tanks on onset of wet season flows (delays) and on biological cues	Hydropower generation on wetland ecosystems	Water harvesting on runoff/flood and downstream groundwater recharge (reduced)
Sediment load	Large-scale deforestation on reservoirs	Overgrazing, or erosion in smallholder agriculture on reservoir (siltation)	Dam retaining silt on fertilization of downstream floodplains	Diffuse deforestation on silt load and delta fanning

a. Qanat is an ancient system of tunnels and wells built to capture water in a mountain and channel it to a lower level.
Source: Based on Molle 2008.

 Point, large-scale user or intervention

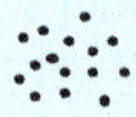 Diffuse, scattered users or interventions.

nested nature of these scales also means that the modes of governance will have to be consistent and interrelated.

Sectoral conflicts. Sectoral conflicts oppose users from different sectors (domestic, hydropower, irrigation, industries, recreation and so on), including ecosystems, whose sustainability depends on environmental flows. These conflicts are both economic (the return per cubic metre differs greatly across these uses) and political (the social importance and the political clout of each sector also varies). Box 9.2 illustrates the case of conflict between agriculture and industry in Orissa, India.

Perhaps the most common conflict is between agriculture and cities. Half the world lives in cities – and this share is increasing – while agriculture is generally the largest user of water. Moving water from agriculture to uses with higher economic value is frequently proposed, for several reasons. Agriculture gets by far the largest share of diverted water resources and also consumes the most water through plant evapotranspiration. Cities are also thirsty. The value-added of water in non-agricultural sectors is usually far higher than in agriculture. This apparent misallocation is often attributed to government failure to allocate water rationally.[5]

Box 9.2 Conflict between agriculture and industry over water in Orissa, India

The Hirakud Dam in Orissa, India, was the first multipurpose dam to become operational after India's independence in 1947. Built across the Mahanadi River, it is the longest and largest earthen dam in the world, and its reservoir is the largest artificial lake in Asia. The dam helps control floods in the Mahanadi, provides irrigation to 155,600 hectares of land and generates up to 307.5 megawatts of electricity through its two power plants. Thanks to irrigation provided by the dam, Sambalpur District is referred to as the rice bowl of Orissa.

With new state development policies based on industrialization, the reservoir started supplying water to industrial plants pumping from the reservoir. In 2006 the state government signed memorandums of understanding with 17 companies to provide them water from the reservoir. Meanwhile, 50 years after the dam's construction, many downstream areas had yet to receive irrigation water, and tension was building between reservoir authorities on one side and local governments and farmers associations on the other on water releases from the dam. In June 2006, 25,000 farmers, fearing that diversion of water could deprive more than 20,000 hectares of irrigation water, formed an 18 kilometre-long human chain near Sambalpur to protest the provision of water to industries. Five months later, in November 2007, 30,000 farmers gathered at the reservoir to protest. This large turnout surprised even the protest organizers and demonstrated the desperation of farmers over their water supply. Both events were covered by the media. Under pressure by the opposition party, Orissa's chief minister assured farmers' representatives that not a single drop of the farmers' share would be diverted to industries and announced a 20 billion rupees package for canal repair work in the Hirakud area.

Source: Thierry Facon, Food and Agriculture Organization of the United Nations, Bangkok regional office, adapted from *Kalinga Times* 2007 and South Asia Network on Dams, Rivers and People 2006.

Another common intersectoral conflict is between hydropower and other sectors, especially agriculture and fisheries. Because the energy production of hydropower plants follows consumer demand, the dams may release water when downstream irrigators do not need it. Real-time management of stored water can result in better outcomes because it enables water to be released when needed for multiple users. Dams may also harm fisheries by impeding fish migration and reducing productivity by altering the water regime. Famous conflicts have occurred on the Columbia River, in the northwestern United States, where intensive river damming has affected salmon and other species. Some dams have been decommissioned to restore ecosystem connectivity. Conflict is looming between dams planned or under construction in the Mekong River basin and the river-abundant fisheries. It is feared that the cumulative effects of these dams, notably those planned on the mainstream river, will have a deleterious impact on Tonle Sap, Cambodia's great lake, and on the fisheries that provide 60% of the basin population's protein intake.

Dams, irrigation schemes and cities consume water or change flow pathways. The poor and the environment, the residual user, bear a disproportionate share of the negative consequences. Massive upstream diversions have typically affected downstream lakes or deltas, such as in the Colorado and the Indus basins (box 9.3; see also figure 10.2 in chapter 10 and box 11.1 in chapter 11). Diversion of the lower Ganges River by the Farraka Dam damaged the ecology of the Sunderbands wetlands, and the project for interlinking northern and southern rivers in India could dramatically compound these impacts on the Ganges-Brahmaputra delta. Excess use of groundwater in many large coastal cities (such as Chennai, Jakarta, Lima and Tel Aviv) has led to the depletion of local aquifers and allowed seawater to intrude and salinize the aquifers.

Meeting water needs during dry seasons and ensuring security of supply require water storage. Climate change will intensify climate irregularity, so that more storage will be needed to ensure the same level of security. More water will have to be kept in reservoirs as reserves for dry spells, leaving less water for use on average. And this increased need for storage is occurring at a time when pressure from users is forcing water managers to take risks and reduce carryover stocks. In many regions of the world the need for more storage is not taken into account, resulting in a growing frequency of local crises during extreme drought.

Integrated management of reservoirs in a river basin is a realistic solution. Part 4 illustrates possible responses and approaches.

Transboundary competition. When a river or aquifer crosses a political boundary and there is competition between sectors or countries, problems become more complex

Box 9.3 Competition for water and downstream impacts in the Indus River basin

Degradation of the Indus delta environment, in the downstream reaches of the Indus River basin, has a long and complex history. Gradual increases in irrigation demands and cultivated areas throughout the basin, punctuated by years of drought and the construction of reservoirs, have resulted in progressive reductions in freshwater flows to the delta over the last 40 years and contention over water diversions.

This environmental degradation is widely acknowledged as stemming from both local threats, such as unsustainable fisheries exploitation and industrial and urban pollution from nearby Karachi, and external threats related to competition for upstream irrigation diversions and storage water management (ineffective drainage, low irrigation efficiency and inadequate farming practices, and reduced freshwater flows).

The poor are bearing the brunt of the consequences, including water-logging and increased salinization of land, aquifers and surface water as reduced freshwater flows are unable to prevent seawater intrusion and land erosion, and reductions in their livelihood assets and opportunities as a result of declining fisheries, deteriorating grazing grounds and reduced agricultural outputs and related revenues. Drinking water shortages have led to an increase in water-related diseases, forcing households to purchase water from tankers at great expense and women and children to spend more time fetching water from sources farther away.

Source: Brugère and Facon 2007.

Box 9.4 Fisheries and hydropower competing in the Mekong River basin

After years of being undisturbed by humans, the Mekong River basin has undergone rapid change in recent years. Populations have been displaced by dams in Thailand, and there has been protracted conflict over the impact of the Pak Mun Dam on fisheries. In Cambodia the loss of lives due to the release of water from dams on the upper Se San in Viet Nam have stirred public awareness of the social and environmental costs of conventional infrastructure development.

A major challenge for the basin is to design hydropower facilities with minimum impact on fisheries. Mekong fisheries account for 17% of the world freshwater fish catch, and numerous studies have shown the importance of fish to the diets and incomes of populations in the basin.

Recent announcements of bilateral agreements between Lao PDR and Thailand and between Lao PDR and Viet Nam for dams on the main stem of the river together with numerous private contracts agreed by Cambodia, Lao PDR and Viet Nam (mostly with companies in China and South-East Asia) have raised concerns about whether these new projects will benefit from the lessons learned from past mistakes. The marginalization of regional international players (such as the Asian Development Bank, the Mekong River Commission and the World Bank), the lack of transparency of the planning processes and the abruptness of official declarations about the signed agreements have left little room for discussion of the economic soundness and impacts of the projects. The central concern remains the fate of fisheries as new dams are planned on the main stream, an issue on which specialists at the Mekong River Commission, the World Fish Centre and elsewhere have issued severe warnings.

Source: Molle, Foran, and Käkönen forthcoming; Mekong River Commission 2008.

Figure 9.2 **Three types of response to water scarcity and competition**

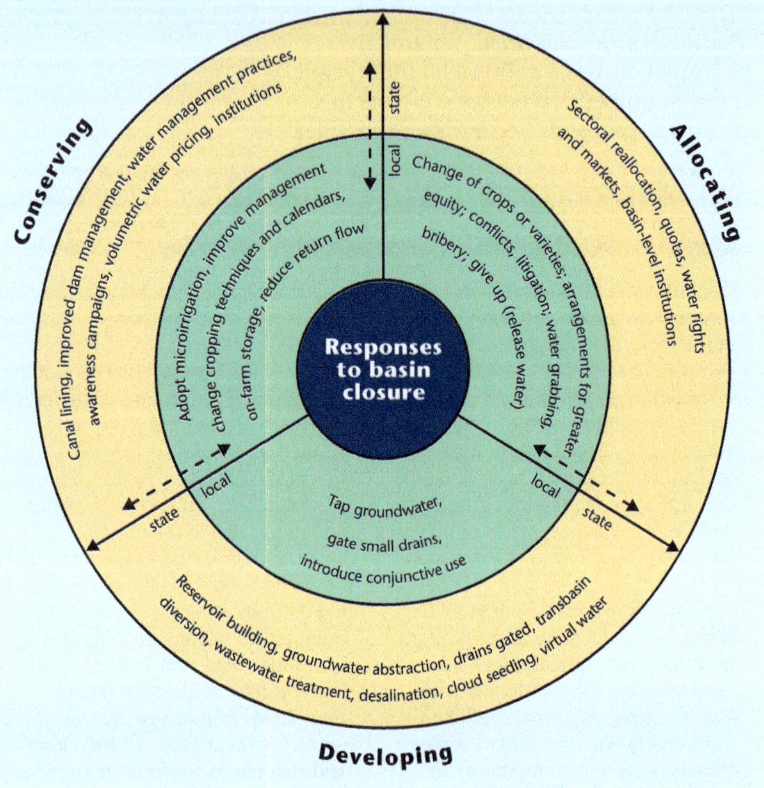

Source: Based on Comprehensive Assessment of Water Management in Agriculture 2007.

and can lead to conflict. The Mekong River basin has been an exception, with concerns around water arising only recently. Partly because of conflicts unrelated to water the river has long-remained undisturbed, but dam development to meet the growing need for energy in most of the riparian countries is putting other downstream uses at risk – particularly fisheries (box 9.4).

Despite competing demands and conflict, however, there is little historic evidence that water itself has led to international warfare or that a war over water would make strategic, hydrographic or economic sense.[6] At the international level water appears to provide reasons for transboundary cooperation rather than war, often preventing instead of causing escalation.[7] Many multilateral treaties on freshwater resources have stressed multiple objectives – economic development, joint management and water quality – rather than just water quantity and hydropower (see appendix 2).[8] The way Mexico and the United States resolved their dispute over the allocation of water from the Rio Grande River, which included a cost-sharing arrangement for water conservation measures, offers interesting lessons for the peaceful resolution of water disputes (see box 15.22 in chapter 15). A recent shift in emphasis from water sharing to benefit sharing promises greater transboundary cooperation.

Managing competition through supply and demand management and reallocation

There are many shortcomings in how water is managed today in a context of increased scarcity: low efficiency, environmental degradation, and inequity. Despite some improvements competition is increasing and water use efficiency remains low in most sectors. But the answer is not just more efficient allocation mechanisms and more emphasis on greater yields and productivity, because these alone may lead to further losses in equity and environmental sustainability. Rather, a combination of supply and demand management measures is needed.

Three common responses to competition

The responses to increased competition for water are supply augmentation, conservation and reallocation (figure 9.2). The most conventional response is to develop new resources. For the state this typically means building new reservoirs or desalination plants or interbasin transfer. For users this means more wells or farm ponds or gating drains to store water. Conserving water includes increasing the efficiency of use by reducing losses. Changes in allocation, to ease competition or to maximize water use, are based on economic, social, environmental or other criteria. Augmentation is a supply management strategy, while conservation and reallocation are demand management strategies, roughly defined as 'doing better with what we have'.[9]

Supply augmentation is typically constrained by the availability of storage sites, the social and environmental costs and the rising financial cost of water. With needs outstripping available stocks in many basins, transfers between basins have become more frequent. Amman, Athens, Bangkok, Kathmandu, Los Angeles and Mexico City are procuring water further afield. The massive transfer of water now under way in China (from the Yangtze River to the Yellow River) is being emulated in Brazil, India, Jordan and Thailand. While this trend is likely to continue, its potential will gradually be exhausted and its costs will spiral upwards. Other small-scale options, such as farm ponds in Asia or wells, have also been widely developed. Desalination is an option in specific locations (islands and coastal cities), but its cost is likely to remain high (though it is declining) and its

Box 9.5 The untapped potential of marginal-quality water

Non-conventional water resources, especially marginal-quality water (urban wastewater, agricultural drainage water and saline surface water or groundwater), are an important source of water that is still undervalued.

Urban wastewater use in agriculture remains limited, except in a few countries with very meagre water resources (40% of uses in the Gaza Strip, 15% in Israel and 16% in Egypt with the reuse of drainage water). Elsewhere, even where water is scarce, wastewater use accounts for less than 4% of all uses (2.3% in Cyprus, 2.2% in Syria, 1.1% in Spain and 1.0% in Tunisia). The use of urban wastewater – treated or not – is growing, particularly for farming around cities, often because higher quality water resources are not available.

Desalination based on brackish water sources (48%) and seawater (52%) is increasingly affordable as a result of new membrane technology ($0.60-$0.80 per cubic metre). It is used mostly for drinking water (24%) and industrial supplies (9%) in countries that have reached the limits of their renewable water resources (such as Cyprus, Israel, Malta and Saudi Arabia; figure 1). Little is used for agriculture (1%), but its use for high-value crops in greenhouses is gradually increasing. Desalination accounted for only 0.4% of water use in 2004 (nearly 14 cubic kilometres a year; figure 2), but production should double by 2025.

Source: Blue Plan, MAP, and UNEP 2007.

Figure 1 Desalination capacity in selected countries, 2002

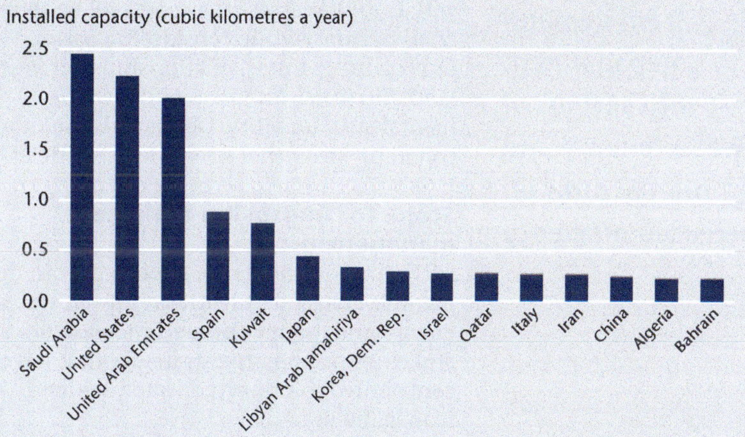

Source: Based on Maurel 2006.

Figure 2 Rapid growth of global installed capacity for desalination, 1966-2004

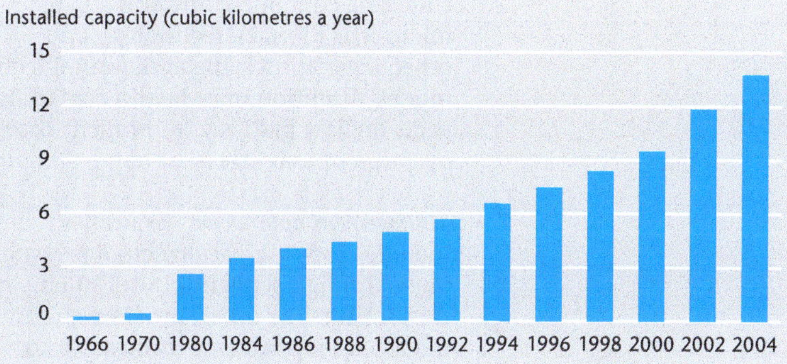

Source: Based on Maurel 2006.

use limited to urban supply. Other non-conventional sources of supply include wastewater, secondary sources (such as treated irrigation drainage) and the mining of fossil (non-renewable) aquifers (box 9.5). Figure 9.3 shows the relative importance of these sources in selected Middle East and Mediterranean region countries.

Because of reuse of water in basins and users' adjustments to scarcity, fully developed basins or aquifers tend to have much less 'slack' than is often thought, and the potential for net water savings at the basin level is often overstated. When limits are reached and improved efficiency and demand management possibilities are exhausted, there are often no win-win solutions to meet additional demands. Rather, resources must be reallocated from one source to another. These demand management options are discussed in more detail in the following section.

Countries rarely resort to all three options at once unless pressure over the resource is severe, as in Tunisia (see chapter 15).

Figure 9.3 Importance of non-conventional sources of water for selected Middle East and Mediterranean region countries, 2000-06

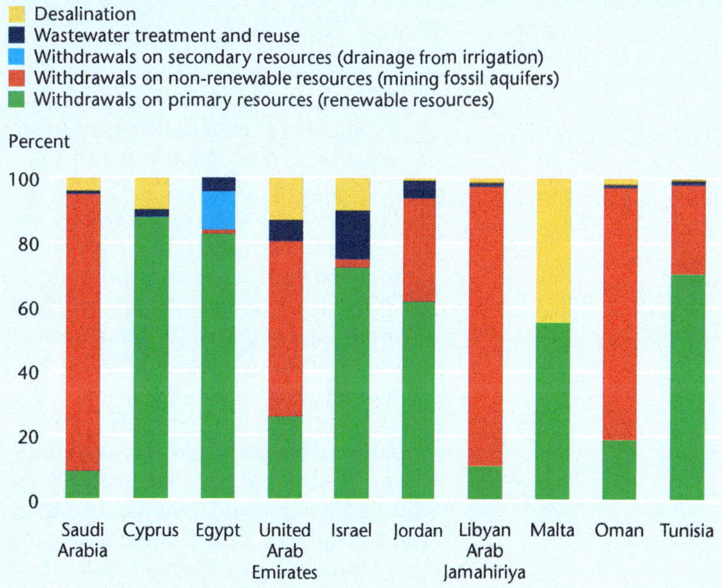

Source: FAO-AQUASTAT; Benoit and Comeau 2005.

Large improvements are expected from demand management, with savings in water, energy and money through increased efficiency

Mechanisms are needed to make optimal choices along the spectrum of options. Optimal choices should emerge from informed processes of consultation and negotiation that assess the costs and benefits of all options, while considering basin interconnectedness, relationships between land and water resources and environmental sustainability. Decisions should be coherent with other government policies.

Scope for improving demand management

Large improvements are expected from demand management, with savings in water, energy and money through increased efficiency. Among the strategies that have contributed to improved water demand management are:

- *Technological improvements.* These include reducing leakage in urban networks, changing equipment and shifting to micro-irrigation, biotechnology and other water-conserving agricultural techniques. Attention must be directed to impacts on flow pathways to properly assess overall water savings (see chapter 3).
- *Management approaches.* Examples include cropping-pattern change, water reuse through sequential uses in irrigation schemes or urban processes, reuse in closed-loop systems (industry and energy sectors) and reallocation across sectors.
- *Economic incentives.* Using water pricing, taxes and fees for demand management and allocation of water has proved effective in domestic and industrial sectors, but these measures are not a workable option for most irrigation schemes in developing countries.[10] Payment for environmental services has been found to be a useful economic restraint in some cases (see chapters 4 and 14).
- *Legal and regulatory approaches.* 'Polluter pays' and 'user pays' principles have reduced both water use and pollution in industry, and participatory management has increased user participation by controlling individual water demands (see chapter 4).

Urban distribution networks and irrigation schemes lose large amounts of water through leakage and percolation. Among the 23 countries of the Mediterranean Action Plan, in a region where water stakes are high, an estimated 25% of water is lost in urban networks and 20% in irrigation canals (map 9.1). Realistically, only part of the water lost can effectively be recovered technically and at a reasonable cost. Cities such as Rabat and Tunis have cut their losses to 10%.[11] Even when the water is returned to the water system, these losses and leakages constitute a failure of the supply infrastructure as they result in significant financial costs (for producing drinking water and pumping and transporting water) and additional environmental and health risks. Technology (canal lining, micro-irrigation) can often solve part of the problem, but a large part of the losses are due to management or regulatory flaws.

While irrigation losses and inefficiency appear high, with only a third of the water supplied reaching plant roots, most of the losses become return flows, which are tapped by other users elsewhere in the basin or serve important environmental functions. There may be little water to be saved in fully developed basins, and conservation interventions can often end up as reallocation.[12]

Localized irrigation (micro-irrigation), for example, has a limited impact on water depleted by evapotranspiration in the fields and chiefly reduces return flows. Thus water 'saved' by upstream irrigators can come at the expense of downstream users, allowing upstream irrigators to expand their cultivation. This may be desirable from the perspective of the upstream farmer, but the result is increased water depletion.

Price-based regulation was emphasized in the 1990s for limiting water use, but its benefits have failed to materialize, most notably in the irrigation sector.[13] Water can also be saved by changing behaviours – through awareness campaigns, quotas or water pricing (box 9.6).

In the industrial sector a combination of subsidies, higher water prices and environmental regulations have encouraged industries to improve processes and reduce withdrawals (see chapter 7). It is hard to get a consolidated picture of how industries manage water worldwide, but there are global indications that the business community is devoting growing attention to water management,[14] as a result of increased efforts to improve water management. Industries can realize major savings in natural and financial resources by raising awareness through environmental audits and by investing modest amounts. In agricultural and emerging market economies the scope for progress through clean processes is even greater, since production processes are generally well below world standards. Multi-national

Map 9.1 **Difference between water withdrawn and water effectively used in Mediterranean region countries, all uses, 2000-05**

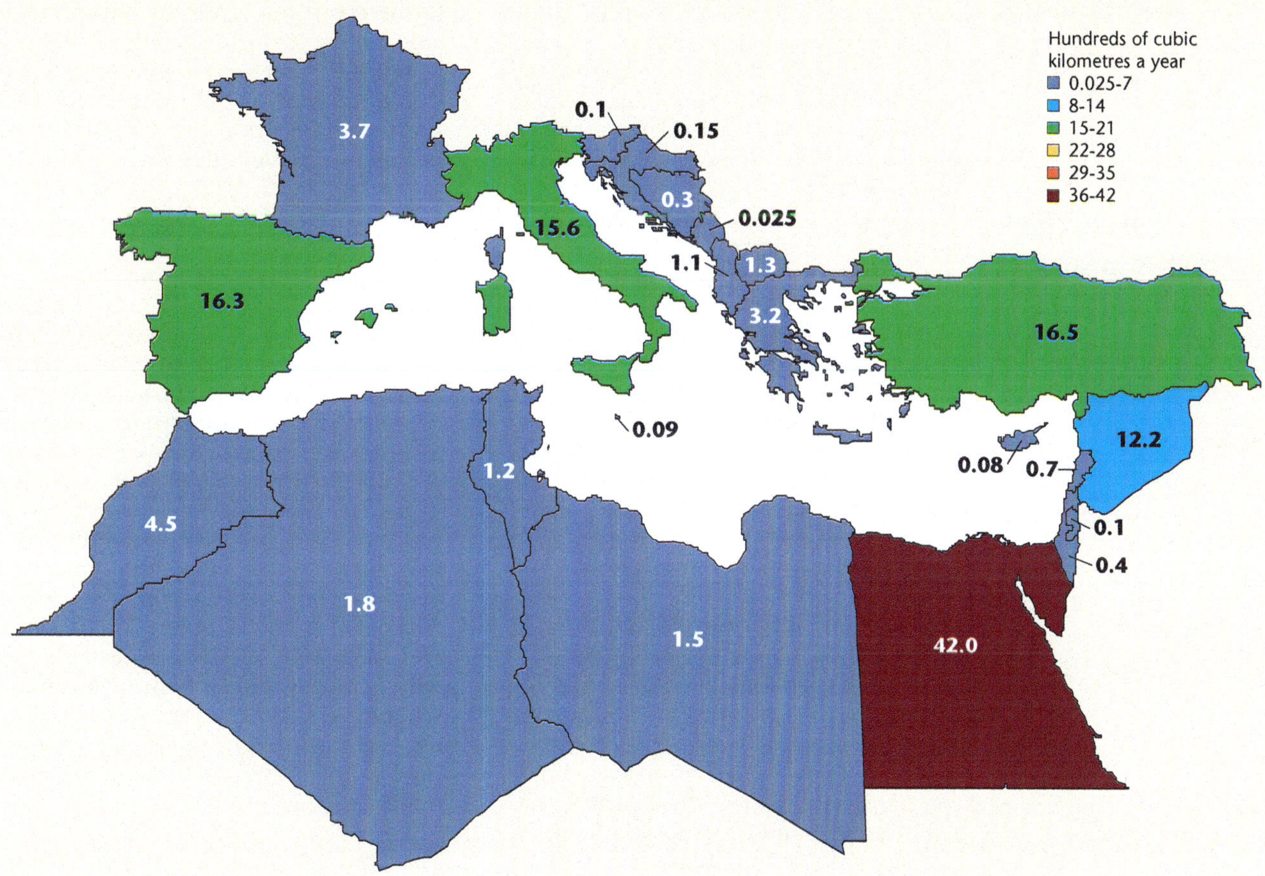

Source: Jean Margat, adapted from Benoit and Comeau 2005; Blue Plan, MAP, and UNEP 2007.

companies can play a key role. In some countries public intervention through subsidies or more stringent enforcement are necessary. The international competitiveness of companies in the global market is enhanced by a commitment to best environmental practices, which reduce pollution and improve the efficiency of water use.

At the national level a growing number of companies are introducing clean production processes – often for pollution reduction – that result in substantial water savings. These efforts are supported by various UN programmes (United Nations Environment Programme, United Nations Industrial Development Organization) through a network of cleaner production centres in 27 countries.

Chapters 14 and 15 present several examples of approaches to demand management.

Reallocation, efficiency and equity issues

Water, like any resource, when it is scarce or requires scarce resources to supply

Box 9.6 **Signs of progress in urban areas: examples in Asia and Australia**

Water withdrawals increased in Asia and Australia in all sectors until the 1980s, when withdrawals for agriculture declined and growth in overall water withdrawals slowed. There is still considerable potential for improvement. Some recent efforts have been stimulated by the United Nations Economic and Social Commission for Asia and the Pacific, which has promoted the concept of eco-efficiency of water use and water infrastructure development, and by complementary research.

Implementation of demand management measures has been uneven across the region, but interest in improving water use efficiency is growing in many countries. Singapore has reduced urban domestic water demand from 176 litres per person a day in 1994 to 157 litres in 2007 as a result of additional and targeted public expenditure for improved demand management.

In Bangkok and Manila leak detection programmes have lowered estimated unaccounted-for water and allowed postponing the development of new infrastructure. Effluent charges have been an important instrument for stimulating efficient water uses in households and commercial establishments.

In 2008 Sydney Water in Australia began providing homes in the Hoxton Park area with two water supplies – recycled water and drinking water (dual reticulation). Recycled water is to be used for gardens and other outdoor needs, toilet flushing and potentially as cold water in washing machines and for certain non-residential purposes. The recycled water taps, pipes and plumbing are coloured purple to distinguish recycled water from drinking water.

Source: www.sydneywater.com.au; UNESCAP 1997, 2004; Kiang 2008.

Once basic human and environmental water needs have been met, the remainder should ideally go to where water has the highest value to society

it, rises in economic value. Once basic human and environmental water needs have been met, the remainder should ideally go to where water has the highest value to society. Since much water is used for productive or 'lifestyle' purposes, it is appropriate to apply economic criteria to its allocation. But water pricing alone will not produce the necessary reallocation, since prices in many sectors do not reflect underlying economic values, and there are many cases of market or service failure. In several Eastern European countries price increases resulted in reductions in urban water consumption to half the level of two decades ago.[15] Flow reductions can lead to secondary water quality problems in supply networks (increase of water residence time), odour problems in sewerage systems and added burdens at wastewater treatment plants, which become hydraulically underloaded and have to treat much denser raw wastewater than before.

Reallocation from lower- to higher-value uses can be achieved by enabling the traditional markets as well as by applying administrative measures, creating water markets or trading water rights. In each case society should set appropriate limits on transfers to protect third parties, the environment and wider social interest. Subject to these conditions, competition for water can be healthy.

In countries that recognize water trading rights, many cities have met their growing water needs by purchasing farms or properties with water rights and taking over the rights. Some non-governmental organizations 'compete' on behalf of the environment by purchasing the rights to a certain volume of water in a river or lake, which they then leave in the water body. These are examples of one-off transactions. But in certain regions (Chile, parts of Australia, some western states of the United States) the conditions have been created for regular water trading (see box 4.2 in chapter 4). There, water markets are commonly used by farmers wanting supplementary water for valuable crops during drought conditions or by cities to create reserves in anticipation of impending droughts. Prices set in these markets signal the marginal values of water in these different uses, which are usually much higher than average values.[16]

These 'efficiency' criteria need to be reconciled with society's desire for equity (the satisfaction of basic needs) and environmental sustainability. Such balancing of water needs can be achieved by a combination of administrative allocation, tariff structures with adequate provisions to protect the poor and other relevant measures. There is a role for subsidies in water services, but they should be carefully targeted to specific functions. Poor people and other disadvantaged groups without sustainable access to safe water and adequate sanitation are usually willing to pay within their means for reliable access to service because improving access (through standpipes or household connections) yields large financial dividends.

The Comprehensive Assessment of Water Management in Agriculture argues for reforms to enable more efficient use of water.[17] Policy-makers need to recognize the incentives and resource constraints confronting small farmers, but it would be a mistake to assume that farmers do not respond to market incentives (food prices have an impact on cropping patterns). Farmers will invest in inputs and irrigation technology (meaning higher water costs) if they believe that they will achieve higher returns. There is no reason for efficiency, equity and environmental sustainability to be out of alignment in that case.

Notes

1. OECD 2008.
2. OECD 2008.
3. Molle, Wester, and Hirsch 2007.
4. Barreteau, Bousquet, and Attonaty 2001; Barbier and Thompson 1998; Neiland et al. 2000.
5. Molle and Berkoff 2005.
6. Wolf, Yoffe, and Giordano 2003; Gleick 2008.
7. van der Molen and Hildering 2005.
8. Wolf, Yoffe, and Giordano 2003.
9. Winpenny 1994.
10. Molle and Berkoff 2008.
11. Blue Plan, MAP, and UNEP 2007.
12. Molle, Berkoff, and Barker 2005; Molle, Wester, and Hirsch 2007; Molle et al. 2008.
13. Molle and Berkoff 2008.
14. WBCSD n.d.; World Economic Forum 2008.
15. Somlyódy and Varis 2006.
16. Winpenny 1994; Molle and Berkoff 2005.
17. Comprehensive Assessment of Water Management in Agriculture 2007.

References

Barbier, E. B., and J. R. Thompson. 1998. The Value of Water: Floodplain versus Large-Scale Irrigation Benefits in Northern Nigeria. *Ambio* 27 (6): 434-40.

Barreteau O., F. Bousquet, and J.-M. Attonaty. 2001. Role-Playing Games for

Opening the Black Box of Multi-Agent Systems: Method and Lessons of Its Application to Senegal River Valley Irrigated Systems. *Journal of Artificial Societies and Social Simulation* 4 (2).

Benoit, G., and A. Comeau, eds. 2005. *A Sustainable Future for the Mediterranean: The Blue Plan's Environment and Development Outlook.* London: Earthscan.

Blue Plan, MAP (Mediterranean Action Plan), and UNEP (United Nations Environment Programme). 2007. Water Demand Management, Progress and Policies: Proceedings of the3rd Regional Workshop on Water and Sustainable Development in the Mediterranean Zaragoza, Spain, 19-21 March. MAP Technical Reports Series 168. Athens: United Nations Environment Programme.

Brugère, C., and T. Facon. 2007. Doomed Delta? Challenges to Improved Governance, Planning and Mitigation Measures in the Indus Delta. Paper presented at the International Delta 2007 Conference on Managing the Coastal Land-Water Interface in Tropical Delta Systems, 7-9 November, Bang Sean, Thailand.

Comprehensive Assessment of Water Management in Agriculture. 2007. *Water for Food, Water for Life: A Comprehensive Assessment of Water Management in Agriculture.* London: Earthscan, and Colombo: International Water Management Institute.

Courcier, R., J. P. Venot, and F. Molle. 2005. *Historical Transformations of the Lower Jordan River Basin: Changes in Water Use and Projections (1950 – 2025). Comprehensive Assessment of Water Management in Agriculture Research Report 9.* Colombo: Comprehensive Assessment Secretariat, International Water Management Institute.

Gleick, P. H. 2008. Water Conflict Chronology. Pacific Institute, Oakland, CA. www.worldwater.org/conflictchronology.pdf.

***Kalinga Times.* 2007.** Stop Corporate Control over Water: Farmers Warn Govt. www.kalingatimes.com/orissa_news/news/20070111_farmers_warn_govt.htm.

Kiang, Tay Teck. 2008. Singapore's Experience in Water Demand Management. Paper presented at the 13th International Water Resources Association World Water Congress, 1-4 September, Montpellier, France.

Maurel, A. 2006. *Dessalement de L'eau de Mer et des Eaux Saumâtres: et Autres Procédés Non Conventionneles d'approvisionnement en Eau Douce.* 2eme édition. Cachan, France: Tec & Doc Lavoisier.

Mekong River Commission. 2008. *Hydrological, Environmental and Socio-Economic Modelling Tools for the Lower Mekong Basin Impact Assessment. Water Utilisation Programme Final Report Part 2: Research Findings and Recommendations.* Helsinki and Espoo, Finland: Finnish Environment Institute, Environmental Impact Assessment Centre of Finland Ltd., and Helsinki University of Technology.

Molle, F. 2008. Why Enough Is Never Enough: The Societal Determinants of River Basin Closure. *International Journal of Water Resources Development* 24 (2): 217-26.

Molle, F., and J. Berkoff. 2005. *Cities versus Agriculture: Revisiting Intersectoral Water Transfers, Potential Gains and Conflicts. Comprehensive Assessment of Water Management in Agriculture Research Report 10.* Colombo: Comprehensive Assessment Secretariat.

———, eds. 2008. *Irrigation Water Pricing: The Gap between Theory and Practice. Comprehensive Assessment of Water Management in Agriculture Series 4.* Wallingford, UK: Centre for Agricultural Bioscience International.

Molle, F., J. Berkoff, and R. Barker. 2005. Irrigation Water Pricing in Context: Exploring the Gap between Theory and Practice. Paper presented at the workshop Comprehensive Assessment of Water Management in Agriculture, International Water Management Institute, Montpelier, France, 6-10 June 2005.

Molle, F., T. Foran, and M. Käkönen, eds. Forthcoming. *Contested Waterscapes in the Mekong Region: Hydropower, Livelihoods and Governance.* London: Earthscan.

Molle, F., P. Jayakody, R. Ariyaratne, and H. S. Somatilake. 2008. Irrigation vs. Hydropower: Sectoral Conflicts in Southern Sri Lanka. *Water Policy* 10 (S1): 37-50.

Molle, F., P. Wester, and P. Hirsch. 2007. River Basin Development and Management. In *Water for Food, Water for Life: A Comprehensive Assessment of Water Management in Agriculture.* London: Earthscan, and Colombo: International Water Management Institute.

Neiland A. E., J. Weeks, S. P. Madakan, and B. M. B. Ladu. 2000. Inland Fisheries of North East Nigeria Including the Upper River Benue, Lake Chad and the Nguru–Gashua Wetlands II: Fisheries Management at Village Level. *Fisheries Research* 48 (3): 245-61.

OECD (Organisation for Economic Co-operation and Development). 2008. *OECD Environment Outlook to 2030.* Paris: Organisation for Economic Co-operation and Development.

Somlyódy, László, and Olli Varis. 2006. Freshwater Under Pressure in Groundwater Mangement and Policy: Its Future Alternatives. *International Review for Environmental Strategies* 6 (2): 181-204.

South Asia Network on Dams, Rivers and People. 2006. Protest against Diversion of Hirakud Water. *Dams, Rivers and People* 4 (9-10): 8.

UNESCAP (United Nations Economic and Social Commission for Asia and the Pacific). 1997. Guidelines for the Establishment of Pricing Policies and Structure for Urban and Rural Water Supply. ST/ESCAP/1738. Manila: United Nations Economic and Social Commission for Asia and the Pacific.

———. 2004. Guidelines on Strategic Planning and Management of Water Resources. ST/ESCAP/2346. New York: United Nations.

van der Molen, I., and A. Hildering. 2005. Water: Cause for Conflict or Co-operation? *ISYP Journal on Science and World Affairs* 1 (2): 133-43.

WBCSD (World Business Council for Sustainable Development). n.d. Global Water Tool. Geneva. www.wbcsd.org/templates/TemplateWBCSD5/layout.asp?type=p&MenuId=MTUxNQ&doOpen=1&ClickMenu=LeftMenu.

Winpenny, James. 1994. *Managing Water as an Economic Resource.* London and New York: Routledge.

Wolf A. T., S. B. Yoffe, and M. Giordano. 2003. International Waters: Identifying Basins at Risk. *Water Policy* 5 (1): 29-60.

World Economic Forum. 2008. Water as Critical as Climate Change. Press Release, 23 January. Davos, Switzerland. www.weforum.org/en/media/Latest%20Press%20Releases/WaterCritical.

State of the resource

PART 3

Chapter

Chapters 10-12 Coordinator
Andras Szöllösi-Nagy (UNESCO), with the support of Anil Mishra (UNESCO) and Siegfried Demuth (UNESCO)

Facilitator
Denis Hughes

Chapter 13 Coordinator
Andras Szöllösi-Nagy (UNESCO)

Associate coordinator
Wolfgang Grabs (WMO)

Facilitator
Denis Hughes

Author: Denis Hughes

Part 3 of the Report highlights the key issues associated with the Earth's natural water cycles (chapter 10) and identifies the changes occurring in the water cycle (chapter 11) and how they affect hazards and opportunities for using water resources (chapter 12). It concludes with a discussion of the available information – and the information that is lacking – to quantify the resource and patterns of change (chapter 13).

Water resources are made up of many components associated with water in its three physical states (liquid, solid and gas). Under natural conditions water results from complex interactions between atmospheric, land surface and subsurface processes that affect its distribution and quality. The components of the water cycle (rainfall, evaporation, runoff, groundwater, storage and others) therefore all differ in their chemical and biochemical qualities, spatial and temporal variability, resilience, vulnerability to pressures (including land use and climate change), susceptibility to pollution and capacity to provide useful services and to be used sustainably. A consequence of this variability is that while human pressures have resulted in large modifications to the global cycle, the directions and degrees of change are complex and difficult to ascertain.

The uneven distribution over time and space of water resources and their modification through human use and abuse are sources of water crises in many parts of the world. Broad overviews at the global scale are difficult because of the spatial and temporal complexity arising from the conjunction of land, water and atmospheric elements – made worse by incomplete monitoring and fragmented availability of data to quantify the resource and its changes.

The main components of water have come to be designated as blue, green and white water:

- **Blue water** is liquid water moving above and below the ground and includes surface water and groundwater. As blue water moves through the landscape, it can be reused until it reaches the sea.

- **Green water** is soil moisture generated from rainfall that infiltrates the soil and is available for uptake by plants and evapotranspiration. Green water is non-productive if evaporated from soil and open water.

- **White water** (sometimes considered the non-productive part of green water) is water that evaporates directly into the atmosphere without having been used productively and includes losses from open water and soil surfaces.

In addition, grey and black water refer to the quality of the resource.

- **Grey water,** usually wastewater, may be poor in quality, but usable for some purposes.

- **Black water** is so heavily polluted (usually with microbes) as to be harmful (to humans and ecosystems) or at least economically unusable.

About 60% of total global water fluxes (flows, movements and transfers between physical states) are attributable to green water flow, the component of the water cycle most susceptible to land use and land cover changes, and to changes in atmospheric conditions that control evaporative demand, such as temperature, solar radiation and atmospheric vapour pressure deficit. In addition, there is the Earth's 'solid' water, the contribution that snow and ice make to the global resource and how these components may be affected by changes in climate conditions.

Detecting and quantifying changes involve separating natural variability in climate and hydrologic processes from the variability and trends caused by other factors that influence the hydrologic landscape and then measuring and quantifying this variability. This hydrologic landscape has already been highly modified by land use changes, which can both accelerate (for example, through urbanization and vegetation degradation) and dampen (for example, through afforestation) hydrologic responses. It is also modified by engineered systems, including in-stream effects (such as dams, direct abstractions, return flows and interbasin transfers) and off-stream effects (such as irrigation). Climate change is being superimposed on an already complex hydrologic landscape, making its signal difficult to isolate, and yet making its influence felt throughout the water supply, demand and buffering system. There are already detectable changes in some parts of the world in the first-order climate parameters of temperature and rainfall. But this signal is not yet discernable in many parts of the world in second-order parameters of importance to water resources managers, such as changes in runoff and groundwater.

Chapter 11 reviews the scientific evidence of change in the components of the water cycle, focusing on impacts related to climate change. While trends in precipitation have been noted in some parts of the world, in other areas precipitation patterns have remained about the same within the period of observed data. There is evidence of changes in seasonality and frequency of heavy precipitation events in some areas. Despite the evidence of temperature changes, there is little evidence of detectable changes in evaporation and evapotranspiration. Part of the reason may stem from decreased solar radiation due to increased aerosol use or cloud cover, but difficulties in obtaining representative measurements could be obscuring any trends. Similarly, detecting trends in soil moisture is constrained by limited observations, while model studies are subject to uncertainties about the model and data input assumptions.

Despite the limitations of global datasets, many studies have shown changes in runoff and streamflow. Many have focused on low (drought) or high (flood) extremes (chapter 12). Except in regions with flows affected by glacier meltwater, the general conclusion is that global trends are not present or cannot be detected at this stage, although climate change–related trends are evident in some regions. Many rivers have been altered by engineered systems, including dams, diversions, return flows and interbasin transfers and by land use modifications. The same is true of groundwater resources, which have been heavily used for human supply and agriculture for many years. While many groundwater abstraction schemes access fossil water (water unrelated to current conditions), renewable groundwater resources depend on highly variable recharge volumes. It is thus realistic to expect future recharge regimes to reflect changes in the driving hydrologic processes (such as precipitation and evapotranspiration) that might result from anticipated climate changes. That these are not yet detectable could be related to the buffering capacity of groundwater (slower processes), as well as to the lack of a suitably integrated global database. While changes have been observed in natural lakes and wetlands in recent decades in many parts of the world, the primary factors driving these changes are region specific.

The strongest evidence for the effects of climate change on water resources comes from areas where the rate at which the solid phase of water (snow and ice) is converted into liquid is important. In permafrost regions changes have been detected in the depth of frozen ground and in the duration, thickness and areal extent of the seasonal freeze and thaw within seasonally frozen regions. Potential impacts include surface settlement, swamping, landslides and greater sediment loads. More than 15% of the world's population live where water resources availability depends heavily on snowmelt from ephemeral snowpacks or perennial glaciers. Changes have been observed in snow cover extent and snow water equivalent and in the frequency with which precipitation falls as snow.

Despite observations that the snow cover season has shortened and that this change appears to have accelerated in recent decades, some inconsistencies remain in the data sources. There is considerable evidence to suggest that glaciers have retreated globally since the mid-19th century and that this retreat has accelerated since the mid-1970s in response to rising air temperature and changes in the amount and kind of precipitation.

Among the consequences of a changing hydrologic cycle is its interaction with the terrestrial carbon cycle, because of positive feedbacks to climate change. The terrestrial biosphere is thought to have taken up roughly 25% of anthropogenic carbon emissions during the last century, and how long this can continue is unclear. Observations suggest that the rate of carbon uptake depends on hydrologic and climate conditions, as well as land use. But long-term observations are sparse, making trend detection difficult.

Most climate scientists agree that global warming will result in an intensification, acceleration or enhancement of the global hydrologic cycle, and there is some observational evidence that this is happening already. Data limitations in length of record, continuity and spatial coverage contribute to the uncertainty, while natural climate variability and multiyear variability associated with large-scale atmospheric circulation patterns influence the interpretation of many trends in ways that are not yet fully understood. Improving data collection and reducing the uncertainties associated with modelling studies are important for future impact assessments.

Hazards (chapter 12) can result from too much water (floods, erosion, landslides and so on) or too little (droughts and loss of wetlands or habitat) and from the effects of chemical and biological pollution on water quality and in-stream ecosystems. Water-related hazards can be naturally occurring or anthropogenic. The chapter states that the natural variability of water resources and changes, whatever the cause, can also provide positive opportunities – with careful management. In many areas hydrologic extremes have increased. Deaths and material damage from extreme floods can be high, and more intense droughts, affecting increasing numbers of people, have been observed in the 21st century. Such droughts have been linked to higher temperatures and decreased precipitation, but are also frequently a consequence of resource mismanagement.

Changes in cultivation and interruptions in sediment delivery through the construction of dams can lead to changes in erosion and sediment transport. In some developing countries rapidly expanding population has driven land clearing and rapid expansion of cultivated land. The water quality and ecology of many of the world's rivers have been altered partly by changes in flow and partly by inputs of chemical and biological waste from human activities. Global warming is expected to lead to changes in water temperature, with substantial effects on energy flows and matter recycling, resulting in algal blooms, increases in toxic cyanobacteria bloom and reductions in biodiversity, among other impacts.

The increased exposure to potential climate change hazards has led to more awareness of water resources management. The response of management strategies to potential climate change threats is an opportunity to implement more resource-sustainable policies and practices. For example, in areas of increasing water stress, groundwater is often an important buffer resource, capable of responding to increased water demands or compensating for the declining availability of surface water. There are significant opportunities for both mitigation and adaptation strategies, such as stronger observation networks (chapter 13), increased integration of groundwater and surface water supplies (including artificial recharge), improved early warning and forecasting systems for hazardous events, improved risk-based approaches to management and greater community awareness of sustainable water resources use.

Management of global water resources requires reliable information about the state of the resource and how it is changing in response to all the drivers that affect it. Worldwide, water observation networks are inadequate for current and future management needs and risk further decline (chapter 13). There are insufficient data to understand and predict the current and future quantity and quality of the resource. Political protocols and imperatives for sharing data are also inadequate. While new technologies in satellite remote sensing and modelling present opportunities, their value is limited by our inability to ground truth and validate much of the simulated information. To improve monitoring and to use data more effectively and efficiently, countries need to place observations and continual assessments of water resources higher on the political and development agendas. The financial and human resources that countries can commit to achieving these improvements will differ greatly. But unless a worldwide effort is made to improve our knowledge and understanding of changes in the global water resource, future management will be undertaken in an environment of greater uncertainty and high risk.

Chapter 10

The Earth's natural water cycles

Author: Charles J. Vörösmarty

Contributors: Daniel Conley, Petra Döll, John Harrison, Peter Letitre, Emilio Mayorga, John Milliman, Sybil Seitzinger, Jac van der Gun and Wil Wollheim

Coordinator: Andras Szöllösi-Nagy (UNESCO)

Facilitator: Denis Hughes

Key messages

- The uneven distribution of water resources over time and space and the way human activity is affecting that distribution today are fundamental sources of water crises in many parts of the world.
- Climate change is superimposed on the complex hydrologic landscape, making its signal difficult to isolate and its influence felt throughout the water supply, demand and buffering system.

The contemporary water cycle – and so freshwater resources – is defined by the interaction of natural and human factors. This chapter highlights key components of the global water cycle most directly relevant to the state of the water resources base. Such components include the land-based water system and the world's oceans, circulation of water in the atmosphere, surface and subsurface water associated with the continental land mass, accessible and virtual water and people's role in stabilizing and redirecting this resource – and limiting it through mismanagement, including overabstraction and pollution.

Overview of the global hydrologic cycle

Water unifies the climate, biosphere and chemo-lithosphere of the planet. The physical state of water and its transformations are linked to energy exchanges called *atmospheric teleconnections* (such as El Niño) and to feedbacks in the climate system. Water movement is the largest flow of any kind through the biosphere and is the primary vehicle for erosion and dissolution of continents. Freshwater strongly determines the productivity of biomass and supports critical habitats and biodiversity. Humans struggle to stabilize and make available adequate water, despite an unforgiving climate, failed governance and mismanagement, which lead to depletion and pollution.

Components of the global water cycle – defining the water resources base

The land-based hydrologic cycle is the fundamental building block of water resources. Freshwater is but a small fraction – about 2.5% – of the total water on Earth, the 'blue planet'. All human enterprise requires water. It is required for food production, industry, drinking water, inland water transport systems, waste dilution and healthy ecosystems.

Water links Earth's atmosphere, land mass and oceans through the global water cycle – circulating through each of these domains, changing phase between solid, liquid and gas; supporting the biosphere and humans; wearing away the continents and nourishing coastal zones. Water also serves as a conveyance system for bioactive chemicals (including poisons) that eventually find their way from continental source areas into the world's oceans.

Precipitation is the ultimate source of freshwater. After losing water back to the atmosphere through evaporation and evapotranspiration, precipitation recharges groundwater and provides surface and subsurface runoff. This runoff ultimately flows downstream through river corridors and into groundwater aquifers and constitutes an important regional source of water (table 10.1). Variation in precipitation and atmospheric demand for evapotranspiration thus geophysically limit water availability. Globally, about two-thirds of precipitation is returned to the atmosphere.

Latin America is the most water-rich region, with about a third of global runoff. Asia is next, with a quarter of global runoff, followed by the countries of the Organisation for Economic Co-operation and Development (mainly North America, Western Europe and Australasia) (20%), and sub-Saharan Africa and Eastern Europe, the Caucasus and Central Asia, each with about 10%. The Middle East and North Africa is the most water-limited region, with only 1% of global runoff. Over 85% of precipitation is evaporated or transpired. In regions where runoff is scarce or ill-timed, rainwater is an important component of the resource picture.

Table 10.2 describes the land-based water cycle as it relates to water resources systems. A substantial portion of the water associated with the renewable land-based hydrologic cycle is inaccessible to humans due to remoteness or an inability to store seasonal flows.[1] Accessibility is also affected by political preferences and unequal distribution of wealth and technological resources, which can prevent the delivery of water even when its physical presence is confirmed (the concept of economic water scarcity). Further, some groundwater stocks are accessible but not renewable, such as ancient aquifers in regions that today lack a replenishment source. An appropriate conceptual framework is thus required that combines physical and human dimensions. Map 10.1 contrasts the two perspectives and shows the portion of the global water cycle that is accessible to humans. Some additional supplies can be made available through non-conventional means, such as desalination, often at substantial cost in infrastructure and operations.

Spatial and temporal variability

Map 10.1 clearly shows an uneven spatial distribution of water supply. Variations in seasonality and the episodic nature of rainfall, snowfall, snowmelt and evapotranspiration all contribute to temporal incongruities that show up as flooding, seasonal low flows and longer-term drought, challenging water managers to forecast conditions and specify water allocations under a cloud of uncertainty.

Major interconnections are established through circulation patterns in the global atmosphere, which lead to the

Globally, about two-thirds of precipitation is returned to the atmosphere

Table 10.1 **Estimates of renewable water supplies, access to renewable supplies and population served by freshwater, 2000**

Indicator	Asia	Eastern Europe, the Caucasus and Central Asia	Latin America	Middle East and North Africa	Sub-Saharan Africa	OECD	Global total
Area (millions of square kilometres)	20.9	21.9	20.7	11.8	24.3	33.8	133.0
Total precipitation (thousands of cubic kilometres a year)	21.6	9.2	30.6	1.8	19.9	22.4	106.0
Evaporative returns to atmosphere (percent of precipitation)	55	27	27	86	78	64	63
Total renewable water supply (blue water flows; thousands of cubic kilometres a year) [% of global runoff]	9.8 [25]	4.0 [10]	13.2 [33]	0.25 [1]	4.4 [11]	8.1 [20]	39.6 [100]
Renewable water supply (blue water flows accessible to humans; thousands of cubic kilometres a year) [percent of total renewable water supply]	9.3 [95]	1.8 [45]	8.7 [66]	0.24 [96]	4.1 [93]	5.6 [69]	29.7 [75]

Note: Means computed based on methods in Vörösmarty, Leveque, and Revenga (2005). Estimates are based on climate data for 1950-96, computed using estimates of population living downstream of renewable supplies in 2000.
Source: Fekete, Vörösmarty, and Grabs 2002.

Table 10.2 Definitions of key components of the land-based hydrologic cycle and examples of their reconfiguration by humans

Water system element	Space and time variability	Typical roles in water resources systems	Management challenges, vulnerabilities and opportunities
Green water • Soil moisture (non-productive green water is evaporated from soil and open water surfaces)	• Very high over both dimensions	• Direct support to rainfed cropping systems	• Highly sensitive to climate variability (both drought and flood); limited capacity to control • Can be augmented by rainfall-harvesting techniques (many traditional and widely adopted) • Weather and climate forecasts help in scheduling planting, harvest, supplemental irrigation and other activities • Performance improved or compromised by land management • Selection of improved crop strains for climate-proofing
Blue water (natural and altered) • Net of local groundwater recharge and surface runoff, streamflow	• High over both dimensions	• Farm ponds and check dams augment green water in rainfed cropping systems • Source waters and entrained constituents delivered downstream within watersheds	• Highly sensitive to climate variability (both drought and flood) and ultimately climate change • Some capacity to control • Habitat management highly localized • Many small engineering works can propagate strong cumulative downstream effects • Poor land management heightens possiblities of flash flooding followed by dry streambeds
• Inland water systems (lakes, rivers, wetlands)	• Decreased variability with increased size	• Key resource over district, national, and multinational domains • Important role in transport, waste management, and domestic, industrial and agricultural sectors	• Water losses through net evaporation occur naturally and through human use • Legacy of upstream management survives downstream (e.g., irrigation losses, pollution) • Multiple sector management objectives may be difficult to attain simultaneously • Potential upstream-downstream conflicts (human to human; human to nature), including international
• Ground water (shallow)	• Moderate over both dimensions; links to streams	• Locally distributed shallow well systems serving drinking water and irrigation needs	• Intimate connection to weather and climate means water yields subject to precipitation extremes • Easily polluted • Easily overused, resulting in temporary depletion; some loss of regional importance to oceans
• Fossil groundwater (deep)	• Extremely stable	• Critical (and often sole) source of water in arid and semi-arid regions	• Large repositories of water but with limited recharge potential • Use typically non-sustainable, leading to declining water levels and pressure, increasing extraction costs • Low replenishment rates mean pollution often effectively becomes permanent
Blue water (engineered) • Diversions, including reservoirs and interbasin transfers • Reused waters	• Stable to very stable	• Critical (and often sole) source of water in arid and semi-arid regions • Altered blue water balance as flows stabilized or redirected from water-rich times and places to water poor times and places • Multiple uses: hydropower, irrigation, domestic, industrial, recreational, flood control • Secondary reuse as effluents in irrigation	• Large quantities of water with high recharge potential • Modified flow regime, with positive and negative impacts on humans and ecosystems • Can destroy river fish habitat while creating lake fisheries by fragmenting habitat • Natural ecosystem 'cues' for breeding and migration removed • Water supplies stabilized for use when needed most by society • Sediment trapping, leading to downstream inland waterway, coastal zone problems • Potential for introduction of exotic species • Greenhouse gas emission from stagnant water • Health problems (e.g., schistosomiasis) from stagnant water • Social instability due to forced resettlement

(continued)

Table 10.2 **Definitions of key components of the land-based hydrologic cycle and examples of their reconfiguration by humans** (continued)

Water system element	Space and time variability	Typical roles in water resources systems	Management challenges, vulnerabilities and opportunities
Virtual water (not an additional water system element)	• Stable, but linked to fluctuations in global economy	• Water embodied in production of goods and services, typically with crops traded on the international market • Not explicitly recognized as a water resources management tool until recently	• Can implicitly off-load water use requirements from more water-poor to more water-rich locations • Particularly important where rainfed agriculture is restricted and irrigation relies on rapidly depleting fossil groundwater sources
Desalination	• Stable	• Augmentation in water-scarce areas	• Costly, special use water supply; technologies rapidly developing for cost-effectiveness

Source: Author's compilation.

Map 10.1 **Contrasts between geophysical and human-dimension perspectives on water, most recent year available**

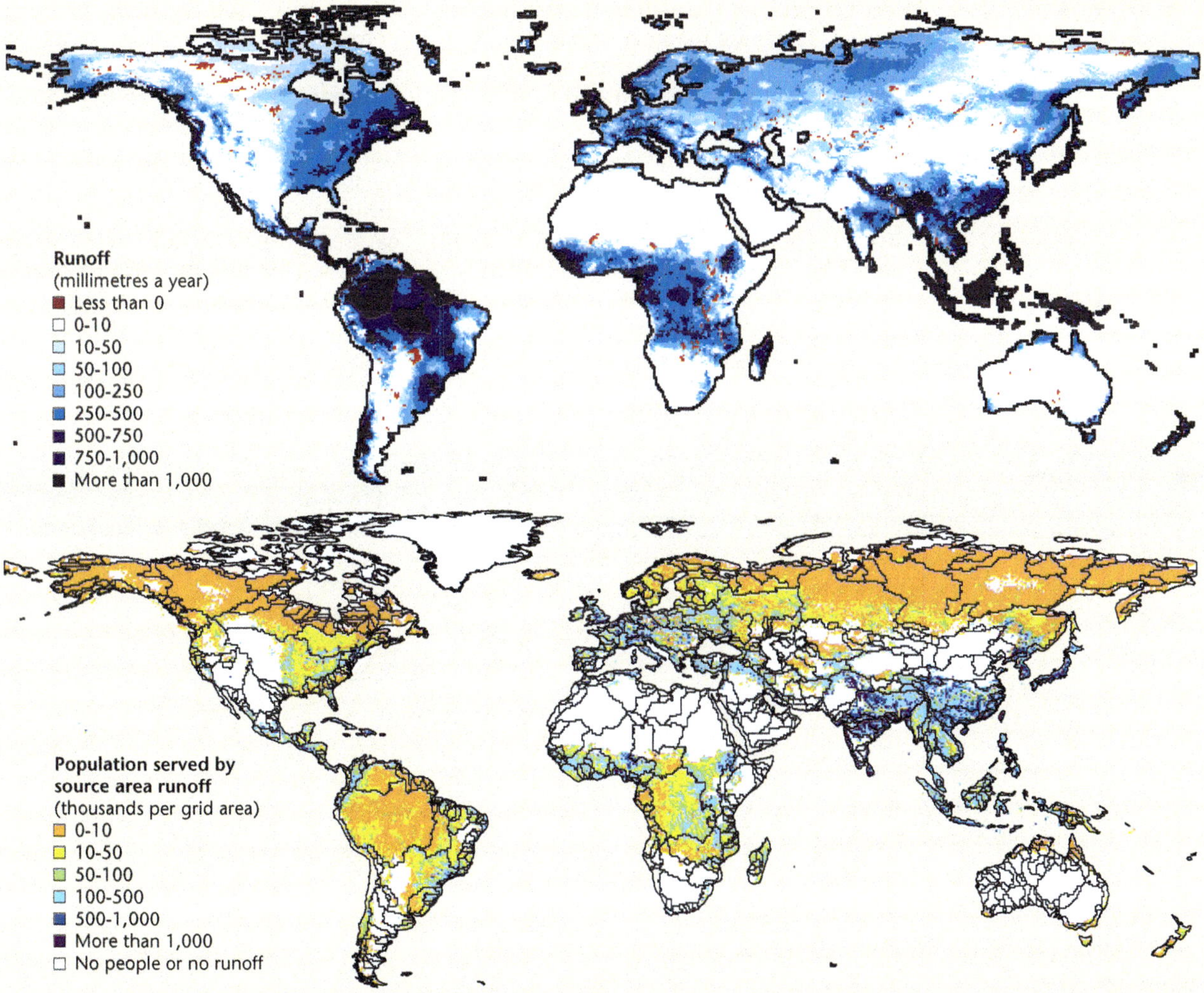

Note: The top map shows runoff-producing areas in absolute terms, with darker blue indicating areas that generate intense local-scale runoff. This is the traditional view of the global distribution of the renewable water resources base. The bottom map shows the importance of all of the world's runoff-producing areas, as measured by the human population served. Thus, runoff produced across a relatively unpopulated region like Amazonia, while a globally significant source of water to the world's oceans, is much less critical to the global water resources base than runoff produced across a region like South Asia.
Source: Vörösmarty, Leveque, and Revenga (2005), updated from Fekete, Vörösmarty, and Grabs (2002).

The global distribution of freshwater must be considered together with its accessibility

redistribution of water to oceans and land masses. The position of mountain ranges across the continents can be used to divide the Earth into two domains – exorheic zones, where water flows to the oceans, and internally draining endorheic zones, many in the 'rain shadow' of the world's main precipitation belts (figure 10.1). Much of the endorheic land is positioned mid-continent and distant from the ocean, resulting in a characteristically dry environment. Here, 10%-15% of the global land mass generates only 2% of global renewable freshwater resources. Mountain systems are important as the world's 'water towers' and generate a substantial share of the global water resources base for the billions of people who live downstream. By contrast, the margins of the continents (exorheic plains), because of their intimate connection with ocean-derived moisture, generate about half of all renewable freshwater resources, collectively greater than all mountain water towers. There are many concerns about the impact of climate change on this geography of precipitation and runoff-producing areas.[2]

The global distribution of freshwater must be considered together with its accessibility. With about 75% of total annual runoff accessible to humans (see table 10.1) and with slightly more than 80% of the world's population (4.9 billion people) served by renewable and accessible water,[3] almost 20% of people are unserved by naturally occurring renewable resources and must take their supply from ancient aquifers (aquifer mining), interbasin transfers and desalinized seawater. Except in the exorheic mountain regions, where water is relatively abundant, most of the population has only a small share of the global freshwater resource. Using high resolution global maps of population and water supply, a study showed that 85% of the world's population resides in the drier half of the Earth.[4] More than 1 billion people living in arid and semi-arid parts of the world have access to little or no renewable water resources. However, there is still uncertainty surrounding estimates of the renewable supply, water use and derived statistics (table 10.3; see also chapter 7).

There is great variation in flow reliability, with variability greatest in regions with the lowest levels of runoff (the drier regions; map 10.2). Patterns of reliable monthly river flows also confirm the sensitivity of arid and semi-arid regions, defined from a hydrologic (local runoff and river corridor supplies) rather than a climatic (rainfall variability alone) perspective.[5] This variability reduces projections of average annual GDP growth rates by as much as 38%, and even a single drought event within a 12-year period can reduce growth rates over the period by 10%. Flooding can also have devastating effects, particularly in areas with high population density and without adequate early warning and emergency response systems (map 10.3). During 1992-2001 floods accounted for 43% of recorded

Figure 10.1 **Distribution of global runoff to the oceans (exorheic) or internal receiving waters (endorheic) and the corresponding distribution of contemporary population served**

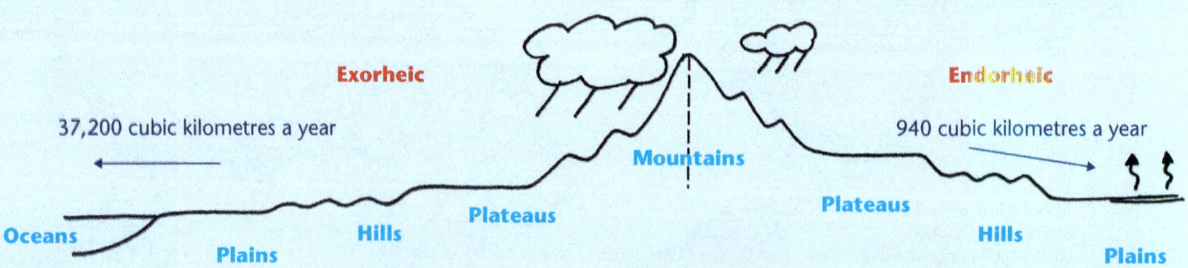

	Plains	Hills	Plateaus	Mountains	Mountains	Plateaus	Hills	Plains
Area (millions of square kilometres)	63	11	14	28	6	2.5	0.4	9
Depth change (millimetres a year)	293	445	153	424	86	38	102	35
Total resources (thousands of cubic kilometres a year)	18.4	4.9	2.1	11.8	0.5	0.09	0.04	0.3
Population served (bilions)	3.3	0.8	0.4	1.5	0.2	0.03	0.02	0.2

Source: Updated from Vörösmarty and Meybeck (2004); land form categories from Meybeck, Green, and Vörösmarty (2001).

disasters and affected more than 1.2 billion people.[6]

People have responded to such variability in the water cycle with investments in engineered water stabilization, such as reservoirs, interbasin transfers and deep groundwater pumping. These stabilization arrangements bring new patterns of hydrograph variability (see box 11.1 in chapter 11). Figure 10.2 shows an example of a typical effect of a series of river flow regulations. While such changes may stabilize flows and thus optimize water availability for a variety of human uses, they also create substantial distortions in flows that stress downstream aquatic biota (see chapter 9).[7]

A more stable and reliable source of freshwater resides below ground. Groundwater reservoirs are recharged directly by surplus rainfall percolating through soil or indirectly by surface water losses to the subsurface and infiltration of excess irrigation water or water from other uses. Approximately 90% of the world's groundwater discharge feeds into streams, accounting for almost 30% of global runoff.[8] Most groundwater systems have large storage volumes and high storage to throughput ratios (known as residence time, or average time that inflow volumes remain in storage). Because of these characteristics, groundwater resources are much less affected by short-term fluctuations in climate than are surface water resources (table 10.4). Groundwater reservoirs thus add persistency and stability to the terrestrial hydrologic system and enable humans, fauna and flora to survive extended dry periods. This underlines the potential of groundwater for coping with

Table 10.3 **Indicative range of uncertainty in recent assessments of renewable water supply, most recent year available**

Region	Renewable water supply (cubic kilometres a year)	Mean water crowding (people per million cubic metres a year)
Asia	7,850-9,700	320-384
Former Soviet Union	3,900-5,900	48-74
Latin America	11,160-18,900	25-42
North Africa and Middle East	300-367	920-1,300
Sub-Saharan Africa	3,500-4,815	115-160
Organisation for Economic Co-operation and Development	7,900-12,100	114-129
Global total	38,600-42,600	133-150

Note: Supply here refers to global total renewable runoff, both accessible and remote from human population and croplands.

Source: The ranges reported here are from three global-scale water resources models, two of which were used in the *Millennium Ecosystem Assessment*: Vörösmarty, Federer, and Schloss (1998); Fekete, Vörösmarty, and Grabs (2002) and Federer, Vörösmarty, and Fekete (2003) for the Condition and Trends Working Group assessment and Alcamo et al. (2003) and Döll, Kaspar, and Lehner (2003) for the Scenarios Working Group. A third model from Dirmeyer, Gao, and Oki (2002) and Oki et al. (2003) was also compared.

Map 10.2 **Global variations in the relationship between low flows and mean flows (percentage deviation of 1 in 10 year low flows relative to mean flows measured over 1961-90)**

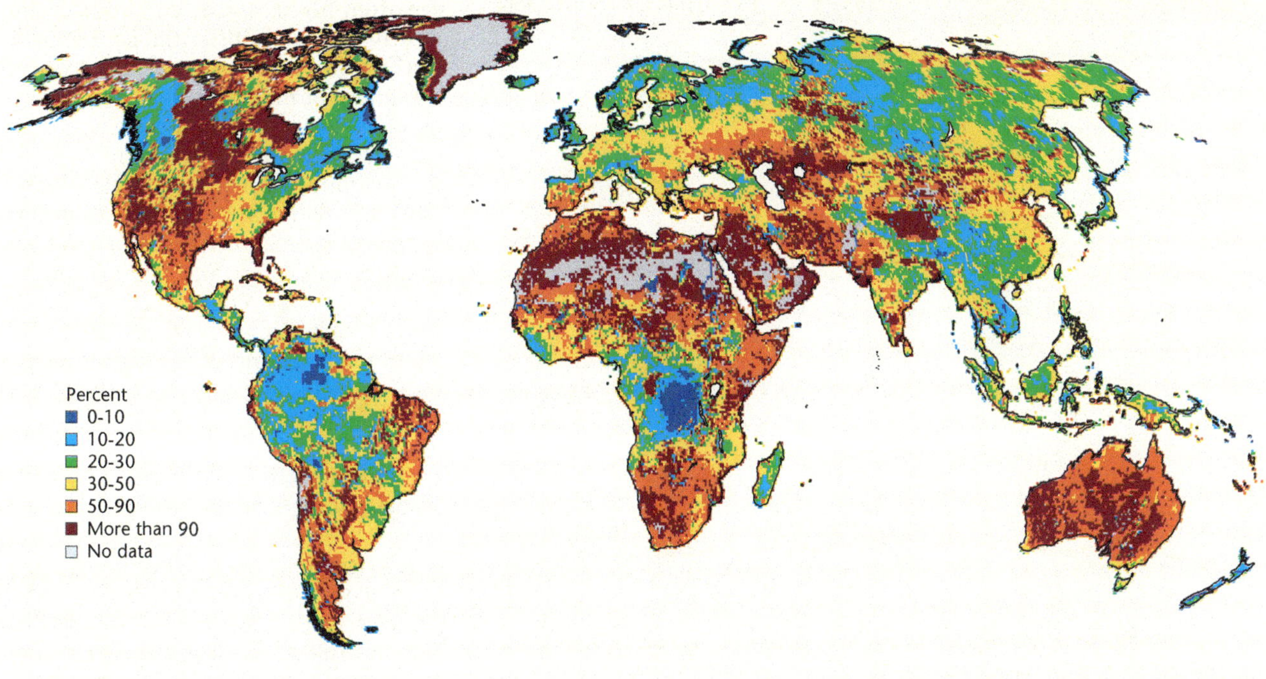

Source: Based on Döll, Kaspar, and Lehner 2003.

Map 10.3 **Impact of flood losses (comparative losses based on national GDP)**

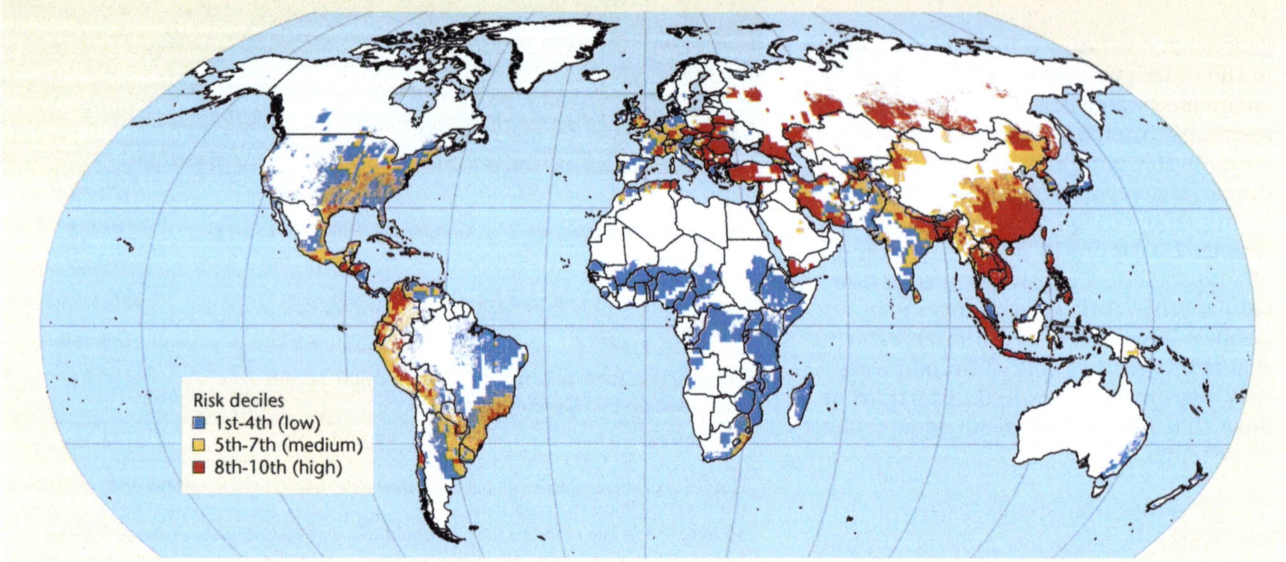

Note: Deciles refer to the level of risk, normalized for comparing 10 categories.
Source: Based on Dilley et al. 2005.

Figure 10.2 **Impact of the Davis Dam on the Colorado River hydrograph**

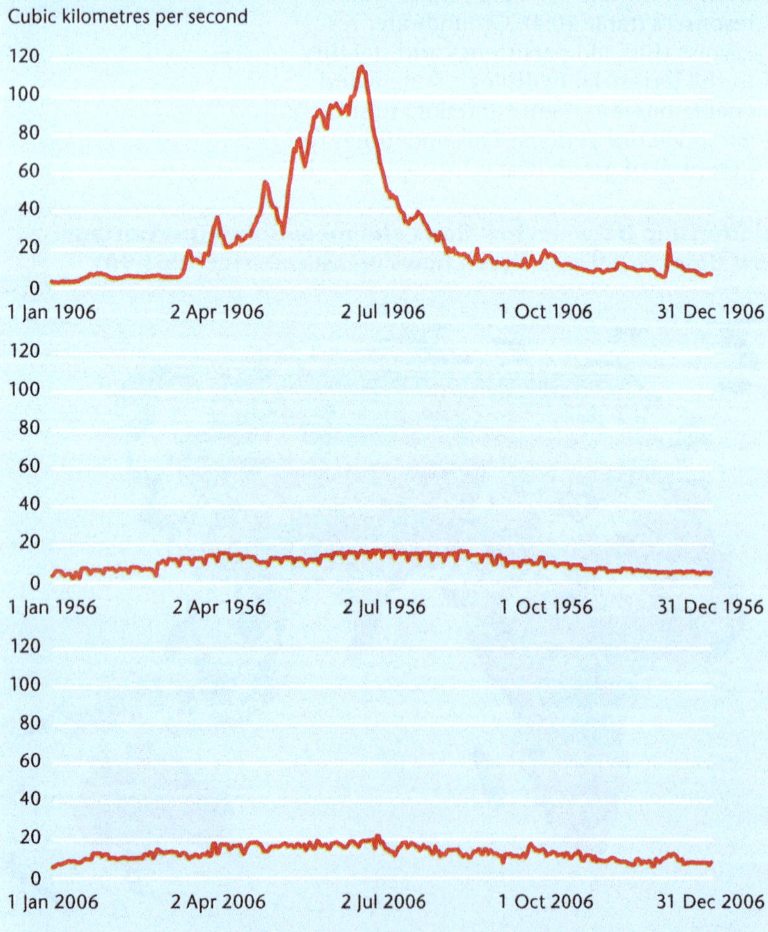

Note: The dam was completed in 1950, has a maximum capacity of 2 cubic kilometres and residence time change (average time that inflows remain in storage) of about 0.5 year.
Source: US Geological Survey station records.

increasing water scarcity due to climate change. At the same time, because of the strong interdependence between groundwater and surface water, the overall resource is difficult to quantify, and there is a risk of double counting available water resources.

Recent estimates put the mean renewable groundwater resource at 2,091 cubic metres per person a year, or about a third of total renewable resources per capita.[9] Although groundwater systems are often highly localized, groundwater clearly makes a substantial contribution to the water resources base, constituting 20%-50% of municipal water supply.[10] As much as 60% of groundwater withdrawal is used to irrigate crops in arid and semi-arid regions. Information on groundwater storage is scarce and not very accurate because of the enormous effort and cost required to explore and assess groundwater reservoirs. The geographic distribution of long-term average diffuse groundwater recharge and of the known larger groundwater reservoirs are in maps 10.4 and 10.5.

Relationship of water to global biogeochemical cycles

A growing body of evidence indicates that human activities are affecting river water chemistry on a global scale. It is estimated that less than 20% of the world's drainage basins exhibit nearly pristine water quality and that the riverine transport of inorganic nitrogen and phosphorus has increased severalfold over the last 150-200 years.[11] Monitoring and analysis are needed to understand the effects of these changes on water resources, but observed information is lacking (see chapter 13), and much of our understanding is based on modelling and inferences.

Water as the conveyor for particulates and dissolved materials, linking land, ocean and atmosphere

Water mobilizes and transports materials essential for life in terrestrial and aquatic ecosystems. For example, nitrogen, phosphorus and silica are important nutrients that limit maximum plant and algal biomass, while organic carbon from land is an important energy source in downstream freshwater and marine systems. Water also transports natural materials that directly influence the health of organisms (for example, through conductivity and pH) and habitat structure (for example, through sediments). Under natural conditions these materials originate in atmospheric transport and deposition, biologic activity and erosion or weathering from bedrock and soils. Multiple human activities lead to additional sources of such elements (figure 10.3) as well as to material not naturally present in water, such as pesticides and synthetic chemicals that can mimic or block hormones and their natural functions ('endocrine disrupters') (see section on water pollution as a constraint to supply).

Table 10.4 **Estimated mean residence times (storage to throughput) and stored water volumes of the main components of the Earth's hydrosphere**

Component	Mean residence time	Total water stored (thousands of cubic kilometres)	Freshwater stored (thousands of cubic kilometres)
Permafrost zone, ground ice	10,000 years	300	300
Polar ice	9,700 years	24,023	24,023
Oceans	2,500 years	1,338,000	na
Mountain glaciers	1,600 years	40.6	40.6
Groundwater (excluding Antarctica)	1,400 years	23,400	10,530
Lakes	17 years	176.4	91.0
Swamps	5 years	11.5	11.5
Soil moisture	1 year	16.5	16.5
Streams	16 days	2.1	2.1
Atmosphere	8 days	12.9	12.9
Biosphere	Several hours	11.2	11.2
Total		1,385,985	35,029

na is not applicable.
Note: Components may not sum to total because of rounding. Reservoirs of water that respond slowly to change have long residence times. The atmosphere exhibits huge variability, its dynamics changing over very short space and time scales, whereas permafrost is sluggish and would be expected to respond slowly to forced changes such as those associated with global warming. Residence time also has an enormous impact on water quality. Streams and river waters, with their generally short residence times, are able to respond relatively quickly to pollution control measures, whereas groundwater can remain polluted – and taken out of the resource supply pool – for centuries unless costly remediation measures are applied.
Source: Based on Shiklomanov and Rodda 2003.

Rivers have traditionally been considered simple transporters of materials, but it is increasingly acknowledged that chemical

Map 10.4 **Patterns of long-term average diffuse groundwater recharge, 1961-90**

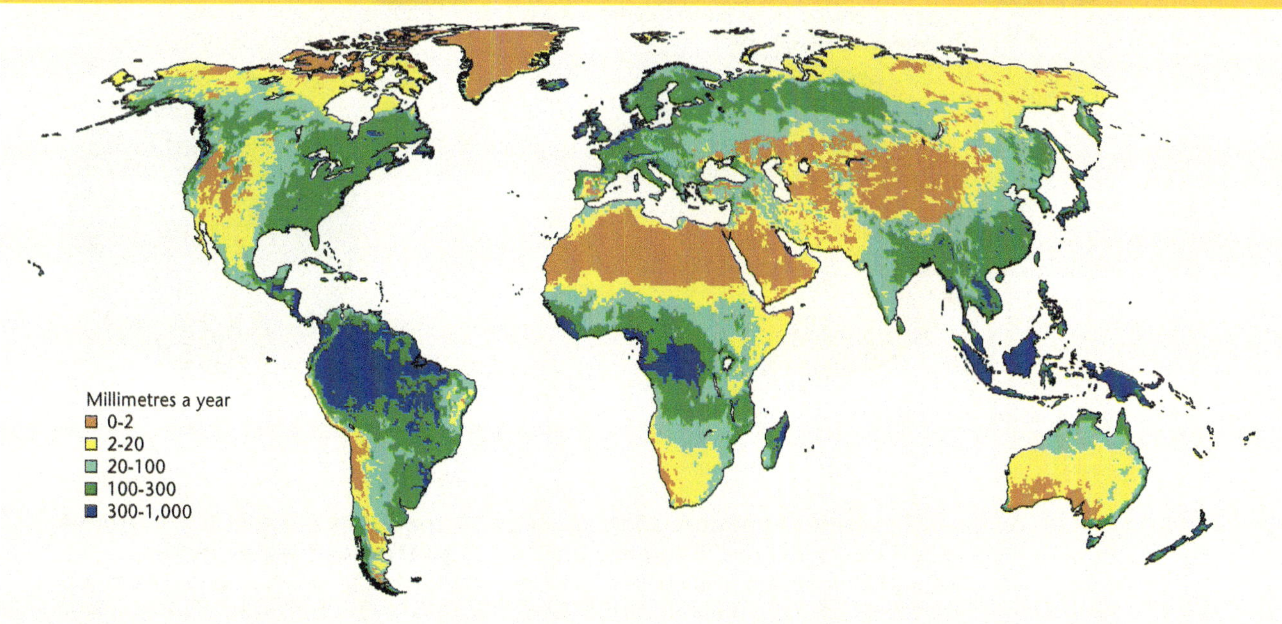

Source: Based on Döll and Fiedler 2007.

Map 10.5 **Global groundwater recharge, most recent year available**

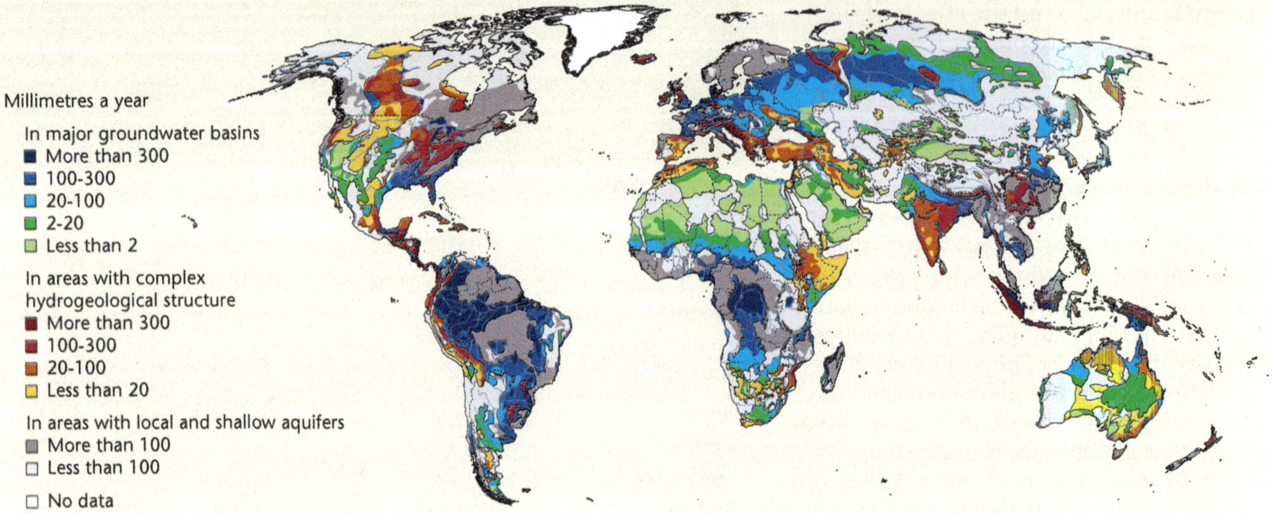

Source: Based on WHYMAP 2008.

Figure 10.3 **Human activities are sources for dissolved inorganic nitrogen, organic nitrogen, inorganic phosphorus and organic phosphorus in coastal zones**

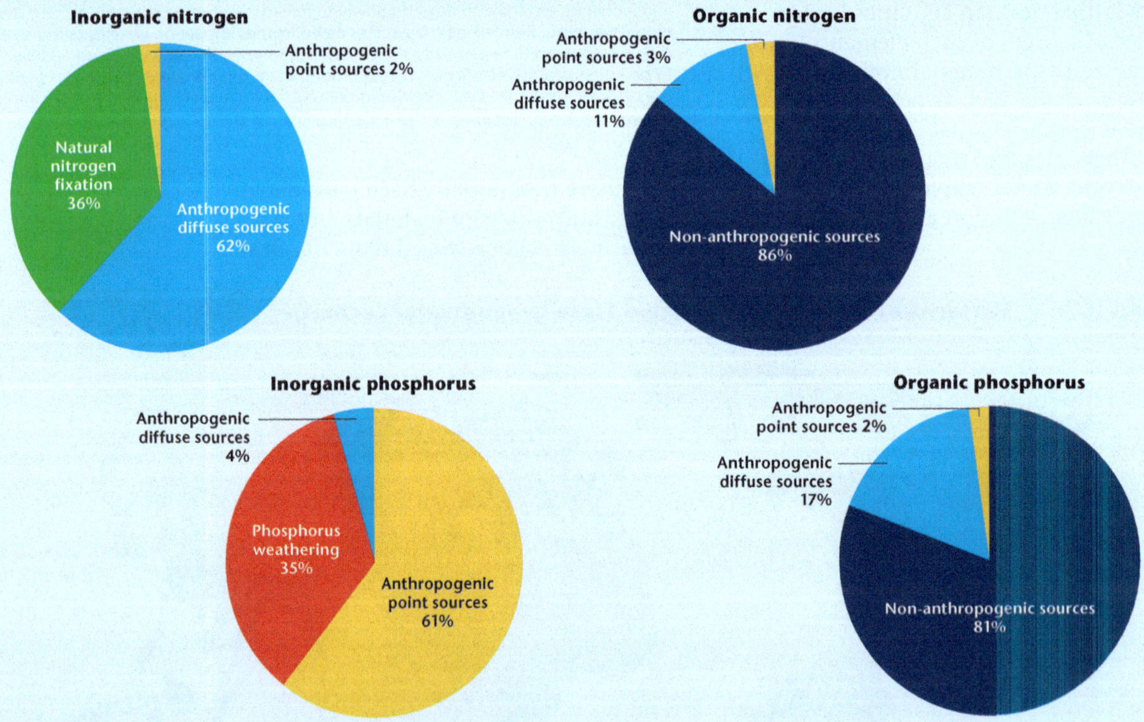

Note: Human activity dominates among the inorganic source terms.
Source: Seitzinger et al. 2005.

and biochemical transformations occurring during water transit through basins can have important influences on materials transport and pollutant loads.[12] The quantity and timing of water flows influence the mobility of pollutant sources and their dilution potential.

Spatial heterogeneity in global patterns of water quality

Making broad statements about water quality is difficult because of spatial and temporal biogeochemical complexity as well as definitional problems and incomplete monitoring (see chapter 13). The multiple factors that influence material fluxes through aquatic systems (sources, hydrology, geomorphology and biology) suggest an inherent complexity in characterizing water quality at regional and global scales. Uncertainties are associated with each factor.

The situation is exacerbated by the uneven distribution over space and time of field observations of material fluxes and water quality. The European Union and the United States have relatively high densities of water quality measurement stations, while the rest of the world has a much sparser set of stations, with low frequency of sampling and observations over shorter periods.

Observations over short periods can fail to capture changes in human activities across the multiple time scales over which hydrologic variability occurs – from individual storms to seasonal, annual and multidecadal cycles. This variability is also expressed spatially, influencing the development of source and reactivity hot spots (such as erosion and wetland development). This interaction can make it difficult to attribute changes to anthropogenic or natural phenomena.

Humans are accelerating – and decelerating – the constituent cycles

Human pressures have greatly modified the behaviour of both hydrology and constituent transports, particularly over the last century.[13] This will likely remain true well into the future, but the direction of the evolution of these transports, and so the quality of inland water, is a complex function of four major changes that must be considered simultaneously:

- Human activities have greatly accelerated the biogeochemical cycles and the global transfer of materials, including sediment from increased erosion associated with poor land management, construction and other activities.
- Fluvial system filters have been greatly modified and in the case of artificial impoundments have increased in importance.
- River water discharge to oceans is controlled and reduced by water engineering and irrigation, with irretrievable losses on the order of 200 cubic kilometres (km^3) a year for reservoir evaporation and 2,000 km^3 a year for agriculture.[14]
- New and esoteric engineered compounds, many long-lived, are appearing in waterways. We have simultaneously increased and decreased the levels of various constituents in our waterways, but the exact nature of the acceleration or deceleration is complex and ambiguous.

Human population and economic growth lead to increased demand for land and commodities for food production, housing and fuel (see chapters 2 and 4). The natural capacity for land to support human populations is insufficient, so people enhance food production with fertilizer and intensive agriculture.[15] Agricultural activities are also accelerating the elemental cycles of nitrogen and phosphorus, since more must be added to the landscape in biologically usable forms. Globally, nitrogen inputs have more than doubled, with similar increases in transport to the ocean.[16] Similar results have been documented for phosphorus.[17]

Changes in hydrologic factors associated with human activity have unintentionally exacerbated other changes related to land use. Erosion associated with changes in land use is increasing sediment delivery to aquatic systems, but much of this material is captured by increased reservoir capacity, with only a small net change at the basin mouth.[18] A third of sediment destined for coastal zones no longer arrives because of sediment trapping and water diversion,[19] resulting in a net increase in erosion of deltas and other sensitive coastal settings that require a steady supply of land-derived sediment.[20] Reservoir construction appears to have attenuated silica, nitrogen and phosphorus fluxes, though the role appears to be less than for sediments.[21]

Net increases in nutrient loads (particularly nitrogen and phosphorus) have resulted in the eutrophication (excessive plant growth and decay) of lakes, rivers and receiving coastal waters and subsequent degradation of ecosystems, fisheries and human health. Anoxic dead zones result from excess nutrient inputs from agriculture, as from the Mississippi River basin into the Gulf of Mexico[22] and from the Yangtze River plume.[23] Alterations to inputs and 'fluvial filters' of nutrients (nitrogen, phosphorus, silicon) are changing elemental ratios in freshwater and downstream coastal waters. In addition, as different nutrient forms (such as organic and inorganic nitrogen) are shown to have contrasting watershed sources and human pressures, ratios among nutrient forms

Human pressures have greatly modified the behaviour of both hydrology and constituent transports, particularly over the last century

New modelling approaches now include mechanisms that hold promise for predicting global patterns of water quality

may also be changing.[24] Shifting nutrient ratios alter the composition of biological assemblages in freshwater and coastal systems, including the occurrence and recurrence of harmful algal blooms.[25]

Recent progress in describing patterns in global water quality

Despite limitations in characterizing human processes that determine the chemical characteristics of freshwater resources, syntheses of river observations and process studies have substantially advanced our ability to quantify the transformation of watershed-derived inputs into river loads and exports to coastal zones.[26] Recent models predict mean annual nutrient status based on geospatial datasets defining watershed inputs (natural and human), hydrologic and physical properties and biological processing potentials within rivers. They rely on global calibration of basin-scale parameters with river mouth observations to provide a consistent, spatially explicit picture of worldwide nutrient exports to coastal zones. Submodels of Global-NEWS (Nutrient Export from Watersheds), organized under the United Nations Educational, Scientific and Cultural Organization's Intergovernmental Oceanic Commission, apply a consistent framework and datasets to calculate river exports of carbon, nitrogen and phosphorus (dissolved and particulate, inorganic and organic), enabling an integrated assessment of impacts of a range of human pressures on receiving waters.[27] But such models are static and have limited process representation, so they are unable to account for variability over relatively short time scales (less than 1 month), which is critical in characterizing water quality.

Recently developed spatially distributed, mechanistic models of nutrient fluxes through river systems have been applied to numerous basins and are important for understanding the mechanisms controlling material fluxes.[28] Such efforts have also been applied globally to integrate spatially distributed controlling mechanisms. For example, global terrestrial nitrogen models now account for within-basin patterns of nitrogen loading, hydrologic conditions, land surface characteristics and ecosystem processes to predict nitrogen export fluxes.[29]

Recently, a spatially distributed modelling approach was applied to global inland aquatic systems to estimate the relative importance of small rivers, large rivers, lakes and reservoirs in the global aquatic nitrogen cycle by integrating the spatial distribution of inputs from land, discharge conditions, network geomorphology, position of various water bodies and rates of biological activity (figure 10.4). The relative importance of different types of water bodies varies with latitude because of the distribution of channel bottom surface area relative to the position of nitrogen inputs, which are increasingly determined by human inputs of fertilizers, sewage and animal wastes and by atmospheric deposition.[30] These approaches now include key mechanisms that hold promise for predicting global patterns of water quality and thus of water supply compromised by pollution, changes in runoff and streamflow, runoff variability, temperature and hydraulic modification.

Figure 10.4 The spatial distribution of surface area and nitrogen inputs and removal by types of water bodies differ by latitude, most recent year available

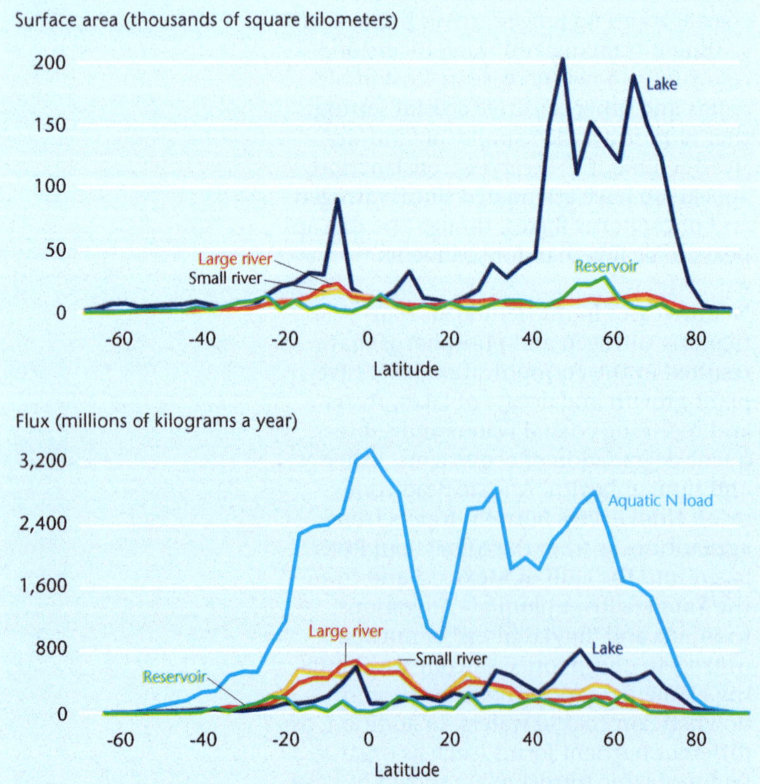

Source: Based on Wollheim et al. 2008.

Water pollution as a constraint on water supply

Good water quality is important to human health and the health of aquatic ecosystems. Increasing pressures from development lead to deteriorating surface water and groundwater quality (table 10.5), rising human health challenges, growing requirements for water treatment and greater likelihood of deteriorating ecosystem function.[31] Pollutants (including excess nutrients) may be retained in soils, aquifers and aquatic sediments for extended periods. This persistence can lead to continued mobilization long after human inputs have been suppressed or when affected sediments are later disturbed.[32]

Humans have long relied on dilution and transport by aquatic systems to manage pollution – and the water quality of freshwater resources. In some cases aquatic systems permanently remove pollutants to the atmosphere, as in the denitrification of excess nitrogen. These are important ecosystem services that depend on characteristics of the water cycle. Changes in the water cycle will lead to changes in the capacity of natural ecosystems to provide these services.[33] Because aquatic systems are highly connected, local changes in aquatic ecosystem services often cause impacts far downstream (see chapter 8).

Pollutants can be categorized as those that directly affect human health and those that affect ecosystems. Pollutants affecting human health include fæcal coliform contamination, residual pesticides and metals. Examples include contamination of groundwater by arsenic in Bangladesh[34] (see box 8.3 in chapter 8) and by mercury in the northeastern United States[35] and nitrogen in drinking water supplies (see chapter 6).[36]

In developed countries many pollution issues have been addressed and ameliorated over the last 40 years, especially those pertaining to point sources. But in developing countries pollution remains among the most important water resources problems. These include lack of sewage treatment and point source controls and contamination with pathogens, combined with poor access to clean water.[37] In developed countries non-point-source pollution remains a critical issue, in part because management requires multijurisdictional approaches that make implementation difficult. But successful policies addressing acidification of surface water by atmospheric deposition in Europe and North America are leading to the recovery of many surface waters from acidification, providing a hopeful model for multijurisdictional landscape management.[38]

Good water quality is important to human health and the health of aquatic ecosystems

Table 10.5 **Principal symptoms of human-river system interactions and human pressures on water use**

Symptoms	Land use change	Mining and smelting	Industrial transformation	Urbanization	Reservoirs	Irrigation	Other water management
Organic matter	+ + –		+	+ + +	+ – –	–	+
Salts	+	+ + +	+	+	+	+ + +	
Acids							
Direct inputs		+ +	+				
Atmospheric changes		+	+ +	+ +			
Metal							
Direct inputs		+ +	+ +	+	– – –		
Atmospheric inputs changes		+ + +	+ +	+			
Historical		+ + +	+	+	– – –		
Total suspended solids	+ + +	+ +	+	+	– – –	-	-
Nutrients	+ + +		+	+ +	– –	+ –	
Water-borne diseases	+ –			+ + +	+	+	+
Persistent organic pollutants							
Direct inputs	+ +		+ +	+ +	– –		
Atmospheric inputs	+		+	+			
Historical			+ + +	+ +			
Mean runoff	+ –		–	+	–	– – –	
Flow regime	×			×	×××	×	×
Aquatic habitat changes	×	××		××	×××	×	×××

Note: Amplitude of change ranges from + to + + + (increase) and – to – – – (decrease). The × symbols refer to magnitude of change without an indication of direction.
Source: From Meybeck 2003.

Notes

1. Postel, Daily, and Ehrlich 1996.
2. IPCC 2007.
3. Vörösmarty, Leveque, and Revenga 2005.
4. Vörösmarty, Leveque, and Revenga 2005.
5. Vörösmarty et al. 2005.
6. CRED 2002.
7. Olden and Poff 2003.
8. Margat 2008.
9. Döll and Fiedler 2007; Vörösmarty, Leveque, and Revenga 2005.
10. Morris et al. 2003; Zekster and Margat 2003.
11. Vörösmarty and Meybeck 2004.
12. Seitzinger et al. 2006; Cole et al. 2007; Battin et al. 2008.
13. Meybeck and Vörösmarty 2005.
14. Shiklomanov and Rodda 2003.
15. Imhoff et al. 2004; Haberl et al. 2007.
16. Green et al. 2004; Bouwman et al. 2005.
17. Harrison, Caraco, and Seitzinger 2005; Seitzinger et al., 2005.
18. Syvitski et al. 2005.
19. Vörösmarty et al. 2003.
20. Ericson et al. 2006.
21. Dumont et al. 2005; Harrison et al. 2005.
22. Rabalais et al. 2007.
23. Wang 2006.
24. Seitzinger et al. 2005.
25. Wang 2006.
26. Seitzinger et al. 2006; Seitzinger et al. 2005; Cole et al. 2007; Green et al. 2004; Boyer et al. 2006; Smith et al. 2003; Wollheim et al. 2008.
27. See Seitzinger et al. 2005 and recent special issue of *Global Biogeochemical Cycles*.
28. Ball and Trudgill 1995; Lunn et al. 1996; Alexander et al. 2002.
29. Bouwman et al. 2005.
30. Wollheim et al. 2008.
31. MEA 2005.
32. Meybeck 2003; Meybeck and Vörösmarty 2005.
33. Hinga and Batchelor 2005.
34. Mukherjee et al. 2006.
35. Driscoll et al. 2007.
36. Townsend et al. 2003.
37. WHO/UNICEF 2004.
38. Driscoll et al. 2003; Warby, Johnson, and Driscoll 2005.

References

Alcamo, J., P. Döll, T. Henrichs, F. Kaspar, B. Lehner, T. Rösche, and S. Siebert. 2003. Global Estimates of Water Withdrawals and Availability under Current and Future "Business-As-Usual" Conditions. *Hydrological Sciences Journal* 48 (3): 339-48.

Alexander, R. B, P. J. Johnes, E. W. Boyer, and R. A. Smith. 2002. A Comparison of Models for Estimating the Riverine Export of Nitrogen from Large Watersheds. *Biogeochemistry* 57/58: 295-339.

Ball, J., and S. T. Trudgill. 1995. Overview of Solute Modeling. In *Solute Modelling in Catchment Systems*, ed. S. T. Trudgill. Chichester, UK: John Wiley and Sons.

Battin, T. J., L. A. Kaplan, S. Findlay, C. S. Hopkinson, E. Marti, A. I. Packman, J. D. Newbold, and F. Sabater. 2008. Biophysical Controls on Organic Carbon Fluxes in Fluvial Networks. *Nature Geosciences* 1 (2): 95-100.

Bouwman, A. F., G. Van Drecht, J. M. Knoop, A. H. W. Beusen, and C. R. Meinardi. 2005. Exploring Changes in River Nitrogen Export to the World's Oceans. *Global Biogeochemical Cycles* 19, GB1002, doi:10.1029/2004GB002314.

Boyer, E. W., R. W. Howarth, J. N. Galloway, F. J. Dentener, P. A. Green, and C. J. Vörösmarty. 2006. Riverine Nitrogen Export from the Continents to the Coasts. *Global Biogeochemical Cycles* 20, GB1S91, doi:10.1029/2005GB002537.

Cole, J. J., Y. T. Prairie, N. F. Caraco, W. H. McDowell, L. J. Tranvik, R. G. Striegl, C. M. Duarte, P. Kortelainen, J. A. Downing, J. J. Middelburg, and J. Melack. 2007. Plumbing the Global Carbon Cycle: Integrating Inland Waters into the Terrestrial Carbon Budget. *Ecosystems* 10 (1): 171-84.

CRED (Centre for Research on the Epidemiology of Disasters). 2002. EM-DAT: Emergency Events Database. www.cred.be/emdat.

Dilley, M., R. S. Chen, U. Deichmann, A. L. Lerner-Lam, M. Arnold, with J. Agwes, P. Buys, O. Kjekstad, B. Lyon, and G. Yetman. 2005. *Natural Disaster Hotspots: A Global Risk Analysis: Synthesis Report.* New York: Columbia University.

Dirmeyer, P., X. Gao, and T. Oki. 2002. *GSWP-2: The Second Global Soil Wetness Project Science and Implementation Plan.* IGPO Publication Series No. 37, International GEWEX (Global Energy and Water Cycle Experiment) Project Office.

Döll, P., and K. Fiedler. 2007. Global-Scale Modelling of Groundwater Recharge. *Hydrology and Earth System Sciences* 4 (6): 4069–4124.

Döll, P., F. Kaspar, and B. Lehner. 2003. A Global Hydrological Model for Deriving Water Availability Indicators: Model Tuning and Validation. *Journal of Hydrology* 270 (1): 105–134.

Driscoll, C. T., K. M. Driscoll, K. M. Roy, and M. J. Mitchell. 2003. Chemical Response of Lakes in the Adirondack Region of New York to Declines in Acidic Deposition. *Environmental Science and Technology* 37: 2036-42.

Driscoll, C. T., Y. J. Han, C. Y. Chen, D. C. Evers, K. F. Lambert, T. M. Holsen, N. C. Kamman, and R. K. Munson. 2007. Mercury Contamination in Forest and Freshwater Ecosystems in the Northeastern United States. *BioScience* 57 (1): 17-28.

Dumont, E., J. A. Harrison, C. Kroeze, E. J. Bakker, and S. P. Seitzinger. 2005. Global Distribution and Sources of Dissolved Inorganic Nitrogen Export to the Coastal Zone: Results from a Spatially Explicit, Global Model. *Global Biogeochemical Cycles* 19, GB4S02, doi:10.1029/2005GB002488.

Ericson, J. P., C. J. Vörösmarty, S. L. Dingman, L. G. Ward, and M. Meybeck. 2006. Effective Sea-Level Rise in Deltas: Causes of Change and Human-Dimension Implications. *Global & Planetary Change* 50 (1): 63-82.

Federer, C. A., C. J. Vörösmarty, and B. Fekete. 2003. Sensitivity of Annual Evaporation to Soil and Root Properties in Two Models of Contrasting Complexity. *Journal of Hydrometeorology* 4 (1): 1276-90.

Fekete, B. M., C. J. Vörösmarty, and W. Grabs. 2002. High Resolution Fields of Global Runoff Combining Observed River Discharge and Simulated Water Balances. *Global Biogeochemical Cycles* 6 (3), doi:10.1029/1999GB001254.

Green, P. A., C. J. Vörösmarty, M. Meybeck, J. N. Galloway, B. J. Peterson, and E. W. Boyer. 2004. Pre-Industrial and Contemporary Fluxes of Nitrogen through Rivers: A Global Assessment Based on Typology. *Biogeochemistry* 68 (1): 71-105.

Haberl, H., K. H. Erb, F. Krausmann, V. Gaube, A. Bondeau, C. Plutzar, S. Gingrich, W. Lucht, and M. Fischer-Kowalski. 2007. Quantifying and Mapping the Human Appropriation of Net Primary Production in Earth's Terrestrial Ecosystems. *Proceedings of the National Academy of Sciences of the United States of America* 104:12942-45.

Harrison, J. A., N. Caraco, and S. P. Seitzinger. 2005. Global Patterns and Sources of Dissolved Organic Matter Export to the Coastal Zone: Results from a Spatially Explicit Global Model. *Global Biogeochemical Cycles* 19, doi 10.1029/2005gb002480.

Harrison, J. A., S. P. Seitzinger, A. F. Bouwman, N. F. Caraco, A. H. W. Beusen, and C. J. Vörösmarty. 2005. Dissolved Inorganic Phosphorus Export to the Coastal Zone: Results from a Spatially Explicit, Global Model. *Global Biogeochemical Cycles* 19, GB4S03, doi:10.1029/2004GB002357.

Hinga, K. R., and A. Batchelor. 2005. Waste Detoxification. In *Millennium Ecosystem Assessment, Volume 1: Conditions and Trends Working Group Report*, ed., R. Hassan, R. Scholes, and N. Ash. Washington, DC: Island Press.

Imhoff, M. L., L. Bounoua, T. Ricketts, C. Loucks, R. Harriss, and W. T. Lawrence. 2004. Global Patterns in Human Consumption of Net Primary Production. *Nature* 429 (6994): 870-73.

IPCC (Intergovernmental Panel on Climate Change). 2007. *Climate Change 2007: The Physical Science Basis. Summary for Policymakers.* Contribution of Working Group I to the Fourth Assessment Report of the Intergovernmental Panel on Climate Change, ed., S. Solomon, D. Qin, M. Manning, Z. Chen, M. Marquis, K. B. Averyt, M. Tignor, and H. L. Miller. Cambridge, UK: Cambridge University Press.

Lunn, R. J., R. Adams, R. Mackay, and S. M. Dunn. 1996. Development and Application of a Nitrogen Modeling System for Large Catchments. *Journal of Hydrology* 174 (1): 285-304.

Margat, J. 2008. *Les Eaux Souterraines: Une Richesse Mondial.* Paris: United Nations Educational, Scientific, and Cultural Organization

MEA (Millennium Ecosystem Assessment). 2005. *Ecosystems and Human Well-being: Wetlands and Water Synthesis.* Washington, DC: World Resources Institute.

Meybeck, M. 2003. Global Analysis of River Systems: From Earth System Controls to Anthropocene Syndromes. *Philosophical Transactions of the Royal Society of London* B 358: 1,935-55.

Meybeck, M., P. Green, and C. J. Vörösmarty. 2001. A New Typology for Mountains and Other Relief Classes: An Application to Global Continental Water Resources and Population Distribution. *Mountain Research and Development* 21 (1): 34-45.

Meybeck, M., and C. J. Vörösmarty. 2005. Fluvial Filtering of Land-to-Ocean Fluxes: From Natural Holocene Variations to Anthropocene. *Comptes Rendus Geoscience* 337: 107-23.

Morris, B. L, A. R. Lawrence, P. J. Chilton, B. Adams, R. Calow, and B. A. Klinck. 2003. Groundwater and Its Susceptibility to Degradation: A Global Assessment of the Problem and Options for Management. Early Warning and Assessment Reports Series, RS.03–3, United Nations Environment Programme, Nairobi, Kenya.

Mukherjee, A., M. K. Sengupta, M. A. Hossain, S. Ahamed, B. Das, B. Nayak, D. Lodh, M. M. Rahman, and D. Chakraborti. 2006. Arsenic Contamination in Groundwater: A Global Perspective with Emphasis on the Asian Scenario. *Journal of Health, Population and Nutrition* 24 (2): 142-63.

Oki, T., Y. Agata, S. Kanae, T. Saruhashi, and T. Musiake. 2003. Global Water Resources Assessment under Climatic Change in 2050 Using TRIP. *IAHS Publication* 280: 124–33.

Olden, J. D., and N. L. Poff. 2003. Redundancy and the Choice of Hydrologic Indices for Characterizing Streamflow Regimes. *River Research and Applications* 19: 101-21.

Postel, S. L, G. C. Daily, and P. R. Ehrlich. 1996. Human Appropriation of Renewable Fresh Water. *Science* 271: 785-88.

Rabalais, N. N., R. E. Turner, B. K. Sen Gupta, E. Platon, and M. L. Parsons. 2007. Sediments Tell the History of Eutrophication and Hypoxia in the Northern Gulf of Mexico. *Ecological Applications* 17: S129-S143.

Seitzinger, S., J. A. Harrison, J. K. Bohlke, A. F. Bouwman, R. Lowrance, B. J. Peterson, C. R. Tobias, and G. van Drecht. 2006. Denitrification across Landscapes and Waterscapes: A Synthesis. *Ecological Applications* 6: 1051-76.

Seitzinger, S. P., J. A. Harrison, E. Dumont, A. H. W. Beusen, and A. F. Bouwman. 2005. Sources and Delivery of Carbon, Nitrogen, and Phosphorus to the Coastal Zone: An Overview of Global Nutrient Export from Watersheds (NEWS) models and their application. *Global Biogeochemical Cycles* 19: GB4S01, doi:10.1029/2005GB002606.

Shiklomanov, I. A., and J. Rodda, eds. 2003. *World Water Resources at the Beginning of the 21st Century.* Cambridge, UK: Cambridge University Press.

Smith, S.V., D. Swaney, L. Talaue-McManus, J. Bartley, P. Sandhei, C. McLacghlin, V. Dupra, C. Crossland, R. Buddemeier, B. Maxwell, and F. Wulff. 2003. Humans, Hydrology, and the Distribution of Inorganic Nutrient Loading to the Ocean. *BioScience* 53 (3): 235-45.

Syvitski, J. P. M., C. J. Vörösmarty, A. J. Kettner, and P. Green. 2005. Impact of Humans on the Flux of Terrestrial Sediment to the Global Coastal Ocean. *Science* 308 (5720): 376-80.

Townsend, A. R., R. W. Howarth, F. A. Bazzaz, M. S. Booth, C. C. Cleveland, S. K. Collinge, A. P. Dobson, P. R. Epstein, D. R. Keeney, M. A. Mallin, C. A. Rogers, P. Wayne, and A. H. Wolfe. 2003. Human Health Effects of a Changing Global Nitrogen Cycle. *Frontiers in Ecology and the Environment* 1: 240-46.

Vörösmarty, C. J., E. M. Douglas, P. A. Green, and C. Revenga. 2005. Geospatial Indicators of Emerging Water Stress: An Application to Africa. *Ambio* 34: 230-36.

Vörösmarty, C. J., C. A. Federer, and A. Schloss. 1998. Potential Evaporation Functions Compared on U.S. Watersheds: Implications for Global-Scale Water Balance and Terrestrial Ecosystem Modeling. *Journal of Hydrology* 207 (3-4): 147-69.

Vörösmarty, C. J., C. Leveque, and C. Revenga. 2005. Fresh Water. In *Millennium Ecosystem Assessment,* Volume 1, *Conditions and Trends Working Group Report.* Washington, DC: Island Press.

Vörösmarty, C. J., M. Meybeck, B. Fekete, K. Sharma, P. Green, and J. Syvitski. 2003. Anthropogenic Sediment Retention: Major Global Impact from Registered Impoundments. *Global and Planetary Change* 39 (1-2): 169-190.

Vörösmarty, C. J., and M. Meybeck. 2004. Responses of Continental Aquatic Systems at the Global Scale: New Paradigms, New Methods. In *Vegetation, Water, Humans and the Climate*, ed., P. Kabat, M. Claussen, P. A. Dirmeyer, J. H. C. Gash, L. Bravo de Guenni, M. Meybeck, R. A. Pielke Sr., C. J. Vörösmarty, R. W. A. Hutjes, and S. Lutkemeier. Berlin: Springer.

Wang, B. D. 2006. Cultural Eutrophication in the Changjiang (Yangtze River) Plume: History and Perspective. *Estuarine Coastal and Shelf Science* 69: 471-77.

Warby, R. A. F., C. E. Johnson, and C. T. Driscoll. 2005. Chemical Recovery of Surface Waters across the Northeastern United States from Reduced Inputs of Acidic Deposition: 1984-2001.

Environmental Science & Technology 39 (17): 6,548-54.

WHO (World Health Organization)/ UNICEF (United Nations Children's Fund). 2004. *Meeting the MDG Drinking Water and Sanitation Target: A Mid-term Assessment of Progress*. Geneva: World Health Organization and New York: United Nations Children's Fund.

WHYMAP. 2008. *Groundwater Resources of the World*. Hannover: BRG and Paris: United Nations Educational, Scientific, and Cultural Organization.

Wollheim, W. M., C. J. Vörösmarty, A. F. Bouwman, P. A. Green, J. Harrison, E. Linder, B. J. Peterson, S. Seitzinger, and J. P. M. Syvitski. 2008. Global N Removal by Freshwater Aquatic Systems: A Spatially Distributed, within-Basin Approach. *Global Biogeochemical Cycles* 22, GB2026, doi:10.1029/2007GB002963.

Zektser, I., and J. Margat. 2003. *Groundwater Resources of the World and Their Use*. Paris: United Nations Educational, Scientific, and Cultural Organization–International Hydrological Programme.

The Earth's natural water cycles

Chapter 11

Changes in the global water cycle

Author: Dennis P. Lettenmaier

Contributors: Vladimir Aizen, Abou Amani, Theodore Bohn, Filippo Giorgi, Stephan Harrison, Tom Huntington, Rick Lawford, Peter Letitre, Harry Lins, Jun Magomi, Gi-Kyun Park, Igor Severskiy, W. James Shuttleworth, Pratap Singh, Soroosh Sorooshian, Willi Struckmeier, Kuni Takeuchi, Lena Tallaksen, Charles Vörösmarty, Tandorig Yan and Tinjun Zhang

Coordinator: Andras Szöllösi-Nagy (UNESCO)

Facilitator: Denis Hughes

Key messages

- Among the consequences of a changing hydrologic cycle is its interaction with the terrestrial carbon cycle. The terrestrial biosphere may have taken up roughly 25% of anthropogenic carbon emissions during the last century; it is unclear how long this can continue.
- Most climate scientists agree that global warming will result in an intensification, acceleration or enhancement of the global hydrologic cycle, and there is some observational evidence that this is already happening.
- It is increasingly clear that the assumption of statistical stationarity is no longer a defensible basis for water planning.

Water is essential to life, sustainable economic growth and the functioning of ecosystems. These are all affected by variations in water storage and fluxes at the land surface – storage in soil moisture and groundwater, snow, and surface water in lakes, wetlands and reservoirs – and precipitation, runoff and evaporative fluxes to and from the land surface. Water planners have generally seen these key elements in the land surface water cycle as at least approximately statistically stationary, their challenge being to characterize and buffer against natural variability. Over the last several decades, however, studies of changes in streamflow, snowpack and evapotranspiration have made it increasingly clear that the assumption of statistical stationarity is no longer defensible. Arguing that 'stationarity is dead', at least for water planning, some researchers are making the case for a new initiative in water management to deal with nonstationarity problems.[1] Meeting this challenge requires understanding the nature of observed changes in the land surface water cycle – the objective of this chapter.

Changes in the water cycle

This section reviews current knowledge about changes in the land surface water cycle, both in surface fluxes (precipitation and evapotranspiration) and storage (soil moisture, groundwater, and lakes and reservoirs). It also discusses interactions between the water and carbon cycles as well as some aspects of current hydrologic observations.

Observed trends in precipitation

Contributors: Filippo Giorgi with Abou Amani

Precipitation is a key component of the Earth's hydrologic cycle and one of the most difficult to observe and model accurately. It is determined by large-scale processes (such as the location of major storm tracks) as well as local processes (such as topographical uplift) and therefore exhibits pronounced variability at a wide range of spatial and temporal scales. In addition, the mode of precipitation – its frequency, intensity and occurrence in liquid or solid form – is often more important than the total amount in affecting the hydrologic cycle and the

Observations of precipitation do not show a continuous trend throughout the century but rather variable trends at multidecadal scales

human and natural systems that depend on it (see chapter 7 for the implications of climate change and changes in the hydrologic cycle on uncertainty in agriculture).

Another critical problem is that precipitation observations can be subject to considerable measurement error. Measurement station densities in many mountainous regions and remote areas are low and uneven, precluding proper estimates of regionally averaged values. While remotely sensed measurements offer a global and regularly spaced picture, they are based on uncertain conversions of radiometric data into precipitation rates. Furthermore, precipitation variability at multidecadal scales can mask long-term trends. Because this variability increases as the spatial scale decreases, identifying regional and local trends is especially difficult.[2]

The Intergovernmental Panel on Climate Change recently completed a thorough assessment of observed precipitation trends in the 20th century at global and regional scales based on six station and satellite observation datasets.[3] Over global land areas precipitation shows an increasing trend superimposed on large interdecadal oscillations from the beginning of the century to the 1950s.[4] This is followed by a decreasing trend until the end of the century, still superimposed on large interdecadal variability. In other words, observations of precipitation do not show a continuous trend throughout the century but rather variable trends at multidecadal scales.

Map 11.1 shows the geographic distribution of annual precipitation for 1901-2005 and 1979-2005. During the second period anthropogenic greenhouse gas emissions are estimated to be dominant in determining global warming.[5] The two precipitation patterns differ greatly due to the large multidecadal variability of regional precipitation.

Map 11.1 **Trend in annual precipitation rate, 1901-2005 and 1979-2005**

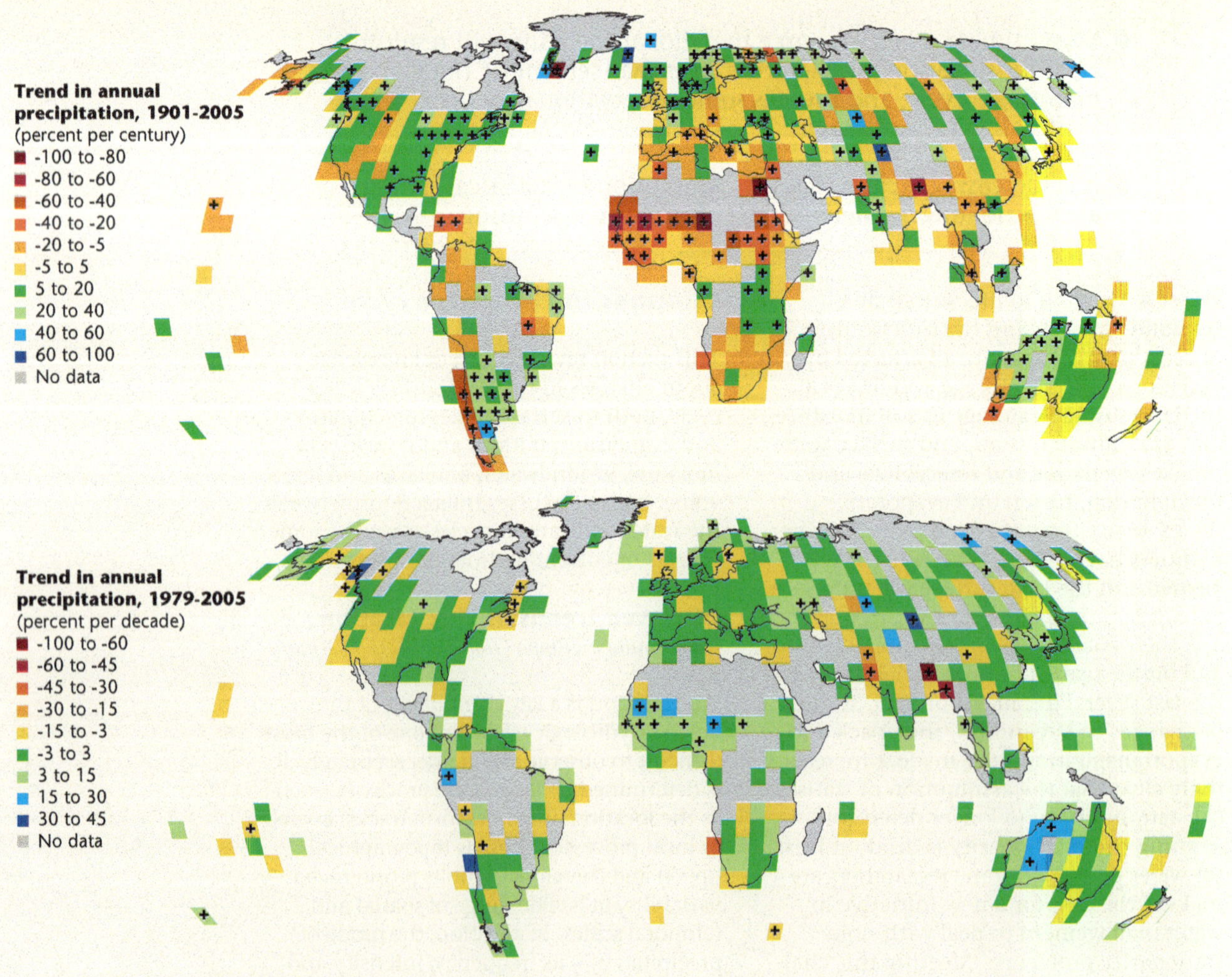

+ is a trend that is statistically significant at the 5% level.
Source: IPCC 2007, chapter 3.

The most striking example of this difference is the Sahel region in Africa, where the 1901-2005 trend is negative, but the 1979-2005 trend is positive. One study finds large multidecadal fluctuations in annual rainfall anomalies for the Sahel for 1895-2000, with largely positive anomalies for 1925-70 and pronounced drought conditions for 1970-2000.[6] In addition, while the precipitation trend between 1901 and 2000 is negative, after about 1985 it becomes positive. This example illustrates the difficulty of identifying long-term changes in precipitation trends, especially at the regional scale.

During 1901-2005 there were positive precipitation trends over eastern North America, southern South America, Northern Europe, Central Asia and Western Australia and negative trends over the Sahel, Southern Africa, the Mediterranean and Southern Asia (see map 11.1). During 1979-2005 additional negative trends emerged, such as over portions of the western United States, northwestern India and Pakistan; however, the precipitation trends are generally noisy.

Variables that integrate components of the hydrologic cycle may be more indicative for identifying precipitation trends. An example is the Palmer Drought Severity Index (PDSI), an index of drought that measures the cumulative deficit in soil moisture derived from precipitation and temperature budgets. Positive values imply more severe drought conditions. Map 11.2 shows that the PDSI has increased globally during the 20th century but that geographic variability in the PDSI trend became more pronounced. Areas increasingly subject to more severe drought conditions include the Sahel and West Africa, Southern Africa, the Mediterranean, Central America, South and South-East Asia, Eastern Australia, North and Northeast China, Southeastern Brazil and the Northern

Variables that integrate components of the hydrologic cycle may be more indicative for identifying precipitation trends

Map 11.2 **Geographic distribution of the trend in the Palmer Drought Severity Index (PDSI) and annual variations in the globally averaged PDSI, 1900-2000**

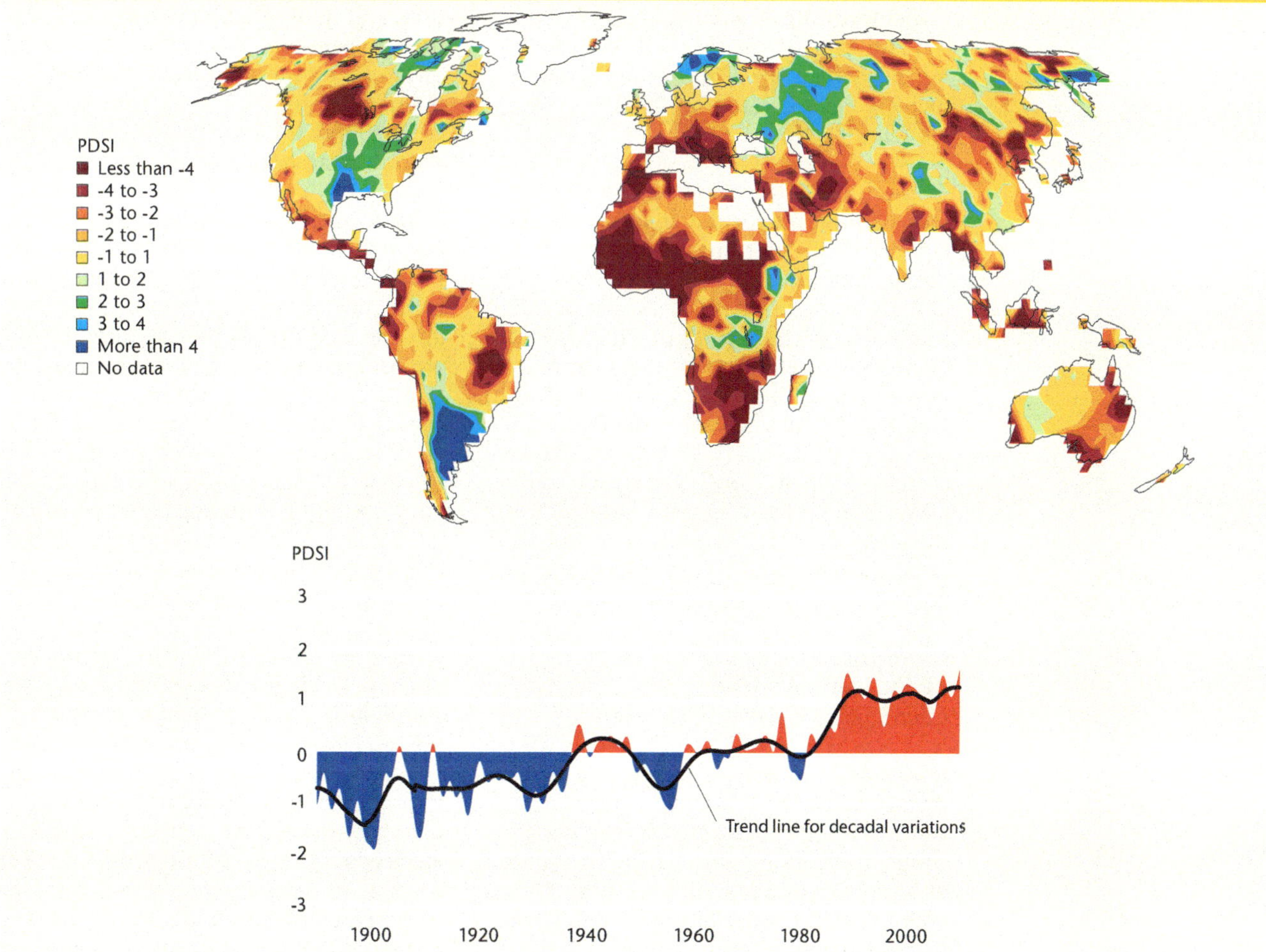

Note: The PDSI is a measure of drought severity based on precipitation and temperature information. Positive values imply more severe drought conditions.
Source: Based on IPCC 2007, chapter 3.

> **Current methods of measuring evaporation are either too recent or too inaccurate to be used in analyses of long-term gradual change in evaporation**

Territories of North America. Areas with reduced drought trends include southern South America, the eastern United States, Northeastern Europe and the Ethiopian highlands.

Several similarities are evident when these 20th century trends in the PDSI are compared with projections of annual precipitation change from 22 coupled atmosphere-ocean general circulation models at the end of the 21st century prepared for IPCC scenario A1B (in the middle of the IPCC scenario range).[7] The Mediterranean Basin, Central America, Southern Africa, Southeastern Brazil and Eastern Australia, where the PDSI trend shows increased drought conditions, are projected to receive substantially less precipitation. These areas thus appear most at risk of water stress, presumably due to anthropogenic climate change. West Africa and the Sahel, where the positive PSDI trend is at a maximum, are not projected to be affected by large precipitation changes. This might be an indication that the Sahel drought of the last part of the 20th century was a natural climate fluctuation not directly tied to global warming. Similarly, areas of Central and East Asia where the PDSI trend was positive in the 20th century show increased precipitation in the 21st century model projections.

As mentioned, the mode of precipitation may be more important than average precipitation in determining hydrologic impacts. Widespread increases in heavy precipitation events have been observed in some places where total precipitation has decreased, and more precipitation now falls as rain rather than snow in northern regions.[8] At the same time, the length, frequency and intensity of heat waves have increased widely. Consistent with a warmer atmosphere with a greater water-holding capacity, these changes are also found in IPCC climate projections under different greenhouse gas emission scenarios.[9]

Changes in evaporation and evapotranspiration

Contributor: W. James Shuttleworth

Actual evaporation can be measured either by integrating the water transferred into the atmosphere using micrometeorological techniques or by measuring the liquid water loss from representative sample volumes of the soil-atmosphere interface.[10] All available methods, however, are either too recent or too inaccurate to be used in analyses of long-term gradual change in evaporation.[11] Consequently, there are little or no direct plot scale data on actual evaporation that are long-term or accurate enough.

In the absence of sufficient data, alternative methods have been used. Some studies have diagnosed changes as the residual in the area-average water balance,[12] but most studies have used the measured rate of evaporation from evaporation pans, rates calculated using estimation equations, or both. Because many of these studies used the language of early evaporation theory instead of that of the physics and physiology of evaporation, confusion has resulted, generating much controversy.[13]

Two main explanations have been advanced for observed reductions in pan evaporation. The first is that open water evaporation has decreased because the net radiation available to support evaporation at the ground has declined, whether because of higher concentrations of atmospheric aerosols, increasing cloud cover, climate change or a combination of these. If a decline in net radiation were the only reason for the reduction in pan evaporation, and if all other meteorological, soil moisture and plant physiology influences on actual evaporation had remained unchanged, that would indicate an associated reduction in area-average actual evapotranspiration.[14]

Enhancement of the hydrologic cycle is an important feature of climate change projections, not only because water vapour is the most important greenhouse gas but also because the additional water in the atmosphere may alter cloud cover and thus affect surface solar radiation. The extent to which cloud cover may increase is not yet clear. Some observational studies have calculated reductions in solar radiation of a few percent per decade based on calculated changes in open water and reference crop evaporation using historical data on hours of sunshine, implying cloud cover may be increasing in some regions.[15] Meanwhile, studies reporting significant changes in observed solar radiation have confidently ascribed them to local or regional changes in atmospheric aerosol concentration.[16] Regionally varying, but nonetheless widespread, impacts of increasing atmospheric aerosols and associated reductions in surface solar radiation are now well documented[17] and modelled in general circulation models.[18] One study estimates reductions in solar radiation of about 2.75% a decade,[19] while the IPCC estimates the change in global radiative forcing due to sulphate aerosols since 1750 at between -0.2 and -0.8 watts per square metre (m^2).[20]

Thus, the evidence is now strong of a real but regionally dependent reduction in surface solar radiation over the last few decades, a result of either increasing cloud cover or increased aerosol concentration. This is likely to be at least one contributory factor in the observed reduction in pan evaporation. Recently, however, a study has used a physically based model of pan evaporation in the semi-arid climates of Australia to demonstrate that decreased wind speed is the dominant cause of reduced pan evaporation there.[21]

A second widely reported explanation for reduced pan evaporation is that the average near-surface vapour pressure deficit has shrunk or that the average near-surface wind speed has slowed, or a combination of the two. Most discussion of this potential mechanism has assumed the validity of the hypothetical complementary relationship between actual and potential evaporation proposed by Bouchet.[22] Modelling studies, however, do not support the numerical accuracy of this complementary relationship.[23] Nonetheless, the basic mechanism for a shrinking of the vapour pressure deficit through atmospheric conditioning by enhanced regional actual evaporation is plausible, and it has been shown that, when surface-atmosphere coupling processes are considered, Bouchet's complementary relationship is approximately correct.[24]

In fact, there is substantial evidence that open water evaporation and reference crop evapotranspiration rates are increasing and that there are regional changes in the vapour pressure deficit and wind speed.[25] In addition, some studies have used historical climate data to investigate the relative importance of the meteorological controls on open water and reference crop evapotranspiration.[26] These studies offer no clear evidence that change in surface radiation or regional wind speed is consistently the dominant – let alone the sole – cause of observed changes in pan evaporation and some evidence of regional difference in the relative importance of various meteorological variables. In addition, several studies provide evidence that on average for large areas of Asia and North America actual evapotranspiration is increasing, even though pan evaporation is decreasing.[27]

Interpreting observed pan evaporation data in terms of actual regional evaporation is therefore complex. One study argues that two distinct influences are involved: type A processes related to large-scale changes in atmospheric concentrations and circulation, which modify surface evaporation rates in the same direction, and type B processes related to coupling between the surface and the atmospheric boundary layer at landscape scale, which usually modify area-average evaporation and pan evaporation in different directions.[28] Evidence suggests that both influences are present, with their relative strengths depending on the regional climate aridity. In semi-arid and arid regions the effect of decreased wind speed, higher aerosol concentrations and increasing cloud cover may be the dominant cause of the observed reduction in pan evaporation. However, because actual evaporation is so strongly determined by available precipitation in these regions, it does not necessarily follow that time-average actual evapotranspiration is decreasing. In humid regions evidence suggests that actual evaporation has increased despite an aerosol- or cloud-cover-induced decrease in solar radiation.

Soil moisture

Contributors: Soroosh Sorooshian with Jialun Li and Gi-Kyun Park

Understanding and predicting variations in the water cycle require knowledge of soil moisture variations. These can be investigated using direct measurement, remote-sensing, meteorological-based methods and model simulations. Each method has shortcomings. In-situ soil moisture measurements are too sparse to draw conclusions about multidecadal soil moisture trends at a global scale. Satellite sensors are suboptimal in microwave frequency, and record lengths are too short to provide meaningful trend information. Physically based model reconstructions using precipitation and temperature to estimate soil wetness provide the best current insights into long-term soil moisture trends, but studies using these methods are not comprehensive enough to fully explain the uncertainties and to generate unambiguous results.

In-situ measurements are crucial for climatological analysis, model evaluation and ground truthing remote sensing data. Ground observations are limited at both spatial and temporal scales and can represent only a small portion of the Earth's land surface. Gravimetric soil moisture content has been measured in the Russian Federation and Ukraine for about 40 years and in China for about 20 years.

There is substantial evidence that open water evaporation and reference crop evapotranspiration rates are increasing and that there are regional changes in the vapour pressure deficit and wind speed

Inconsistencies in national hydrologic monitoring networks, data access and data quality prevent systematic and comprehensive global assessments of trends in runoff and streamflow

Advanced soil moisture sensors such as neutron probes, time domain reflectrometry, frequency domain reflectometry and tensiometers provide more continuous measurements. The Soil Climate Analysis Network of the US Department of Agriculture National Resources Conservation Service reports real-time soil moisture observations using meteor burst communication, with some records going back to the early 1990s.[29]

Gravimetrically observed soil moisture data for the Russian Federation and Ukraine for about 40 years show that summer soil moisture increased significantly from 1958 to the mid-1990s. Researchers suggest that solar dimming, a decrease in evaporation and an increase in carbon dioxide could be responsible, although contrary to model predictions based on global warming related to greenhouse gases.[30]

A study found strong correlations between the PDSI for 1870-2002 based on global gridded precipitation and surface temperature records and observed soil moisture content to 1 metre in depth during warm season months in China, Mongolia, the former Soviet Union and the United States (Illinois).[31] The study found that global land areas in both very dry (PDSI less than -3.0) and very wet (PDSI more than +3.0) conditions have increased from approximately 20% to 38% since 1972, suggesting surface warming as the primary cause after the mid-1980s.

Land surface models are a more sophisticated method than the PDSI for reconstructing long-term soil moisture variations. While the ability of models to faithfully reproduce observations is always an issue, studies using observations over Illinois[32] and a transect over Eurasia[33] found that the variable infiltration capacity model accurately reproduced observed interseasonal and interannual variability. Another study used the variable infiltration capacity model to reconstruct soil moisture globally for 1950-2000 and found a small wetting trend at a global scale in the reconstructed soil moisture.[34] The study attributes the wetting trend primarily to increasing precipitation. The reconstructed trend in soil wetness in the United States is consistent with other studies.[35]

An alternative to land service models (which use observed records of surface climate variables) is coupled land-atmosphere model reanalysis output and coupled model climate reconstructions. These are reconstructions of a past period using a fixed version of a numerical weather prediction model, with assimilation of atmospheric and other data as though the model were being run in real time. Analyses of the variability in soil moisture in output from several recent global reanalyses found no long-term linear trends at the global scale.[36] An analysis of output from climate simulations of 25 general circulation models found that the models were unable to reproduce observed trends in soil moisture over the Russian Federation and Ukraine in the second half of the 20th century.[37]

Remote sensing estimates of soil moisture offer another alternative to in-situ observations. Surface soil moisture can be retrieved from low-frequency microwave satellite data. Current satellite-based estimates of soil moisture (using mostly shorter wavelength sensors) reflect the water content of only the upper 1-2 centimetres of the soil column, because vegetation can mask the signature of the soil below. Data from the Gravity Recovery and Climate Experiment (GRACE) satellite have been used to represent terrestrial water storage variability in several studies since its launch in 2002.[38] Major drawbacks to GRACE data are the low spatial resolution (hundreds of kilometres) and a wide 'visual' field that encompasses the total signature of variations in moisture storage from the centre of the Earth to the top of the atmosphere, so that sources of variations other than soil moisture (such as atmospheric moisture, snow water storage, groundwater, lakes and reservoirs) must be removed to analyse soil moisture trends and variability.

Runoff and streamflow trends

Contributors: Harry Lins with Lena Tallaksen

Inconsistencies in national hydrologic monitoring networks, data access and data quality prevent systematic and comprehensive global assessments of trends in runoff and streamflow (see chapter 13). Sparse stream gauge networks in developing countries, often coupled with national data policies that restrict dissemination to regional or international data centres, limit the comprehensiveness of worldwide trend assessments.[39] Several studies of trends in streamflow have nevertheless been published. Most used stream gauge records that are 'climate sensitive', meaning that the data are from stations thought to be minimally affected by such confounding anthropogenic influences as upstream regulation, diversions, groundwater extractions and land use change. As a result, the trends reported in these studies are most directly attributable to variability or change in climate.

This synthesis of streamflow trends draws largely on national and continental analyses, especially for Europe and North America, where the most comprehensive work has been done. Few studies have been published for Australia or for countries in Asia (outside of Siberia), Africa or South America. However, results of a global synthesis that includes some trend information for these continents are included at the end of this section.[40]

A study for Canada found that annual mean streamflow generally decreased between 1947 and 1996, with significant decreases in the southern part of the country, particularly southern Alberta and British Columbia.[41] The study also found reduced monthly mean streamflow except in March and April, when significant increases were observed. In southern Canada significant decreases were also observed in all percentiles of daily flow, along with the annual mean. Over northern British Columbia and the Yukon Territory, however, significant increases were identified in the lower flow percentiles.

In the United States the most extensive studies of trends analysed long-term streamflow records for 395 stations across the continental United States with continuous daily records for 1944-93 (updated to 435 stations for 1940-99).[42] The study found a preponderance of upward trends in all but the highest flows, which were less numerous and for which upward and downward trends were about equal. Another study, working with a similar dataset, found that the streamflow increases appeared as a step change around 1970 rather than as a gradual trend.[43] A step change implies that the climate system has shifted to a new regime that may remain stable until a new shift occurs, whereas a gradual trend is likely to continue into the future.

In Europe much of the published work on streamflow trends has focused on flooding. A study evaluating high-flow and flood records in a network of climate-sensitive catchments in the United Kingdom applied trend tests to indicators of flood magnitude and frequency and of high-flow magnitude (10- and 30-day maximums) and duration (prevalence of high flows).[44] Over the 30- to 40-year period ending in 2003 positive trends were identified in all indicators. The trends were found primarily in upland maritime-influenced catchments in northern and western areas of the United Kingdom. The results suggest a trend towards more protracted high flows in northern and western areas, but trends in flood magnitude were less prevalent. Only a few trends were found in the English lowlands, and these were influenced by a sequence of notable flood events through the exceptionally wet winter of 2000/01. The study shows that the recent increases in northern and western catchments were probably caused by a shift towards a more prevalent positive phase in the North Atlantic Oscillation since the 1960s. Moreover, fluctuations in the records suggest that recent trends may be influenced by multidecadal variability.

An analysis of systematic flood records for winter and summer in Central Europe since the middle of the 19th century and longer-term historical records of major floods since the 16th century found a decrease in winter flooding on the Elbe and Oder Rivers in Germany over an 80- to 150-year period and no trend in summer flooding.[45] The study attributed the winter decrease in part to a decline in strong freezing events that reduce late winter ice jamming and consequent higher flood peaks. The study also detected long-term changes in flood frequency between the 16th and 19th centuries but concluded that reductions in river length, construction of reservoirs and deforestation had minor influences. These results are consistent with findings for the Elbe and Vltava Rivers in the Czech Republic of reduced frequency and intensity of flooding during the 20th century.[46]

Analysis of annual runoff volumes and annual and seasonal flood peaks in Sweden found that average runoff increased 5% over the 20th century and even more during the 19th century, when temperatures were lower.[47] In Switzerland a study of annual streamflow found a general increase over the same time period, due mostly to increases in winter, spring and autumn runoff.[48] Winter increases were observed across the entire flow distribution but were especially marked for maximum flows. Increases in spring and autumn runoff were concentrated primarily in moderate and low-flow percentiles. Farther southeast, in Turkey, analysis of monthly mean streamflow records for 26 basins for 1964-94 found that streamflow had generally decreased in western and southern Turkey, while not changing significantly in eastern Turkey.[49] A study of a pan-European dataset of more than 600 daily streamflow records covering 1962-90,

Few studies of streamflow trends have been published for Australia or for countries in Asia (outside of Siberia), Africa or South America

Groundwater is an often neglected part of the global water cycle, partly because it is invisible and difficult to monitor and partly because the data for understanding groundwater flows are limited

1962-95, 1930-95 and 1911-95 tested for trends in hydrologic drought and detected no significant changes for most stations.[50] The study concluded that there was no evidence that drought conditions had become more severe or frequent.

Several recent studies have investigated trends in Siberian river discharge. Annual discharge from the six largest Russian Arctic rivers (Kolyma, Lena, Ob, Pechora, Severnaya and Yenisei) increased 7% over 1936-99, mostly in winter.[51] These observations indicate a general increase in freshwater flux to the Arctic Ocean. There is concern that this increase may influence the global climate system, through both ocean circulation and the carbon cycle (see next section). Another study found significant positive trends in minimum flows in many smaller rivers across European and Siberian Russian Federation, suggesting an increasing contribution of groundwater to the hydrologic cycle of northern Eurasia.[52]

Two recent studies assessed global high and low flow trends by analysing trends in annual maximum flow and in peaks over threshold and annual low flows. One study analysed records from 195 stations on six continents, most of them (92%) in North America, Europe and Australia (table 11.1). The records for 70% of the stations showed no trend in the annual maximum flow, 14% an increasing trend and 16% a decreasing trend (box 11.1). The trends for each continent were generally consistent with the aggregate totals. Using a much smaller sample of stations (21), another study found a mixed pattern of trends in peaks over threshold, with approximately 30% of stations exhibiting a trend and with more downward than upward trends.[53] A very different pattern was apparent in low flows, with about 52% of stations exhibiting statistically significant trends in seven-day low flows, all of them increasing. This implies a reduction in the incidence of hydrologic drought, consistent with the results of a study of global trends in soil moisture and drought.[54] These two broad-scale results are generally consistent with those from regionally specific investigations, and both studies conclude that the results do not support the hypothesis that global warming has as yet caused an increase in hydrologic extremes – more floods or hydrologic droughts.

Groundwater trends

Contributors: Willi Struckmeier with Peter Letitre

Groundwater is an often neglected part of the global water cycle, partly because it is invisible and difficult to monitor and partly because the data for understanding groundwater flows are limited. Thus, despite its importance for river baseflow and wetlands, groundwater is frequently ignored in water balance calculations and in water resources planning and management. As long as groundwater levels are roughly stable – or the annual variations lie within certain intervals – the groundwater resource is regarded as constant. However, for longer-term evaluations of global change, trends in groundwater resources are crucial, since groundwater has a buffer function in short-term climate variations and is a key element of adaptation strategies.

Groundwater flow processes are usually much slower than atmospheric or surface water processes, often by two or three orders of magnitude. In large aquifer systems containing most of their groundwater in stock, sometimes more than 90%

Table 11.1 Trends in annual maximum streamflow, by continent, for 195 stream gauging stations worldwide, various years

Region	Number of stations	Stations with increasing trend		Stations with no trend		Stations with decreasing trend	
		Number	Percent	Number	Percent	Number	Percent
Africa	4	1	25	1	25	2	50
Asia	8	0	0	5	63	3	38
South America	3	0	0	3	100	0	0
North America	70	14	20	44	63	12	17
Australia-Pacific	40	1	3	34	85	5	13
Europe	70	11	16	50	71	9	13
Total	195	27	14	137	70	31	16

Note: The analysis was based on the Mann-Kendall test (two-sided, significance level 10%). Values may not sum to 100% because of rounding.
Source: Kundzewicz et al. 2004.

of groundwater resources are made up of 'dead storage' formed in times when rainfall and recharge conditions were more favourable. While not part of the current water cycle, it is an extremely valuable asset, chiefly in arid and semi-arid regions such as the deserts of North Africa, the Arabian Peninsula, Central Asia and Australia. These fossil groundwater resources are increasingly used for agricultural, industrial and domestic water supplies, although they are almost never recharged and will be depleted one day. In many areas fossil groundwater is the only reliable water resource.[55]

Observed groundwater levels indicate that recharge volumes may vary considerably from year to year in response to climate variations. It is therefore realistic to expect changes in groundwater recharge as a consequence of climate change. However, a recent US Climate Change Science Program report notes that 'the ability to predict the effects of climate and climate change on groundwater systems is nowhere near as advanced as for surface water systems.'[56] An examination of the sensitivity of the Ellensburg, Washington, groundwater basin to climate and land cover change found that the sensitivities of groundwater recharge and crop water use to climate change were closely intertwined.[57] In particular, for the native grassland vegetation groundwater recharge was projected to increase, whereas for the mostly irrigated pasture vegetation, recharge was projected to decrease. Seasonal patterns of evapotranspiration are substantially different for natural vegetation and for irrigated crops.

Another study evaluated the sensitivity of two unconfined aquifers in northwestern North America to climate change, one in a humid area west of the Cascade Mountains and the other in a much drier climate east of the Cascades.[58] For the

Observed groundwater levels indicate that recharge volumes may vary considerably from year to year in response to climate variations

Box 11.1 Changes in discharge of major global rivers

Although considerable attention has been given to trends in discharge of unregulated rivers, most studies find that there are considerably more rivers with no statistically significant trends that could be attributed to long-term climate variations and change than rivers with significant trends.

For instance, at the 10% significance level, 70% of the records analysed showed no statistically significant change. However, the discharge of many large rivers has indisputably been affected by water management, especially dam construction but also within-basin diversions for other beneficial uses such as irrigation and municipal and industrial water supply, as well as transbasin diversions.

The figure shows the observed flow of the Columbia River after construction of large reservoirs in Canada and the United States (totalling about 30% of the mean annual discharge), as well as 'naturalized' discharge (the discharge that would have occurred in the absence of the dams) and projected effects of climate change on the naturalized flows by 2050 (the orange band reflects the range of climate model projections). Although the projected effects of climate change are substantial, they are much less than the observed effects of water management (see chapter 9).

The figure shows other observed effects of water management globally. Since construction of the High Aswan Dam almost all the Nile River's discharge is now either diverted for irrigation or lost to reservoir evaporation. Flows of the Syr Darya River have declined greatly because of irrigation diversions upstream (the Syr Darya is one of the major tributaries of the Aral Sea, whose declining levels are closely related to reductions in the river's flows). In contrast, the flow of the Burntwood River increased by a factor of almost four following an upstream diversion into the river basin in the 1970s for hydropower production.

Source: Lins and Slack 2005; Kundzewicz et al. 2005; GWSP 2005.

Changes in the flow of major global rivers associated with water management

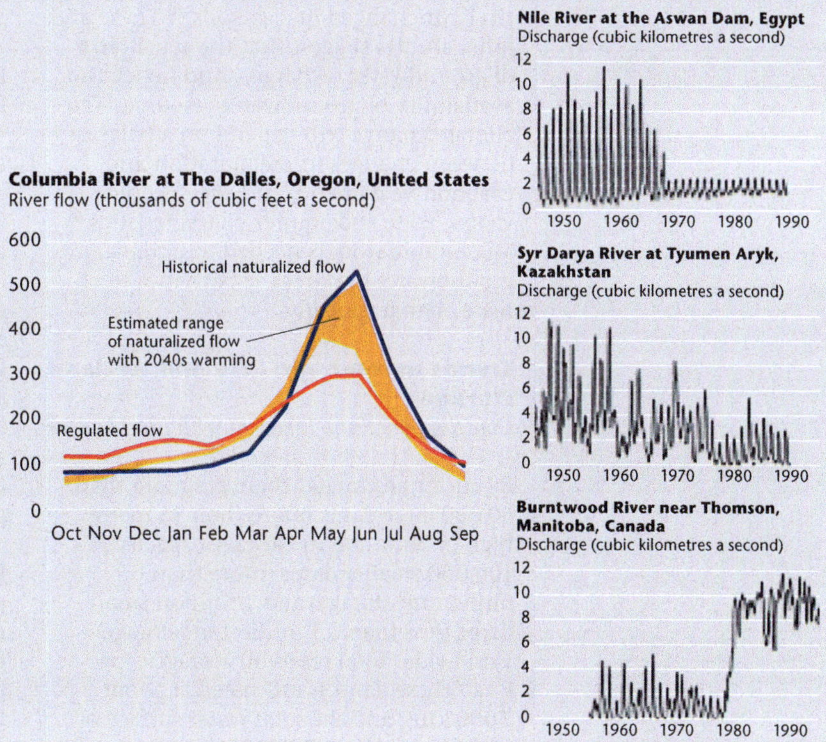

Source: Based on GWSP 2005.

> **By current estimates there are more than 50,000 large dams, 100,000 smaller dams and 1 million small dams worldwide**

aquifer in the humid region recharge was closely related to projected changes in precipitation and its seasonal pattern, and for the particular climate scenario examined groundwater levels were predicted to decline slightly. For the aquifer in the drier climate river discharge dominated aquifer variations. Projected changes in groundwater level therefore closely followed projected changes in river discharge, which were higher in winter and early spring and lower in summer and fall.

A modelling study of two sites in Australia, one with a Mediterranean climate and one with a subtropical climate, found that in the Mediterranean climate changes in evaporative demand related to rising temperatures dominated the hydrologic response, whereas in the subtropical climate changes in rainfall characteristics dominated.[59] A spatially distributed hydrologic simulation model – used to evaluate the combined effects of projected increases in precipitation, temperature and potential evapotranspiration in a future climate – predicted that groundwater recharge and subsurface storage and discharge would increase in sandy soils but remain relatively unchanged in clay soils.[60] These and other studies suggest that the sensitivity of groundwater recharge – and hence the availability of groundwater resources – to climate change will depend on a balance between changes in precipitation and evaporative demand, and site-specific vadose zone and aquifer characteristics.[61] Much more work is needed to understand groundwater resources sensitivities to climate change globally.

Trends in reservoir, lake and wetland storage

Contributors: Kuni Takeuchi with Jun Magomi

By current estimates there are more than 50,000 large dams (more than 15 metres high or 3 million m^3 storage capacity), 100,000 smaller dams (more than 0.1 million m^3 storage) and 1 million small dams (less than 0.1 million m^3 storage) worldwide. Total reservoir storage capacity of these dams is estimated at about 7,000 km^3, and the total water surface of reservoirs is about 500,000 square kilometres (km^2). Although there are a huge number of reservoirs, 95% of total reservoir capacity is accounted for by about 5,000 large reservoirs (dams more than 60 metres high), and more than 80% are used for hydropower generation (see chapter 7).[62]

Reservoirs. In the past 100 years, but mainly during 1950-90, many reservoirs were constructed in North America, the southeast coast of South America, Australia, China and the Russian Federation. About 350 large reservoirs are currently under construction in China, India, Iran and Turkey and countries in the Middle East and South-East Asia.[63] Since the late 20th century reservoirs have experienced various changes, such as dam removals in the United States; conflicts over reservoir water use between upstream and downstream countries along the Euphrates, Mekong and Syr Darya Rivers; and reservoir sedimentation. More than 25% of global suspended sediment discharge is thought to be trapped by reservoirs.[64] Although storage of impounded water is known to have increased greatly over the main period of dam construction during the 20th century, trends in global reservoir storage over the last 20 years of reduced dam construction are less clear. One study suggests only a modest change in reservoir storage over the last decade,[65] and there have been suggestions that a prevalence of drought in key areas of the world may have reduced global reservoir storage over the last decade.

Lakes. Several studies have provided extensive global data on natural lakes.[66] Lakes store the largest volume of fresh surface water (about 90,000 km^3), more than 40 times that in rivers and streams and about 7 times that in wetlands. Together with reservoirs, they cover an estimated 2.7 million km^2, or about 2% of surface area outside the polar regions.[67] Most lakes are small, but the 145 largest lakes are estimated to contain more than 95% of lake freshwater. Lake Baikal (Russian Federation), the world's largest, deepest and oldest lake, alone contains 27% of all lake freshwater. Lake water serves commerce, fishing, recreation and transport, and supplies water for much of the world's population.

While changes in lake coverage over the past few decades have been observed in many parts of the world, the primary factors driving these changes are specific to each region. The surface area of Lake Chad shrank from 23,000 km^2 in 1963 to less than 2,000 km^2 by the mid-1980s, due largely to drought. The Aral Sea has also shrunk dramatically, and the volume of water in the basin has plummeted 75% since 1960 – changes attributable primarily to diversions of inflows for irrigation. The level of the Caspian Sea fell 3 metres between 1929 and 1977 but then rose 3 metres by 1995.[68] In Siberia changes in total lake area over the last three decades of the 20th century have been correlated with the state of the underlying permafrost. Total lake area has increased in the continuous permafrost zone (12% in western Siberia[69] and 14.7% in eastern Siberia),[70] while decreasing in the discontinuous (-13%), sporadic (-12%) and isolated (-11%) permafrost areas of western Siberia (map 11.3).[71] The changes in Siberian lake extent are seen as symptomatic of permafrost degradation and havc important consequences for global climate through their influence on the carbon cycle (see section titled 'Links between the terrestrial carbon and water cycles').

Wetlands. Water-saturated environments, wetlands are commonly characterized as swamps, bogs, marshes, mires and lagoons. Although they contain only 10% of the water in lakes and other surface waters, wetlands cover an area about 3-4 times greater than do the world's lakes[72] and play important roles in flood protection, groundwater recharge, food production, water quality, wildlife habitat and

Map 11.3 **Example of decline in lake abundance and total lake area in the discontinuous permafrost zone of western Siberia, 1973-97**

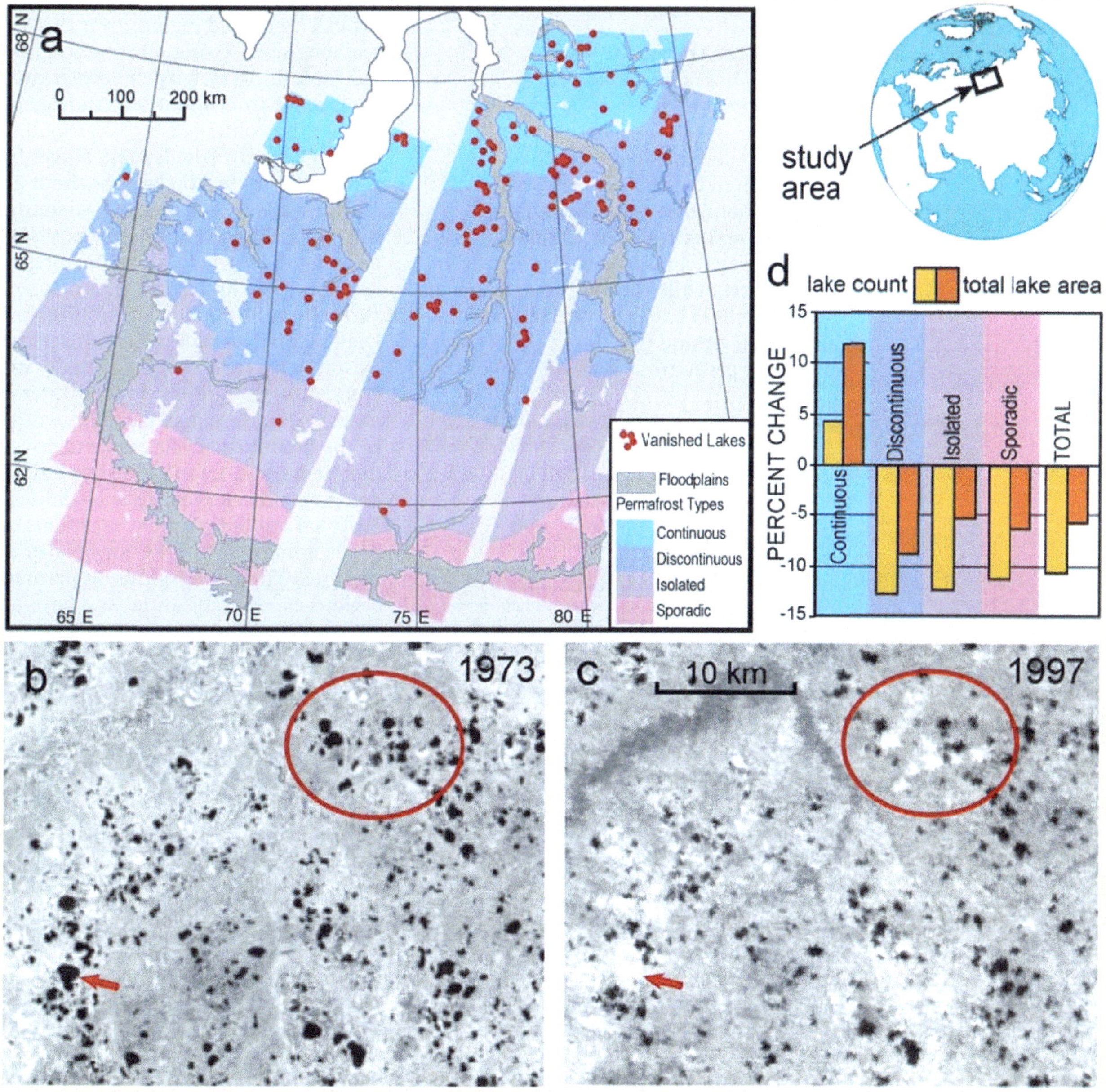

Note: Changes such as those in this map are thought to be symptomatic of permafrost degradation. Net increases in lake abundance and area have occurred in continuous permafrost, suggesting an initial but ephemeral increase in surface ponding. Decadal scale variations in lake, wetland and reservoir storage are natural characteristics of the dynamics of these water bodies and cannot necessarily be ascribed to climate, land cover or other anthropogenic causes.
Source: Based on Smith, Sheng, et al. 2005.

Both current measurements and climate model projections show that climate changes in cold-land regions are among the largest changes over the entire globe and will continue to be so

biogeochemical cycling.[73] During the last century numerous wetlands were destroyed. Currently, extensive work is being done through the 'Wise Use' campaigns sponsored by Ramsar, WWF and the United Nations Environment Programme to maintain critical services in water and related livelihood and food production areas of wetlands. Roughly half of the world's wetlands occur in high latitudes, and many of these owe their existence at least in part to the drainage impedance of permafrost. There is concern that permafrost degradation may cause some of these wetlands to drain and be replaced by grasslands, with serious implications for the global carbon cycle and possible feedbacks to global climate change.

Permafrost trends

Contributors: Tinjun Zhang with Vladimir Aizen

Frozen ground includes soils affected by short-term freeze-thaw cycles, seasonally frozen ground and permafrost. Permafrost regions occupy approximately 24% of the exposed land area of the Northern Hemisphere, while the long-term average maximum area extent covered by seasonally frozen ground (including the active layer over permafrost) is about 51% of the Northern Hemisphere land area.[74] Permafrost exists mainly in high latitudes and high elevation regions (map 11.4). Permafrost in Eurasia occurs over the entire Arctic and boreal forest areas and includes the mountainous regions of Central Asia (Tien Shan and Pamir), the Tibetan Plateau and high elevated areas of the Himalayas. Over North America permafrost is distributed mainly over Alaska and the Canadian Arctic, with the southern boundary of the latitudinal permafrost varying from 50°N to 57°N.[75] Due to the effect of the Rocky Mountains, mountain permafrost can extend as far south as 37°N.

Map 11.4 Circumpolar permafrost extent, 2000

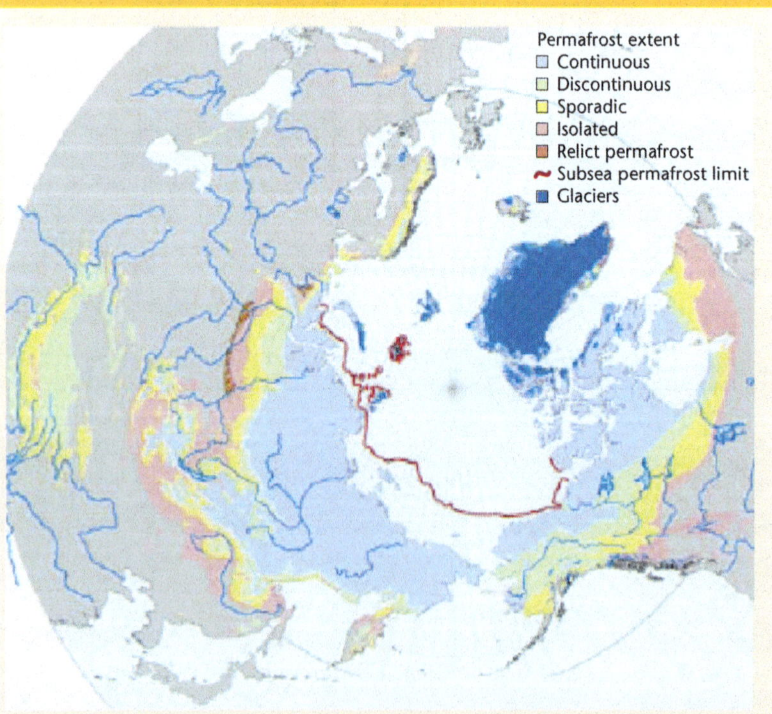

Note: Permafrost occupies the entire area of the continuous permafrost zone (except beneath large rivers and deep lakes) and underlies 10%-90% of the surface in the discontinuous permafrost zone. In the isolated permafrost zone permafrost occupies less than 10% of the area.
Source: Based on Brown et al. 1997.

Changes in the regime of ground ice in permafrost directly regulate the hydrologic cycle of cold regions over the short and long term. Using information from the Circum-Arctic Map of Permafrost and Ground Ice Conditions,[76] a study estimated that the volume of excess ground ice in the Northern Hemisphere ranges from 10,800 km^3 to 35,460 km^3, or about 2.7-8.8 metres sea-level equivalent.[77] Assuming the average porosity of permafrost soil to be about 40%, the total volume of ground ice (both pore and excess ground ice) varies from 54,000 km^3 to 177,000 km^3. Under global warming scenarios permafrost is expected to degrade rapidly in the 21st century.[78] As a result melt-water of excess ground ice may participate directly in the hydrologic cycle, while melt-water of pore ground ice may become a significant groundwater resource in cold regions. Seasonal and interannual variations of soil water storage within the active layer and seasonally frozen layer in non-permafrost regions can be substantial and have a significant impact on the hydrologic cycle in cold seasons and cold regions.

Both current measurements and climate model projections show that climate changes in cold-land regions are among the largest changes over the entire globe and will continue to be so. The most important changes that affect permafrost are increases in air temperatures and intensification of the hydrologic cycle. These climate changes inevitably change the energy and mass fluxes at the land surface and the near-surface and subsurface physical conditions in cold-land regions. These changes in the physical environment are forcing changes in permafrost conditions and permafrost degradation: increase in permafrost temperature, thickening of the active layer, thermokarst and talik development, decrease in permafrost area and eventually complete disappearance of permafrost at local, regional and global scales.

Permafrost degradation is initiating many natural processes that will intensify changes in the cold-land regions and through these changes will affect the entire Earth system. Some of these processes could be fast developing and very destructive for the northern and high-altitude ecosystems and infrastructure. Examples include surface settlement, swamping, changes in the extent of thaw lakes, landslides and slope failures, thermal erosion of river banks and deep gully formation, dramatic increase in river sediment loads and desertification. The release of carbon stored in thawing permafrost soils may have significant consequences for global climate (see section titled 'Links between the terrestrial carbon and water cycles').

Observations of changes in permafrost conditions include mainly increased permafrost temperatures, thickening of the active layer and thermokarst and talik development (including changes in the extent of thaw lakes) over the Northern Hemisphere permafrost regions. Ground-based measurements indicate that permafrost temperature increases are greatest in continuous permafrost regions and are lower (or there is no change) in discontinuous and sporadic permafrost regions. For example, permafrost temperature in northern Alaska increased about 4°-7°C during the 20th century, almost half of it during the last 20 years.[79] Increases in permafrost temperatures in the Alaskan interior have ranged from about 0.5°C to 1.5°C since the 1980s.[80] Data from the northern Mackenzie Valley in the continuous permafrost zone of Canada show a 1°C increase in permafrost temperatures at depths of 20-30 metres since the 1990s, with smaller changes in the central Mackenzie Valley and no trend in the southern Mackenzie Valley where permafrost is very thin.[81]

Field measurements in the northern European part of the Russian Federation show that temperature increase was greatest within the continuous permafrost zone for the last 20 years at 1.6°-2.8°C, at a depth of 6 metres,[82] while within the discontinuous permafrost zone the increase was 1.2°C during 1970-95.[83] In the continuous permafrost zone over the Lena River basin of Siberia permafrost temperature has increased more than 3°C since the early 1960s, while in the discontinuous permafrost zone over the Yenesei River basin the rate of temperature increase has fallen substantially. In most other discontinuous permafrost zones over the past 30 years, such as in Central Mongolia,[84] the Qinghai-Tibetan Plateau[85] and the Swiss Alps,[86] the increase in permafrost temperature was less than 1°C. Changes in air temperature alone cannot account for the increase in permafrost temperature in continuous permafrost zones, but changes in the insulation provided by snow may be partly responsible.[87] The smaller change and the absence of a trend in discontinuous and sporadic permafrost zones are likely due to the absorption of latent heat required to melt ice.

Changes in seasonally frozen ground have occurred mainly in the timing, duration, thickness and area of seasonal soil freeze and thaw. Based on soil temperature measurements in the active layer and upper permafrost up to 3.2 metres from 31 hydrometeorological stations in the Russian Federation, the active layer exhibited a statistically significant deepening of about 0.25 metre from the early 1960s to 1998.[88] The International Permafrost Association started a network of Circumpolar Active Layer Monitoring (CALM) stations in the 1990s to monitor the response of the active layer and upper permafrost to climate change that currently incorporates more than 125 sites in the Arctic, the Antarctic and several mid-latitude mountain ranges.[89] The results from northern high-latitude sites in North America demonstrate substantial interannual and interdecadal fluctuations, but with no significant trend in active layer thickness in response to air temperature variations. Evidence from the CALM European monitoring sites indicates that active layer thickness was greatest in the summers of 2002 and 2003.[90] Active layer thickness has increased by up to 1.0 metre over the Qinghai-Tibetan Plateau since the early 1980s.[91]

The thickness of seasonally frozen ground in non-permafrost regions has decreased more than 0.34 metre across the Russian Federation since the middle of the 1950s[92] and up to 0.2 metre over the northern Tibetan Plateau from 1967 through 1997.[93] The main driving forces for the decrease are significant winter warming and changes in snow cover. The duration of seasonally frozen ground decreased by more than 20 days during 1967-97 over the Qinghai-Tibetan Plateau, due mainly to the earlier onset of the spring thaw.[94] The estimated maximum area of seasonally frozen ground has decreased by about 7% in the Northern Hemisphere during the 20th century. Evidence from satellite remote sensing data shows that the onset of the spring thaw and autumn

The estimated maximum area of seasonally frozen ground has decreased by about 7% in the Northern Hemisphere during the 20th century

Most studies suggest that over the Northern Hemisphere the snow cover season has shortened and spring melt has occurred earlier in the last 50-100 years

freeze advanced 5-7 days in Eurasia over 1988-2002, leading to an earlier start to the growing season but no changes in its length.[95] Over North America the onset of the autumn freeze was delayed 5 days, which was partly responsible for lengthening the growing season 8 days over 1988-2001.[96]

Snow trends

Contributors: Stephan Harrison with Dennis P. Lettenmaier

More than one-sixth of the world's population lives in areas where surface water is derived mainly from snowmelt, either seasonally ephemeral snowpacks or perennial glaciers.[97] These areas also account for more than a quarter of global gross domestic product. Changes in the seasonal patterns of runoff or permanent changes in runoff volume resulting from changes in snow cover (map 11.5) are therefore of great concern.

Most studies suggest that over the Northern Hemisphere the snow cover season has shortened and spring melt has occurred earlier in the last 50-100 years. Some studies suggest that these changes may have accelerated in the last several decades, but inconsistencies in data sources complicate the picture. The mountainous areas of North America have exhibited downward trends in snow water equivalent that seem related primarily to increased temperature. At some high-latitude locations, however, mid-winter snow water equivalent has increased, possibly in response to increased precipitation in some generally cold continental interior climates.

Analyses of long-term changes in snow cover extent and snow water equivalent over parts of the Northern Hemisphere have used station data, available in some cases since early in the 20th century, and satellite estimates of snow cover extent, available for almost 40 years. Using reconstructed snow cover extent for much of the Northern Hemisphere, one study found that winter snow water equivalent increased about 4% per decade, as did winter snow depth over the Russia Federation.[98] In contrast, over Eurasia and North America spring snow cover extent and snow water equivalent decreased substantially beginning about 1980. A study using the more recent period of satellite data found substantial declines in snow cover extent over the Northern Hemisphere during 1972-2006, especially in spring.[99] Spring declines appeared to be amplified poleward and were larger for North America than Eurasia. Another study using the same satellite data inferred trends of about 3-5 days per decade earlier snow melt and about the same increase in length of the snow-free season over the Northern Hemisphere.[100] A study using reconstructed snow cover records for North America reached similar conclusions.[101]

A number of regional studies have also been conducted. One analysed manual observations of snow depth and newly fallen snow over the Swiss Alps during 1931-99 and found that both the number of days of continuous snow cover and the number of days with snowfall increased gradually until about 1980 and declined thereafter.[102] Trends were progressively more pronounced with decreasing elevation. Increased temperature was the main cause of reductions in the number of days with snow on the ground at low elevation in the Swiss Alps.[103] Analysis of Russian Federation snow depth data for 1936-83 from 119 stations showed a statistically significant increasing trend in winter snow depth.[104] While annual snow cover duration over southern Canada generally has not changed, winter snow cover has increased and spring snow cover has decreased, although changes in snow cover were modest.[105] Snowfall over Canada north of 55°N increased about 20% during 1950-90, associated primarily with increased winter precipitation.[106] Snowfall changes in southern Canada are highly correlated with winter temperature, with snowfall decreasing in the warmer areas

Map 11.5 Changes in the duration of spring snowcover, 1978-2006

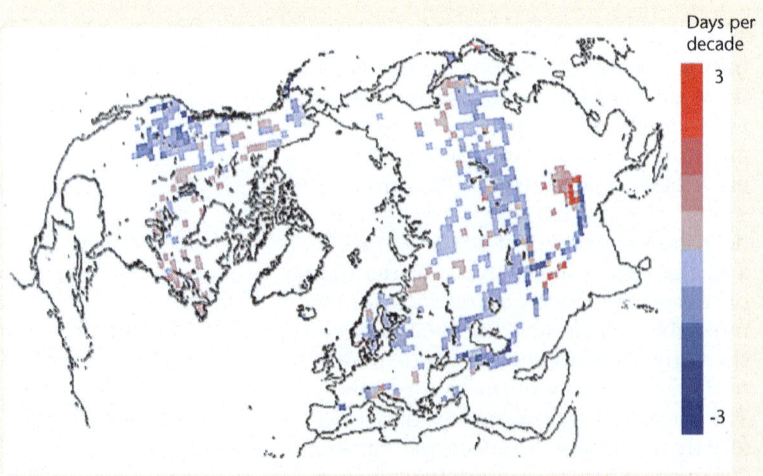

Source: US National Snow and Ice Data Center (NSIDC) from the US National Oceanic and Atmospheric Administration weekly snow cover maps.

and increasing in colder areas, and overall annual precipitation increasing.

Several recent studies have analysed long-term changes in snow-related variables over the Western United States. One analysed 230 snow water equivalent time series in the Pacific Northwest for 1950-2000 and found a strong preponderance of downward trends, especially in the Cascade Mountains, where winter temperatures generally are higher than elsewhere in the region.[107] The declines in snow water equivalent were generally larger in absolute values at lower elevations. An expansion of the analysis to the Western United States and to 1915-2003 found generally similar results.[108] Analysis of changes in the timing of spring snowmelt runoff across the Western United States for 1948-2002 found that for snowmelt-dominated (mostly mountainous) river basins the centroid of the annual hydrograph was consistently shifting to an earlier date but that for coastal basins without a substantial snowmelt component, the date changed little or runoff was later.[109]

Other studies have analysed changes in snow-related variables elsewhere in the United States, mostly in the northeast. One report, based on data for 1952-2005 in the Catskill region of New York State, found that peak snowmelt generally shifted 1-2 weeks earlier, apparently due to an increasing trend in maximum March air temperature.[110] Similar studies for New England found 1-2 week advances of the centre of the volume of runoff.[111] Analysis of the ratio of snow to precipitation for Historical Climatology Network sites in New England found a general decline in the ratio and decreasing snowfall amounts.[112] And 18 of 23 snow course sites in and near Maine with records spanning at least 50 years had decreases in snowpack depth or increases in snowpack density.[113]

Trends in glaciers

Contributors: Vladimir Aizen with Stephan Harrison, Xin Li, Igor Severskiy, Pratap Singh and Tandong Yao

Glaciers cover 11% of the Earth's land surface and store about 75% of its freshwater. Alpine glaciers supply water and generate river flow vital to the millions of people living downstream and to forestry, agriculture, industry and urban areas in adjacent lowlands. Glacier meltwater is particularly important when the lower water courses flow through semi-arid and arid regions with high evaporation and high demands for irrigation water during the vegetation period. Glaciers integrate climate variations over a wide range of time scales, preserving past climatic signatures and making them natural sensors of climate variability. Climate change is having a significant impact on the snow and ice glacier water resources and river water supply in mountain regions.

There is considerable evidence that glaciers have retreated globally since the middle of the 19th century, after the 'little ice age',[114] with the rate of retreat accelerating from the mid-1970s in response to rapid increases in air temperature and changes in the amount of precipitation and its composition (rain or snow).[115] Tropical glaciers are more sensitive than glaciers at higher latitudes to changes in climate. In the tropical Andes the trend in air temperatures has been about a 0.1°C gain per decade since 1939, and the rate has tripled over the last 25 years. The Andes contain 99% of the world's tropical glaciers, most of which are undergoing considerable recession, with many reduced in volume by 30% since 1980.[116] The development of glaciers is also limited by the low levels of precipitation and continuous rise in air temperatures. In the tropical Andes runoff during the dry season (May-September) is often fed solely by glaciers, so that glacier retreat has major implications for seasonal water supplies.[117]

Changes in air temperature and precipitation may have different impacts in different mountain regions at macro- and mesoscales and even in small catchments. In arid regions of western Argentina, central and northern Chile and much of western Peru climate change has resulted in warming and decreasing precipitation during the 20th century, and the glaciers have retreated much faster in South America than in Central Asia.

The coexistence of elevated cold and arid areas creates unique climate and hydrologic regimes not only in the tropical Andes but also at mid-latitudes of Central Asia and the Tibetan Plateau. The large Aral-Caspian and Tarim closed-drainage basins and the great Asian rivers, such as the Ob, Yenisei, Huang He, Yangtze, Mekong and Brahmaputra, are fed by glaciers in the Altai, Pamir, Tibetan and Tien Shan Plateaus. Central Asia glaciers cover 81,500 km^2 and contain approximately 8,000 km^3 of freshwater.[118]

Changes in global and regional air temperature and the frequency of major atmospheric circulation processes regulating

Climate change is having a significant impact on the snow and ice glacier water resources and river water supply in mountain regions

Interactions between a changing hydrologic cycle and a changing terrestrial carbon cycle can be correlated with climate change

moisture flow over Central Asia are major driving forces of glacier mass balance and variability in river discharge. Glaciologic observations conducted on the Tibetan Plateau revealed that between the 1950s and 1960s 50% of glaciers retreated, 30% advanced and 20% were stable. During the 1970s glaciers were relatively stable, but since then recessions have accelerated, and in the 1990s up to 95% of 620 glaciers studied were retreating. The recession rate was 4 metres a year at the north Tibetan Plateau, intensifying up to 65 metres a year to the southeast.[119] The total glacier area of the Tibetan Plateau has shrunk 5.5% in the last 45 years.[120]

Altai Mountain ridges define the northern periphery of the Central Asian mountains and the southern periphery of the Asian Arctic Basin. Altai glaciers cover 2,040 km^2 in southern Siberia, Mongolia and northwestern China. The Pamir glacier, which extends to the most western periphery of the Central Asian mountains (Tajikistan and northwestern China), covered 12,100 km^2 in the late 1970s. In the last half century annual precipitation increased by 3.2 millimetres (mm) a year in Altai, notably in spring and summer months, while no significant change in precipitation occurred in the adjacent lowlands. In the northwestern and central Pamir at elevations above 3,000 metres annual precipitation increased 8.1 mm a year over the last 17 years.[121] Despite the increase in precipitation, glacier recession occurred in Altai and Pamir due to increases (0.03°C a year) in spring and summer air temperatures,[122] which have intensified snow and glacier melt and increased discharge to the Ob and Yenisei Rivers by 7% and to the Pamir River by 13.5%. The Altai glacier area shrank 7.2% on average between 1952 and 2006.[123]

The glacier area of the Tien Shan, a mountain range in Central Asia, has decreased by 1,620 km^2 (10.1%) during the last 30 years. The rate of glacier recession varied between 3.5% in central Tien Shan, with large high-elevated glacier massifs, and 14.1% in the low western Tien Shan, with small sparse glaciers. The rate of recession was three times faster during 1977-2003 than during 1943-77. The surface of some glaciers dropped 100 metres during 1977-2000 (map 11.6).[124] Annual runoff of the major Tien Shan rivers averages 67 km^3 a year, including glacial melt of about 14 km^3 a year, or 20%. During droughts the proportion of glacial runoff increases to 30%. The duration of snow melt (from maximum snow cover to its disappearance) declined from 168 days to 138 days in 2007. The area of seasonal snow cover decreased 15% (by 120,000 km^2), and the date of maximum snow cover has come later. Further decline may be accelerated by increased rainfall instead of snowfall in the early spring at high elevations and, consequently, a lesser heat expenditure for the snowmelt.

Mathematical simulations of the current state of the Tien Shan glacier area and forecasts of the potential impact of global and regional climate change on the glaciers and glacier river runoff in the Tien Shan estimate that an increase in air temperature of 1°C at equilibrium line elevation must be compensated for by a 100 mm increase in precipitation to maintain glaciers in their current state. Glaciers are predicted to decline to 94% of their current number, 69% of current covered area and 75% of current volume. Glacier runoff is predicted to be 75% of its current value.[125]

While the Tien Shan glacier area has been continuously declining, the annual river discharge has been growing over the last decade due mainly to increased precipitation. One of the main predictors of the current year's river discharge in Tien Shan is the volume of river runoff the previous year, which could be replenished by groundwater. The possible sharp change in river runoff indicates the non-linear system response caused mainly by the non-linear response of evapotranspiration to changes in temperature and precipitation. Thus, a precipitation surplus accelerates evapotranspiration when air temperature rises, while a precipitation deficit slows this process even with rising air temperature, which increases the albedo of glacier surface in summer and reduces potential snow and ice melt. Current glacier recession, while initially boosting river flows, eventually causes runoff to decrease.

Glaciers in the Himalayas and European Alps are receding and disappearing faster than glaciers in Central Asia.[126] Recent studies revealed that large Himalayan glaciers are retreating at a rate of more than 30 metres a year, resulting in a 21% reduction in glacier area since the 1980s.[127]

Links between the terrestrial carbon and water cycles

Contributors: Theodore Bohn with Dennis P. Lettenmaier and Charles Vörösmarty

Interactions between a changing hydrologic cycle and a changing terrestrial

Map 11.6 Tien Shan, Akshiirak glacier massif

Glacier area recessions, 1943-2003

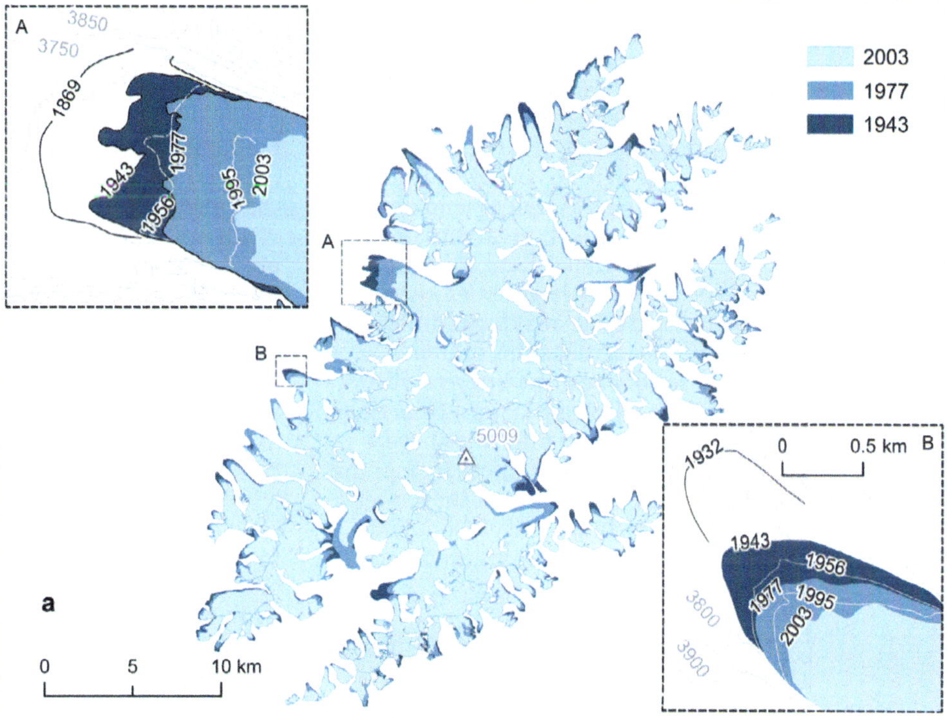

Glacier surface degradation, 1977-2000

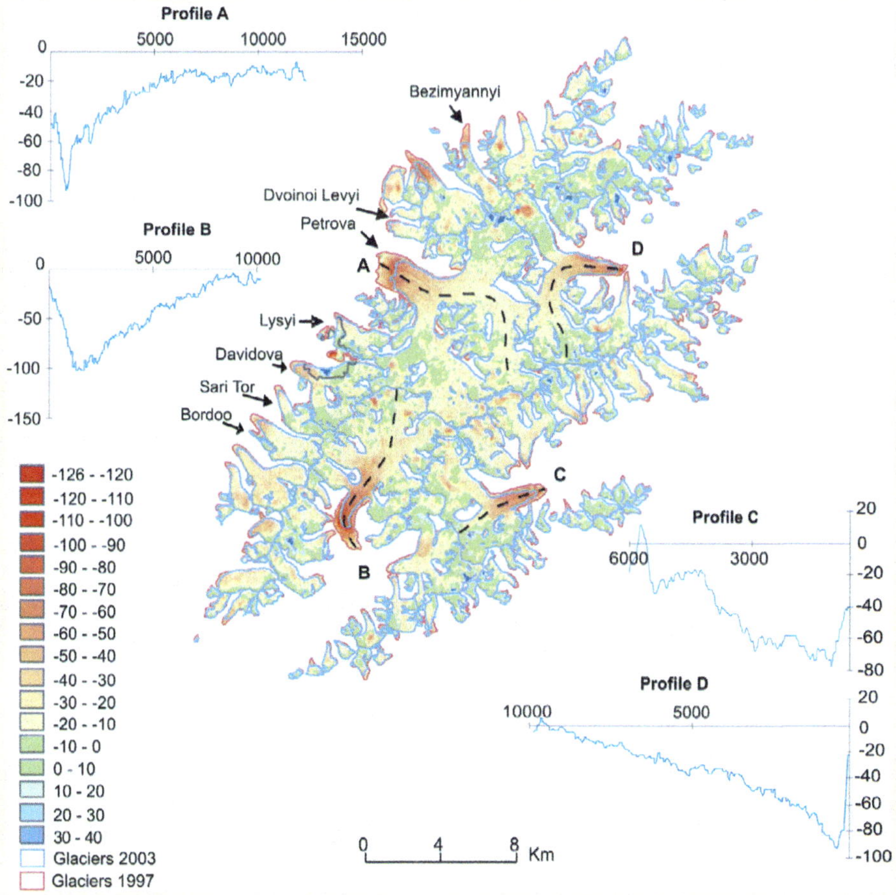

Note: Top maps are aerial photogrammetry and Advanced Spaceborne Thermal Emission and Reflection Radiometer (ASTER) images. Bottom map is aerial photogrammetry and Shuttle Radar Topography Mission (SRTM) data.
Source: Based on Aizen et al. 2006.

Estimated net primary biomass productivity increased significantly over 25% of the global vegetated area and decreased significantly over 7% of the area

carbon cycle can be correlated with climate change. The terrestrial biosphere plays an important role in the global climate system, having taken up roughly 25% of anthropogenic carbon emissions during the last century.[128] It is not clear how long the biosphere can continue to absorb atmospheric carbon at this rate. Observations suggest that the rate of carbon uptake depends on hydrologic and climate conditions as well as land use. However, long-term observations are much sparser for terrestrial carbon storage and flux, especially over large scales, than for the hydrologic cycle, making it difficult to discern the relationships of trends in global or regional carbon budgets with climate and hydrologic factors. Nonetheless, some of these relationships have become apparent, while in other cases the strong relationships between shorter-term hydrologic variability and carbon fluxes have important implications for observed hydrologic trends.

Because water plays different roles in each stage of the terrestrial carbon cycle, it is useful to consider each stage separately. The vast majority of terrestrial carbon is stored in soil, litter and above-ground vegetative biomass. Carbon enters these environmental compartments primarily through plant photosynthesis (productivity). Carbon may leave these compartments through respiration (by plants or through decomposition in the soil), export of carbon into lakes and stream networks and disturbances such as fire (or changes in land use). Some of these stages lend themselves to large-scale monitoring more easily than others.

Productivity exhibits the tightest coupling between the hydrologic and terrestrial carbon cycles, in the link between photosynthesis and transpiration. Accordingly, productivity is subject to many of the same climate limitations as evaporation, including soil moisture, temperature, incident solar radiation and humidity. Attribution of changes in productivity to trends in climate factors is not entirely straightforward, in part because more than one of these factors can be colimiting. Nonetheless, because of its manifestation in the form of biomass, productivity (equivalent to photosynthesis minus plant respiration, or 'net productivity') lends itself more easily to large-scale measurement than other carbon fluxes and is often estimated on global scales using satellite-based data.[129]

An analysis of global trends in net primary biomass productivity (map 11.7) by the geographic distribution of climate factors limiting productivity and trends in those factors during 1982-99 found that estimated net primary productivity increased

Map 11.7 **Interplay of climate factors and net primary productivity**

Trends in estimated net primary productivity, 1982-99

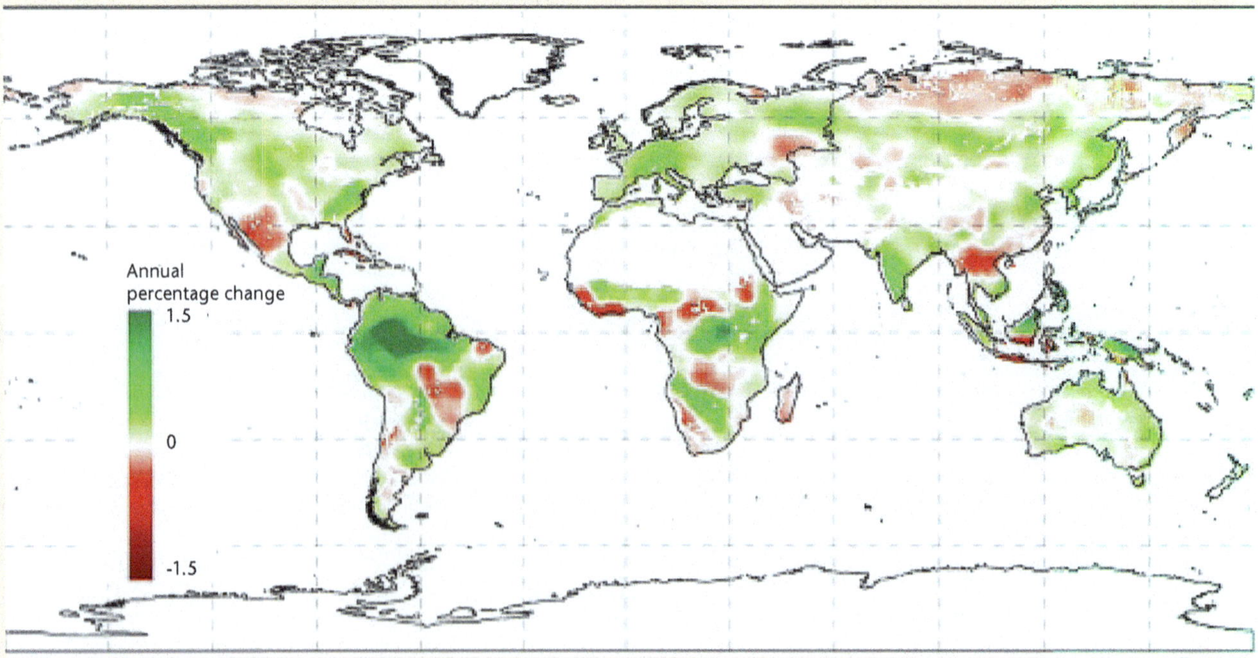

Source: Based on Nemani et al. 2003.

significantly over 25% of the global vegetated area and decreased significantly over 7% of the area, for a net global increase of 6.17% (3.42 petagrams of carbon). Trends in these climate factors were estimated to explain roughly 40% of the trends in net primary productivity, with the remainder attributed to changes in vegetation (for example, land use changes). Trends in net primary productivity over regions where productivity is limited or colimited by solar radiation, such as the Amazon Basin and the northern and southern fringes of tropical Africa and South-East Asia, correlated strongly with trends in cloudiness. Increases in growing season precipitation over water-limited areas such as the Sahel, Southern Africa and Northern and Western Australia led to increases in net primary productivity, while decreases in growing season precipitation in water-limited areas such as Northern Mexico and Central Australia led to decreases in net primary productivity. Temperature-limited areas such as northwest North America and Siberia experienced increased or decreased net primary productivity in accordance with trends in average temperature over the growing season.

Smaller-scale analyses of China[130] and North America[131] agree for the most part with these general conclusions. Both studies cite lengthening of the growing season (essentially a function of temperature) as the primary driver behind increases in net primary productivity over most of their respective domains. However, the North America study found that increased summer precipitation was primarily responsible for the increases in net primary productivity in the central plains, while land use changes may have been responsible for increases over the southeast.

Complicating matters is the seasonal interplay among climate factors. While increased air temperatures in late spring have led to the earlier onset of the growing season in some areas such as the western United States, the lack of a concomitant increase in summer precipitation can result in an earlier exhaustion of soil moisture and reduced productivity in autumn, at least partially negating the gains from increased annual temperature.[132] Thus, it is possible that as temperatures rise, productivity in these regions will become more water-limited and increasingly correlated with annual precipitation.

Hydrology also exerts a strong influence on carbon fluxes out of terrestrial ecosystems. Soil respiration is the largest efflux, approximately equal annually to productivity on a global scale. Field studies have shown that soil respiration has a complex dependence on hydroclimate factors, such as soil temperature and moisture, as well as biogeochemical factors; in particular, soil moisture determines the proportions of carbon released to the atmosphere as carbon dioxide and methane.[133]

While strong global correlations between trends in soil respiration and trends in hydroclimate factors have not been found,[134] some hydrologic trends are expected to have serious implications for soil respiration. One example is permafrost degradation, and the associated changes in lake extent in Siberia (see discussion earlier in this chapter). Much of the permafrost in Siberia contains tremendous reservoirs of carbon-rich soil, termed *yedoma* (500 gigatons of carbon, or roughly the carbon content of the atmosphere), which has been protected from respiration by frozen conditions since the last ice age.[135] In the continuous permafrost zone many lakes have expanded over the last few decades,[136] actively thawing the surrounding and underlying yedoma. Because the newly thawed soil around and under the lakes is saturated with water, respiration of the carbon produces methane, a much stronger greenhouse gas than carbon dioxide. Strong methane emissions were observed from the active margins of thaw lakes in eastern Siberia, and the recent expansion of these thaw lakes (estimated at 14.7% in area between 1974 and 2000) may have resulted in a 58% increase in methane emissions, or 1.4 million metric tonnes a year. Because the recent expansion of these thaw lakes is a result of permafrost degradation, there is concern that methane emissions from the lakes could exacerbate climate warming.

Fires constitute another large carbon efflux, but their sporadic occurrences have made precise measurements of long-term carbon budgets for small areas difficult to attain. Long-term trends can be assessed for large regions, however, through satellite imagery, which captures trends in fire frequency. Similar to productivity, fire frequency exhibits a dependence on multiple climate factors, including temperature, precipitation and soil moisture. One study found that annual wildfire frequencies in the Western United States increased significantly during 1970-2003.[137] Fire frequency over this period exhibited high correlations with both the timing of spring snow melt and average summer temperatures (to

Hydrology also exerts a strong influence on carbon fluxes out of terrestrial ecosystems

There is a consensus among climate scientists that climate warming will intensify, accelerate or enhance the global hydrologic cycle

which snow melt timing is sensitive). Conversely, several studies have found strong negative correlations between trends in fire frequency and summer storm frequency.[138] As mentioned, warmer summers and earlier snow melts tend to result in earlier exhaustion of soil moisture and onset of water limited conditions, while more frequent summer storms would be expected to delay the onset of water-limitation. Thus, two major carbon fluxes, productivity and fires, depend strongly on the annual duration of water limitation.

The final major efflux of carbon from terrestrial ecosystems is the export of carbon from soils to aquatic systems as particulate organic carbon (POC) and dissolved organic carbon (DOC). Because much of this carbon can be subsequently respired and returned to the atmosphere, either in streams or the ocean, this flux is an important loss term in the terrestrial carbon budget. A substantial amount of POC consists of organic carbon sorbed onto soil sediment particles, entering streams through erosion and mass wasting (see section on erosion and sediment transport), so trends in sediment transport would be expected to have strongly correlated with trends in POC export (although other factors such as soil carbon content and in-stream chemistry would also exert an influence on POC export). One study estimated that, globally, 0.4-1.2 petragrams of carbon are transported to the oceans as POC each year. However, no global trends in POC export have yet been assessed.[139]

Several studies have observed marked increases in the annual fluxes of DOC in many temperate and boreal streams around the world. While it has been difficult to attribute all the observed trends to a single cause,[140] hydrology appears to play a role in some cases, through changes in groundwater drainage. For example, in the Arctic, several studies have found strong correlations between daily river discharge and DOC concentrations.[141] In this context, the increases in the annual discharge of the six major rivers of the Russian Federation Arctic[142] and especially the recently discovered increase in minimum flows across Northern Eurasian pan-Arctic[143] discussed in the section on runoff and streamflow trends may have important consequences for the carbon cycle.

Because minimum flows generally reflect the influence of groundwater, the cause of these trends has been speculated to be a reduction in the intensity of seasonal soil freezing, allowing more connectivity in subsurface drainage networks. If this process is indeed occurring, it is conceivable that the increased flushing of the soils by groundwater, accompanied by longer growing seasons and greater microbial activity during seasons in which the soils historically have been frozen, could lead to greater mobility and loss of soil carbon. Establishing a link between these phenomena will require further research. In addition, the thawing of permafrost in Siberia may release large amounts of soil carbon into streams in the future,[144] but this is based on a space-for-time substitution rather than a direct observation of a trend.

In summary, while hydrologic processes are important in all stages of the carbon cycle, trends in the carbon cycle are only sometimes strongly correlated with trends in the hydrologic cycle – namely, when water availability is the dominant limiting factor. For some ecosystems water availability may become a limiting factor as other climate factors change. Direct observations of carbon fluxes and storages have rarely been made over a long enough time period or a large enough region for significant trends to be detected, but strong evidence can still exist to indicate potential impacts of hydrologic trends on the carbon cycle.

Is the hydrologic cycle accelerating?

Contributor: Tom Huntington

There is a consensus among climate scientists that climate warming will intensify, accelerate or enhance the global hydrologic cycle.[145] Intensification could be evidenced or caused by increasing rates of evaporation, evapotranspiration, precipitation and streamflow (in some areas). Associated changes in atmospheric water content, soil moisture, ocean salinity and glacier mass balance (seasonal) may also be implicated. The mechanism most often cited is that warmer air temperatures result in higher saturation vapour pressure (about 7% higher per degree Kelvin) and hence atmospheric water vapour content.[146] Some argue that recent satellite observations do not support subdued sensitivity, and report increases in water vapour content, precipitation and evaporation of about 6% per degree Kelvin.[147] The intensification response to future warming remains a critical question in assessing hydrologic response to climate warming.

The IPCC has found global average increases in surface air temperatures over land of 0.74°C ± 0.18°C for 1906–2005.[148] There are fewer long-term, continuous and quality-assured records for precipitation than for air temperature and spatial heterogeneity. Trends in precipitation have been more variable spatially and temporally than trends in temperature (see section on precipitation trends), but increases have been noted in most of North America, southern South America, northern Eurasia and western Australia and decreases in western Africa and the Sahel and Chile during 1901-2005. Current estimates of global average long-term trends in precipitation do not show significantly increasing precipitation over the period of observation as earlier assessments found,[149] possibly because of recent decreases and different methodologies and observations. There is also some evidence that snowfall has increased in northern high latitudes[150] and over mountain and subpolar glaciers.[151]

Recent studies or assessments have reviewed various components of the hydrologic system and have concluded that the evidence supports an ongoing intensification of the hydrologic cycle,[152] though with substantial variability across regions and time. One frequent concern is that data are usually incomplete both temporally and spatially.

Evapotranspiration cannot be directly measured over large land areas, but indirect measurements from long-term river basin water balance studies provide reasonable estimates based on precipitation minus runoff (assuming no net change in storage). Increases in precipitation were substantially higher than increases in runoff, suggesting that during the 20th century evapotranspiration had increased in the Mississippi River basin, other large and smaller river basins in the continental United States and the La Plata River basin in South America.[153] In most of Canada streamflow has been stable or decreasing (see section on runoff and streamflow trends), but precipitation has been increasing.[154] Together, these observations suggest that evapotranspiration has been increasing.[155]

As discussed in the section on 'Changes in evaporation and evapotranspiration', pan evaporation has decreased over much of the continental United States and the former Soviet Union, despite indications of increased actual evapotranspiration. Although the apparent difference in direction of trends in pan and actual evapotranspiration has been termed the 'evaporation paradox',[156] the reasons are likely to be more straightforward. In arid regions increasing actual evapotranspiration may result from increasing precipitation because evaporative demand greatly exceeds supply. In humid areas increasing actual evapotranspiration may result from decreases in net radiation and surface humidity, and perhaps some change in surface wind.

Atmospheric water vapour content (specific humidity) has increased in recent decades because warmer air can hold more moisture.[157] This aspect of an intensifying hydrologic cycle is profoundly important because water vapour is also a radiatively active gas.[158] Because accumulation of other radiatively active gases induces climate warming, this feedback mechanism increasing atmospheric water vapour content will amplify the warming.

As summarized earlier, analysis of streamflow trends in northern temperate and high latitudes and in southern temperate latitudes shows more upward than downward trends, while parts of West Africa, Southern Europe and southernmost South America have seen decreased runoff.[159] River basins in the continental United States show trends towards increasing mean and low flows but no clear evidence of trend direction in high flows.[160] Most rivers draining to the Arctic Ocean show increasing flow, although the cause of these trends, which are mostly in winter discharge, remain unclear.[161]

On balance, there is some observational evidence that surface air temperature warming has intensified the global hydrologic cycle. Data limitations in length of record, continuity and spatial coverage contribute uncertainty to this broad conclusion, and especially for changes in hydrologic extremes (such as floods). Natural climatic and multiyear variability associated with large-scale atmospheric circulation patterns such as the El Niño-Southern Oscillation, the Pacific Decadal Oscillation and the North Atlantic Oscillation influence many of the observed trends in ways not yet fully understood.

Assessing future impacts of climate change

Contributor: Rick Lawford

An accelerated global water cycle associated with global warming could have important consequences for the world's water resources. Focusing on projected

An accelerated global water cycle associated with global warming could have important consequences for the world's water resources

Many of the more indirect impacts of climate change on surface water are not fully known

future changes, this section considers the consequences for water management as well as adaptation strategies for coping.

The latest IPCC report describes water cycle effects associated with projected global warming ranging from a warmer atmosphere that holds more water vapour to more severe regional water shortages in semi-arid and arid regions.[162] These projections result mostly from general circulation models of the coupled land-atmosphere-ocean climate system. These models represent the climatic forcings and response of the Earth's systems to these forcings and to anthropogenic changes in the composition of the atmosphere, typically at spatial resolutions of 2°-4° latitude by longitude. Although the reliability of these projections is improving, a number of inherent uncertainties remain because of poorly defined initial conditions, natural and human processes and feedbacks, inadequate process representation in climate models, scale mismatches, extremes of climate and long-term climate variability, among other reasons. Furthermore, accurate projections of future changes in land hydrologic processes based on general circulation model simulations are complicated by subgrid spatial heterogeneity and the highly non-linear nature of the processes. Projected changes based on these simulations are generally more accurate at large scales and less so at the regional scales where mitigation and adaptation must take place.

Due to the localized response of water resources to large-scale forcings, global projections from climate models are of limited value for water resources applications unless accompanied by downscaling. Hydrologic models are a central component of the tools used to assess the possible implications of a change in climate for the hydrology of a location or a basin. Climate projections must be transformed into values at points or for small areas for use in impact assessments. Downscaling can be conducted by statistical methods or dynamic regional climate models.[163] Regional models can use global model outputs as their boundary conditions to provide much higher resolution outputs that account more fully for topography and more accurately represent critical physical processes. Statistical downscaling methods are easier to apply and hence more widely used; they involve 'training' a statistical model that relates large-scale climate model output to local conditions using historical observations (for example, of physical climate variables like precipitation and temperature) and historical general circulation models, and then applying the same relationships to general circulation model simulations of future large-scale conditions.

Hydrologic impact studies assess the effects of climate change on individual processes to infer how the process may change with a shifting climate. They typically use deterministic hydrologic models to estimate changes in water availability and related impacts. Hydrologic sensitivity studies involve running models with and without a particular process to determine its contributions.

Ensemble simulations provide another technique for assessing the uncertainty in impact of changes due to climate and other factors. On a global basis and for the Colorado River basin the nature of runoff changes is projected by a suite of general circulation models using this ensemble information to estimate uncertainty in projections of future hydrologic change.[164]

Approaches to incorporating climate change information in decision-making can be either direct or indirect. Direct approaches incorporate climate change information directly into decision-making – for example, climate scientists interacting with partnering utilities to find space and time scales appropriate for adaptations to reduce the risk of climate extremes. Indirect approaches involve potentially affected people in studies of the readiness of societies to adapt to climate change. Although the indirect approach has dominated to date, as water managers and decision-makers become more serious about adaptation to climate change, the direct approach will likely begin to predominate.

Summary

Table 11.2 summarizes the key findings of a literature review of ongoing changes in the land surface water cycle. The picture is incomplete – both because a brief summary such as this cannot be exhaustive and because the most comprehensive studies have been conducted where the highest-quality and lengthiest datasets are available, resulting in non-uniform coverage globally. Furthermore, while confidence in projections of the thermal aspects of climate change is growing, many of the more indirect impacts of climate change on surface water are not fully known. Improving this knowledge by acquiring better and more comprehensive data is critical

Table 11.2 **Summary of key findings relative to trends in land surface water cycle components**

Hydrologic variable	Key findings
Precipitation	The mode of precipitation may be more important than the average precipitation in determining hydrologic impacts. Widespread increases in heavy precipitation events have been observed in some places where total precipitation has decreased. At the same time the length, frequency and intensity of heat waves have widely increased. In addition, more precipitation now falls as rain rather than snow in northern regions. All these changes are consistent with a warmer atmosphere with a greater water-holding capacity.
Evaporation and evapotranspiration	Several studies provide evidence that on average for large areas of Asia and North America, actual evapotranspiration is increasing, even though pan evaporation is decreasing.
Soil moisture	In-situ soil moisture measurements are too sparse to draw conclusions about multidecadal soil moisture trends at the global scale, and current satellite sensors are suboptimal in terms of microwave frequency and have record lengths that are too short to provide meaningful information about trends. Studies using physically based model reconstructions with precipitation and temperature to estimate soil wetness are not comprehensive enough to fully understand the uncertainties and generate unambiguous results.
Runoff and streamflow	Two recent studies that assess high and low flow trends on a worldwide basis concluded that their results do not support the hypothesis that global warming has, to date, caused an increase in hydrologic extremes, such as more floods and (hydrologic) droughts. Where long streamflow records support century-scale trend analysis, there is evidence of increases in low flows and mean annual flows, but not floods. These changes appear to be generally consistent with observed precipitation increases over the same period.
Groundwater	A recent US Climate Change Science Program report states that '. . . the ability to predict the effects of climate and climate change on groundwater systems is nowhere near as advanced as for surface water systems.' Much more work is needed to understand the sensitivity of this critical resource to climate change globally.
Reservoir, lake and wetland storage	Changes in lake extent have been observed in many parts of the world over the past few decades, but the primary factors behind these changes are regionally specific. Decadal scale variations in lake, wetland and reservoir storage are natural characteristics of the dynamics of these water bodies and cannot necessarily be ascribed to climate, land cover or other anthropogenic causes.
Permafrost	Changes in the physical climate at high latitudes, primarily increasing air temperature, are forcing changes in permafrost conditions and permafrost degradation. These include increased permafrost temperatures, thickening of the active layer, and thermokarst and talik development (including changes in the extent of thaw lakes) over the Northern Hemisphere permafrost regions. Ground-based measurements indicate that the magnitude of permafrost temperature increase is greatest in continuous permafrost regions and transitions to no change in discontinuous and sporadic permafrost regions.
Snow	Most studies suggest that over the Northern Hemisphere the length of the snow cover season has decreased and spring melt has occurred earlier, over the last 50-100 years. Some studies suggest that these changes may have accelerated in the last several decades; however inconsistencies in data sources complicate such a conclusion.
Glaciers	There is strong evidence that glaciers have retreated globally since the middle of the 19th century, after the 'little ice age', and this retreat has accelerated from the mid-1970s as a response to rapid increases in air temperature, and changes in precipitation amount and rain and snow partitioning. Although there is evidence of glacier retreat globally, tropical glaciers are more sensitive than those at higher latitudes, and have shown the most rapid changes.

to adapting water management to global change. In addition to these requirements, an end-to-end analysis (from data through models to decision-making) is needed to identify the most significant sources of uncertainty and misunderstanding.

A deficiency of this review is that the published literature is heavily biased towards the 'old world' because these countries are much more likely to have long-term hydrologic observations that support rigorous statistical trend analyses. In almost all cases the hydrologic time series that are the basis for the studies summarized in this chapter are in-situ observations. Ongoing efforts, such as the International Hydrological Programme's Flow Regimes from International Experimental Network and Data (FRIEND), are helping rebuild hydrologic networks in developing countries but have not existed long enough to provide stable multidecadal record lengths for trend analysis. This should also be alleviated as satellite records of key hydrologic variables become long enough to support trend analyses, and as methods of merging multisatellite data sources are improved. The Global Earth Observation System of Systems (GEOSS) effort, fostered by the Group on Earth Observations, has improvement of Earth, and specifically hydrologic observations, as a central goal.

Notes

1. Milly et al. 2008.
2. Giorgi 2002.
3. IPCC 2007.
4. IPCC 2007, figure 3.12.
5. IPCC 2007.
6. L'hote et al. 2002.
7. IPCC 2007, chapter 10.
8. IPCC 2007, chapter 3.
9. IPCC 2007, chapter 9.
10. Shuttleworth 2008.
11. Farahani et al. 2007. An exception is accurate-weighing lysimeters, which are very expensive and are mainly used in short-term research studies to measure evapotranspiration.
12. Gedney et al. 2006.
13. See, for example, Brutsaert and Parlange 1998; Ohmura and Wild 2002; and Roderick and Farquhar 2002, 2004, 2005.
14. Shuttleworth et al. forthcoming.
15. For example, Chattopadhyay and Hume 1997; Thomas 2000; Shenbin, Yvnfeng, and Thomas 2006; Xu et al. 2006.
16. For example, Askoy 1997; Omran 1998; Cohen, Ianetz, and Stanhill 2002.
17. Stanhill and Cohen 2001.
18. IPCC 2007.
19. Stanhill and Cohen 2001.
20. IPCC 2007.
21. Roderick et al. 2007.
22. Bouchet 1963.
23. De Bruin 1983, 1989; McNaughton and Spriggs 1986, 1989.
24. Shuttleworth et al. forthcoming.
25. Chattopadhyay and Hume 1997; Thomas 2000; Xu et al. 2006; Shenbin, Yunfeng, and Thomas 2006.
26. Chattopadhyay and Hume 1997; Thomas 2000; Xu et al. 2006.
27. Lawrimore and Peterson 2000; Golubev et al. 2001; Hobbins and Ramirez 2004.
28. Shuttleworth et al. 2009.
29. www.wcc.nrcs.usda.gov/scan/.
30. Robock et al. 2005; Li, Robock, and Wild 2007.
31. Dai, Trenberthy, and Qian 2004.
32. Maurer et al. 2002.
33. Nijssen, Schnur, and Lettenmaier 2001.
34. Sheffield and Wood 2008.
35. Groisman et al. 2004; Andreadis and Lettenmaier 2006.
36. Lu et al. 2005 and Li et al. 2005.
37. Li, Robock, and Wild 2007.
38. Tapley et al. 2004; Rodell et al. 2006; Yeh et al. 2006; Swenson and Wahr 2006; Syed et al. 2005.
39. Lins 2008.
40. Kundzewicz et al. 2005; Svensson, Kundzewicz, and Maurer 2005.
41. Zhang et al. 2001.
42. Lins and Slack 1999, 2005.
43. McCabe and Wolock 2002.
44. Hannaford and Marsh 2007.
45. Mudelsee et al. 2003.
46. Yiou et al. 2006.
47. Lindstrom and Bergstrom 2004.
48. Birsan et al. 2005.
49. Kahya and Kalayci 2004.
50. Hisdal et al. 2001.
51. Peterson et al. 2002.
52. Smith et al. 2007.
53. Svensson, Kvndzewicz, and Maurer 2005.
54. Sheffield and Wood 2008.
55. Foster and Loucks 2006.
56. Backlund, Janetos, and Schimel 2008, p. 145.
57. Vaccaro 1992.
58. Scibek and Allen 2006.
59. Green et al. 2007.
60. van Roosmalen, Christensen, and Sonnenborg 2007.
61. Backlund, Janetos, and Schimel 2008.
62. ICOLD 2003; Lempérière 2006.
63. Lempérière 2006.
64. Vörösmarty et al. 1997.
65. Chao, Wu, and Li 2008.
66. Meybeck 1995; Shiklomanov and Rodda 2003; Lehner and Döll 2004.
67. Lehner and Döll 2004.
68. www.caspage.citg.tudelft.nl.
69. Smith, Sheng et al. 2005.
70. Walter et al. 2006.
71. Smith, Sheng et al. 2005.
72. Lehner and Döll 2004.
73. Mitsch and Gosselink 2000.
74. Zhang et al. 1999, 2003.
75. Brown and Goodison 1996, Zhang et al 1999.
76. Brown and Goodison 1996.
77. Zhang et al. 1999.
78. Lawrence and Slater 2005.
79. Lachenbruch and Marshall 1986; Osterkamp 2005.
80. Osterkamp 2005.
81. Smith, Burgess et al. 2005.
82. Pavlov 1996.
83. Oberman and Mazhitova 2001.
84. Sharkhuu 2003.
85. Wu and Zhang 2008.
86. Vonder Muhll et al. 2004.
87. Zhang 2005.
88. Zhang et al. 2005.
89. Brown, Hinkel, and Nelson 2000.
90. Harris et al. 2003.
91. Zhao et al. 2004.
92. Frauenfeld et al. 2004.
93. Zhao et al. 2004.
94. Zhao et al. 2004.
95. Smith, Saatchi, and Randerson 2004.
96. McDonald et al. 2004.
97. Barnett, Adam, and Lettenmaier 2005.
98. Brown 2000.
99. Déry and Brown 2007.
100. Dye 2002.
101. Frei, Robinson, and Hughes 1999.
102. Laternser and Schneebeli 2003.
103. Scherrer and Appenzeller 2004.
104. Ye 2000.
105. Brown and Goodison 1996.
106. Groisman and Easterling 1994.
107. Mote 2003.
108. Mote et al. 2005.
109. Stewart, Cayan, and Dettinger 2005.
110. Burns, Klaus, and McHale 2007.
111. Hodgkins, Dudley, and Huntington 2003; Hodgkins and Dudley 2006a.
112. Huntington et al. 2004.
113. Hodgkins and Dudley 2006b.
114. Mayewski and Jeschke 1979.
115. Liu et al. 2006; Aizen et al. 2006.
116. Francou et al. 2003.
117. Juen, Kaser, and Georges 2007.
118. Shi 2005.
119. Yao et al. 2004; Yao et al. 2007.
120. Kang et al. 2004.
121. Finaev 2007.
122. Aizen et al. 2005; Finaev 2007.
123. Surazakov, Aizen, and Nikitin 2007.
124. Aizen et al. 2006.
125. Aizen, Aizen, and Kuzmichenot 2007.
126. Ageta et al. 2001; Paul et al. 2004.
127. Srivastava, Gupta, and Mukerji 2003; Kulkarni et al. 2007.
128. IPCC 2007.
129. Net primary productivity is generally estimated using satellite-derived a normalized difference vegetation index (NDVI).
130. Fang et al. 2003.
131. Hicke et al. 2002.
132. Baldocchi et al. 2001; Nemani et al. 2003.
133. See, for example, Dise et al. 1993.
134. Kirschbaum 2006.

135. Walter et al. 2006.
136. Smith, Sheng et al. 2005; Walter et al. 2006.
137. Westerling et al. 2006.
138. Holden et al. 2007; Knapp and Soule 2007.
139. Stallard 1998.
140. Worrall and Burt 2007.
141. Holmes et al. 2008; Raymond et al. 2007.
142. Peterson et al. 2002.
143. Smith et al. 2007.
144. Frey and Smith 2005.
145. Del Genio, Lacis, and Ruedy 1991; Loaiciga et al. 1996; Trenberth 1999; Held and Soden 2000; Arnell and Liu 2001.
146. Allen and Ingram 2002.
147. Wentz et al. 2007.
148. Trenberth et al. 2007.
149. Dai, Fung, and Del Genio 1997; New et al. 2001.
150. Trenberth et al. 2007.
151. Dyurgerov 2003; see also the section on trends for glaciers.
152. Dai 2006; Held and Soden 2006; Huntington 2006; Dirmeyer and Brubaker 2006; Holland et al. 2007; Trenberth et al. 2007.
153. Milly and Dunne 2001; Walter et al. 2004; Szilagyi, Katul, and Parlange 2002; Berbery and Barros 2002.
154. Zhang et al. 2000.
155. Fernandes, Korolevych, and Wang 2007.
156. Bouchet 1963.
157. Dai 2006; Willett et al. 2007; Wentz et al. 2007; Allen and Ingram 2002; Held and Soden 2006.
158. Held and Soden 2000.
159. Milly et al. 2005.
160. Lins and Slack 1999, 2005.
161. Peterson et al. 2002; Adam and Lettenmaier 2008.
162. IPCC 2007.
163. See, for example, Wood et al. 2004 for a comparison.
164. Milly, Dunne, and Vecchia 2005.

References

Adam, J. C., and D. P. Lettenmaier. 2008. Application of New Precipitation and Reconstructed Streamflow Products to Streamflow Trend Attribution in Northern Eurasia. *Journal of Climate* 21 (8): 1807-28.

Ageta, Y., N. Naito, M. Nakawo, K. Fujita, K. Shankar, A. P. Pokhrel, and D. Wangda. 2001. Study Project on the Recent Rapid Shrinkage of Summer-Accumulation Type Glaciers in the Himalayas, 1997-1999. *Bulletin of Glaciological Research* 18: 45-9.

Aizen, V. B., E. M. Aizen, K. Fujita, S. Nikitin, K. Kreutz, and N. Takeuchi. 2005. Stable-Isotope Time Series and Precipitation Origin from Firn Cores and Snow Samples, Altai Glaciers, Siberia. *Journal of Glaciology* 51 (175): 637-54.

Aizen, V. B., E. M. Aizen, and V. A. Kuzmichenok. 2007. Glaciers and Hydrological Changes in the Tien Shan: Simulation and Prediction. *Environmental Research Letters* 2 (October-December): 1-10. http://dx.doi.org/10.1088/1748-9326/2/4/045019.

Aizen, V. B., E. M. Aizen, A. B. Surazakov, and V. A. Kuzmichenok. 2006. Assessment of Glacial Area and Volume Change in Tien Shan (Central Asia) During the Last 150 Years Using Geodetic, Aerial Photo, ASTER and SRTM Data. *Annals of Glaciology* 43: 202-13.

Allen, M. R., and W. J. Ingram. 2002. Constraints on Future Changes in Climate and the Hydrologic Cycle. *Nature* 418 (6903): 224-32.

Andreadis, K., and D. Lettenmaier. 2006. Trends on 20th Century Drought over the Continental United States. *Geophysical Research Letters* 33: L10403. doi:10.1029/2006GL025711.

Arnell, N. W., and Liu C. 2001. Hydrology and Water Resources. In *Climate Change 2001: Impacts, Adaptation and Vulnerability*, eds. J. J. McCarthy, O. F. Canziani, N. A. Leary, D. J. Dokken, and K. S. White. Cambridge, UK: Cambridge University Press.

Askoy, B. 1997. Variations and Trends in Global Solar Radiation for Turkey. *Theoretical and Applied Climatology* 58 (3-4): 71-7.

Backlund, P., A. Janetos, and D. Schimel, eds. 2008. *The Effects of Climate Change on Agriculture, Land Resources, Water Resources, and Biodiversity in the United States.* Washington, DC: U.S. Climate Change Science Program.

Baldocchi, D., E. Falge, Gu L., R. Olson, D. Hollinger, S. Running, P. Anthoni, C. Bernhofer, K. Davis, R. Evans, J. Fuentes, A. Goldstein, G. Katul, B. Law, Lee X., Y. Malhi, T. Meyers, W. Munger, W. Oechel, K.T. Paw U, K. Pilegaard, H.P. Schmid, R. Valentini, S. Verma, T. Vesala, K. Wilson, and S. Wofsy. 2001. FLUXNET: A New Tool to Study the Temporal and Spatial Variability of Ecosystem-Scale Carbon Dioxide, Water Vapour, and Energy Flux Densities. *Bulletin of the American Meteorological Society* 82 (11): 2415-34.

Barnett T. P., J. C. Adam, and D. P. Lettenmaier. 2005. Potential Impacts of a Warming Climate on Water Availability in Snow-Dominated Regions. *Nature* 438 (7066): 303-9.

Berbery, E. G., and V. R. Barros. 2002. The Hydrologic Cycle of the La Plata Basin in South America. *Journal of Hydrometeorology* 3 (6): 630-45.

Birsan, M.-V., P. Molnar, P. Burlando, and M. Pfaundler. 2005. Streamflow Trends in Switzerland. *Journal of Hydrology* 314: 312-29.

Bouchet, R. J. 1963. Evapotranspiration Reele, Evapotranspiration Potentielle, et Production Agricole. *Annals of Agronomy* 14: 543-824.

Brown, J., O. J. J. Ferrians, J. A. Heginbottom, and E. S. Melnikov. 1997. International Permafrost Association Circum-Arctic Map of Permafrost and Ground Ice Conditions. US Geological Survey Circum-Pacific Map Series Map CP-45. Boulder, CO: National Snow and Ice Data Center and World Data Center for Glaciology.

Brown, R. D. 2000. Northern Hemisphere Snow Cover Variability and Change, 1915–97. *Journal of Climate* 13: 2339-55.

Brown, R. D., and B. E. Goodison. 1996. Interannual Variability in Reconstructed Canadian Snow Cover, 1915-1992. *Journal of Climate* 9 (6): 1299-1318.

Brown, J. K., M. Hinkel, and F. E. Nelson. 2000. The Circumpolar Active Layer Monitoring (CALM) Program: Research Design and Initial Results. *Polar Geography* 24 (3): 166-258.

Brutsaert, W., and M. P. Parlange. 1998. Hydrological Cycle Explains the Evaporation Paradox. *Nature* 396 (5): 284-5.

Burns, D. A., J. Klaus, and M. R. McHale. 2007. Recent Climate Trends and Implications for Water Resources in the Catskill Mountain Region, New York, USA. *Journal of Hydrology* 336 (1-2): 155-170.

Chao B. F., Wu Y. H., and Li Y. S. 2008. Impact of Artificial Reservoir Water Impoundment on Global Sea Level. *Science* 320 (5873): 212-14.

Chattopadhyay, N., and M. Hume. 1997. Evaporation and Potential Evaporation in India under Conditions of Recent and Future Climate Change. *Agricultural and Forest Meteorology* 87: 55-73.

Cohen, S., A. Ianetz, and G. Stanhill. 2002. Evaporative Climate Changes at Bet Dagan, Israel, 1964-1998. *Agricultural and Forest Meteorology* 111: 83-91.

Dai, A. 2006. Recent Climatology, Variability, and Trends in Global Surface Humidity. *Journal of Climate* 19 (15): 3589-606.

Dai, A., I. Y. Fung, and A. D. Del Genio. 1997. Surface Observed Global Land Precipitation Variations during 1900-88. *Journal of Climate* 10: 2943-62.

Dai, A., K. Trenberth, and T. Qian. 2004. A Global Dataset of Palmer Drought Severity Index for 1870-2002: Relationship with Soil Moisture and Effects of Surface Warming. *Journal of Hydrometeorology* 5 (6): 1117-30.

De Bruin, H. A. R. 1983. A Model of the Priestley-Taylor Parameter, α. *Journal of Applied Meteorology* 22: 572-8.

———. **1989.** Physical Aspects of the Planetary Boundary Layer. In *Estimation of Areal Evaporation*, eds. T. A. Black, D. L. Spittlehouse, M. D. Novak, and D. T. Price. Wallingford, UK: International Association of Hydrological Sciences .

Del Genio, A. D., A. A. Lacis, and R. A. Ruedy. 1991. Simulations of the Effect of a Warmer Climate on Atmospheric Humidity. *Nature* 351: 382-5.

Déry, S. J., and R. D. Brown. 2007. Recent Northern Hemisphere Snow Cover Extent Trends and Implications for the Snow-Albedo Feedback. *Geophysical Research Letters* 34: L22504. doi:10.1029/2007GL031474.

Dirmeyer, P. A., and K. L. Brubaker. 2006. Evidence for Trends in the Northern Hemisphere Water Cycle. *Geophysical Research Letters* 33: L14712. doi:14710.11029/12006GL026359.

Dise, N. B., E. Gorham, and E. S. Verry. 1993. Environmental Factors Controlling Methane Emissions from Peatands in Northern Minnesota. *Journal of Geophysical Research-Atmospheres* 98 (D6): 10583-94.

Dye, D. G. 2002. Variability and Trends in the Annual Snow-Cover Cycle in Northern Hemisphere Land Areas, 1972–2000. *Hydrological Processes* 16: 3065-77.

Dyurgerov, M. 2003. Mountain and Subpolar Glaciers Show an Increase in Sensitivity to Climate Warming and Intensification of the Water Cycle. *Journal of Hydrology* 282: 164-76.

Fang J. Y., Piao S., C. B Field, Pan Y., Guo Q., Zhou L., Peng C., and Tao S. 2003. Increasing Net Primary Production in China from 1982 to 1999. *Frontiers in Ecology and the Environment* 1 (6): 293-97.

Farahani, H. J., T. A. Howell, W. J. Shuttleworth, and W. C. Bausch. 2007. Evapotranspiration: Progress in Measurement and Modeling in Agriculture. *Transactions of the American Society of Agricultural and Biological Engineers* 50(5): 1627-38.

Fernandes, R., V. Korolevych, and S. Wang. 2007. Trends in Land Evapotranspiration over Canada for the Period 1960-2000 Based on in Situ Climate Observations and a Land Surface Model. *Journal of Hydrometeorology* 8: 1016-30.

Finaev, A. F. 2007. Impact of Climate and Water Resources Changes on Land Degradation in Tajikistan. Proceedings of the NASA LCLUC Workshop, 16-21 September 2007, Urumqi, China.

Foster, S., and D. P. Loucks, eds. 2006. *Non-Renewable Groundwater Resources. A Guidebook on Socially-Sustainable Management for Water-Policy Makers. IHP-VI, Series on Groundwater No.10.* Paris: United Nations Children's Fund.

Francou, B., M. Vuille, P. Wagnon, J. Mendoza, and J.-E. Sicart. 2003. Tropical Climate Change Recorded by a Glacier in the Central Andes during the Last Decades of the Twentieth Century: Chacaltaya, Bolivia, 16S. *Journal of Geophysical Research* 108 (D5): 4154. doi:10.1029/2002JD002959.

Frauenfeld, O., Zhang T., Roger G. Barry, and David G. Gilichinsky. 2004. Interdecadal Changes in Seasonal Freeze and Thaw Depths in Russia. *Journal of Geophysical Research* 109: D05101. doi:10.1029/2003JD004245.

Frei, A., D. A. Robinson, and M. G. Hughes. 1999. North American Snow Extent: 1900-1994, *International Journal of Climatology* 19: 1517-34.

Frey, K. E., and L. C. Smith. 2005. Amplified Carbon Release from Vast West Siberian Peatlands by 2100. *Geophysical Research Letters* 32 (9).

Garcia, N. O., and C. R. Mechoso. 2005. Variability in the Discharge of South American Rivers and in Climate. *Hydrological Sciences Journal* 50: 459-78.

Gedney, N., P. M. Cox, R. A. Betts, O. Boucher, C. Huntingford, and P. A. Stott. 2006. Detection of a Direct Carbon Dioxide Effect in Continental River Runoff Records. *Nature* 439 (16): 835-7. doi:10.1038.

Giorgi, F. 2002. Dependence of Surface Climate Interannual Variability on Spatial Scale. *Geophysical Research Letters* 29: 2101.

Golubev, V. S., J. H. Lawrimore, P. Y. Groisman, N. A. Speranskaya, S. A. Zhuravin, M. J. Menne, T. C. Peterson, and R. W. Malone. 2001. Evaporation Changes over the Contiguous United States and the Former USSR: a Reassessment. *Geophysical Research Letters* 28 (13): 2665-68.

Green, T. R., B. C. Bates, S. P. Charles, and P. M. Fleming. 2007. Physically Based Simulation of Potential Effects of Carbon Dioxide–Altered Climates on Groundwater Recharge. *Vadose Zone Journal* 6: 597-609. doi:10.2136/vzj2006.0099.

Groisman, P. Y., and D. R. Easterling. 1994. Variability and Trends in Total Precipitation and Snowfall over the United States and Canada. *Journal of Climate* 7: 184-205.

Groisman, P., R. Knight, T. Karl, D. Easterling, B. Sun, and J. Lawrimore. 2004. Contemporary Changes of the Hydrological Cycle over the Contiguous United States: Trends Derived from In-Situ Observations. *Journal of Hydrometeorology* 5: 64-85.

GWSP (Global Water System Project). 2005. The Global Water System Project Science Framework and Implementation Activities. Earth System Science Partnership Report No. 3. www.gwsp.org/fileadmin/downloads/GWSP_Report_No_1_Internetversion_01.pdf.

Hannaford, J., and T. J. Marsh. 2007. High-Flow and Flood Trends in a Network of Undisturbed Catchments in the U.K. *International Journal of Climatology* 27. doi:10.10002.joc.1643.

Harris, C., D. Vonder Muhll, K. Isaksen, W. Haeberli, J. L. Sollide, L. King, P. Holmlund, F. Dramis, M. Guglielmin, and D. Palacios. 2003. Warming Permafrost in European Mountains. *Global Planet Change* 39: 215-25.

Held, I. M., and B. J. Soden. 2000. Water Vapour Feedback and Global Warming. *Annual Review of Energy and the Environment* 25: 441-75.

———. **2006.** Robust Responses of the Hydrological Cycle to Global Warming. *Journal of Climate* 19: 5686-99.

Hicke, J. A., G. P. Asner, J. T. Randerson, C. Tucker, S. Los R. Birdsey, J. C. Jenkins, and C. Field. 2002. Trends in North American Net Primary Productivity Derived from Satellite Observations, 1982-1998. *Global Biogeochemical Cycles* 16 (2).

Hisdal, H., K. Stahl, L. M. Tallaksen, and S. Demuth. 2001. Have Streamflow Droughts in Europe Become More Severe or Frequent? *International Journal of Climatology* 21 (1): 317-33.

Hobbins, M. T., and J. A. Ramirez. 2004. Trends in Pan Evaporation and Actual Evapotranspiration across the Conterminous U.S.: Paradoxical or Complementary? *Geophysical Research Letters* 31: L3503. doi:10, 1029/2004GL019846.

Hodgkins, G. A., and R. W. Dudley. 2006a. Changes in Late-Winter Snowpack Depth, Water Equivalent, and Density in Maine, 1926-2004. *Hydrological Processes* 20: 741-51.

———. **2006b.** Changes in the Timing of Winter-Spring Streamflows in Eastern North America, 1913–2002. *Geophysical Research Letters 33:* L06402. doi:10.1029/2005GL025593.

Hodgkins, G. A., R. W. Dudley, and T. G. Huntington. 2003. Changes in the Timing of High River Flows in New England over the 20th Century. *Journal of Hydrology* 278: 244-52.

Holden, Z. A., P. Morgan, M. A. Crimmins, R. K. Steinhorst, and A. M. S. Smith. 2007. Fire Season Precipitation Variability Influences Fire Extent and Severity in a Large Southwestern Wilderness Area, United States. *Geophysical Research Letters* 34 (16).

Holland, M. M., J. Finnis, A. P. Barrett, and M. C. Serreze. 2007. Projected Changes in Arctic Ocean Freshwater Budgets. *Journal of Geophysical Research* 112: G04S55. doi:10.1029/2006JG000354.

Holmes, R. M., J. W. McClelland, P. A. Raymond, B. B. Frazer, B. J. Peterson, and M. Stieglitz. 2008. Lability of DOC Transported by Alaskan Rivers to the Arctic Ocean. *Geophysical Research Letters* 35 (3). doi: 10.1029/2007GL032837.

Hu C., Liu C., Zhou Z., and R. Jayakumar. 2008. Changes in Water and Sediment Loads of Rivers in China. International Research and Training Center on Erosion and Sedimentation, Beijing.

Huntington, T. G. 2006. Evidence for Intensification of the Global Water Cycle: Review and Synthesis. *Journal of Hydrology* 319: 83-95.

Huntington, T. G., G. A. Hodgkins, B. D. Keim, and R. W. Dudley. 2004. Changes in the Proportion of Precipitation Occurring as Snow in New England (1949 to 2000). *Journal of Climate* 17: 2626-36.

ICOLD (International Commission of Large Dams). 2003. *World Register of Dams 2003.* Paris: International Commission of Large Dams.

IPCC (Intergovernmental Panel on Climate Change). 2007. *Climate Change 2007: The Physical Science Basis. Summary for Policymakers.* Contribution of Working Group 1 to the Fourth Assessment Report of the Intergovernmental Panel on Climate Change. Cambridge, UK: Cambridge University Press. http://ipcc-wg1.ucar.edu/wg1/wg1-report.html.

Juen, I., G. Kaser, and C. Georges. 2007. Modelling Observed and Future Runoff from a Glacierized Tropical Catchment (Cordillera Blanca, Perú). *Global and Planetary Change* 59: 37-48.

Kahya, E., and S. Kalayci. 2004. Trend Analysis of Streamflow in Turkey. *Journal of Hydrology* 289: 128-44.

Kang E.-S., Shen Y.-P., Li X., Liu C.-H., Xie Z.-C., Li P.-J., Wang J., Che T., and Wu L.-Z. 2004. Assessment of the Glacier and Snow Water Resources in China, a Report to the Ministry of Water Resources of China. CAREERI/CAS, Lanzhou.

Kirschbaum, M. U. F. 2006. The Temperature Dependence of Organic-Matter Decomposition – Still a Topic of Debate. *Soil Biology and Biochemistry* 38: 2510-18.

Knapp, P. A., and P. T. Soule. 2007. Trends in Midlatitude Cyclone Frequency and Occurrence during Fire Season in the Northern Rockies: 1900-2004. *Geophysical Research Letters* 34: L20707. doi:10.1029/2007GL031216.

Kundzewicz, Z. W., D. Graczyk, T. Maurer, I. Pinskwar, M. Radziejewski, C. Svensson, and M. Szwed. 2004. *Detection of Change in World-Wide Hydrological Time Series of Maximum Annual Flow. GRDC Report Series Report 32.* Koblenz, Germany: Global Runoff Data Centre.

———. **2005.** Trend Detection in River Flow Series: 1. Annual maximum flow. *Hydrological Science Journal* 50 (5): 797-810.

Kulkarni, A. V., I. M Bahuguna, B. P. Rathore, S. K. Singh, S. S. Randhawa, R. K. Sood, and S. Dhar. 2007. Glacial Retreat in Himalaya Using Indian Remote Sensing Satellite Data. *Current Science* 92 (1): 1-10.

Lachenbruch, A. H., and B. V. Marshall. 1986. Changing Climate: Geothermal Evidence from Permafrost in the Alaskan Arctic. *Science* 234: 689-96.

Lal, R., M. Griffin, J. Apt, L. Lave, and N. Granger Morgan. 2004. Managing Soil Carbon. *Science* 304 (5669): 393.

Laternser, M., and M. Schneebeli. 2003. Long-Term Snow Climate Trends of the Swiss Alps (1931–99). *International Journal of Climatology* 23: 733-50.

Lawrence, D. M., and A. G. Slater. 2005. A Projection of Severe Near-Surface Permafrost Degradation during the 21st Century. *Geophysical Research Letters* 32: L24401. doi:10.1029/2005GL025080.

Lawrimore, J. H., and T. C. Peterson. 2000. Pan Evaporation Trends in Dry and Humid Regions of the United States. *Journal of Hydrometeorology* 1 (6): 543-46.

Lehner, B., and P. Döll. 2004. Development and Validation of a Global Database of Lakes, Reservoirs and Wetlands. *Journal of Hydrology* 296: 1-22.

Lempérière, F. 2006. The Role of Dams in the XXI Century: Achieving a Sustainable Development Target. *International Journal on Hydropower & Dams* 13 (3): 99-108.

L'hote, Y., G. Mahe, B. Some, and J. P. Triboulet. 2002. Analysis of a Sahelian Annual Rainfall Index Updated from 1896 to 2000. The Drought Still Goes on. *Hydrological Sciences Journal* 47: 121-36.

Li H., A. Robock, Liu S., Mo X., and P. Viterbo. 2005. Evaluation of Reanalysis Soil Moisture Simulations Using Updated Chinese Soil Moisture Observations. *Journal of Hydrometeorology* 6: 180-93.

Li H., A. Robock, and M. Wild. 2007. Evaluation of IPPC Fourth Assessment Soil Moisture Simulations for the Second Half of the Twentieth Century. *Journal of Geophysical Research* 112: D06106. doi:10.1029/2006JD007455.

Lindstrom, G., and S. Bergstrom. 2004. Runoff Trends in Sweden 1807-2002. *Hydrological Sciences Journal* 49: 69-83.

Lins, H. F. 2008. Challenges to Hydrological Observations. *WMO Bulletin* 57 (1): 55-8.

Lins, H. F., and J. R. Slack. 1999. Streamflow Trends in the United States. *Geophysical Research Letters* 26: 227-30.

———. **2005.** Seasonal and Regional Characteristics of U.S. Streamflow Trends in the United States from 1940 to 1999. *Physical Geography* 26: 489-501.

Liu Shiyin, Ding Yongjian, Shangguan Donghui, Zhang Yong, Li Jing, Han Haidong, Wang Jian, and Xie Changwei. 2006. Glacier Retreat as a Result of Climate Change Due to Warming and Increased Precipitation in the Tarim River Basin, Northwest China. Annals of Glaciology 43 (1): 91-6.

Loaiciga, H. A., J. B. Valdes, R. Vogel, J. Garvey, and H. Schwarz. 1996. Global Warming and the Hydrologic Cycle. *Journal of Hydrology* 174: 83-127.

Lu C., M. Kanamitsum, J. Roads, W. Ebisuzaki, K. Mitchell, and D. Lohmann. 2005. Evaluation of Soil Moisture in the NCEP-NCAR and NCEP-DOE Global Reanalyses. *Journal of Hydrometeorology* 6: 391-408.

Maurer, E. P., A. W. Wood, J. C. Adam, D. P. Lettenmaier, and B. Nijssen. 2002. A Long-Term Hydrologically-Based Data Set of Land Surface Fluxes and States for the Conterminous United States. *Journal of Climate* 15 (22): 3237-51.

Mayewski, P. A., and P. A. Jeschke. 1979. Himalayan and Trans-Himalayan Glacier Fluctuations since AD 1812. *Arctic and Alpine Research* 119 (3): 267-87.

McCabe, G. J., and D. M. Wolock. 2002. A Step Increase in Streamflow in the Conterminous United States. *Geophysical Research Letters* 29. doi:10.1029/2002GL015999.

McDonald, K. C., J. S. Kimball, E. Njoku, R. Zimmermann, and M. Zhao. 2004. Variability in Springtime Thaw in the Terrestrial High Latitudes: Monitoring a Major Control on the Biospheric Assimilation of Atmospheric CO_2 with Spaceborne Microwave Remote Sensing. *Earth Interactions* 8 (Paper No. 20): 1-23.

McNaughton, K. G., and T. W. Spriggs. 1986. A Mixed-Layer Model of Regional Evaporation. *Boundary Layer Meteorology* 34:243-62.

———. **1989.** An Evaluation of the Priestley-Taylor Equation. In *Estimation of Areal Evaporation*, eds. T. A. Black, D. L. Spittlehouse, M. D. Novak, and D. T. Price. Wallingford, UK: International Association of Hydrological Sciences.

Meybeck, M. 1995. Global Distribution of Lakes. In *Physics and Chemistry of Lakes*, eds. A. Lerman, D. M. Imboden, and J. R. Gat. Berlin: Springer.

Milly, P. C. D., J. Betancourt, M. Falkenmark, R. M. Hirsch, Z. W. Kundzewicz, D. P. Lettenmaier, and R. J. Stouffer. 2008. Stationarity Is Dead: Whither Water Management? *Science* 319: 573-74.

Milly, P. C. D., and K. A. Dunne. 2001. Trends in Evaporation and Surface Cooling in the Mississippi River Basin. *Geophysical Research Letters* 28: 1219-22.

Milly, P. C. D., K. A. Dunne, and A. V. Vecchia. 2005. Global Pattern of Trends in Streamflow and Water Availability in a Changing Climate. *Nature* 438: 347-50.

Mitsch, W. J., and J. G. Gosselink. 2000. *Wetlands.* New York: John Wiley and Sons.

Montgomery, D. R. 2007. Soil Erosion and Agricultural Sustainability. *Proceedings of the National Academy of Sciences* 104 (33): 13268-72.

Mote, P. W. 2003. Trends in Snow Water Equivalent in the Pacific Northwest and Their Climatic Causes. *Geophysical Research Letters* 30. doi:10.1029/2003GL017258.

Mote, P. W., A. F. Hamlet, M. P. Clark, and D. P. Lettenmaier. 2005. Declining Mountain Snowpack in Western North America. *Bulletin of American Meteorological Society* 86: 39-49.

Mudelsee, M., M. Börngen, G. Tetzlaff, and U. Grünewald. 2003. No Upward Trends in the Occurrence of Extreme Floods in Central Europe. *Nature* 425: 166-9.

Nemani, R. R., C. D. Keeling, Hashimoto H., W. M. Jolly, S. C. Piper, C. J. Tucker, R. B. Myneni, and S.W. Running. 2003. Climate-Driven Increases in Global Terrestrial Net Primary Production from 1982 to 1999. *Science* 300 (5625): 1560-3.

New, M., M. Todd, M. Hulme, and P. D. Jones. 2001. Precipitation Measurements and Trends in the Twentieth Century. *International Journal of Climatology* 21: 1899-1922.

Nijssen, B. N., R. Schnur, and D. P. Lettenmaier. 2001. Global Retrospective Estimation of Soil Moisture Using the VIC Land Surface Model, 1980-1993. *Journal of Climate* 14 (8): 1790-1808.

Oberman, N. G., and G. G. Mazhitova. 2001. Permafrost Dynamics in the Northeast of European Russia at the End of the 20th Century. *Norwegian Journal of Geography* 55: 241-4.

Ohmura, A., and M. Wild. 2002. Is the Hydrological Cycle Accelerating? *Science* 298 (5597): 1345-6.

Omran, M. A. 1998. Analysis of Solar Radiation over Egypt. *Theoretical and Applied Climatology* 67: 225-40.

Osterkamp, T. E. 2005. The Recent Warming of Permafrost in Alaska. *Global Planet Change* 49: 187-202. doi: 10.1016/j.gloplacha.2005.09.001.

Paul, F., A. Kaab, M. Maisch, T. Kellenberger, and W. Haeberli. 2004. Rapid Disintegration of Alpine Glaciers Observed with Satellite Data. *Geophysical Research Letters* 31: L21402. doi:10.1029/2004GL020816.

Pavlov, A. V. 1996. Permafrost-Climate Monitoring of Russia: Analysis of Field Data and Forecast. *Polar Geography* 20 (1): 44-64.

Peterson, B. J., R. M. Holmes, J. W. McClelland, C. J. Vörösmarty, R. B. Lammers, A. I. Shiklomanov, and S. Rahmstorf. 2002. Increasing River Discharge to the Arctic Ocean. *Science* 298: 2171-3.

Raymond, P. A., J. W. McClelland, R. M. Holmes, A. V. Zhulidov, K. Mull, B. J. Peterson, R. G. Striegl, G. R. Aiken, and T. Y. Gurtovaya. 2007. Flux and Age of Dissolved Organic Carbon Exported to the Arctic Ocean: A Carbon Isotopic Study of the Five Largest Arctic Rivers. *Global Biogeochemical Cycles* 21 (4). doi: 10.1029/2007GB002934.

Robock, A., M. Mu, K. Vinnikov, I. Trofimova, and T. Adamenko. 2005. Forty-Five Years of Observed Soil Moisture in the Ukraine: No Summer Desiccation (Yet). *Geophysical Research Letters* 32: L03401. doi:10.1029/2004GL021914.

Rodell, M., Chen J., H. Kato, J. Famiglietti, J. Nigro, and C. Wilson. 2006. Estimating Groundwater Storage Changes in the Mississippi River Basin (USA) Using GRACE. *Hydrogeology Journal.* doi:10.1007/s10040-006-0103-7. ftp://ftp.csr.utexas.edu/pub/ggfc/papers/Rodell2006_HJ.pdf.

Roderick, M. L., and G. D. Farquhar. 2002. The Cause of Decreased Pan Evaporation over the Past 50 Years. *Science* 298: 1410-11.

———. **2004.** Changes in Australian Pan Evaporation from 1970 to 2002. *International Journal of Climatology* 24: 1077-90.

———. **2005.** Changes in New Zealand Pan Evaporation from the 1970s. *International Journal of Climatology* 25: 2013-39.

Roderick, M. L., L. D. Rotstayn, G. D. Farquhar, and M. T. Hobbins. 2007. On the Attribution of Changing Pan Evaporation. *Geophysical Research Letters* 34: L17403. doi:10.1029/2007GL031166.

Scherrer, S. C., and C. Appenzeller. 2004. Trends in Swiss Alpine Snow Days: The Role of Local- and Large-Scale Climate Variability. *Geophysical Research Letters* 31. doi:10.1029/2004GL020255.

Scibek, J., and D. M. Allen. 2006. Comparing Modeled Responses to Two High-Permeability, Unconfined Aquifers to Predicted Climate Change. *Global and Planetary Change* 50: 50-62.

Sharkhuu, N. 2003. Recent Changes in the Permafrost of Mongolia. In *Proceedings of the 8th International Conference on Permafrost, 21-25 July 2003, Zurich, Switzerland,* eds. M. Phillips, S. M. Springman, and L. U. Arenson. Lesse, The Netherlands: A.A. Balkema.

Sheffield, J., and E. Wood. 2008. Global Trends and Variability in Soil Moisture and Drought Characteristics, 1950-2000, from Observation Driven Simulations of the Terrestrial Hydrologic Cycle. *Journal of Climate* 21: 432-58.

Shenbin, C., L. Yunfeng, and A. Thomas. 2006. Climatic Change on the Tibetan Plateau: Potential Evapotranspiration Trends from 1961-2000. *Climate Change* 76: 291-319.

Shi Y. F. 2005. *Glacial Inventory of China (Synthesis volume).* Shanghai: Science Popularization Press.

Shiklomanov, I. A., and J. Rodda. 2003. *World Water Resources at the Beginning of the 21st Century.* Cambridge, UK: Cambridge University Press.

Shuttleworth, W. J. 2008. Evapotranspiration Measurement Methods 2008. *Southwest Hydrology* 7 (1): 22-3. www.swhydro.arizona.edu/archive/V7_N1/.

Shuttleworth, W. J, A. S. Capdevila, M. L. Roderick, and R. Scott. Forthcoming. On the Theory Relating Changes in Area-Average and Pan Evaporation. *Quarterly Journal of the Royal Meteorological Society.*

Smith, S. L., M. M. Burgess, D. Riseborough, and F. M. Nixon. 2005. Recent Trends from Canadian Permafrost Monitoring Network Sites. *Permafrost and Periglacial Processes* 16: 19-30.

Smith, L. C., T. M. Pavelsky, G. M. MacDonald, A. I. Shiklomanov, and R. B. Lammers. 2007. Rising Minimum Flows in Northern Eurasian Rivers: A Growing Influence of Groundwater in the High-Latitude Hydrologic Cycle. *Journal Geophysical Research* 112: G04S47. doi:10.1029/2006JG000327.

Smith, N. V., S. S. Saatchi, and T. Randerson. 2004. Trends in High Latitude Soil Freeze and Thaw Cycles from 1988 to 2002. *Journal of Geophysical Research* 109: D12101. doi:10.1029/2003JD004472.

Smith, L. C., Sheng Y., G. M. MacDonald, and L. D. Hinzman. 2005. Disappearing Arctic Lakes. *Science* 308: 1429.

Srivastava, D., K. R. Gupta, and S. Mukerji, eds. 2003. *Proceedings: Workshop on Gangotri Glacier: Lucknow 26-28 March 2003: Special Publication No. 80. Geological Survey of India.* New Delhi: Vedams eBooks.

Stallard, R. F. 1998. Terrestrial Sedimentation and the Carbon Cycle: Coupling Weathering and Erosion to Carbon Burial. *Global Biogeochemical Cycles* 12: 231-57.

Stanhill, G., and S. Cohen. 2001. Global Dimming: A Review of the Evidence for A Widespread and Significant Reduction in Global Radiation with Discussion of its Probable Causes and Possible Agricultural Consequences. *Agriculture and Forest Meteorology* 107: 255-78.

Stewart, I. T., D.R. Cayan, and M. D. Dettinger. 2005. Changes toward Earlier Streamflow Timing across Western North America. *Journal of Climatology* 18: 1136-55.

Surazakov, A. B., V. B. Aizen, and S. A. Nikitin. 2007. Glacier Changes in the Siberian Altai Mountains, Ob River Basin (1952-2006) Estimated with High-Resolution Imagery. *Environmental Research Letters* 2 04 5017. doi:10.1088/1748-9326/214/045017. http://dx.doi.org/10.1088/1748-9326/2/4/045017.

Svensson, C., Z. W. Kundzewicz, and T. Maurer. 2005. Trend Detection in River Flow Series: 2. Flood and low-flow index series. *Hydrological Science Journal* 50: 811-24.

Swenson, S., and J. Wahr. 2006. Post-Processing Removal of Correlated Errors in GRACE data. *Geophysical Research Letters* 33: L08402. doi:10.1029/2005GL025285.

Syed, T. H., J. S. Famiglietti, Chen J., M. Rodell, S. I. Seneviratne, P. Viterbo, and C. R. Wilson. 2005. Total Basin Discharge for the Amazon and Mississippi River Basins from GRACE and a Land-Atmosphere Water Balance. *Geophysical Research Letters* 32: L24404. doi:10.1029/2005GL024851.

Szilagyi, J., G. G. Katul, and M. B. Parlange. 2002. Evapotranspiration Intensifies over the Conterminous United States. *Journal of Water Resources Planning and Management* 127: 354-62.

Tapley B. D., S. V. Bettadpur, M. Watkins, and C. Reigber. 2004. The Gravity Recovery and Climate Experiment: Mission Overview and Early Results. *Geophysical Research Letters* 31: L09607. doi:10.1029/2004GL019920.

Thomas, A. 2000. Spatial and Temporal Characteristics of Potential Evapotranspiration Trends over China. *International Journal of Climatololgy* 20: 381-96.

Trenberth, K. E. 1999. Conceptual Framework for Changes of Extremes of the Hydrological Cycle with Climate Change. *Climatic Change* 42: 327-39.

Trenberth, K. E., L. Smith, T. Qian, A. Dai, and J. Fasullo. 2007. Estimates of the Global Water Budget and Its Annual Cycle Using Observational and Model Data. *Journal of Hydrometeorology* 8 (4): 758.

Vaccaro, J. 1992. Sensitivity of Groundwater Recharge Estimates to Climate Variability and Change, Columbia Plateau, Washington. *Journal of Geophysical Research* 97: 2821-33.

van Roosmalen, L., B. S. B. Christensen, and T. O. Sonnenborg. 2007. Regional Differences in Climate Change Impacts on Groundwater and Stream Discharge in Denmark. *Vadose Zone Journal* 6: 554-71. doi:10.2136/vzj2006.0093.

Vonder Muhll, D., J. Notzli, K. Makowski, and R. Delaloye. 2004. Permafrost in Switzerland in 2000/2001 and 2001/2002. Glaciological Report (Permafrost) No. 2/3, Glaciological Commission of the Swiss Academy of Sciences, Zurich.

Vörösmarty, C. J., M. Meybeck, B. Fekete, and K. Sharma. 1997. The Potential Impact of Neo-Cartorization on Sediment Transport by the Global Network of Rivers. In *Proceedings of the Fifth IAHS Scientific Assembly at Rabat, Morocco, April-May 1997*, eds. D. E. Walling and J.-L. Probst. Wallingford, UK: International Association of Hydrological Sciences.

Walter, M. T., D. S. Wilks, J.-Y. Parlange, and R. L. Schneider. 2004. Increasing Evapotranspiration from the Conterminous United States. *Journal of Hydrometeorology* 5: 405-08.

Walter, K. M., S. A. Zimov, J. P. Chanton, D. Verbyla, and F. S. Chapin. 2006. Methane Bubbling from Siberian Thaw Lakes as a Positive Feedback to Climate Warming. *Nature* 443. doi:10.1038/nature05040.

Wentz, F. J., L. Ricciardulli, K. Hilburn, and C. Mears. 2007. How Much More Rain Will Global Warming Bring? *Science* 317: 233-5.

Westerling, A. L., H. G. Hidalgo, D. R. Cayan, and T.W. Swetnam. 2006. Warming and Earlier Spring Increase Western U.S. Forest Wildfire Activity. *Science* 313. doi:10.1126/science.1128834.

Willett, K. M., N. P. Gillett, P. D. Jones, and P. W. Thorne. 2007. Attribution of Observed Surface Humidity Changes to Human Influence. *Nature* 449: 710-13.

Wood, A. W., L. R. Leung, V. Sridhar, and D. P. Lettenmaier. 2004. Hydrologic Implications of Dynamical and Statistical Approaches to Downscaling Climate Model Outputs. *Climatic Change* 62: 189-216.

Worrall, F., and T. P. Burt. 2007. Trends in DOC Concentration in Great Britain. *Journal of Hydrology* 346: 81-92.

Wu Q. B., and Zhang T. 2008. Recent Permafrost on the Qinghai-Tibetan Plateau. *Journal of Geophysical Research* 113: D13108. doi:10.1029/2007JD009539.

Xu C. -Y., G. Lebing, Jiang T., Chen D., and V. P. Singh. 2006. Analysis of the Spatial Distribution and Temporal Trend of Reference Evaporation and Pan Evaporation in Changjiang (Yangtze River) Catchment. *Journal of Hydrology* 327: 81-93.

Yao T. D., Pu J. C., Tian L. D., Yang W., Duan K. Q., Ye Q. H., and L. G. Thompson. 2007. Recent Rapid Retreat of the Naimona'nyi Glacier in Southwestern Tibetan Plateau (in Chinese with English Abstract). *Journal of Glaciology and Geocryology* 29 (4): 503-8.

Yao T. D., Wang Y. Q., Liu S. Y., Pu J. C., Shen Y. P., and Lu A. X. 2004. Glacial Retreat High Asia and Its Influence on Water Resource in Northwest China. *Science in China* 47 (12): 1065-75.

Ye H. 2000. Decadal Variability of Russian Winter Snow Accumulation and Its Associations with Atlantic Sea Surface Temperature Anomalies. *International Journal of Climatology* 20: 1709-28.

Yeh P. J. -F., S. C. Swenson, J. S. Famiglietti, and M. Rodell. 2006. Groundwater Storage Changes Inferred from the Gravity Recovery and Climate Experiment (GRACE). *Water Resources Research* 42: W12203. doi:10.1029/2006WR005374.

Yiou P., P. Ribereau, P. Naveau, M. Nogaj, and R. Brazdil. 2006. Statistical Analysis of Floods in Bohemia (Czech Republic) since 1825. *Hydrological Sciences Journal* 51: 930-45.

Zhang T. 2005. Influence of the Seasonal Snow Cover on the Ground Thermal Regime: An Overview. *Reviews of Geophysics* 43: RG4002. doi:10.1029/2004RG000157.

Zhang T., R. G. Barry, K. Knowles, F. Ling, and R. L. Armstrong. 2003. Distribution of Seasonally and Perennially Frozen Ground in the Northern Hemisphere. In *Proceedings of the 8th International Conference on Permafrost, Zurich, Switzerland, 21-25 July 2003*, eds. M. Phillips, S. M. Springman, and L. U. Arenson. Lesse, The Netherlands: A. A. Balkema.

Zhang T., O. W. Frauenfeld, M. C. Serreze, A. Etringer, C. Oelke, J. McCreight, R. G. Barry, D. Gilichinsky, Yang D., Ye H., Ling F., and S. Chudinova. 2005. Spatial and Temporal Variability of Active Layer Thickness over the Russian Arctic Drainage Basin. *Journal of Geophysical Research* 110: D16101. doi:10.1029/2004JD005642.

Zhang X., K. D. Harvey, W. D. Hogg, and T. R. Yuzyk. 2001. Trends in Canadian Streamflow. *Water Resources Research* 37: 987-98.

Zhang T., Roger G. Barry, K. Knowles, J. A. Heginbottom, and J. Brown. 1999. Statistics and Characteristics of Permafrost and Ground Ice Distribution in the Northern Hemisphere. *Polar Geography* 23 (2): 147-69.

Zhang X., L. A. Vincent, W. D. Hogg, and A. Niitsoo. 2000. Temperature and Precipitation Trends in Canada during the 20th Century. *Atmosphere-Ocean* 38: 395-429.

Zhao L., Ping C. -L., Yang D., Cheng G., Ding Y., and Liu S. 2004. Changes of Climate and Seasonally Frozen Ground over the Past 30 Years in Qinghai-Xizang (Tibetan) Plateau, China, Global Planet. *Change* 43: 19-31.

Changes in the global water cycle

Chapter 12

Evolving hazards – and emerging opportunities

Author: Bruce Stewart

Contributors: Yoganath Adikari, Vladimir Aizen, Ali Chavoshia, Hu Chunhong, Pablo García, M. Gopalakrishnan, Hege Hisdal, A.W. Jayawardena, Zbigniew Kundzewicz, Peter Letitre, Robert Naiman, Julian Olden, Roberto Pizzaro, Erich Plate, Carmen Revenga, James Robertson, Léna Salamé, Kyoji Sassa, Pratap Singh, Vladimir Smakhtin, Lena M. Tallaksen, Jos G. Timmerman, Carel De Villeneuve, Des Walling and Maciej Zalewski

Coordinator: Andras Szöllösi-Nagy (UNESCO)

Facilitator: Denis Hughes

Key messages

- In many places climate-related events have become more frequent and more extreme. In developing countries extreme floods can result in many deaths, while in developed countries they can result in billions of dollars in damages. More intense droughts in the past decade, affecting an increasing number of people, have been linked to higher temperatures and decreased precipitation but are also frequently a consequence of the mismanagement of resources and the neglect of risk management.
- Changes in flow and inputs of chemical and biological waste from human activity have altered the water quality and ecological functioning of many of the world's rivers. Global warming is expected to have substantial effects on energy flows and matter recycling through its impact on water temperature, resulting in algal blooms, increases in toxic cyanobacteria bloom and reductions in biodiversity.
- In areas of increasing water stress groundwater is an important buffer resource, capable of responding to increased water demands or of compensating for the declining availability of surface water.

A review of recent changes in the global water cycle that analysed more than 100 studies (based on observations) found rising global and regional trends in runoff, floods and droughts, and other climate-related events and variables in the second half of the 20th century that together support the perception of an intensification of the hydrologic cycle.[1] Meanwhile, substantial uncertainty remains about trends of hydroclimate variables due to differences in responses by variables and across regions, major spatial and temporal limitations in data (see chapter 13) and the effects of modifications in water resources development (withdrawals, reservoirs, land use changes and so on) on flow regimes.

Hazards vary with climate regions

Just as regions are experiencing different degrees of change related to climate variations and population and development pressures, so are they responding differently to changes in hydrologic extremes. This chapter identifies the areas that are most sensitive to changes in extremes and hazards and those that are likely to experience the most negative impacts on water resources.

- *Deserts* face conflicting influences under climate change: potentially seeing more vegetation with higher carbon dioxide levels, but overall facing increases in drought and

Regions in the transition zone between major climate zones are susceptible to drought and thus to potential changes in climate

temperatures. With an already fragile environment, desert ecosystems could experience severe impacts.

- *Grasslands* are influenced by precipitation, both its total amount and its variability. Changing seasonal variability is important even when total precipitation is rising, and declining summer rainfall could damage grassland fauna.

- *Mediterranean ecosystems* are diverse and vulnerable, susceptible to changes in water conditions. Even with a temperature rise of 2°C, the Southern Mediterranean may lose 60%-80% of species.

- *Tundra and Arctic regions* face the loss of permafrost and the potential for methane release with greater warming at the poles.

- *Mountains* are seeing shortened and earlier snow and ice melt and related changes in flooding. At higher altitudes increased winter snow can lead to delayed snow melt.

- *Wetlands* will be negatively affected where there is decreasing water volume, higher temperatures and higher-intensity rainfall.

Some studies have used climate models and greenhouse gas emission scenarios from the recent assessment by the Intergovernmental Panel on Climate Change (IPCC) to forecast differences between climate zones today and in 2100.[2] They have found that under both high- and low-emission scenarios many regions would experience biome-level changes, suggesting that rainforest, tundra or desert areas may no longer have the same type of vegetation by 2100 because of climate shifts (box 12.1).[3] By the end of the 21st century large portions of the Earth's surface may experience climates not found today, and some 20th century climate characteristics may disappear.

Regions in the transition zone between major climate zones (particularly between temperate and dry climates) are susceptible to drought and thus to potential changes in climate. A shift in climate may create a new transition zone, with unknown feedback mechanisms. A northward shift is observed in Southern Europe, causing a decline in summer precipitation in Central and Eastern Europe. Climate models consistently predict an increase in summer temperature variability in these areas and attribute it mainly to strong land-atmosphere interactions. This could cause more droughts and heat waves in this and other mid-latitude regions. Regional climate models suggest that towards the end of the 21st century about every second summer could be as warm as or warmer (and as dry as or dryer) than the summer of 2003 in Europe.

Snow cover has decreased in most regions, especially in spring and summer. Snow cover in the Northern Hemisphere observed by satellite from 1966 to 2005 decreased in every month except November and December, with a step-wise drop of 5% in annual mean in the late 1980s. In the Southern Hemisphere the few long records or proxies show mostly decreases or no changes in the past 40 years or more.[4]

The Himalayan region is highly vulnerable to climate change because its major river drainage systems depend on substantial contributions from snow and glacier melt. In India the river systems originating from the Himalayas (Ganges-Brahmaputra and Indus) contribute more than 60% to the total annual runoff for all the rivers of India. These river systems hold immense potential as a future water source and drain the major plains of the country. Some Himalayan rivers receive more than half their flow from snow and glacier melt runoff near the foothills of the Himalayas. Melting of glaciers and a reduction in solid precipitation in mountain regions would directly affect water resources for domestic supplies, irrigated agriculture, hydropower generation and other water-dependent activities.

Changes in average streamflow

While hazards are normally associated with hydrologic extremes, changes in average streamflow, especially in already water-stressed areas, could cause substantial risks to human activities. The IPCC report suggests that by 2050 annual average runoff will have increased by 10%-40% at high latitudes and decreased by 10%-30% over some dry regions at mid-latitudes and semi-arid low latitudes.[5] However, in many water-scarce regions land use change and increasing water resources development and use could mask the effects of climate change. At high latitudes, where an increase in annual flow is predicted, the impact on low flow and drought depends on the seasonal distribution of precipitation, the storage capacity of the catchment (ability to take advantage of higher winter precipitation), changes in evapotranspiration and the length of the growing season.

Box 12.1 A global perspective on regional vegetation and hydrologic sensitivities to climate change

The Mapped Atmosphere-Plant-Soil System (MAPSS), a biogeographic model, predicts changes in vegetation leaf area index, site water balance, runoff and biome boundaries. Global equilibrium impacts on these ecosystem properties were simulated under five general circulation model (GCM) potential climate scenarios with doubled carbon dioxide concentration.

Leaf area index is the ratio of the vegetation's leaf surface area per unit of ground area. The greater the leaf surface area, the more rapidly the vegetation will extract soil water. Most ecosystems will grow as much leaf area as can be supported by the water available during an average growing season. Thus, under normal conditions many ecosystems are very near a drought threshold. Warming lengthens the growing season, and evaporative demand increases exponentially with rising temperature. Consequently, entire landscapes can extract all soil moisture before the end of the growing season and become susceptible to sudden decline under rapid warming, especially if coincident with a short-term drying trend. Regional increases in precipitation and benefits to plant water use efficiency from elevated carbon dioxide concentrations can offset the increased drought stress in some ecosystems. However, at global scales most ecosystem models show that the rapid increases in evaporative demand can overwhelm these benefits over large areas, possibly within the next few decades.

Average simulated change in vegetation leaf area index from five general circulation model scenarios

With no change in water use efficiency

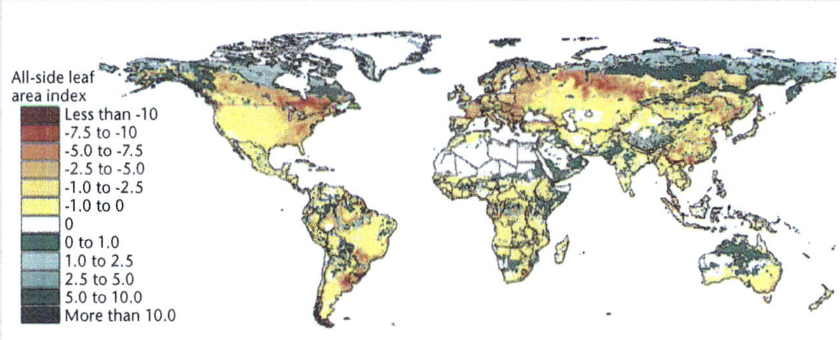

With increase in water use efficiency

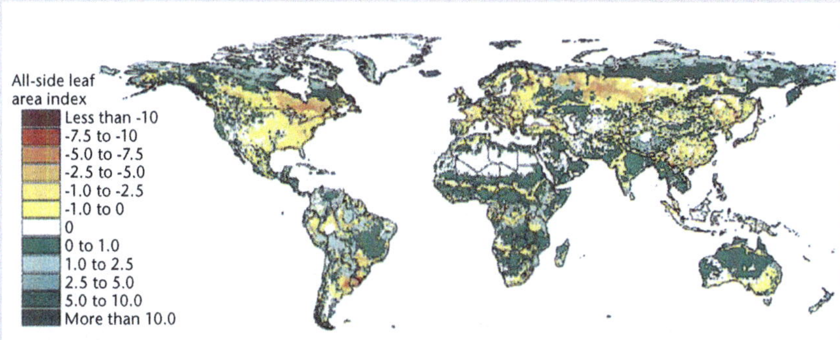

Note: All-side leaf area index is the leaf area (all-sides) per unit ground area; it is a non-dimensional (unitless) measure.
Source: Neilson and Marks 1994.

Regional patterns of vegetation change and annual runoff are surprisingly consistent across the five GCM scenarios, considering the relative lack of consistency in predicted changes in regional precipitation patterns (see bottom map). Eastern North America and Eastern Europe to western Russia show particular sensitivity to drought-induced forest decline. Uncertainties about potential evapotranspiration and vegetation water use efficiency can alter the sign of the simulated regional responses, but the relative responses of adjacent regions appear to be a function largely of the background climate rather than of the vagaries of the GCMs and are intrinsic to the landscape. Thus, spatial uncertainty maps can be drawn even under the current generation GCMs.

Source: Ronald P. Neilson, Department of Forest Ecosystems and Society, Oregon State University; Neilson et al. 1998; Scholze et al. 2006.

The IPCC report notes increased annual runoff and earlier spring peak discharge in many glacier- and snow-fed rivers, indicating a regime shift for some rivers. This trend is projected to continue in response to increasing temperatures, resulting in increased summer streamflow in downstream regions receiving melt water from major mountain ranges, followed eventually by reduced streamflow.

Changes in extreme events

Extreme water-related events can have positive as well as negative impacts. They recharge natural ecosystems, providing more abundant water for food production, health and sanitation (box 12.2). In the lower Mekong River delta, for example, Cambodians trap water and nutrients carried down by the sediment during flood periods and use them to replenish rice paddies. Floods can be important to the aquatic and riparian ecology, as demonstrated during the artificial flooding of the Colorado River at the Grand Canyon in the United States. But extreme water-related events also destroy lives and property. The most common extreme events are floods and droughts.

Floods and flooding

With global climate change and projected increases in global temperature, scientists generally agree that the hydrologic cycle will intensify and that extremes will become more common.

Box 12.2 Managing urban stormwater in association with land use and land-cover planning can convert a nuisance into a resource

Urban areas cause substantial changes in stormwater hydrology, increasing runoff volumes and peak discharges and altering water quality. In the traditional view urban runoff is a nuisance to be removed as quickly and effectively as possible. However, new studies and initiatives in several countries around the world have shown that stormwater can become a resource, not merely a nuisance. Optimal management of runoff is determined by local conditions and can include recharge of aquifers, retention and detention to improve water quality and reduce downstream flood impacts and the cost of drainage systems, and on-site use of the water (rainwater-harvesting) to irrigate and enhance the urban environment. Stormwater management can be exercised from the household level to the entire built area of a city.

Traditionally planners decide on land uses and land cover and then give engineers the task of designing drainage systems to remove the runoff. In a framework known as water-sensitive planning, water considerations are incorporated in land use and land-cover planning from the outset, following best management practices on the distribution of pervious and impervious land cover and on constructed facilities for capturing, detaining, storing and infiltrating runoff. Efficient water use and conservation, recycling of grey water and possible use of treated wastewater should also be incorporated in planning.

With some effort, water-sensitive designs can also be applied to existing urban areas. By connecting roof drains to pervious spaces, constructing low walls around properties to make them into mini-detention basins that store and infiltrate runoff and directing excess water to playgrounds, planners can make better use of stormwater runoff and reduce flooding damages in open public spaces and parks. In the Coastal Plain of Israel, for example, where annual rainfall is just 500 millimetres, it has been estimated that the aquifer recharge could be increased by about 25,000-77,000 cubic metres per square kilometre of urban area by connecting roof drains to a 15% pervious area on the property and surrounding the property with a low (approximately 20 centimetres) wall.

Source: Carmon and Shamir 1997.

The moisture-holding capacity of the atmosphere has been increasing at a rate of about 7% per 1°C of warming, creating the potential for heavier precipitation. There have likely been increases in the number of heavy precipitation events in many land regions – consistent with a warming climate and the observed increase in atmospheric water vapour, even where total precipitation has declined.

In developing countries extreme floods can result in many deaths, while in developed countries extreme floods cause material damage in the billions and tens of billions of dollars (see table 12.1 for examples). Destructive floods in the last decade across the globe have led to record high material damage. Other extreme high-impact water-related events not listed in table 12.1 are floods in Europe in 1997 and 2002 and floods in China in 1996 ($26 billion in material damage) and 1998 ($30 billion in material damage). Yearly economic losses from extreme events rose tenfold between the 1950s and 1990s in inflation-adjusted dollars.[6]

The question is whether the frequency and magnitude of flooding are also increasing and, if so, whether that is in response to climate variability and change. Disaster losses, mostly weather and water related, have grown much more rapidly than population or economic growth, suggesting a climate change factor. Globally, the number of great inland flood catastrophes was twice as large per decade between 1996 and 2005 as between 1950 and 1980, and economic losses were five times as great. The dominant drivers of these upward trends are socioeconomic factors, such as population growth, land use change and greater use of vulnerable areas.

Documented trends in floods show no evidence for a globally widespread change. One study identified an apparent increase in the frequency of 'large' floods (exceeding 100-year return period levels) in 16 large basins across much of the globe during the 20th century.[7] Analyses of long time series of monthly river flow data showed that seven of eight 100-year floods occurred in the more recent half of the records.

However, subsequent studies have provided less widespread evidence. A global change detection study does not support the hypothesis of a global increase of annual maximum river flows.[8] The study found increases in 27 cases, decreases in 31 cases and no trend in the remaining 137 cases of the 195 catchments examined worldwide. Of the 70 time series for Europe only 20 show statistically significant changes (11 increases and 9 decreases). However, the overall maxima for the period (1961-2000) occurred more frequently (46 times) in the later subperiod, 1981-2000, than in the earlier subperiod (24 times), 1961-80. Evidence is stronger for changes in the timing of floods, with increasing late autumn and winter floods. Fewer ice-jam related floods have been observed in Europe.

Low flows and droughts

Climate change is expected to influence precipitation, temperature and potential evapotranspiration and, through their combined effects, to influence the

Table 12.1 **Examples of major floods and flooding worldwide, 1860-2008**

Date	Location	Meteorological conditions	Peak discharge (cubic metres per second)	Impact material damage (US$ millions)	Human losses
January 2008	Zambezi River, Mozambique	Heavy torrential precipitation in Mozambique and neighbouring countries	3,800	2	20 dead, 113,000 displaced
May 2006	Lower Yukon, United States	Snowmelt, ice-jam break-up	na	na	na
April-May 2003	Santa Fe, Argentina	Saturated soil due to heavy precipitation in summer 2002 and April 2003	4,100	na	22 dead, 161,500 displaced
February 2000	Limpopo River, Mozambique	Extreme precipitation in tropical depression, enforced with torrential rain of three cyclones	10,000	na	700 dead, 1,500,000 displaced
July 1997	Czech Republic	Saturated ground after extreme long-lasting precipitation and extreme precipitation	3,000	1.8	114 dead, 40,000 displaced
June 1997	Brahmaputra River, Bangladesh	Torrential monsoon rains during monsoon season	10,200	400	40 dead, 100,000 displaced
March-April 1997	Red River, United States	Heavy rains and snowmelt	3,905	16,000	100,000 homes flooded, 50,000 displaced
November 1996	Subglacial Lake Grímsvötn, Iceland	Jökulhlaup flood	50,000	12	na
February 1996	West Oregon, United States	Extreme spring snowmelt and heavy spring precipitation	na	na	9 dead, 25,000 displaced
July 1995	Athens, Greece	Storm of a short duration and extreme intensity	650	na	50,000 displaced
November 1994	Po River, Italy	Cold front associated with cyclonic circulation and heavy rainfall	11,300	na	60 dead, 16,000 displaced
February 1994	Meuse River, Europe	Heavy rain due to low pressure system	3,100	na	na
September 1993	Mississippi River, United States	Heavy precipitation in June and July; saturated soil due to extremely high precipitation	na	15,000	50 dead, 75,000 displaced
November 1988	Hat Yai City, Thailand	Brief torrential monsoon rain	na	172	664 dead, 301,000 displaced
January 1983	Northern Peru	El Niño situation with heavy rains	3,500	na	380 dead, 700,000 displaced
August 1979	Machu River, India	Exceptionally heavy rainfall, swollen river, resulting in collapse of the Matchu Dam	16,307	100	1,500 dead, 400,000 displaced
June-September 1954	Yangtze River, China	Intensive rainfall over several months	66,800	na	30,000 dead, 18,000 displaced
January 1953	North Sea, Netherlands	High spring tide and a severe European windstorm	na	504	1,835 dead, 100,000 displaced
January 1910	Seine River, France	Very wet period for six months followed by heavy rains in January	460	na	200,000 displaced
May 1889	Johnstown, Pennsylvania, United States	Extremely heavy rainfall due to storm followed by breach of dike	na	17	2,200 dead
July 1860	Eastern Norway	Frost and heavy snowfall followed by snowmelt and heavy precipitation	3,200	na	12 dead

na is not available or not applicable.

Source: Compiled by Siegfried Demuth, International Hydrological Programme, UNESCO.

In the past three decades droughts have become more widespread, more intense and more persistent

occurrence and severity of droughts. But it is difficult to disentangle the impacts of climate change from those of other human influences (engineered effects and land use changes) and multidecadal climate variability. More intense droughts, affecting more people and linked to higher temperatures and decreased precipitation, have been observed in the 21st century, in Europe and globally.[9] A similar pattern is found for heat waves. The high pressure system that developed over Western Europe in 2003 blocked moist air masses from the west and allowed warm, dry air masses from Northern Africa to move northwards. The result was large precipitation deficits and record-breaking temperatures across most of Central and Southern Europe, with drought conditions lasting from March to September.

Several summaries of observed and predicted impacts of climate change on hydrologic droughts have been published.[10] A study of spatial and temporal changes in streamflow droughts using a dataset of more than 600 daily European streamflow records from the European Water Archive of the UNESCO Flow Regime from International Experimental Data (FRIEND) detected no significant changes for most stations.[11] However, distinct regional differences were found. For 1962-90, examples of increasing drought were found in Spain, the eastern part of Eastern Europe and large parts of the United Kingdom, whereas examples of decreasing drought were found in large parts of Central Europe and the western part of Eastern Europe.

These trends in streamflow drought could be explained largely by changes in precipitation or artificial hydrologic influences in the catchment. However, the period analysed and the selection of stations can also influence regional patterns. Recent trend studies of long time series in the Czech Republic show that, following a catastrophic flood in 2002, extreme hydrologic drought occurred in 2003 and 2004 as a result of an increasing trend in air temperature and a long-term decline in precipitation, especially during the summer months.[12] Extreme droughts occurred in Europe in 2003 and 2006. All these recent droughts could well change the spatial pattern of droughts found for Europe in the earlier study.

The heat wave and drought between June and mid-August 2003 in Europe were accompanied by annual precipitation deficits of up to 300 millimetres. Vegetation and ecosystems suffered heat and drought stress, and record wildfires were experienced (more than 5% of the forest area of Portugal burned). Gross primary production of terrestrial ecosystems across Europe fell to an estimated 30% of normal. Damages resulting from agricultural crop losses and higher production costs were estimated at more than €13 billion. Major rivers, including the Danube, Loire, Po and Rhine, were at record low levels, disrupting inland navigation, irrigation and power-plant cooling. Extreme glacier melt mitigated the effects of low rainfall and high evaporation on streamflows in rivers partly fed by glaciers, such as the Danube and the Rhine.

In the past three decades droughts have become more widespread, more intense and more persistent due to decreased precipitation over land and rising temperatures, resulting in enhanced evapotranspiration and drying. The occurrence of droughts seems to be determined largely by changes in sea surface temperatures, especially in the tropics, through associated changes in atmospheric circulation and precipitation. In the western United States diminishing snow pack and resultant reductions in soil moisture also appear to be factors. In Australia (box 12.3) and Europe the extremely high temperatures and heat waves accompanying recent droughts have implied direct links to global warming. Sahelian droughts have led to severe losses of livestock, with losses as high as 62% observed in part of Ethiopia in 1998-99.[13] Globally, very dry areas (land areas with a Palmer Drought Severity Index of 3.0 or less) have more than doubled since the 1970s (from about 12% to 30%), with a large jump in the early 1980s due to an El Niño Southern

Box 12.3 **Drought in Australia, 1996-2007**

For large parts of southern and eastern Australia dry conditions have persisted since October 1996. For some areas the rainfall deficit over this period exceeded a full year's normal rainfall. In the agriculturally important Murray-Darling River basin October 2007 marked the sixth year of lower than average rainfall, with November 2001-October 2007 being the driest such six-year period on record.

The recent drought in Australia has contributed to changes in Australia's management of water resources. Acknowledging that too much water has been taken from rivers and aquifers, particularly in the Murray-Darling basin, Australia has decided that it must make better use of its water resources. This means improved efficiency and productivity of water use and better use of water markets to optimize the economic benefits that water brings (see box 4.2 in chapter 4). Australia understands that it must secure water supplies for current and future needs, including from a range of new sources that rely less on rainfall given the clear threat climate change poses to traditional water sources.

Water restrictions have been put in place in all major cities in response to the severe drought. These include restrictions on watering lawns, using sprinkler systems, washing vehicles, hosing in paved areas, refilling swimming pools and others. Restrictions can be adjusted to current conditions. In some cities water inspectors monitor water use and can impose fines or shut off water supplies for water use infractions.

Source: www.mdbc.gov.au/ and www.bom.gov.au/.

Oscillation-related precipitation decrease over land and subsequent increases due primarily to surface warming.[14]

Climate change and other global trends such as increasing population and increasing deforestation demand better risk assessment in the management of vulnerable water resources, balancing social, environmental and economic requirements, to achieve a sustainable water supply system.

Changes in groundwater

The groundwater portion of the water cycle has been subjected to massive changes, particularly during the past hundred years as humans learned to dig or drill wells and abstract groundwater using pumps. Use of groundwater for irrigated agricultural has increased enormously in the past 50 years, and some 70% of global groundwater abstraction is now estimated to be used in irrigation. Particularly in areas associated with the green revolution, heavy groundwater pumping has led to unsustainable conditions, with falling water levels, degraded groundwater aquifers and increased salinization. Pollution of shallow aquifers became widespread four or five decades ago and has triggered water quality protection measures in many countries (see chapter 8).

Groundwater abstractions have also contributed to the development of rural economies,[15] indicating that a balance has to be found in groundwater development. Groundwater serves a water shortage buffer function during short-term climate variations and is important to adaptation strategies. In many places groundwater mining from fossil aquifers is the only reliable means of obtaining water (see chapter 11). These groundwater resources are increasingly being used for agricultural, industrial and domestic water supplies, although they are almost never recharged.

Changing land use and water infrastructure have also greatly modified groundwater regimes, with groundwater pumping from deep aquifers now a worldwide phenomenon. In many places groundwater is pumped with no understanding of its source or its annual recharge and therefore of how much may be used sustainably. Results include falling water levels, desiccated wetlands, dewatered rock sequences and land subsidence (box 12.4). More data are urgently needed to quantify groundwater resources worldwide as a basis for improved groundwater management.

Heavy groundwater pumping has led to unsustainable conditions, with falling water levels, degraded groundwater aquifers and increased salinization

Changes in erosion, landslides, river morphology and sedimentation patterns

A more vigorous hydrologic cycle would imply greater water extremes, which could affect the relationships between hydrology

Box 12.4 Controlled exploitation and artificial recharge as effective measures against detrimental subsidence

Industrial and agricultural development in the last century, accompanied by an exponential growth of cities, led to concentrated pumping of groundwater resources worldwide. During the 1960s and 1970s subsidence occurred in many parts of the world, with widespread damage to property and infrastructure. Large cities built on highly compressible sediments in coastal areas increasingly experienced flooding and salt intrusion. Controlled pumping schemes and artificial recharge measures have managed to slow and even reverse subsidence. But with sea levels projected to rise as a result of global climate change, maintaining these schemes and minimizing the contribution of subsidence has become even more urgent.

Even as Venice experienced a relative sea level rise of 23 centimetres over the last century, the subsidence associated with aquifer depletion increased exponentially until 1961. With curtailing of overexploitation since 1970, subsidence has stabilized at the rate of natural subsidence, at less than 0.5 millimetre a year, and the artesian aquifers have begun to rebound.

The effects of rapid urbanization and industrialization are especially apparent in China, where increasing subsidence has led to extensive environmental and economic damage in more than 45 cities, more than 11 of which have experienced cumulative subsidence of more than 1 metre. Tianjin experienced related economic losses from 1959 to 1993 estimated at $27 billion. Shanghai took drastic measures in 1965, as total subsidence since 1920 had reached as much as 2.63 metres. Pumping has been reduced by 60%, and users are requested to inject the same quantity of water into aquifers in winter as is used in summer. While pumping-related subsidence has been controlled, drainage for construction and compaction of foundation layers have been causing subsidence rates of up to 10 millimetres a year since 1990.

Many groundwater basins in the arid and semi-arid United States experienced subsidence in response to heavy pumping, effectively reducing aquifer storage. As a result of rapid growth of Las Vegas, Nevada, pumping rates have exceeded natural recharge since about 1960, despite imports of Colorado River water. In the late 1980s the Las Vegas Water Valley District initiated an artificial recharge programme, injecting Colorado River water into the principal aquifers. Net annual pumpage has now been reduced to the level of natural recharge. The water level drawdown has recovered from a maximum of 90 metres to as few as 30 metres. Subsidence has also decreased considerably, although the depressurized aquifer continues to compact, evidence that the detrimental effects of overpumping can continue long after control measures have been taken.

Source: Ger de Lange, the Netherlands Organisation for Applied Scientific Research, Built Environment and Geosciences; Poland 1984; Carbognin, Teatini, and Tosi 2005; Hu et al. 2004; Chai et al. 2004; Wang 2007; Bell et al. 2008.

> **Increased erosion rates have important implications for the sustainability of the global soil resource, food security and environment**

and geomorphology. More intense rainfall could lead to more water-induced erosion, while drier climates could result in wind-induced erosion. And changes in the seasonal distributions of rainfall can have significant implications for patterns of vegetation growth and thus for soil erosion. Climate and erosion are interdependent components of the Earth's hydrologic cycle and of the environment. In addition to being affected by shifts in climate, soil erosion can affect climate. Desertification processes are intertwined with soil degradation and vegetation changes. These changes, possibly exacerbated by erosion, result in the loss of soil carbon and the release of carbon dioxide into the atmosphere, contributing to global warming. Changes in vegetative growth and land use that are driven by accelerated erosion can also influence the hydrologic cycle and thus the climate.

Changes in the key hydrologic drivers, such as rainfall amount and intensity, surface runoff and river discharge, caused by climate change and changes in land cover and land use, can be expected to cause significant increases in global soil erosion and in the sediment loads transported by rivers. Changes in sediment load could reflect both changing rates of sediment mobilization and supply to the river system and changes in transport capacity caused by changes in discharge and the impact of reservoirs and other human-made sinks and stores in reducing downstream fluxes. In turn, changes in the sediment regimes of rivers affect the storage capacity of reservoirs and the yield of water resources systems. Although data are limited, it is possible to assess the general magnitude and direction of changes in erosion and sediment transport over the past decades.

Erosion rates

The conversion of native vegetation to agriculture has been shown to increase soil erosion rates 10- to 100-fold.[16] With agricultural land now occupying about 37% of the ice-free area of the continents, it is clear that agriculture has had an enormous impact on global erosion rates. Increased erosion rates have important implications for the sustainability of the global soil resource, food security and the environment.

Much of the world's farmland has been cultivated for centuries and in some regions for millennia. Major increases in soil erosion rates are unlikely to have occurred within these areas in the recent past. But in other areas, particularly in developing countries, a rapidly expanding population has led to recent land clearance and rapid expansion of cultivated land. Since 1960 world population has approximately doubled, and cropland has increased by more than 10%.[17]

From a global perspective, however, such recent increases in soil loss are likely to have been at least partially offset by reduced erosion in other regions following implementation of soil conservation programmes and improved land management during the 20th century. In the United States soil conservation and related measures promoted by the Food Security Act of 1985 have reduced total annual erosion from cropland by an estimated 40%, from 3.4 gigatonnes (Gt) a year to 2.0 Gt.[18] In China erosion control measures in the loess region of the Middle Yellow River basin after 1978 helped reduce the annual sediment load of the Middle Yellow River from about 1.6 Gt in the mid-20th century to 0.7 Gt at the end of the 20th century.[19] Elsewhere, the progressive introduction of no-till and minimum till practices – now implemented on an estimated 5% of world cropland[20] – has also reduced erosion rates on cultivated land. Such measures typically reduce erosion rates by more than one order of magnitude.[21]

While an accurate assessment of the relative importance of these opposing trends for the contemporary global soil erosion budget is still not possible, it is clear that significant changes are occurring. Furthermore, there is increasing recognition that, with the greater variability of rainfall and the higher frequency of extreme storm events accompanying future climate change, erosion rates in many areas of the world are likely to rise. A recent study in the Midwestern United States, combining the output from general circulation models with erosion models that also took into account the likely impact of climate change on crop management and crops grown, suggested that erosion rates would increase in 10 of the 11 study area regions. Increases relative to 1990-99 were predicted to range from 33% to 274% by 2040-59.

Sediment loads

A river's sediment load is sensitive to a range of environmental controls related to both the supply of sediment to the river and its ability to transport that sediment. Long-term sediment measurements are unavailable for most of the world's rivers, precluding detailed analysis of global trends, but available data show that important changes are occurring.[22] For many rivers there is evidence of reduced sediment loads in recent years,[23] primarily because of the construction of dams, which trap a large proportion of the sediment load previously transported by the river. One estimate suggests that more than 40% of global river discharge is intercepted by large dams (dams with a storage capacity of more than 0.5 cubic kilometre).[24] Dams on the Colorado and the Nile Rivers have reduced the sediment load of those rivers to near zero.

Changes in sediment load have important implications for global geochemical cycling and sediment-associated carbon fluxes. From a more local perspective increased sediment loads frequently cause degradation of water quality and aquatic habitats and increased siltation of reservoirs, river channels, canal systems and harbours. In many contexts reduced sediment loads bring benefits, but in deltas and coastal areas reduced sediment inputs can remove important nutrient sources and lead to shoreline recession. Landslides, the most dramatic erosion process, present the gravest hazard to human communities. Until the International Geotechnical Society UNESCO Working Party on World Landslide Inventory was established during the International Decade for Natural Disaster Reduction of 1990-2000, there was no common definition of landslides. The international unified definition was agreed as 'the movement of a mass of rock, debris or earth down a slope'.[25] Landslides are classified by a combination of the type of material (rock, debris and earth) and type of movement (fall, slide, flow, topple and spread).

While the definition has been approved internationally, no unified map of the distribution of landslides has ever been prepared. However, a global-scale landslide susceptibility map was produced by the International Programme on Landslides with World Bank funding.[26] Landslide susceptibility is calculated from topography, earthquake and rainfall data. The model does not include the shear strength of soils, which is difficult to quantify at the global scale. Recently, the US National Aeronautics and Space Administration (an International Consortium on Landslides member) also compiled a landslide susceptibility map using Shuttle Radar Topography Mission data, the Food and Agriculture Organization's digital soil map and other information.

Changes in precipitation or precipitation-causing phenomena (such as cyclones and typhoons) can lead to increased severity or frequency of landslides. If these changes are accompanied by seismic activities, there is a strong potential for an increase in landslide-triggering events. For example, in 2004 Japan experienced the greatest number of typhoons in its history. After one of these typhoons the 2004 Niigata-ken Chuetsu earthquake occurred. The earthquake triggered more large-scale landslides than had events of a similar magnitude, such as the 1995 Hyogo ken-Nanbu (Kobe) earthquake and the 2005 Fukuoka-ken Saiho-oki offshore earthquake. In February 2006 a small earthquake, which occurred five days after three days of continuous rainfall, triggered a huge landslide in Leyte Island, Philippines, killing more than 1,000 people.

Climate change will affect water quality and ecosystem health through higher water temperatures, lower water levels, more flooding and changes in lake stratification patterns. Aquatic ecosystem dynamics are driven by temperature and water availability, which determine energy flow; the primary production, composition, structure and biological diversity of ecosystems; the range of global biomass; pattern of ecosystem succession and the type of climax biome.

Climate change will affect water quality and ecosystem health through higher water temperatures, lower water levels, more flooding and changes in lake stratification patterns

Increased water temperatures promote algal blooms and increase toxic cyanobacteria bloom. A toxin produced by microcistis (*Microcistina*-LR) is 10 times more toxic than strychnine. Toxic cyanobacteral blooms, already present on all continents, may intensify, requiring restrictions on people's use of water resources.

Relatively small increases in temperature also accelerate energy flow and matter cycling: a 1°C warming enhances ecosystem productivity by 10%-20% at all trophic levels. An increase in zooplankton consumption may reduce the density of this food source, resulting in a decline in the food base for fish, inhibiting growth and favouring small species over large (an insufficient food base for larger species). An overlapping of changing abiotic conditions, such as rising temperature and declining dissolved oxygen content, may be an additional stressor, contributing to a lowering of biodiversity and ecosystems function. This could mean shifts in dominant species, a destabilization of the ecosystem equilibrium and a shift to another steady-state. Rising water temperatures and related changes in ice cover, salinity, oxygen levels and water circulation have already contributed to global shifts in the range and abundance of algae,

If hazards become more severe (in intensity or magnitude), countries will face new challenges, requiring additional cooperation with other concerned countries in mitigating hazards

zooplankton and fish in high-latitude oceans and high-altitude lakes, as well as to earlier migrations of fish in rivers.

The effects of increased temperatures and the acceleration of biological processes will differ depending on hydrologic type and the characteristics and complexity of aquatic ecosystems. In colder regions, for example, rising water temperatures can improve water quality during winter and spring, with earlier ice breakup increasing oxygen levels and reducing winter fish-kills.[27]

The response of river ecosystems to climate change will depend on their location within the river basin. Longitudinal linkages are important to the functioning of river ecosystems. Upper sections of rivers are usually driven by abiotic factors (flow and water quality), and the biotic structures are better adapted to high abiotic (hydrologic) variability, resistant to rapid and unexpected changes and better able to recover from stress. Down the river course, with more stable abiotic characteristics, biotic processes determine ecosystem dynamics, and ecosystems are more vulnerable to global warming.

Modification of precipitation patterns due to climate change will directly influence runoff and the timing and intensity of nutrient and pollutant supplies to rivers and lakes. Greater changes are expected in catchments with degraded vegetation cover, landscape drainage and wetland loss. Open nutrient cycles in the terrestrial ecosystems due to reduced nutrient retention in biomass and mineralization of organic matter in soils will intensify nutrient loss to freshwater. More intense rainfall events will also lead to greater fluvial erosion and increases in suspended solids loads (turbidity) in lakes and reservoirs.[28] Extension of the growing season due to global warming may increase the duration of agricultural activities, which may cause more nutrient leaching from agricultural areas.[29] All these processes will contribute to intensification of eutrophication, a common problem in lakes and rivers all over the world and a serious hazard for human activities (drinking water, aquaculture, recreation) and ecosystem functioning.

The expected overall lowering of water levels in rivers and lakes will worsen water quality. Water reserves will become more turbid through the resuspension of bottom sediments,[30] and the reduction in water supply will increase the concentration of pollutants in water resources. Oxygenation of river water – a key indictor of biological water quality – is enhanced under high flow conditions that encourage surface aeration. Simulations of stream conditions under several climate change scenarios found that decreased streamflows resulted in decreased oxygen levels and water quality.[31] Salinity levels will increase with decreasing streamflow in semi-arid and arid areas. Salt concentrations are predicted to increase 13%-19% by 2050 in the upper Murray-Darling River basin in Australia.[32] Salinization of water resources is also predicted to be a major hazard for island nations, where coastal seawater intrusion is expected with rising sea levels.

The higher temperature, change in precipitation patterns and shift in regional wind regimes associated with climate change are likely to alter the thermal stratification of lakes and reservoirs. Higher temperatures are likely to increase thermal stability and alter mixing patterns in lakes, resulting in reduced oxygen concentrations and increased release of phosphorous from sediments.[33] Simulations suggest that lakes in subtropical zones (latitude 30° to 45°) and subpolar zones (latitude 65° to 80°) are subject to greater relative changes in thermal stratification patterns than are mid-latitude or equatorial lakes and that in subtropical zones deep lakes are more sensitive than are shallow lakes.[34]

Simulations also show that winter stratification in cold regions would be weakened and the anoxic zone would disappear.[35] The greatest increases in water temperatures are foreseen in lakes where the duration of ice cover will be substantially reduced. In addition, simulations show a 10° or more northward shift in the boundary of ice-free conditions in the Northern Hemisphere.[36] Observations during droughts in the boreal region of northwestern Ontario, Canada, show that lower inflows and higher temperatures produce a deepening of the thermocline.[37] Changes in wind speed and direction contributing to patterns of lake and reservoir mixing and thermal stratification may alter the biomass cycling in lakes. Countries that share water resources may face additional challenges under conditions of changing hazards. In areas with experience of hazards countries are used to managing such crises. But if hazards become more severe (in intensity or magnitude), countries will face new challenges, requiring additional cooperation with other concerned countries in mitigating hazards.[38]

In new hazard-exposed areas there will be great variations in how countries mitigate hazards affecting international waters. In broad terms OECD countries in Europe, North America and South East Asia would be able to direct their institutional and financial resources towards new cooperative efforts. Developing countries, with limited resources and hazard mitigating experience, would be more exposed. Examples include the Mekong River basin and some of the major basins in West and Central Africa.[39]

Changes in transboundary water resources (through engineered developments or climate change) will present opportunities for international cooperation. The cooperation must be based on a common understanding of the nature of the resource as well as its value to the countries who share it. And it must involve collecting and sharing reliable data and applying compatible data analysis methods.

Attention is coming to transboundary groundwater issues much later than to surface water concerns. Governments, institutions and other stakeholders that have developed strategies for effective groundwater resources management at the local and national levels are learning that coordination is also necessary across administrative boundaries (see box 12.5 on the Guaraní aquifer). As a result, over

Box 12.5 Towards transboundary management of the Guaraní aquifer

The Guaraní aquifer in Argentina, Brazil, Paraguay and Uruguay averages some 250 metres in thickness and covers an estimated area of 1.2 million square kilometres. The aquifer's groundwater resources are contained in sand and sandstone beds deposited since the Mesozoic Era and mostly confined below volcanic rocks. Estimated annual recharge is about 166 cubic kilometres, produced mainly in the zones where there are outcroppings of unconfined parts of the aquifer. The climate in the area is humid to subhumid, with annual precipitation of 1,200-1,500 millimetres.

The quality of the Guaraní's groundwater is generally good, with a few exceptions. About 20 million people live over the aquifer, and some parts are intensely exploited, while others are virtually untouched. There are still many gaps in knowledge about the state and performance of this aquifer system.

Location of the Guaraní transboundary aquifer

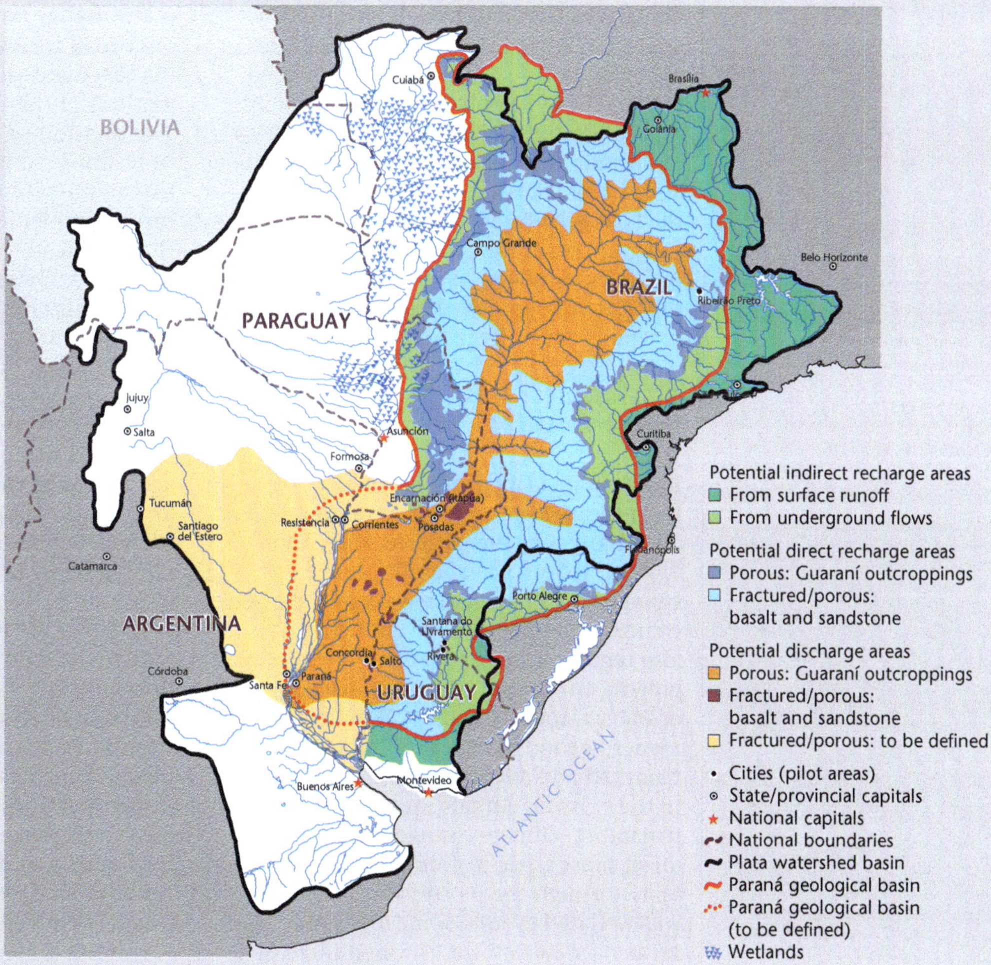

Source: Guaraní Aquifer System Project 2003.

The four countries have cooperated since early 2003 in a project for sustainable management and protection of the aquifer, with support from the Global Environment Facility, the World Bank and the Organization of American States and with participation of the International Atomic Energy Agency and the German Federal Institute for Geosciences and Natural Resources. The project aims to explore and assess the aquifer system in more detail and to develop a framework for coordinated management. In addition, four pilot areas with emerging problems have been selected to gain experience in local management.

Source: www.sg-guarani.org/index; UNESCO/OAS ISARM 2007.

Over many generations the human race has shown an amazing ability to adapt and adjust to climate variability and increasing pressure on resources

the last decade international and regional organizations have developed initiatives related to internationally shared aquifer resources management, in cooperation with countries in the regions concerned.[40]

Mapping and descriptions of transboundary aquifers are the first steps. The United Nations Economic Commission for Europe conducted an inventory of transboundary aquifers in Europe in 1999,[41] and inventories in other regions followed. Recent outcomes are an atlas of the 68 identified transboundary aquifers of the Americas[42] and an assessment report on transboundary rivers, lakes and aquifers that includes 51 transboundary aquifers in South-eastern Europe and 18 in the Caucasus and Central Asia region.[43] International projects are facilitating the exchange of information and experiences and developing improved methodologies and the scientific basis of transboundary aquifer management. A 2007 agreement created the UNESCO Regional Centre for Shared Aquifer Resources Management for Africa in Tripoli. The United Nations International Law Commission, in cooperation with the UNESCO-International Hydrological Programme, drafted articles on the Law on Transboundary Aquifers, which was subsequently approved by the UN General Assembly on 11 December 2008.

Challenges: hazards and opportunities

Based on identified trends, the future will see increased pressure on water resources and changes in the patterns and magnitudes of resource availability related to changing climate patterns. While climate change represents a huge challenge, it also represents an opportunity – for new growth, innovation in the management of water resources and development of a modern economy. Because humans have modified and adapted their lifestyles to the existing climate and its inherent variability, climate change is expected to affect most aspects of human life, notably through the hazardous aspects of water-related events. Some areas may gain greater access to water through increased precipitation, while others may have less or more variable water resources.

Over many generations the human race has shown an amazing ability to adapt and adjust to climate variability and increasing pressure on resources. There are many examples of countries that have managed their scarce resources efficiently and effectively, despite low rainfall and streamflows. Spain, a traditionally dry-climate country, has historically succeeded in managing its water resources through adaptation. An example of potential water gain is South Africa. Analysis of the actual evapotranspiration and yield of five commonly grown crops (beans, groundnuts, maize, millet and sorghum) in two selected districts found that yield increases with evapotranspiration, although the gap remains wide between actual and potential yield and actual and maximum evapotranspiration, especially for rainfed crops.[44] The analysis also showed that a 2°C rise in temperature and a doubling of carbon dioxide concentration in the atmosphere will shorten the growing period of maize, lowering crop water requirements.

The increased exposure to potential climate change hazards has raised awareness of critical issues related to water resources management that require solutions regardless of the impacts of climate change. The revision of management strategies in response to potential climate change threats therefore represents an opportunity to implement policies and practice that will lead to more sustainable use of resources. These could include improved observation networks (see chapter 13), increased integration of groundwater and surface water supplies (including artificial recharge), improved early warning and forecasting systems for hazardous events, improved risk-based approaches to management and the raising of community awareness of sustainable water resources use and individual responses to water-related hazards.

The threat of climate change has led to many developments in the simulation of atmospheric processes, improving the accuracy of climate and weather forecasts. Combined with improved technology for monitoring, collecting and analysing information, these developments should lead to improvements in warning systems for floods and droughts and other major water-related events. If these can be combined with hazard mitigation strategies involving all levels of affected communities, there are enormous opportunities to avoid loss of human life and economic losses.

There are also many specific examples of turning potential hazards into opportunities. These include using increased runoff from glacial melting to develop more reliable water supplies for larger areas and using flood water storage to increase the reliability of water supplies and improve floodplain management and planning (box 12.6).

Small and shallow alluvial aquifers scattered over the Earth's arid and semi-arid regions – preferential zones for human settlement – are probably the most vulnerable groundwater systems to climate change.[45] Yet in areas of increasing water stress groundwater is an important buffer resource, capable of responding to overall increased water demands or of compensating for loss in surface water availability.

This buffer capacity of groundwater systems depends on the ratio between the volume of stored groundwater and the mean annual recharge. Some major aquifers with non-renewable groundwater resources are found in arid environments of North and Southern Africa, the Arabian Peninsula and Australia and under permafrost in Northern Asia.[46] The direct impact of climate change on such resources is negligible, as their stored volumes are usually at least a thousand times the volume of mean annual recharge. Their stocks may be reduced more quickly than before, however, because of larger demands created by climate change and the decline of alternative water sources. Renewable groundwater systems with considerable storage will provide similar buffers in other parts of the world, and an increasingly larger share of total water abstraction is expected to come from groundwater.

The storage capacity of aquifers also offers opportunities for enhancing groundwater storage by artificial recharge or managed aquifer recharge. Managed aquifer recharge, applied at an ever-growing rate,[47] should be part of integrated water and catchment management strategies along with surface water and soil management, erosion and pollution control, demand and environmental management and wastewater reuse. Its role will become increasingly important as the impacts of climate change and variability become more apparent.[48]

Box 12.6 Lake Sarez, Tajikistan – turning a hazard into an opportunity

Lake Sarez, deep in the centre of the Pamir Mountains of Tajikistan, was created in February 1911, when an enormous rock collapsed from the bank of the Murgab River Valley, blocking the river and forming a dam behind which a lake formed. Lake Sarez is 55.8 kilometres long and averages 1.44 kilometres wide, with a maximum depth of 499.6 metres and maximum water volume of 16.074 cubic kilometres. The lake's water level is currently about 50 metres below the top of the dam and rising at a rate of 20 centimetres a year as a result of increased glacial melt, due largely to global warming. At the same time, the permeability of the dam material is changing, and the mineralization of water at the bottom levels is increasing. There is also a real threat of new landslides around the lake.

Multiple hypotheses were developed recently on how the natural barrier would behave in a future earthquake or other catastrophe. Evaluations varied from the categorical 'it will burst' to the no less firm 'it will not'.

While some sources stress that a catastrophic flood is unlikely, no one is dismissive of the risk considering the potential for devastation. A World Bank statement indicated that 'should such a flood occur, the impact on the downstream valleys would be devastating, affecting up to 5 million people'. The impact of a dam break would be felt not only across Tajikistan but also in Afghanistan, Turkmenistan and Uzbekistan.

Tajikistan recently proposed construction of a water pipeline that would serve all of Central Asia with safe drinking water through the regulated drainage of Lake Sarez. This would reduce the risk of a dam break or overflow, while supplying drinking water to the region. Lake Sarez is the largest freshwater reservoir in the upper watershed zone of the Aral Sea basin.

Source: Vefa Mustafaev, UNESCO; Sirodjidin Aslov, Ambassador of Tajikistan to the United Nations; and N. F. Gorelkin Uzbekistan Department of Hydrometerology; World Bank 2005.

Notes

1. Huntington 2006.
2. IPCC 2007.
3. For example, Williams, Jackson, and Kutzbach 2007.
4. IPCC 2007.
5. IPCC 2007.
6. IPCC 2001.
7. Milly et al. 2002.
8. Kundzewicz et al. 2005.
9. Zhang et al. 2007.
10. van Lanen, Tallaksen, and Rees 2007; Huntington 2006.
11. Hisdal et al. 2001.
12. Tallaksen, Demuth, and van Lanen 2007.
13. Easterling et al. 2007.
14. Dai, Trenberth, and Qian 2004.
15. Giordano and Villholth 2007.
16. Montgomery 2007.
17. Wilkinson and McElroy 2007.
18. Uri and Lewis 1999.
19. Hu et al. 2008.
20. Lal et al. 2004.
21. Montgomery 2007.
22. Walling 2006.
23. Walling and Fang 2003.
24. Vörösmarty et al. 2003.
25. Cruden 1991, p. 27.
26. IPL n.d.
27. IPCC 2007.
28. IPCC 2007.
29. Hillbricht-Ilkowska 1993.
30. Atkinson, DePinto, and Lam 1999.
31. Mimikou et al. 2000.
32. IPCC 2007.
33. Bates et al. 2008.
34. Meyer et al. 1999.
35. Fang and Stefan 1997.
36. Hostetler and Small 1999.
37. Schindler and Stainton 1996.
38. Romm 2007.
39. International Crisis Group 2007.
40. For details see www.isarm.net.
41. Almássy and Buzás 1999.
42. UNESCO/OAS ISARM 2007.
43. UNECE 2007.
44. World Bank 2007.

45. van der Gun forthcoming.
46. Foster and Loucks 2006.
47. UNESCO 2005; Fox 2007.
48. Gale 2005.

References

Almássy, E., and Z. Buzás. 1999. *Inventory of Transboundary Groundwaters.* Lelystad, Netherlands: United Nations Economic Commission for Europe.

Atkinson, J. F., J. V. DePinto, and D. Lam. 1999. Water Quality. In *Potential Climate Change Effects on the Great Lakes Hydrodynamics and Water Quality,* eds. D. Lam and W. Schertzer. Reston, VA: American Society of Civil Engineers.

Bates, B., Z. W. Kundzewicz, S. Wu, and J. Palutik, eds. 2008. *Climate Change and Water. Technical Paper of the Intergovernmental Panel on Climate Change.* Geneva: Intergovernmental Panel on Climate Change.

Bell, J. W., F. Amelung, A. Ferretti, M. Bianchi, and F. Novali. 2008. Permanent Scatterer in SAR Reveals Seasonal and Long-Term Aquifer-System Response to Groundwater Pumping and Artificial Recharge. *Water Resources Research* 44 (2): W02407.

Carbognin L., P. Teatini, and L. Tosi. 2005. Land Subsidence in the Venetian Area: Known and Recent Aspects. *Giornale di Geologia Applicata* 1 (1): 5-11.

Carmon, N., and U. Shamir. 1997. Water-Sensitive Urban Planning: Concept and Preliminary Analysis. In *Groundwater in the Urban Environment: Problems, Processes and Management,* eds. J. Chilton, K. Hiscock, P. Younger, B. Morris, S. Puri, S. W. Kirkpatrick, H. Nash, W. Armstrong, P. Aldous, T. Water, J. Tellman, R. Kimblin, and S. Hennings. London: Balkema, Rotterdam, Brookfield.

Chai J. C., Shen S. L., Zhu H. H, and Zhang X. L. 2004. Land Subsidence Due to Groundwater Drawdown in Shanghai. *Géotechnique* 54 (2): 43-147.

Cruden, D. M. 1991. A Simple Definition of a Landslide. *Bulletin of Engineering Geology and the Environment* 43 (1): 27-9.

Dai, A., K. E. Trenberth, and T. Qian. 2004. A Global Data Set of Palmer Drought Severity Index for 1870-2002: Relationship with Soil Moisture and Effects of Surface Warming. *Journal of Hydrometeorology* 5 (6): 1117-30.

Easterling, W. E., P. K. Aggarwal, P. Batima, K. M. Brander, L. Erda, S. M. Howden, A. Kirilenko, J. Morton, J. F. Soussana, J. Schmidhuber, and F. N. Tubiello. 2007. Food, Fibre and Forest Products. In *Climate Change 2007: Impacts, Adaptation and Vulnerability. Contribution of Working Group II to the Fourth Assessment Report of the Intergovernmental Panel on Climate Change,* eds. M. L. Parry, O. F. Canziani, J. P. Palutikof, P. J. van der Linden, and C. E. Hanson. Cambridge, UK: Cambridge University Press.

Fang, X., and H. G. Stefan. 1997. Simulated Climate Change Effects on Dissolved Oxygen Characteristics in Ice-covered Lakes. *Ecological Modelling* 103 (2-3): 209-29.

Foster, S., and D. P. Loucks, eds. 2006. *Non-Renewable Groundwater Resources: A Guidebook on Socially-Sustainable Management for Water-Policy Makers.* Paris: United Nations Educational, Scientific, and Cultural Organization.

Fox, P., ed. 2007. *Management of Aquifer Recharge for Sustainability.* Phoenix, AZ: Acacia Publications.

Gale, I., ed. 2005. *Strategies for Managed Aquifer Recharge (MAR) in Semi-Arid Areas.* Paris: United Nations Educational, Scientific, and Cultural Organization.

Giordano, M., and K. Villholth, eds. 2007. *The Agricultural Groundwater Revolution: Opportunities and Threats to Development. Comprehensive Assessments of Water Management in Agriculture No 3.* Colombo: International Water Management Institute and CABI.

Guaraní Aquifer System Project. 2003. Preliminary Map of the Guaraní Aquifer System. General Secretariat, Guaraní Aquifer System Project, Montevideo. www.sg-guarani.org/index/site/sistema_acuifero/sa001.php.

Hillbricht-Ilkowska, A. 1993. Lake Ecosystems and Climate Changes. *Kosmos* 42 (1): 107-21.

Hisdal, H., K. Stahl, L. M. Tallaksen, and S. Demuth. 2001. Have Streamflow Droughts in Europe Become More Severe or Frequent? *International Journal of Climatology* 21 (1): 317-33.

Hostetler, S. W., and E. E. Small. 1999. Response of North American Freshwater Lakes to Simulated Future Climates. *Journal of the American Water Resource Association* 35 (6): 1625-38.

Hu C., Liu C., Zhou Z., and R. Jayakumar. 2008. Changes in Water and Sediment Loads of Rivers in China. Working Paper prepared as a contribution to UNESCO's *World Water Development Report 3,* International Research and Training Center on Erosion and Sedimentation, Beijing.

Hu R. L., Yue Z. Q., Wang L. C., and Wang S. J. 2004. Review on Current Status and Challenging Issues of Land Subsidence in China. *Engineering Geology* 76 (1-2): 65-77.

Huntington, T. G. 2006. Evidence for Intensification of the Global Water Cycle: Review and Synthesis. *Journal of Hydrology* 319 (1-4): 83-95.

International Crisis Group. 2007. Nigeria: Ending Unrest in the Niger Delta. Africa Report 135, International Crisis Group, Brussels.

IPCC (Intergovernmental Panel on Climate Change). 2001. *Climate Change 2001: Impacts, Adaptation and Vulnerability Contribution of the Working Group II to the Third Assessment Report of the Intergovernmental Panel on Climate Change,* eds. J. J. McCarthy, O. F. Canziani, N. A. Leary, D. J. Dokken, and K. S. White. Cambridge, UK: Cambridge University Press.

———. 2007. *Climate Change 2007: The Physical Science Basis. Summary for Policymakers. Contribution of the Working Group I to the Fourth Assessment Report of the Intergovernmental Panel on Climate Change.* Cambridge, UK: Cambridge University Press.

IPL (International Programme on Landslides). n.d. IPL Leaflet. International Consortium on Landslides, Kyoto University, Disaster Prevention Research Institute, Uji, Kyoto, Japan. http://iclhq.org/IPL-Leaflet-2004.pdf.

Kundzewicz, Z. W., D. Graczyk, T. Maurer, I. Pińskwar, M. Radziejewski, C. Svensson, and M. Szwed. 2005. Trend Detection in River Flow Series: 1. Annual Maximum Flow. *Hydrological Sciences Journal* 50 (5): 797-810.

Lal, R., M. Griffin, J. Apt, L. Lave, and M. N. Granger. 2004. Managing Soil Carbon. *Science* 304 (5669): 393.

Meyer, J. L., M. J. Sale, P. J. Mulholland, and N. L. Poff. 1999. Impacts of Climate Change on Aquatic Ecosystem Functioning and Health. *Journal of the American Water Resources Association* 35 (6): 1373-86.

Milly, P. C. D., R. T. Wetherald, K. A. Dunne, and T. L. Delworth. 2002. Increasing Risk of Great Floods in a Changing Climate. *Nature* 415 (6871): 514-17.

Mimikou, M. A., E. Malta, E. Varanou, and K. Pantazis. 2000. Regional Impacts of Climate Change on Water Resources Quantity and Quality Indicators. *Journal of Hydrology* 234 (1-2): 95-109.

Montgomery, D. R. 2007. Soil Erosion and Agricultural Sustainability. *Proceedings of the National Academy of Sciences* 104 (33): 13268-72.

Neilson, R. P., and D. Marks. 1994. A Global Perspective of Regional Vegetation and Hydrologic Sensitivities from Climatic Change. *Journal of Vegetation Science* 5 (5): 715-30.

Neilson, R. P., I. C. Prentice, B. Smith, T. G. F. Kittel, and D. Viner. 1998. Simulated Changes in Vegetation Distribution under Global Warming. In *The Regional Impacts of Climate Change: An Assessment of Vulnerability,* eds., R. T. Watson, M. C. Zinyowera, R. H. Moss, and D. J. Dokken. Cambridge, UK: Cambridge University Press.

Poland, J. F., ed. 1984. *Guidebook to Studies of Land Subsidence Due to Ground-Water Withdrawal.* Paris: United Nations Educational, Scientific, and Cultural Organization.

Romm, J. 2007. *Hell and High Water: Global Warming – the Solution and the Politics – and What We Should Do.* New York: William Morrow.

Schindler, D. W., and M. P. Stainton. 1996. The Effects of Climatic Warming on the Properties of Boreal Lakes and Streams at the Experimental Lakes Area, Northwestern Ontario. *Limnology and Oceanography* 41 (5): 1004-17.

Scholze, M., W. Knorr, N. W. Arnell, and I. C. Prentice. 2006. A Climate-Change Risk Analysis for World Ecosystems. *Proceedings of the National Academy of Science* 103: 13116-120.

Tallaksen, L. M., S. Demuth, and H. A. J. van Lanen. 2007. Low Flow and Drought Studies – the Northern European (NE) FRIEND Experience (keynote paper). In *Climatic and Anthropogenic Impacts on the Variability of Water Resources International Seminar, 22-24 November 2005. Montpellier, France,* ed. G. Mahé. Paris: United Nations Educational, Scientific, and Cultural Organization, International Hydrological Programme.

UNECE (United Nations Economic Commission for Europe). 2007. *Our Waters: Joining Hands across Borders. First Assessment of Transboundary Rivers, Lakes and Groundwaters.* Geneva and New York: United Nations. www.unece.org/env/water/publications/ assessment/assessmentweb_full.pdf.

UNESCO (United Nations Educational, Scientific, and Cultural Organization). 2005. *Recharge Systems for Protecting and Enhancing Groundwater Resources. Proceedings of the Fifth International Symposium on Management of Aquifer Recharge, Berlin, Germany, 11-16 June 2005.* Paris: United Nations Educational, Scientific, and Cultural Organization.

UNESCO (United Nations Educational, Social, and Cultural Organization)/OAS (Organization of American States) ISARM (International Shared Aquifer Resource Management). 2007. Transboundary Aquifers of the Americas. Final Report of 4th Coordination Workshop, 20-22 November 2006, San Salvador. United Nations Educational, Social, and Cultural Organization/Organization of American States, International Shared Aquifer Resource Management Americas Program, Washington, DC.

Uri, N. D., and J. A. Lewis. 1999. Agriculture and the Dynamics of Soil Erosion in the United States. *Journal of Sustainable Agriculture* 14 (2-3): 63-82.

van der Gun, J. A. M. Forthcoming. Climate Change and Alluvial Aquifers in Arid Regions – Examples from Yemen. In *Climate Change Adaptation in the Water Sector.* London: Earthscan.

van Lanen, H. A. J., L. M. Tallaksen, and G. Rees. 2007. Droughts and Climate Change. In *Commission Staff Working Document Impact Assessment: Accompanying Document to Communication Addressing the Challenge of Water Scarcity and Droughts in the European Union.* Brussels: Commission of the European Communities.

Vörösmarty, C. J., M. Meybeck, B. Fekete, K. Sharma, P. Green, and J. P. M. Syvitski. 2003. Anthropogenic Sediment Retention: Major Global Impact from Registered River Impoundments. *Global and Planetary Change* 39 (1-2): 169-90.

Walling, D. E. 2006. Human Impact on Land-Ocean Sediment Transfer by the World's Rivers. *Geomorphology* 79 (3-4): 192-216.

Walling, D. E., and D. Fang. 2003. Recent Trends in the Suspended Sediment Loads of the World's Rivers. *Global and Planetary Change* 39 (1): 111-26.

Wang P. 2007. The Dragon Head's Story: Water/Land Conflict in Shanghai. *Geophysical Research Abstracts* 9: 11655.

Wilkinson, B. H., and B. J. McElroy. 2007. The Impact of Humans on Continental Erosion and Sedimentation. *Geological Society of America Bulletin* 119 (1-2): 140-56.

Williams, John W., Stephen T. Jackson, and John E. Kutzbach. 2007. Projected Distributions of Novel and Disappearing Climates by 2100 AD. *Proceedings of the National Academy of Sciences of the United States of America* 104 (14): 5738-42.

World Bank. 2005. A Safer Lake Sarez. World Bank, Washington, DC. http://go.worldbank.org/NLINTLWL00.

———. 2007. Actual Crop Water Use in Project Countries: A Synthesis at the Regional Level. Policy Research Working Paper 4288, Sustainable Rural and Urban Development Team, World Bank, Washington, DC.

Zhang X., F. W. Zwiers, G. C. Hegerl, F. H. Lambert, N. P. Gillett, S. Solomon, P. A. Stott, and T. Nozawa. 2007. Detection of Human Influence on Twentieth-Century Precipitation Trends. *Nature* 448: 461-65.

Chapter 13
Bridging the observational gap

Author: Wolfgang Grabs
Coordinator: Andras Szöllösi-Nagy (UNESCO)
Associate coordinator: Wolfgang Grabs (WMO)
Facilitator: Denis Hughes

Key messages

- Worldwide, water observation networks provide incomplete and incompatible data on water quantity and quality for managing water resources and predicting future needs – and these networks are in jeopardy of further decline. Also, no comprehensive information exists on wastewater generation and treatment and receiving water quality on a regional or global scale.
- There is little sharing of hydrologic data, due largely to limited physical access to data, policy and security issues; lack of agreed protocols for sharing; and commercial considerations. This hampers regional and global projects that have to build on shared datasets for scientific and applications-oriented purposes, such as seasonal regional hydrologic outlooks, forecasting, disaster warning and prevention, and integrated water resources management in transboundary basins.
- Improving water resources management requires investments in monitoring and more efficient use of existing data, including traditional ground-based observations and newer satellite-based data products. Most countries, developed and developing, need to give greater attention and more resources to monitoring, observations and continual assessments of the status of water resources.

There is little doubt that global hydrologic data are inadequate in both spatial coverage and frequency of observations. Moreover, hydrologic observation networks are worsening in many countries because of changing national investment priorities and declining human capacity.[1]

Beyond physical inadequacy, there is widespread reluctance to share hydrologic data, largely because of inadequate national administrative procedures and mechanisms, issues related to commercial use of hydrologic data and information, and security concerns and political sensitivities about transboundary resources. Sharing is also hampered by inadequate telecommunications systems. But the main reason for the decline of networks is insufficient awareness of the global value of hydrologic data.

The importance of hydrologic observations

Sound water resources management should be based on a quantitative understanding of the state of the resource. Because the components of the water cycle vary over time, long time series of observational data are essential. Lack of such data compromises the validity of information used for assessments and subsequent decision-making.

When sufficient observational data are lacking, models can be used to generate information for decision-making, provided they have a baseline from which to be calibrated. But synthetically generated data cannot substitute for real-world observations. To keep networks affordable and sustainable, a minimum-density network can effectively coexist with model-derived data and information. Established analytical procedures can offer insight into observational and model uncertainties and provide a basis for thoroughly analysing permissible reduction of information as a result of minimizing networks.

Data are also crucial to improving understanding of the hydrologic cycle for weather- and climate-related science and for water resources management through better assessment methods and improved forecasting services that can reduce disaster damages. One justification for improving hydrologic networks is to minimize uncertainties in hydrologic forecasting and prediction and therefore to minimize decision-making risk. This can be achieved in several ways, including new and better-quality information from improved measurements (quantity, quality, timeliness) and measurement techniques, improved model structures based on better understanding of physical processes, and better mathematical representation and use of available information during model identification and calibration.

Data are crucial to improving understanding of the hydrologic cycle for weather- and climate-related science and for water resources management

Aside from technical considerations, uncertainty in hydrologic data can be attributed to the general inadequacy of observations in spatial and temporal coverage. 'Uncertainty' in this context relates to the adequacy and quality of technical observations for forecasting and assessment – not to the aberrations in technical observations. Uncertainty for assessment and forecasting varies, but is generally high in subtropical and tropical regions as well as polar and mountainous regions. Developing countries, especially the least developed countries, generally have inadequate networks associated with a high uncertainty of hydrologic observations.

There is a critical need for more availability and access to global hydrologic data, information and products for climate and hydrologic research and applications – including the validation and refinement of global circulation models and the quantification of the water balance and its variation over large basins and regions up to the global level. Important scientific issues include the quantification of a postulated accelerated hydrologic cycle as a result of global warming and the contribution of continental runoff to sea level change. In a general sense global hydrologic observations help quantify key environmental and human-induced changes and interactions, identify significant trends, assess variability of freshwater resources and develop adequate response strategies.

While characterizing all components of the water cycle (including water quantity and quality, groundwater and surface water) is important, measuring everything is impractical. The components of surface water measured most frequently are precipitation, streamflow, evaporation and water in storage (reservoirs and natural lakes). Other components (soil moisture, for example) are generally quantified using models that rely on the other measurements. The components of groundwater measured most frequently are inputs (recharge and inflow), aquifer storage capacity and natural outflows. While effective management of groundwater resources depends on measuring these components, none can be measured directly as a single value because they all depend on a large number of variables.

Water quality observation networks are sparse and frequently use spot rather than continuous observations. The numerous water quality variables that determine how water can be used and the treatment technologies required for safe use make it very difficult to design effective and affordable monitoring systems and to identify relationships between water quantity and quality variations. Such information can be extrapolated to sites that have no observations. The problem is exacerbated when natural water quality is modified by both point and diffuse anthropogenic pollution sources. And deteriorating water quality greatly affects the potential use of water, increasing the need for networks of water quality observations.

It is recognized worldwide that management of water resources should be environmentally sustainable and that adequate protection of aquatic ecosystems is extremely important. Aquatic ecosystems are sensitive to changes in water quantity and quality, but the precise manner in which they react to changes in abiotic drivers is poorly understood. Better understanding is vital for allocating water and can be achieved only by monitoring abiotic drivers (flow and water quality) and biotic responses (biomonitoring of biological conditions; see chapter 8).[2]

Recent developments in observation methods, networks and monitoring

The adequacy of any hydrologic observation network is related to the accuracy of

The adequacy of observational networks varies widely by region, but observations for many water cycle variables have inadequate spatial and temporal coverage

measurement systems, the density and representativeness of the network, and monitoring, data retrieval, storage and dissemination practices.

Observational errors are generally minimal when instruments are used in environments and under conditions for which they were designed. But instruments must be well maintained and calibrated, and rigid quality management procedures must be observed from the field to the release of the primary data. Further reducing instrument errors is less critical than expanding operational networks to decrease observation uncertainties. Error bandwidths have not been significantly reduced, due mainly to inadequate network densities, poor quality and control over hydrologic observations and insufficient hydrologic data and information.[3]

The adequacy of observational networks varies widely by region, but observations for many water cycle variables have inadequate spatial and temporal coverage.[4] Continuously and consistently quantifying hydrologic variables at the global, regional and basin levels will require integrated observation systems that use both terrestrial and satellite observations. These systems will need data assimilation products, including models calibrated from the integrated networks and multiplatform observations.

Most of today's data for terrestrial observations and for many satellite observations are funded through national agencies, although satellite observations are increasingly carried out in the framework of multinational agreements. It would be worthwhile to investigate whether selected terrestrial observations could also be carried out under multinational agreements. Given the resource inputs from national agencies, product development and other derived services need to be responsive to national requirements to encourage continued national participation and funding.

A close feedback loop needs to be established between national data providers and the users of global observation systems. For developing countries this requires more participation in global projects. Most information generated from global observation networks is used by developed countries, and much less is used by developing countries. Sharing information across regions, however, becomes increasingly valuable – especially for smaller countries, which can complement their spatially or technologically restricted national observation networks, and for hydrometeorological variables and hydrologic data and information from shared river basins used for forecasting.

For network development and the design of multiplatform hydrologic observation systems, the heterogeneity of research fields makes it difficult to decide what variables should be collected and over what time and space. Network requirements also vary depending on the requirements of the research and application communities. A flexible network architecture with networks as subsets of larger composite networks or observation systems is thus desirable. Minimum operational baseline networks of routine observations (ground- and satellite-based) need to be augmented by research networks on a long-term basis.

Growing demand for precise assessments, forecasting and warnings requires quality management frameworks for all observation systems. The value of data depends on the accuracy and comparability of observations from different entities and instruments (including analytical laboratory procedures). Although regulatory data quality frameworks exist at the national and international levels, their implementation and adherence vary widely and are insufficiently documented from the country level to higher levels. Lack of adherence is frequently related to insufficient technical training of staff involved in measuring and managing data observations. The results are inadequately and irregularly calibrated instruments, few intercomparisons of different observation methods and analytical procedures, poor data quality control measures for consistency and homogeneity (the identification of systematic trends due to environmental or instrumental changes) and a lack of interaction and experience sharing among institutions.

Objective assessment of data quality is rare but nonetheless indispensable for reliably using data in decision-making. For various reasons, including poor quality and security considerations, several data providers are hesitant to share data. In developing countries the attitude exists that collecting some data (of questionable reliability) is better than collecting no data at all. But confidence in good-quality data can be an incentive to share the data with other countries, development programmes and partners.

Changing status of operational data over the recent past

This section examines the status of operational data from terrestrial hydrologic networks, multiple uses of national hydrologic observing systems, integration of multisource observations, the sharing of observations, observations and data on water use, and hydrologic observations from space.

State of terrestrial hydrologic networks

For terrestrial hydrologic observation systems, especially in many developing countries, data collection is inadequate and deteriorating. Many systems lack adequate quality assurance and control standards for calibrating instruments and reducing data. And basic capacity to access, interpret and apply water cycle information from both terrestrial and satellite observation systems is often insufficient.[5]

Many terrestrial hydrologic networks are shrinking for several reasons:

- Available records fulfil present hydrologic information requirements.
- No direct economically justifiable use of hydrologic information is apparent (for example, in pristine basins or stations close to the mouths of rivers and delta areas).
- Logistical problems.
- Budgetary or resource problems.

While the overall number of streamflow stations did not change significantly, stations with a long-term record were most affected by closures. The US Geological Survey reports that from 1980 to 2004, 2,051 stream gauges with 30 or more years of streamflow data were discontinued, leaving 7,360 at the end of 2005.[6] An important source of global hydrologic memory is being lost at a time when such information is needed to characterize the impacts of climate variability and change on hydrology and water resources. Another example is Kyrgyzstan in the Aral Sea basin, one of the best documented environmental disaster cases, where the number of hydrologic stations declined 48% during 1985-2005.[7]

The technical challenges behind the current situation of hydrologic data in Africa can be traced to the low quality and quantity of basic equipment, poor technology, few laboratories for recalibrating equipment, inadequately trained human resources at both professional and technician levels and insufficient funding and capital to sustain current operations or access new technologies.[8] Many challenges in accessing datasets in Africa arise from the reluctance of countries to exchange data freely. One reason: many countries do not feel sufficiently involved in regional or global studies and believe that their data services are not appreciated. Other factors are the absence of protocols and conventions for sharing water in some shared or international basins and aquifers, limited feedback from researchers and studies that use African data and fear of losing ownership of data.[9]

Anecdotal evidence (because no globally representative studies are available) and reviews of funding in the water sector suggest that national agencies and donors are not prepared to invest in multipurpose hydrologic networks expanding to regional or global hydrologic networks. But ongoing projects include new gauging stations at a limited scale, though these stations often operate only for a limited time for specific projects (including scoping projects, such as for proposed irrigation schemes) and rarely have historic records.

One major effort to stem the decline of hydrologic networks is the World Meteorological Organization's World Hydrological Cycle Observation System. Implemented at the regional and transboundary river basin levels, the programme focuses on establishing and operating requirements-driven hydrologic information systems.[10]

There is no systematic monitoring of groundwater, which constitutes 21% of the world's freshwater resources, at the regional or global level.[11] Few of the publicly accessible systems that contain relevant general information on water at the global level store data on groundwater quantity or quality. In many countries little attention is paid to the uniformity of hydrogeological data, precluding cross-comparisons. Few countries have up-to-date groundwater databases from which the current quality and quantity of groundwater resources can be elaborated. Without such systems information is obtained from model-derived data.

Information availability differs across time and space, as do the methods used to process this information. To address this situation, the International Groundwater Resources Centre has established the Global Groundwater Monitoring Network, which uses available data and derived information to periodically assess global groundwater resources.

An important source of global hydrologic memory is being lost at a time when such information is needed to characterize the impacts of climate variability and change on hydrology and water resources

Major investments are needed to reverse the decline of hydrologic observation networks, including surface water and groundwater observations and water quality monitoring

Surface observations reveal only indirect indicators of groundwater system status, such as changes in vegetation patterns, changes in river base flows, appearing or disappearing springs and wetlands, land subsidence and visible changes in the water levels of shallow, large-diameter wells. Subsurface observations are required to quantify groundwater storage through variations in groundwater levels, changing aquifer conditions and alterations in the chemical composition of groundwater. Groundwater levels and water sample collection are carried out using observation wells. Advanced data loggers such as pressure transducers and salinity sensors allow groundwater and salinity levels to be automatically measured at variable intervals. Geophysical methods (well logs and surface studies) help reveal changes in moisture content and salinity and trace some pollutants. There is promising evidence that gravimetric remote sensing methods allow aquifer monitoring on global to subregional levels, especially in sparsely gauged areas.[12]

The availability of usable freshwater is determined not only by the quantity of the resource but increasingly by its quality, which may further reduce the net availability of water resources for different uses and have critical environmental consequences. More than 100 countries contribute to the Global Environment Monitoring System – Water, a programme with more than 3,000 stations operated by Environment Canada under the auspices of the United Nations Environment Programme. The global database that it has built is key for global freshwater quality assessment. But a general lack of institutionalized, continuous dataflow into the database severely hampers further regional and global assessments as well as programmes to improve water quality, especially in transboundary basins.[13]

National data holdings are frequently fragmented, with no metadata catalogue in place that allows a complete picture of a country's water quality situation. Notwithstanding some positive examples, monitoring networks in developing countries are rudimentary, irregularly updated, rarely objective driven and without sufficient quality control to make observations truly useful. It would be a formidable task not only to establish sustainable baseline and specialized water quality observation networks, but also to make them interoperable with surface water and groundwater quantity observations that enable calculating pollutant loads and biogeochemical fluxes, for example.

Observational gaps result from failure to observe and collect data or from lack of access to data. Increasingly, the gaps are due to lost data and information as a result of disasters, social unrest and technological evolution. Data rescue programmes are thus crucial to retaining historical information and to expanding knowledge bases as far back as possible. This is especially important when considering that long records of observations are a prerequisite for detecting climate variability and changes in observations as well as for establishing the baseline hydrologic conditions that existed prior to development activities.

Major investments are needed to reverse the decline of hydrologic observation networks, including surface water and groundwater observations and water quality monitoring. National investment can be mobilized with assistance from development partners, but doing so is very difficult when investments are sought for transboundary hydrologic observation systems or global data collection and monitoring systems. This is partly due to the fact that most development partners, including institutional donors, focus their technical assistance on bilateral needs rather than on regional or global observing systems.

The general trends of in-situ observation methods include:

- Increasing use of automatic logging systems and replacement of instruments with mechanical recording devices.
- Widespread use of motionless observing methods, including measurement of hydrostatic pressure for gauge heights and observation methods without water contact, such as instruments using small radar devices suspended over the water surface to obtain gauge heights.
- Coupling of in-situ stations with automatic data transmission and telecommunications systems, including mobile phone communication using Global System for Mobile/General Packet Radio Service standards.
- Increasing integration of in-situ observation systems with basinwide hydrologic information systems, including forecasting and decision-support systems.

Observation networks for different purposes

National hydrologic observation networks often serve several purposes – such as providing information for water resources assessments and forecasting and serving as least influenced baseline hydrologic stations for climate studies. A clear delineation of national hydrologic observation systems for different uses is the exception rather than the rule. Specialized hydrologic networks are justified, especially when regular network observations do not provide the appropriate specific observations. One example is the need for extreme event data, which are generally less available than regular network observations.

Justifying networks with higher observational density and higher reporting frequency needs to be based on multiple-objective requirements analysis to assess additional economic or scientific value gained. The rational use of national hydrologic networks would be enhanced by clearly delineating priority use of stations based on such criteria as quality of data, length of records, location of the station and reporting cycle. Classifying hydrologic networks this way would allow rapid identification of subnetworks for specialized purposes. Of particular interest would be hydrologic stations recording flows of water to and from national territories to support assessment of the resource at the national level. In addition, observational networks are sometimes established for specific purposes – in particular for environmental assessment studies. Most of these stations are operated for only a limited time, and their data deliveries are generally poorly documented. One recommendation is to document and preserve short-duration hydrologic observations to supplement regular national hydrologic networks.

From data to information – integrating observations

Hydrologic forecasting and assessment products increasingly use multisource observations and complex data assimilation algorithms to improve accuracy, reliability and timeliness. For example, flood forecasts can be based on observations of precipitation from conventional rain gauges, hydrologic radars and satellite precipitation estimates combined with current (real-time) measurements of soil moisture and other variables. Thus, aside from technical opportunities and challenges, integrating multiple observation systems that operate at different spatial and temporal scales requires systems to ensure internally consistent data products, keeping in mind that local, regional and global datasets are equally important because they serve different purposes.

Sharing hydrologic observations

Other than technical obstacles, sharing of hydrologic data can be hampered by limited physical access to data, national data policy and security issues, lack of agreed protocols for sharing and commercial considerations. Whether hydrologic data are a public good or commodity does not have a simple answer. One argument is that data do not have an intrinsic value and therefore are not commercial products. Societal, scientific and commercial value is added as a result of information and service delivery products (such as hydrologic forecasts or assessments) that have identifiable and quantifiable socioeconomic values. For example, flood disaster damages are often less when an effective forecasting service exists. The cost-benefit ratio for hydrologic data used for forecasting ranges from 1:10 to 1:15.[14]

The basis for sharing data and information is defined requirements from national sectors, planning commissions, river basin organizations, hydrologic and climate research communities, and national and regional development partners. Commercial entities, including the public utility sector, are also increasing requirements for shared hydrologic data. A needs-based approach should form the basis of any data-sharing policy. Many data-sharing protocols and agreements already exist at the national, regional and global levels. Data-sharing agreements between riparian countries in transboundary basins where lower riparian countries have a disproportionally larger benefit from upstream observations could also require downstream users to contribute to the maintenance and operation of upstream stations in order to ensure long-term availability of hydrologic observations. Generally, scientific and technological advances and better management of water resources and hydrologic forecasting at all levels should not be hampered by restrictions in data-sharing arrangements.

The United Nations Economic Commission for Europe Convention on the Protection and Use of Transboundary Watercourses and International Lakes requires its parties to exchange data on water quality and quantity and pollution sources as well as environmental conditions of transboundary waters. The first Assessment of Transboundary Rivers, Lakes and Groundwaters in the region,

Integrating multiple observation systems that operate at different spatial and temporal scales requires systems to ensure internally consistent data products, keeping in mind that local, regional and global datasets are equally important because they serve different purposes

The paucity of contributions to the Global Runoff Data Centre is often related not to lack of infrastructure but to general unwillingness to share data in an institutionalized, regular manner

developed under the Water Convention in 2007, describes the hydrologic regime of 140 transboundary rivers, 30 transboundary lakes and 70 transboundary aquifers together with pressure factors in their basins, their status and transboundary impact, as well as trends and envisaged management measures.[15]

Observational gaps also arise when hydrologic observations are not shared. Regional differences in data provided to the Global Runoff Data Centre are apparent in both quantity and timeliness. Hydrologic data from North and Central America, the Caribbean, Europe and Mediterranean Asia are far greater than data from other regions (map 13.1). In other regions few hydrologic stations provide data to the centre, and update intervals are too great (figure 13.1). The paucity of contributions is often related not to lack of infrastructure but to general unwillingness to share data in an institutionalized, regular manner. This regional picture also suggests that most data are shared by a rather constant number of national hydrologic services, with few new services added over time. This hampers regional and global projects that have to build on such datasets for scientific and applications-oriented purposes. Examples include the calibration of models to provide seasonal regional hydrologic outlooks, forecasting, disaster warning and prevention, and water management in transboundary basins.

Inadequate use of current information technologies severely threatens data and information sharing even when basic hydrologic observations are available. Typical inadequacies relate to incompatible information systems, which effectively block the seamless exchange of data and information between systems and different operators and programmes from the national to the global level. These inadequacies lead to an effective blackout of potentially valuable data and information because they are not part of the information management cycle. The situation is especially critical for forecasting purposes when data need to be shared in real or near real time.

Observations and data related to water use

Many models used to extrapolate from observations to areas with no observations available are designed to simulate natural conditions based on hydrometeorological inputs. For model outputs to reflect quantitative changes of water resources availability and use over space and time, model components that simulate anthropogenic development impacts (such as population trends, economic activities and land use changes) need to be integrated. Model calibration and validation likewise rely on information on water storage in lakes and reservoirs, water abstractions and return flows.

Global data on water use exist primarily for the agricultural sector (for example, the AQUASTAT database by the Food and Agriculture Organization of the United Nations). Data on consumptive and non-consumptive use of water resources are not a regular part of many national statistics,

Map 13.1 **Distribution of Global Runoff Data Centre streamflow gauges**

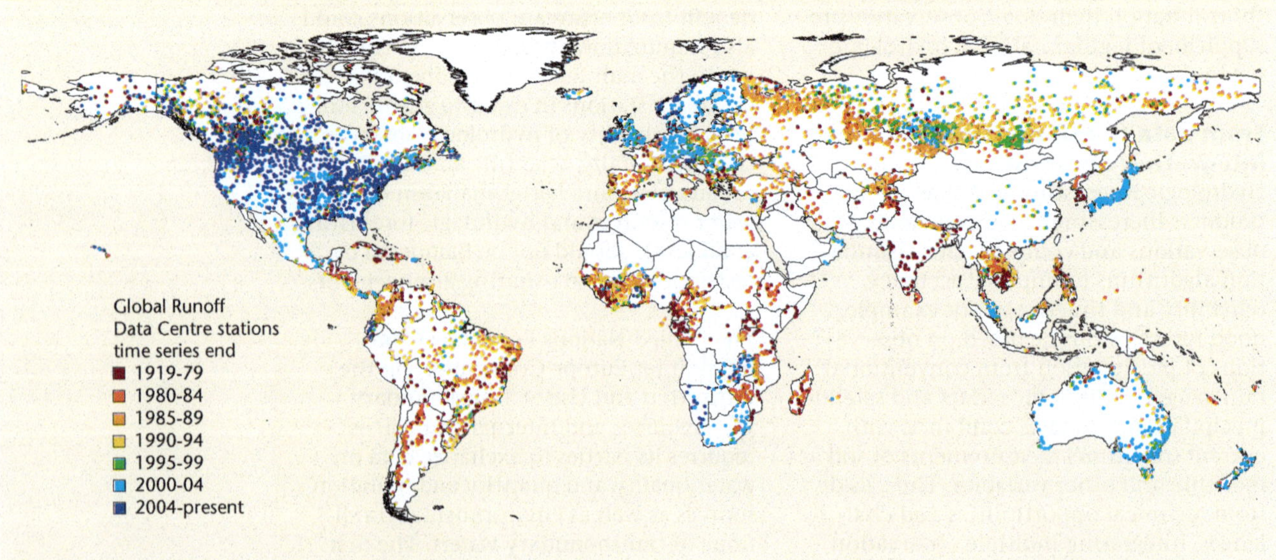

Source: Global Runoff Data Centre (http://grdc.bafg.de/).

Figure 13.1 **Data delivery rate from runoff gauging stations to the Global Runoff Data Centre (number of stations from which data have been received, 2001-07)**

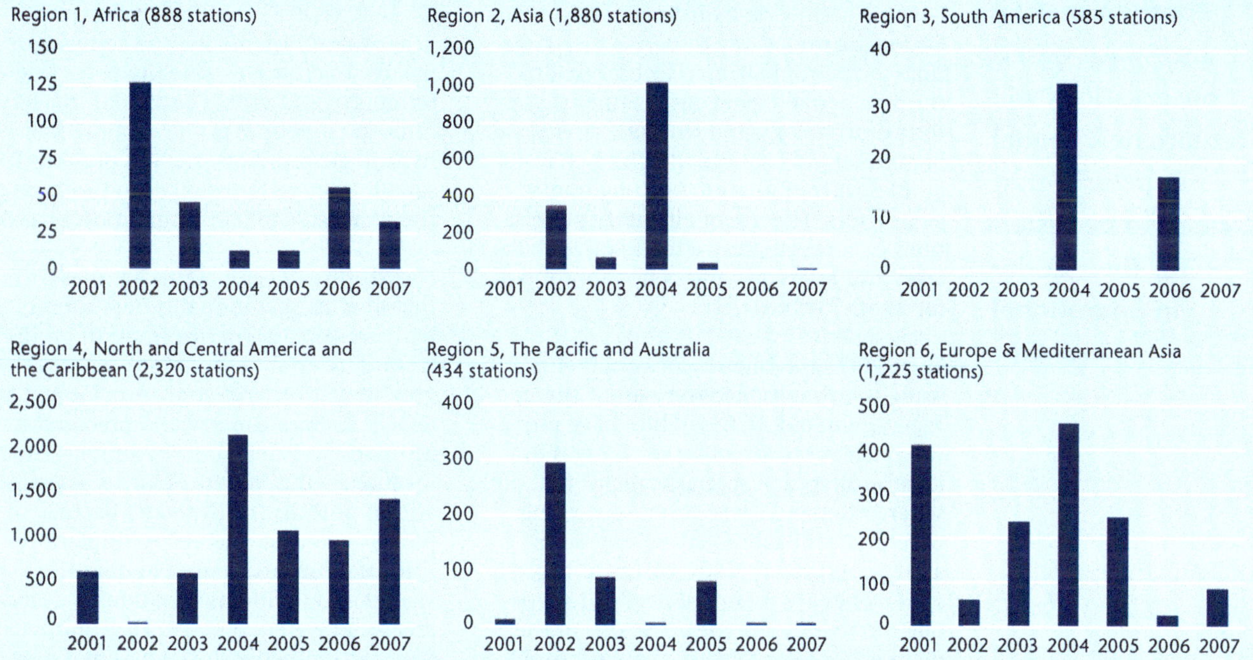

Note: Regions are World Meteorological Organization classifications.
Source: Global Runoff Data Centre (http://grdc.bafg.de/).

although where abstractions are licensed, information on the maximum use of water may be available. This situation is unsatisfactory because it prevents effective management of water demand relative to availability and supply of freshwater.

Despite the importance for the integrated management of water quantity and quality and for understanding water-related health hazards, no comprehensive information exists on the regional or global extent of wastewater generation and treatment and receiving water quality. Even at the national level such information is either inconsistently gathered or unavailable – partly because of ill-defined data collection responsibilities that rest with a multitude of national organizations and commercial entities that rarely share their information.

Hydrologic observations from space

Satellite observations are important means for providing hydrologic data with acceptable spatial and temporal resolution, especially in areas with no or limited infrastructure. However, 'acceptable resolution' depends on what the data are to be used for. In-situ data provide acceptable coverage and temporal resolution mainly in more accessible regions. In-situ data are also used for calibrating and validating space-based information, hydrologic or water resources models and routine real-time forecasting services. For forecasting of runoff and water flows, data must be available within a fraction of an hour up to several hours, depending on the size of the basin and forecasting requirements. For assessment a temporal resolution of several weeks up to one month may suffice.

High priority for additional observations should be focused on data-poor regions, poorly observed hydrologic variables, regions sensitive to change and variables with inadequate spatial resolution. The need for satellite-based observations that complement in-situ observations should be recognized. In-situ and space-based observations for hydrologic applications need to be integrated in a comparable space and time domain and under tight quality control. Such quality control would require increased efforts to assess observation quality through intercomparison and recalibration projects. This is especially important for achieving continuity between historical terrestrial observations and new satellite observations.

Terrestrial water level observations can now be supplemented with sufficient accuracy by radar altimetry instruments flown on the Envisat, Jason and TOPEX satellites. Because the sensor carriers on these satellites are not geostationary, the altimetry observations are taken at virtual gauging

Merging data streams from both terrestrial and space-based observations will require new model structures that need to be tested for their utility in operational services

stations along the path of the satellite at various stretches of a river with an approximately weekly repetition cycle. This allows for basinwide hydrologic assessments. Efforts are under way to derive discharge time series from altimetry observations through actual rating curves that use the river morphology and through virtual calibration curves that use idealized channel profiles in combination with hydraulic parameters. The use of altimetry observations is, however, restricted to large rivers, lakes and reservoirs and is not appropriate for smaller tributaries.

Since 2002 gravimetric measurements using the Gravity Recovery and Climate Experiment (GRACE) satellite have provided the means to observe changes in large aquifers at a spatial scale over 40,000 square kilometres. However, separating water masses (soil moisture, vegetation and groundwater) is still difficult. Inverse approaches are therefore needed to separate the hydrologic contributions of the main water reservoirs (oceans, atmosphere and total continental water storage including snow, soil wetness, groundwater and ice caps) from monthly synthetic GRACE geoïds. Nevertheless, at large scales gravimetric observations of changes in large aquifers from GRACE and next-generation missions like the Gravity Field and Steady-State Ocean Circulation Explorer (GOCE) satellite are nearing a stage where they could be used operationally.

Apart from directly measuring hydrologic variables such as water levels, snow and ice cover extent, soil moisture and groundwater, satellites are invaluable for providing information for hydrologic estimation (modelling) studies. Satellite estimates of precipitation are an obvious example. Several precipitation estimation methods based on microwave and infrared instruments on numerous satellites and associated different data retrieval and interpretation algorithms have reached a semi-operational and, in a few cases, an operational level.[16] Some global or near-global precipitation products have spatial resolutions down to 4 kilometres and temporal resolutions down to three-hourly, making them useful for water management and flood forecasting. Several hydrology-relevant products are associated with the Moderate Resolution Imaging Spectroradiometer instrument operating on both the Terra and Aqua spacecraft: inundation areas, surface changes of lakes and reservoirs, surface reflectance, temperature, land cover and indexes of vegetation change.

Opportunities and challenges

The combination of observations from GRACE, GOCE, satellite altimetry and other space systems such as active and passive microwaves, satellite radar interferometry and visible and radar imagery offer the potential for developing new hydrologic products, such as present-day satellite-derived precipitation products and the planned Global Precipitation Mission.

Near-future developments are likely to include the operational generation of truly multiplatform information products from terrestrial observations merged with gravimetric observations from GRACE and GOCE, radar altimetry and precipitation estimates as inputs into hydrologic models for forecasting, water resources assessment and monitoring of the water balance of basins. Likewise, merged multiplatform information can be used to quantify spatial and temporal variations of flooded areas and water volumes. The improved cooperation between the national meteorological and hydrologic services of many countries with the aim to improve hydrologic forecasting is at the heart of the World Meteorological Organization Flood Forecasting Initiative, launched in 2003. The availability of multiplatform, multivariate data streams in near real time and with high accuracy will be crucial to achieving this goal.

Using the global concept of the World Hydrological Cycle Observing System, building on terrestrial hydrologic information systems and obtaining global hydrologic coverage from satellites to integrate terrestrial and satellite-based observations could close many gaps in hydrologic observations at all levels and support improved hydrologic and water resources assessments, forecasting and research. To make full use of satellite observations, a suite of intercomparison and validation projects is needed to assess the accuracy of satellite-based observations with terrestrial observations in a wide range of environmental conditions. Merging data streams from both terrestrial and space-based observations will require new model structures that need to be tested for their utility in operational services.

Despite current technological and methodological developments, an important challenge remains. To mainstream satellite-based observations that complement terrestrial observations in operational water resources management and forecasting services on a routine basis and for critical assessments, awareness and capacity are needed in national meteorological and hydrologic agencies. Likewise, space organizations need to know the requirements for space-based observations in order to design and operate new and tailor-made missions and to create derived observational and model products for hydrology and water resources management. Space agencies will need to develop front-end tools that allow primary data to be converted into graphics and tables that can be used in models and forecasting routines. This will require an intensified dialogue among space agencies, the science community and hydrologic and meteorologic services to define interinstitutional cooperation and sharing of responsibilities on long-term archiving of satellite observations, access to data and information in support of science and research, and the development of products for operational applications.

Some suggestions for bridging the observational gap

Meta-information systems that promote information rescue and institutional sustainability of water information and knowledge are prerequisites for all levels of water management. Meta-information systems provide generic information about data, information, knowledge sources and data products that are applicable to operations and research. Online dedicated global information systems-based applications can improve access to information. Through a geographic interface and standardized set of water-related attributes, information can be seen in spatial context that shows analogies and patterns. Such applications are indispensable for the global sharing of information and apply to all components of the water cycle.

In addition to the key messages of this chapter, other important steps can be taken to improve the current situation of severe observational gaps in hydrologic observations include:

- At the national, regional and global levels a minimum requirements analysis of long-term, multipurpose observational needs should be undertaken; a new requirement is climate-relevant observations, including those from pristine basins.
- Financing of hydrologic networks, including operation and maintenance, should be based on a multiple-source strategy rather than the prevailing single-source, sector-specific funding arrangements.
- Integrated multiplatform network solutions that combine in-situ and space-based observations and that are affordable for developing countries should be promoted. This would enhance the observational base in spatial and temporal coverage.
- Other hydrologic information – such as in-situ and remotely sensed soil moisture and meteorological data and information including precipitation, evaporation, humidity, temperature and wind fields – needs to be considered to complement hydrologic information and to enhance the information content of hydrologic data through integration in multivariate models and predictions.
- In data-sparse regions in particular, modelling approaches need to be mainstreamed to generate model-derived observation time series. A promising tool is the reconstruction of hydroclimatic data by downscaling. Datasets from the National Centers for Environmental Prediction/National Center for Atmospheric Research for 1948-2007 are a widely used source.
- As observational gaps are often directly related to deficiencies in data transmission and communication, this could be overcome to a large degree by connecting offline operating stations to modern telecommunication systems to increase spatial and temporal availability of data from already existing stations.
- Making maximum use of existing hydrologic observations requires more effort to share hydrologic data and information on all levels, including transboundary river basins and shared aquifer systems.

Online dedicated global information systems-based applications are indispensable for the global sharing of information and apply to all components of the water cycle

Notes

1. GCOS 2003b; US Geological Survey n.d.
2. Personal communication with Richard Robarts, Global Environment Monitoring System–Water.
3. Based on author's discussions with instrument manufactures such as OTT and SEBA.
4. GCOS 2003b.
5. GCOS 2003b.
6. US Geological Survey n.d.
7. Grabs 2007a; and presentation by Kyrgyzstan Hydromet at an ARAL-HYCOS planning meeting, 6-7 December 2006, Almaty.
8. GCOS 2003a.
9. GCPS 2003a.
10. World Hydrological Cycle Observing System (www.whycos.org).
11. Based on data from the International Groundwater Resources Assessment Centre.
12. Grabs 2007b.
13. GEMS-Water 2008.
14. WRI 2004.
15. UNECE 2007.
16. Grabs 2007b.

References

GCOS (Global Climate Observing System). 2003a. *Report of the GCOS/GTOS/HWRP Expert Meeting on Hydrological Data for Global Studies.* WMO/TD-No. 1156. Geneva: World Meteorological Organization.

———. 2003b. *The Second Report on the Adequacy of the Global Observing Systems for Climate in Support of the UNFCCC.* WMO/TD No. 1143. Geneva: World Meteorological Organization.

GEMS-Water (Global Environment Monitoring System–Water Programme). 2008. *Water Quality for Ecosystem and Human Health.* 2nd ed. Nairobi: United Nations Environment Programme.

Grabs, W. 2007a. The State of Hydrological Observation Networks: Technical and Operational Aspects. Overview of Global and Regional Programmes. Presentation for the Third International Conference on Water and Climate, 3-6 September, Helsinki.

———. 2007b. Status of Hydrological In-Situ Networks and Future Developments Including Space-Based Observations. Presentation for the Second Space for Hydrology Workshop, 12-14 November, Geneva.

UNECE (United Nations Economic Commission for Europe). 2007. *Our Waters: Joining Hands across Borders. First Assessment of Transboundary Rivers, Lakes, and Groundwaters.* Geneva and New York: United Nations. www.unece.org/env/water/publications/assessment/assessmentweb_full.pdf.

US Geological Survey. n.d. Trends in the Size of the USGS Streamgaging Network. National Streamflow Information Program, Washington, DC. http://water.usgs.gov/nsip/streamgaging_note.html.

WRI (World Resources Institute). 2004. *World Resources 2002-2004: Decisions for the Earth: Balance, Voice, and Power.* Washington, DC: World Resources Institute.

PART 4

Responses and choices

Chapter 14
Coordinator
Håkan Tropp
(UNDP)

Facilitator
George de Gooijer

Chapters 15-16
Coordinator
Olcay Ünver
(WWAP)

Facilitator
William Cosgrove

PART

4

Authors: George de Gooijer and Håkan Tropp

The Report describes the costs of not investing in water resources development and management – the economic losses and the human suffering and underdevelopment. It also demonstrates the high rates of return generated by such investments.

Part 4 is about response options and how to choose among them. What options are available to decision-makers to respond to the challenges identified in the previous chapters? How can they select among these options? And what are some of the trade-offs in choosing certain options? What can transform competition into synergy?

We have many of the answers. Across the planet we have already shown that it can be done! But there is no one-size-fits-all solution. The best mix of responses to a country's development objectives and policy priorities to meet its water challenges depends on the availability of water in space and in time and the country's technical, financial, institutional and human capacities – its culture, political and regulatory frameworks, and markets.

Responses outside the water domain strongly affect the macro changes that influence how water is used and allocated. They also make water adaptation measures more (or less) effective and less (or more) costly. The Report shows that decisions taken by external actors may have more impact on the state, use and management of water resources than decisions taken by managers within the water sector itself. The demand for and the provision of food and energy, the uneven ability to invest and a changing climate exert strong pressures on water. However, the most important decisions affecting water are often made without water as the primary concern, even though water may play an important role in addressing the issue. Policy decisions on health, food security and energy security, for example, can intensify or alleviate much of the pressure on water resources, affecting both supply and demand.

Key drivers of water use changes – as seen in previous chapters – include demographics, economic development and trade, consumption and climate. These drivers have powerful implications for the options available inside and outside the water domain and for how to mobilize decision-makers and other important actors.

Water resources are strongly affected by climate change and variability. The responses to the challenges posed by climate change, through best practices and low-regret measures, are also specific to each country, especially for vulnerable hot spots such as low-lying islands, deltas, mountain areas and arid regions, where action is not only cost effective in the long term but also urgent.

Options within the water domain are distinct from those outside it. Leaders in the water domain can inform the processes outside their domain and implement decisions for the water domain; but it is the leaders in government, the private sector and civil society who determine the directions that will be taken.

Many countries face multiple challenges but have limited financial and natural resources and implementation capacities. Countries need to fully use synergy opportunities and to make trade-offs and difficult decisions on how to allocate among uses and users to protect their water resources. Improved water management depends on several interrelated factors, including accurate knowledge of the water-related problems to be addressed, their root causes and the management options available to address them; political will; stakeholder participation and cultural acceptance; transparency in management and decision-making; effectiveness of institutional frameworks; and sustainable funding.

An appropriate set of approaches and strategies must be assembled for each country and situation, based on the biophysical characteristics of a water system as well as local, national and sometimes international characteristics and capacities for achieving sustainability of water resources.

The urgency to respond will be especially high for countries already facing severe water challenges and for countries that will face even more severe water challenges if current climate, demographic, socioeconomic and development trends continue.

Responses outside the water domain are paramount in influencing the macro changes in how water is used and allocated and in making adaptation in the water domain more effective, better integrated and less costly. Population growth, urbanization and climate change are forcing the water domain to adapt. Broader policy change and political action are required to change fundamental allocations and uses of water. Global market conditions and trade regimes affect crop prices and choices and thus also have serious implications for agricultural water use and demand.

PART
4

Economic development can improve the water situation for many people, but it can also cause overexploitation of water and the environment.

Traditionally within the water sector the first response to lack of water has been increasing supply. The second response became managing demand, enhancing efficiency and reducing losses. The third response is more drastic and requires decisions outside the water sector on reallocating resources, which can exclude some sectors from further supply. Effective water management combines all three responses and involves all sectors. To achieve results, many actors need to be involved. More often than not this requires convincing fellow decision-makers with well documented arguments. Water professionals who understand the social, economic and political conditions outside the water sector that directly and indirectly affect how water is being used and governed can better inform and participate in decision-making outside the water sector.

Chapter 14 looks at possible responses within the water domain, and chapter 15 looks at responses outside the water domain that affect how water is being used and allocated. The examples of response options are pragmatic and include responses by governments, private sector, civil society and consumers at various levels and scales. The options take into account geographic and hydroclimatic conditions, the level of economic development, water subsectors and the supply and demand side of the water resources equation within the broader socioeconomic framework.

Chapter 16 discusses the need for accelerated investments and informed decision-making through partnerships. It also considers the consequences of increasing uncertainty and how to prepare for decisions under such conditions.

Chapter 14

Options inside the water box

Authors: George de Gooijer, Walter Rast and Håkan Tropp

Contributors: Virginie Aimard, Guy Allaerts, Maggie Black, Rutgerd Boelens, Stefano Burchi, Hein Engel, Joakim Harlin, Molly Hellmuth, Sarah Hendry, Andrew Hudson, Anders Jägerskog, Henrik Larsen, Patrick Moriarty, Eelco van Beek, Peter van der Zaag and James Winpenny

Coordinator: Håkan Tropp (UNDP)

Facilitator: George de Gooijer

Key messages

There are many practical examples of solutions within the water domain. Some options show particular promise:

- Supporting institutional development, to prepare institutions to deal with current and future challenges, through such reforms as decentralization, stakeholder participation and transparency, increased corporatization where feasible and fair, partnerships and coordination (public-private, public-public, public-civil society), and new administrative systems based on shared benefits of water, including when water crosses borders.
- Considering the influence of water law, both formal and customary, including regulations in other sectors that influence the management of water resources.
- Consulting with stakeholders and ensuring accountability in planning, implementation and management as well as building trust within the water and related sectors and fighting corruption and mismanagement.
- Strengthening organization structures and improving the operating efficiency of water supply utilities to improve service quality and increase the coverage and density of connections, while also boosting revenues and creating a more viable financial base to attract further investment.
- Developing appropriate solutions through innovation and research.
- Developing institutional and human capacity, both within the water domain and in areas or sectors outside the water domain. Capacity development can occur through traditional forms of education, on-the-job training, e-learning, public awareness raising, knowledge management and professional networks.
- Creating a favourable investment climate of sound management accountability and good governance within the water sector. This should include new approaches such as payment for environmental services.

Programs and activities are under way around the world that directly address the assessment, allocation or conservation of water resources. Improving water governance includes more efficiently managing available water resources and current and anticipated water uses, and informing water users, stakeholders and decision-

Identifying and implementing effective governance responses to water-related problems remain an elusive goal

makers about the consequences of actions taken (or not taken) to address such issues.

This chapter focuses on what can be done within the domain of water managers to address water problems. Because other publications and reports already deal with the technical and engineering aspects of response options,[1] the main emphasis is on strengthening policy and laws, water resources management, technical capacity, finances and education and awareness.

Undergirding solutions inside the water domain is good water governance, which influences the choices people make about the use (or misuse) of water resources. This focus includes such approaches as water supply and demand management, as well as processes to ensure the collection, analysis and use of the data necessary for making water allocation decisions within a policy framework that is flexible, comprehensive and realistic.

Water governance reform: strengthening policy, planning and institutions

In this Report water governance refers to the political, social, economic, legal and administrative systems that develop and manage water resources and water services delivery at different levels of society while recognizing the role played by environmental services.[2] It encompasses a range of water-related public policies and institutional frameworks and mobilization of the resources needed to support them. Governance issues overlap the technical, environmental and economic aspects of water resources and the political and administrative elements of solving water-related problems (see chapter 4). Identifying and implementing effective governance responses to water-related problems – taking into account the differing contexts in which they may be applied, their integration with other sectors and their impacts on water use equity, efficiency and environmental sustainability – thus remain an elusive goal.

Although many governments have met increasing water demands primarily by augmenting water supplies, many also apply management and technological solutions to address water demands, including more efficient use and conservation. As pressures on water resources continue to rise, countries will also need to consider reallocating water resources from one sector to another, further politicizing water issues within countries and between them. Water reallocation and demand management options include administered solutions (such as allocating less water to agriculture) and economic incentives (providing price signals to decision-makers about the opportunity costs of water). Capital investments in infrastructure, sorely needed in many countries, also require investments in capacity and institutional development to realize and sustain the benefits of increased investments.

The stages of water development and management range from situations in which virtually all water development possibilities have been exhausted (as the Fuyang basin in China, and in the Middle East and North Africa region) to situations in which water development potential still exists (such as the East Rapti River basin in Nepal and many sub-Saharan countries).[3]

Improved planning policies and laws are critical response options. Effective enforcement by government agencies and acceptability and compliance by providers, users and stakeholders are also important to the effectiveness of many water management reforms, as evaluation of the effectiveness of lake basin management interventions in many developing countries has shown.[4]

Integrated approaches to water planning and reform

Both developed and developing countries are reforming their water resources planning policies and laws. European Union members, for example, are implementing the Water Framework Directive. Many middle- and low-income countries in Africa, Asia and Latin America are engaging in reform, focusing on principles of integrated water resources management. A recent United Nations report concluded that implementation of the practices, especially water use efficiency improvements, is lagging (box 14.1).[5]

Implementing integrated water resources management is proving more difficult than envisioned. The approach was meant to facilitate integrating water priorities and related environmental issues into national economic development activities, a goal often considered only after many development activities have already been undertaken (box 14.2). The Sixteenth Session of the Commission on Sustainable Development endorsed integrated water resources management as a framework and essential tool for effectively managing water resources. It recommended that its review on progress in the water and sanitation sector go beyond mere stocktaking of integrated water resources management efforts.[6]

Box 14.1 UN-Water survey on progress towards 2005 targets for integrated water resources management and water efficiency plans

The growing stress on water resources presents managers with increasingly difficult decisions for managing water sustainability. Integrated water resources management assists such decision-making by drawing attention to efficient, equitable and environmentally sound approaches. At the World Summit on Sustainable Development in 2002 countries agreed on a global target to develop integrated water resources management and water efficiency plans by 2005, with support to developing countries throughout all levels.

In 2007/08 UN-Water conducted a survey of government agencies in 104 countries (77 of them developing countries or economies in transition) to assess progress towards the target. A questionnaire was prepared by a UN-Water task force and sent to all Commission on Sustainable Development focal points by the United Nations Department of Economic and Social Affairs. It included questions in the following categories:

- *Main national instruments and strategies that promote integrated water resources management* (policies, laws and integrated water resources management plans).
- *Water resources development,* represented by such items as assessments, regulatory guidelines and basin studies.
- *Water resources management,* as reflected in programs for watershed management, flood control and efficient allocation.
- *Water use,* represented by water use surveys and programs for managing agricultural, industrial and domestic water uses.
- *Monitoring, information management and dissemination,* as reflected in monitoring and data collection networks.
- *Institutional capacity building,* represented by such items as institutional reforms, river basin management institutions and technical capacity-building programs.
- *Stakeholder participation,* illustrated, for example, by decentralized structures, partnerships and gender mainstreaming.
- *Financing,* represented by such items as investment plans, cost recovery mechanisms and subsidies.

Of the 27 developed countries that responded, only 6 have fully implemented national integrated water resources management plans. Another 10 countries have plans in place and partially implemented. Particular areas of improvement are public awareness campaigns and gender mainstreaming.

Of the 77 developing countries that responded 38% had completed plans, with the Americas at 43%, Africa at 38% and Asia at 33%. Africa lags behind Asia and the Americas on most issues, although it is more advanced in stakeholder participation, subsidies and microcredit programs. Asia leads in institutional reform. The survey concluded that to adequately assess the needs for advancing implementation of integrated water resources management, countries needed better indicators and monitoring.

Source: UN-Water 2008.

Box 14.2 Responses to impacts of non-integrated approaches to water resources management

The transboundary Rio Grande River between the United States and Mexico (called Rio Bravo in Mexico) illustrates the negative impacts of non-integrated water resources management. In both countries the region is one of the fastest growing, benefiting from the enhanced economic activities associated with the North American Free Trade Agreement (NAFTA). One result of NAFTA has been a proliferation of product assembly plants (*maquiladoras*) on the Mexican side of the border, making it a magnet for job seekers in Mexico. This growth was accompanied by a proliferation of informal settlements (*colonias*) on both sides of the border. Further, there is extensive agriculture in the lower Rio Grande Valley, making agriculture an important economic sector for Mexico and Texas. Finally, there are seven major paired urban areas along the international portion of the Rio Grande.

Because of the associated water demands, approximately 96% of the average flow of the river is allocated for municipal, agricultural and industrial uses. Although water allocations are governed by several treaties, the river also is subject to the jurisdictional concerns of federal and state agencies in three states in the United States and five states in Mexico. With responsibility for the river's quantity, quality and allocations residing in several international, national and state organizations with differing mandates in two countries, much of the river is overallocated and degraded.

An analysis of 67 EU projects related to integrated water resources management during 1994-2006 provides insight into the practical challenges facing implementation. The analysis finds that integrated water resources management can provide a useful reform and planning framework, even though it has not yet provided unequivocal guidance on implementing national water planning and reforms. To be most effective, the analysis suggests, integrated water resources management must consider policy formulation and implementation as primarily a political process involving government officials, the private sector and civil society.

Source: Moore, Rast, and Pulich 2002; Gyawali et al. 2006.

Too often, only part of the sanitation system is implemented under the guise of being a sanitation solution, and it is later discovered that other components are missing

For example, progress towards meeting the Millennium Development Goal for sanitation is lagging behind expectations partly because of the traditional approach to this issue. 'Sanitation' usually refers to a single technology or instrument designed to handle excreta and wastewater. Septic tanks, pit latrines and composting toilets, among others, are often referred to as 'sanitation systems', when in fact they are technological components. When designed appropriately and linked to a range of other components, they form a robust, sustainable sanitation system.

Too often, only part of the sanitation system is implemented under the guise of being a sanitation solution, and it is later discovered that other components are missing. Examples of such components include provisions for treated effluent (which is often diverted into open drains), fæcal sludge (which is often dumped in open fields) and other side streams that may be generated (such as water from sinks and showers). While the technological components themselves may work, the system as a whole will probably be short-lived.

A sustainable sanitation system includes all the components (physical parts and actions) required to adequately manage human waste. When 'sanitation' is considered as a multistep process and not as a single point, waste products are accounted for from generation to ultimate destination. This concept describes the lifecycle of wastes generated at the household level, which are then processed (stored, transformed and transported) until reaching a final destination. Ideally, waste would be used beneficially, with the nutrients, biogas, soil-conditioner and irrigation water components recovered from wastes and with wastewater used to benefit society in a cycle ('closing the loop'). For example, biogas could be used for cooking gas or electricity, soil fertility could be increased with added soil-conditioner from sludge drying beds and crop production in peri-urban agriculture could be improved through nutrient-rich irrigation water from a constructed wetland.

A more sustainable, holistic sanitation system can be designed by using components that would, in conjunction with existing or innovative new technologies, improve coverage and service while reducing the environmental burden. Several frameworks for a more systematic way of looking at sanitation systems already exist.[7]

Allocating and reallocating water and financial resources is unavoidable in water policy and management. Different

Box 14.3 Water resources management in Tunisia

Water resources management in Tunisia began with development of the supply side, addressing the water demands of various sectors. The country has since established a system of interlinked water sources, making it possible to provide water for multiple purposes, including mixed low- and higher-salinity waters, which makes less usable water more productive. The country developed a national water-savings strategy for both urban and agricultural needs at an early stage of water planning, confirming a cultural 'oasis' tradition of frugal management of scarce water resources.

Several principles underlie the Tunisian water strategy. First is shifting from isolated technical measures to a more integrated water management approach – for example, a participatory approach giving more responsibility to water users. Some 960 water user associations were created, encompassing 60% of the irrigated public areas. Second is the gradual introduction of water reforms and their adaptation to local situations. Third is the use of financial incentives to promote water-efficient equipment and technologies. Fourth is supporting farmer incomes to allow them to plan for and secure agricultural investment and labour. And fifth is a transparent and flexible water pricing system, aligned with national goals of food security, that will gradually lead to the recovery of costs.

Irrigation water demands have been stable for the past six years, despite increasing agricultural development, seasonal peaks in water demands and unfavourable climate conditions (including droughts). The country is now addressing the current water demands of tourism (a source of foreign currency) and of urban areas, to maintain social stability. Wastewater from urban centres is treated and made available for agricultural use. A targeted pricing policy enables full recovery of operating costs of water services, with tourists paying the highest water prices and household users the lowest. Water system monitoring is extensive, including real-time information on all irrigation flows. One result is improved groundwater storage and vegetation recovery in sensitive natural areas. The current plan ends in 2010.

Despite successes, however, Tunisia's water resources are still under considerable stress. The combination of increasing population growth and rising water use in all sectors signals major future threats, providing the impetus for considering scenarios to address future allocation choices.

Source: Blue Plan, MAP, and UNEP 2002; Treyer 2004; UNEP 2008.

stakeholders may see themselves as either winners or losers as a result of the changes arising from reform and planning efforts. Thus, planning should clearly identify the trade-offs among various management options (box 14.3). Further, coping with water deficits and challenges often requires improved engagement between water actors based on water management and allocation issues that extend beyond watershed boundaries.[8] One clear conclusion is that water governance, management and use cannot be considered independently if the goal is water resources sustainability. Both the causes of water problems and their solutions are partly embedded in processes and forces outside the water domain (see chapter 15).

Some countries are developing scenario-based planning tools. The Netherlands has used scenario-based planning to help make decisions on water management options. Its first plan, during the late-1960s, addressed only water quantity, but more recent plans have evolved into a multifaceted water management process, with a main pillar being stakeholder involvement, from other ministries and local authorities to the private sector and the public (box 14.4).

Many Asian countries have essentially exhausted opportunities to further expand irrigated agriculture and other water diversions. The objective in these cases is to increase the productivity of developed water resources and the effectiveness of water-management institutions.[9] But despite progress in improving water management, many countries still face major challenges, a consequence of increasing population, growing water use by all sectors and the approaching impacts of climate change.

Institutional developments: current response options in water reform

The ability of water management agencies to address water variability and to deal with risk and uncertainties such as climate change varies considerably, mainly because few countries have internalized water resources management concerns into their socioeconomic development policies and governance systems. Nor are countries fully including water stakeholders in the decision-making process, particularly at

Few countries have internalized water resources management concerns into their socioeconomic development policies and governance systems

Box 14.4 Integrated water planning in the Netherlands

The Netherlands is preparing its fifth integrated water management plan, with the possible consequences of climate change high on the agenda.

Its first plan, prepared in 1968, was supply-driven and addressed only quantity issues. Deteriorating water quality, and the very dry summer of 1976, led to fundamental changes in the country's approach to water management. The second plan had to be completely different, so the Policy Analysis of Water Management for the Netherlands was carried out before drafting the second plan.

For the second plan, despite a thousand years of experience in water management, the government enlisted the assistance of the RAND Corporation, a U.S.-based think tank with extensive experience in complex policy processes. The company had been involved in an earlier integrated water project in the Netherlands – the storm surge barrier, a multibillion dollar project to protect the southern part of the country – helping them secure close cooperation of other ministries and governmental levels involved.

The plan was expected to achieve three primary ends: develop and apply a methodology for producing alternative water management policies, assess and compare their consequences and help create domestic capability to conduct similar analyses by the related Dutch entities.

Using more than 50 models, the project resulted in a much better operational understanding of the water system. Multiple cost-benefit analyses of options to improve water management led to the identification of implementable local projects and helped avoid large, expensive infrastructure works that proved not to be cost-effective. An important conclusion was that water quality problems cannot be solved at the national level – for example, reallocating water could inflict large losses on some sectors. Tight restrictions on groundwater abstractions were needed to meet desired environmental standards, which would impose large losses on some users. The second plan, published in 1984, reflected this complete change in thinking about how to develop and manage the water system.

Subsequent water management plans continued to develop integrated water resources management. The third plan (1989) added in-depth analysis of the role of ecology in water management, and the fourth plan (1998) focused on specific water systems and themes, facilitating implementation of needed actions and clarifying institutional roles.

The evolution of these five water plans, each building on its predecessor and responding to changing circumstances, facilitated significant shifts in thinking and engendered new approaches to water management. From its origins in a technical, supply-oriented, model-based decision process, the planning process is now multifaceted, with a main pillar being stakeholder involvement (other ministries, local authorities, public and the like) with a focus on sustainability and climate proofing related to anticipated changes.

The lessons learned from developing the five water management plans in an integrated manner are that implementing complete integrated water resources management takes time (more than 30 years in the Netherlands), that external input can facilitate implementation of new concepts and that full involvement of all stakeholders is needed. While cooperation may not always be possible, involvement is essential. By helping stakeholders understand the difficult trade-offs, the participatory practices made it easier to accept the importance of change for the greater good of society.

Source: E. van Beek, H. Engel and G. C. de Gooijer.

Many components of ongoing water reform are part of broader governance reform agendas

the basin level. And still missing are the information, planning tools, management strategies, and human, institutional and system capacities needed to meet local demand for sustainable water development under conditions of climate variability and change.

Many institutional systems are unable to adapt to current and future challenges because of such factors as political power monopolies, unilateral steering by government and bureaucracy, hierarchical control, top-down management and institutional fragmentation. These institutional characteristics also prevent political decision-makers from being fully informed by water sector managers. Still, many developing countries and economies in transition are transforming their water management systems through integrated water resources management approaches. They are incorporating such elements as decentralization (subsidiarity), transparency and stakeholder participation, administrative systems based on river basins and catchments, coordination and integration, partnerships (public-private, public-public, public-community/civil society), use of economic instruments and increased commercialization and privatization.

Decentralization and participation. Many components of ongoing water reform are part of broader governance reform agendas. Uganda, for example, has transferred water responsibilities to district and lower levels, receiving broad and strong political support within the country and illustrating that water reform is an integral part of reform efforts. Most countries have devolved provision of drinking water to the municipal government level. Nevertheless, decentralization and devolution remain problematic. Ethiopia, for example, has transferred important decision-making responsibilities to district and village levels, but has not followed up with capacity development and transfers of funds. Ghana's experience illustrates the importance of user participation and shows how financing can be resolved, demonstrating that decentralization and participation can yield positive outcomes; Bolivia's experience illustrates the beneficial use of cooperatives in such efforts (box 14.5).

Box 14.5 Participatory approaches in decentralized provision of water supply and sanitation services

Enhanced decentralization and participation for rural water supply and sanitation in Ghana. Ghana changed its rural water supply structure, expanding coverage through greater participation and more efficient delivery systems over a period of about 10 years. Water supply coverage rose from 55% in 1990 to 75% in 2004, with most of the increase in rural areas. Decentralization has been a part of broader political reform and improved governance structures.

Responsibility for rural water supply was transferred to local municipal governments and communities, coordinated by the decentralized Community Water and Sanitation Agency. Elected district assemblies are responsible for processing and prioritizing community applications for water and sanitation, awarding contracts for wells and latrine facilities and running latrine subsidy programmes. Village-level participation is part of the new structure. Village water committees plan for local water supply and sanitation facilities and raise funds for investment and operation and management costs. An assessment in 2000 found greater satisfaction with water quality and quantity at the village level. Most community residents contributed financially to these efforts, indicating that they received adequate value for their investments. Members of the village water and sanitation committees have received training and opened bank accounts, and women have played an active role in many communities.

Cooperative for Urban Water and Sanitation Services Delivery in Santa Cruz, Bolivia. Utility cooperatives were initially formed to provide utility services – mainly in rural areas where investor-owned utilities would not expand because of profitability concerns – usually providing services at at-cost prices.

The Cooperative for Urban Water and Sanitation Services in Santa Cruz (SAGUAPAC) provides water and sewerage services to Santa Cruz, Bolivia, a city of 1.2 million inhabitants. The national government approved the autonomous water board's request in 1979 to transform itself into a cooperative, recognizing that a different model was needed to provide services efficiently to a rapidly growing population. Civil opposition to state ownership and recognition that community participation was needed to achieve service improvements contributed to the adoption of a cooperative structure.

SAGUAPAC's service area covers about 63% of the city area and 66% of its population. By 2002 SAGUAPAC was providing water to approximately 95% of the population in its service area and sewerage services to about 50%.

Based on a classical cooperative model, SAGUAPAC has a 27-member Delegate Assembly (three members from each of the nine districts) that elects members of the Administration and Oversight Boards. (Some Bolivian utility cooperative boards have a general assembly instead, open to universal participation.)

SAGUAPAC applies the principles of autonomy, accountability, customer orientation and market orientation. It has become one of the largest urban water cooperatives in the world – serving approximately three-quarters of a million people and billing close to $19 million a year. Assessed against international standards, its performance over the years is considered very good, providing continuous service with good quality water through house connections and maintaining satisfactory financial performance. Most connections are metered (97%), and tariff collection efficiency is 90%.

Source: WSP-AF 2002; UNDP 2006; Ruiz-Mier and Van Ginneken 2006.

River basin management. River basin management integrates physical and administrative boundaries, nesting them within each other at different scales. The aim is to improve coordination in water decision-making. River basin management structures have been installed in many countries, including Australia, Brazil, Kazakhstan, Kenya, South Africa and EU member countries. The European Union Water Framework Directive is a stringent programme for establishing sustainable water resources management. It has had a major impact in new member countries, largely by mobilizing funding for improved water resources management. The government of Québec (Canada) has drafted a water law that identifies river basins as the fundamental water management unit.

Evidence from countries such as South Africa suggests that organizations and catchment bodies smaller than the river basin scale may be ineffective, may be too complex to implement and may offer benefits that are difficult to clearly identify. Several river basin organizations have found implementation challenging, experiencing considerable uncertainty about their roles and functions in implementing integrated approaches to water resources management. Most organizations have limited financial autonomy and depend on money from the central government and the donor community.[10] Despite the spread of river basin organizations at both national and international levels (box 14.6), examples of progress are difficult to find. It is unclear whether this is due to unsatisfactory performance or too short a period to document experiences and results.

Coordination and integration. Coordination with related sectors (agriculture, industry, energy and so on) is vital for improving water resources use and allocation (see chapter 15). But sectoral approaches to water resources management inevitably lead to fragmented, uncoordinated development and management. Fragmented institutional frameworks and overly complex coordination mechanisms in the water sector are common in many countries.[11] When appropriate links are missing, different ministries and agencies deal separately

Sectoral approaches to water resources management inevitably lead to fragmented, uncoordinated development and management

Box 14.6 Integrated management of land-based activities in São Francisco basin, Brazil

The United Nations Environment Programme (UNEP), in cooperation with the National Water Agency (ANA) of Brazil and the Organization of American States (OAS), and with funding from the Global Environment Facility, undertook a project during 1999-2002 to develop a watershed management programme for the São Francisco River basin. The basin traverses five states in Northeastern Brazil before discharging into the Atlantic Ocean.

The basin is strategically important to the economic development of a vast region of Brazil, which has subjected its natural resources to increasing demands. Mining, agricultural, urban and industrial activities contribute large contaminant loads to the system, including organic chemicals, heavy metals and sediment. Environmentally sensitive estuarine wetlands at the river mouth were threatened by unsustainable hydrologic management and land use practices in the basin. The basin's economic development has been haphazard, occurring within a weak institutional framework and resulting in less than optimal use and degradation of its water resources. Regulated flows over large stretches also have altered natural flows, causing changes in the freshwater, estuarine and marine flora and fauna.

The initial project objective was to conduct planning and feasibility studies for formulating an integrated watershed management plan as the basis for environmentally sustainable economic development of the basin. Components included river basin and coastal zone environmental analyses, public and stakeholder participation, development of an organization structure and formulation of a watershed management programme. Concluded in 2002, the environmental analysis provided a sound scientific and technical basis for remedial actions to protect the coastal zone from land-based activities.

Communities were involved in identifying and field testing remedial measures, and a process was established for dialogue among stakeholders and agencies with economic interests in the basin. Basin institutions are being equipped and trained to implement new laws, regulations and procedures for addressing environmental problems. Finally, agencies and individuals both inside and outside government synthesized data and experiences and prepared feasibility assessments and cost analyses for a long-term basin management programme. Some 217 public events were held, including seminars, workshops and plenary sessions. More than 12,000 stakeholders, including more than 400 organizations, universities, non-governmental organizations, unions, associations and federal, state and municipal government organizations, participated in the events. A comprehensive Diagnostic Analysis and Strategic Action Program for the Integrated Management of the São Francisco basin was completed in 2003 and is currently being implemented.

Source: ANA 2004.

Integrated water management approaches must have institutional and legislative governing frameworks to ensure oversight and monitoring of water resources and participation of target groups

with water's many subsectors. Weak water governance encourages economic sectors to compete for larger shares of water resources, to boost economic development or satisfy national production needs. Similarly, where interjurisdictional water governance is weak, riparian countries and jurisdictions sharing a water resource compete to develop their own water infrastructure and use. As this Report shows throughout, it is the leaders in government, business and civil society who make the decisions that determine effective water use policy (see also chapter 15).

Some countries have identified a need for coordination not only among ministries but also with subnational levels in implementing water policy and legislation. Integrated water management approaches must have institutional and legislative governing frameworks to ensure oversight and monitoring of water resources and participation of target groups. Thus, intersectoral coordination of water uses and involvement of water users are necessary at different levels of decision-making.[12]

One approach is to establish water councils, including high-level national water councils, river and basin councils, subnational (regional, governorate, state) councils and water users associations (see box 14.5). Experience with national and subnational water councils is extensive, and their functioning and political influence can vary considerably. Their main purpose is to develop links and structures for managing water resources across sectors and involving water users and stakeholders in planning and strategy development.[13]

Partnerships. Partnerships have been promoted within the water supply sector to improve services. Most have been public-private partnerships, and results have been mixed. Some countries are revising procedures for public-private partnerships (for example, Argentina), while others see no need for change. Some countries (Bolivia, for example) reject any private sector participation in water supply and sanitation.

Other types of partnerships involve civil society, municipalities and the private sector. To be successful, these partnerships require adequate capacity in civil society and private sector organizations and commitment from municipal governments and agencies. Proper incentives and mutual trust are also important.[14] Argentina, Colombia, Honduras, Paraguay and Peru have experience with partnerships going beyond the private sector. The Cartagena Partnership in Colombia, for example, was initially a partnership of municipal authorities, the community and a private water company. It explicitly involved community organizations in mobile payment collection units to collect fees from residents, establishing a clear accountability mechanism for fee collection. In Porto Alegro, Brazil, the municipality, community organizations and the public water company developed a partnership establishing participatory budgetary processes and charging water fees based on consumption rather than property taxes. The partnership improved the financial base for the public water company.

Brazil developed its water resources with eight major coordination mechanisms over a long period and implemented far-reaching institutional and legal changes. The 1988 Constitution specified federal and state government responsibilities and legal authority. The Ministry of Environment and Water Resources was created in 1995, and the National Water Resource Policy Law was passed in 1997. States also passed water laws. The reforms include establishing national, basin and state councils to improve managerial coordination and resolve water conflicts within the federal framework. The public utility (Municipal Department of Water and Sewage) operates

Box 14.7 The right to water

Some countries identify access to water as a human right in their constitution or other high-level legal instrument, thereby opening up constitutional courts and legal mechanisms to individuals and communities seeking to challenge inadequate access. A referendum in Uruguay in 2004, for example, added a human right to water to the Constitution, with more than 64% of the population voting for the amendment. The High Court in South Africa ruled on 30 April 2008 that prepayment meters in Johannesburg were unconstitutional on grounds of discrimination and that Johannesburg Water could afford to supply a minimum amount of water (50 litres a day) to each citizen. Courts in other countries (for example, Argentina, Brazil and India) have also sometimes reversed decisions to disconnect water supplies to poor people who cannot afford to pay. The long-term viability of this approach, however, remains unclear.

Source: COHRE 2007.

under a no-dividend policy, reinvesting all profits and investing at least a quarter of annual revenue in water infrastructure.[15] In Honduras the national water and sanitation operator (SANAA) has long worked with community-based organizations to improve services and procedures. And Malaysia entered into a partnership in an effort to reduce non-revenue losses (box 14.8).

Involving the informal private sector (water vendors) also can improve service delivery and help reduce quality-related problems. New contractual approaches have been developed in Paraguay, for example, to target *aguateros* (mostly small-scale water companies), which have developed piped water supplies in peri-urban areas without public funding. These *aguateros* can now legally take part in public bidding processes, and their performance can be tracked, improving accountability.[16] These examples show that opportunities exist for innovative partnerships and that there is room for new institutional models.

Water institutions and law

Law and policy are connected, with implementation of laws often being a trial and error effort requiring feedback and practical cases that interpret certain aspects of water law (see chapter 4). The government of Kenya, for example, enacted a Water Act in 2002 that established a new policy framework for the water sector. Guided by the legal framework and a draft Zero Investment Plan (2003), Kenya's water sector has undergone radical reform in policies and strategies, with the aim of reducing poverty. Associated goals include efficient service delivery, respect for consumer rights, financial sustainability and service coverage to poor people in both urban and rural areas. Examples from some other countries are illustrated in boxes 14.7 and 14.9.

Local water user groups and communities sometimes acknowledge customary rights to water use and allocation (box 14.10). In rural areas customary water rights often include operational rights and the right to participate in decision-making about operations, inclusion or exclusion of members, water distribution, irrigation schedules, flow rates and organizational positions and responsibilities.[17] African chiefs in charge of villager access to wells and the water court in Valencia, Spain, which regulates irrigation water access, are examples of practices that reflect customary rights.

Another example of customary rights is the traditional *subak* system of irrigation water distribution and use among the traditional rice-growing communities in Bali, Indonesia. The Indonesian Water Act of 2004 recognizes communal rights of local traditional communities as long as they do not contravene legislation and national interests. This is the standard formulation of customary rights protection in water

Local water user groups and communities sometimes acknowledge customary rights to water use and allocation

Box 14.8 Public-private partnership for reducing non-revenue water losses in Malaysia

Many water utilities lose large quantities of water through distribution system leakage or billing weaknesses (so-called 'non-revenue water'), which can undermine their financial viability. The state of Selangor, in Malaysia, experienced a serious water crisis in 1997 attributed to the El Niño weather phenomenon. The distribution leakage rate for the State Waterworks Department was estimated at 25%, or 500,000 cubic metres (m^3) a day, sufficient to serve an estimated 3 million people daily.

To address the problem, the State Waterworks Department employed a locally led consortium, in a joint venture with an international operator. The contract called for reducing non-revenue water losses by 18,540 m^3 a day for a payment equivalent to $243 per cubic metre per day of non-revenue water saved. The contractor was given the flexibility to design and implement activities to reduce losses, with a payment arrangement in place to cover necessary work and materials for detecting and repairing water leaks, identifying illegal connections, replacing customer meters as needed and establishing zones for reducing non-revenue water (district-metered areas). During an initial 18-month phase I, the validity of the concept was tested on a limited portion of the water delivery network.

Phase I exceeded its target, saving 20,898 m^3 a day. Twenty-nine district-metered areas were established, with an average savings of 400 m^3 a day in each area, and some 15,000 water meters were replaced. The cost to the State Waterworks Department was about $215 per cubic metre a day. Phase II (2000-09) has an overall reduction target of 198,900 m^3 a day, for a payment equivalent to $528 per m^3 a day. Based on interim results at the beginning of the sixth project year, 222 district-metered areas were established, more than 11,000 leaks were repaired and 119,000 water meters (of a contractual minimum of 150,000) were replaced. Non-revenue water losses were reduced by 117,000 m^3 a day (20% above the 2009 contract target of 97,500 m^3 a day), and commercial losses were reduced by 50,000 m^3 a day.

Source: Kingdom, Liemberger, and Marin 2006.

A growing number of countries and cities are incorporating water-related adaptations to climate change into planning and policy efforts

Box 14.9 Examples of legal frameworks for managing water

Effectively managing competing water uses requires clear, widely accepted rules on allocating water resources, especially under conditions of scarcity. Water allocation systems should balance equity and economic efficiency.

One means of avoiding conflicts of interests in water legislature is to separate policy, regulation and implementation functions, as Kenya does. The Ministry of Water and Irrigation focuses on policy formulation and guidance, while the Water Services Regulatory Board and Water Resources Management Authority address national and regional regulatory functions. Water service providers (such as community groups, non-governmental organizations, autonomous entities established by local authorities and the private sector under contract to regional water services boards) implement water supply and sanitation services.

Mexico passed the Law on National Waters in November 1992, and implementation regulations were adopted in January 1994. Amendments in 2004 added a package of regulatory, economic and participatory approaches to water resources allocation and pollution control. These include river basin planning, licensing of water abstractions and uses, permitting for wastewater dischargers, charging for water abstractions and wastewater disposal and articulation of federal government administration at the river basin and aquifer levels. The amendments also include provisions for recording legal instruments in a public water rights registry and providing opportunities for community participation through water user organizations and membership in basin councils.

Implementation and enforcement of the new regulatory structure began in 1993 with a survey and registration of abstractions and disposals. It took 10 years, and a series of intermediate regulatory adjustments and massive information campaigns, to complete the process.

Source: Velasco 2003.

Box 14.10 Recognizing customary practices in drafting laws

The framers of the Namibia Water Resources Management Act of 2004 were aware of the potential for deeply rooted customary practices – particularly livestock herding by traditional communities – to clash with the development of large-scale irrigated agriculture or tourism supported by administrative rights for the same waters used by herders. The new act prescribes the processing of abstraction licences and the criteria to inform decisions on licence applications. The law recognizes the existence of a 'traditional community' and the extent of its reliance on a water source affected by a proposed water abstraction (section 35(1)(h)). Accommodation of the 'reasonable requirements' of any traditional community is included among the standard terms and conditions of abstraction licences (section 37(e)).

Another example of customary rights is the traditional *subak* system of irrigation water distribution and use among the rice-growing communities in Bali, Indonesia. The Water Act of 2004 recognizes communal rights of local traditional communities as long as they do not contravene legislation and national interests. This is the standard formulation of customary rights protection in water legislation in countries where customary law is extensively practiced. Although lacking in detail and clarity, such statements can suffice in areas with strong social cohesion and where competition for water from 'outsiders' is limited.

Source: Stefano Burchi, Food and Agriculture Organization of the United Nations.

legislation in countries where customary law is extensively practiced. Although lacking in detail and clarity, such statements can suffice in areas with strong social cohesion and where competition for water from 'outsiders' is limited.[18]

Climate change and water resources

As chapter 5 demonstrates, climate change and variability have many potential impacts, both locally and globally. They may directly affect the quantity and quality of water resources. Climate change and variability also act on the other drivers and thus on water use and demand. The responses to challenges posed by climate change will likely be specific for each country or even parts of each country.

Developments are taking place at policy levels. The National Adaptation Programmes of Action under the United Nations Framework Convention on Climate Change are still in their early phases. Many least developed countries must still coordinate climate- and water-related policies and actions. Bhutan is one example of a country that has coordinated its national water and climate change adaptation policies to meet short- and long-term threats of glacier lake outburst floods resulting from climate change-induced glacier melting.[19]

A growing number of countries and cities are incorporating water-related adaptations to climate change into planning and policy efforts, along with institutional and

Box 14.11 **Water-related responses to climate change**

London and Venice are redesigning their urban stormwater drainage systems to accommodate predicted changes in precipitation frequency and intensity. Tokyo is designing urban holding ponds under roads and parks to temporarily store storm runoff to avoid flash floods. Jakarta recently initiated a programme to construct a major stormwater drainage canal system (East Canal) to provide adequate drainage to its eastern half. Viet Nam has developed an extensive system of dikes, including 5,000 kilometres of river dikes and 3,000 kilometres of sea dikes, to protect from typhoons and rising sea levels.

Countries in the lower Danube River basin in Eastern Europe restored thousands of hectares of aquatic habitat through floodplain restoration.

In Andhra Pradesh, India, removing silt from water tanks allows the capture of more monsoon runoff, resulting in additional benefits of less groundwater pumping, restoration of some dry wells and irrigation of an extra 900 hectares of land. Reconnecting lakes in the Hubei Province in China to the Yangtze River – by opening sluice gates and applying sustainable management techniques – increased wetland areas and wildlife diversity and population and will make the area more resilient to flood flows. There are potential scaling-up possibilities with this approach, as there are hundreds of sluice gates along the Yangtze River that disconnect it from nearby lakes.

Source: World Bank 2008.

technological measures to mitigate such predicted impacts as sea-level rise, more frequent droughts and increased precipitation (box 14.11).

Consulting with stakeholders and avoiding corruption: accountability in planning, implementation and management

Stakeholder engagement is important to improving water resources management through several channels, from direct participation in planning to expanding public awareness. One benefit is reducing corruption, a source of devastating social, economic and environmental impacts, particularly for poor people, and which can increase the investment costs of achieving the Millennium Development Goals.

Engaging stakeholders: benefits and challenges

Stakeholder involvement through public hearings, advisory committees, focus groups, stakeholder forums and the like has often improved water projects, programmes and related human livelihood opportunities. It can also increase public awareness of water issues while informing both the facilitators of change and those involved in it. An example is the study on diffuse water pollution of the North American Great Lakes conducted by the US-Canada International Joint Commission during the 1970s. The commission conducted public hearings throughout the basin, both to educate basin inhabitants about the study goals and to secure their inputs on potential problems and solutions (box 14.12).

Some stakeholders who are left out of decision-making, such as small-scale farmers and poor urban households, may respond by organizing their own activities (box 14.13). Irrigation management transfer in India, Mexico and Turkey, for example, led to investments in new techniques, better collection of water user fees and improved water resources management (box 14.14; see also box 4.4 in chapter 4). In 1998 in the Arwari River catchment in Rajasthan, India, the Arwari River Parliament, with 2,055 members in 70 villages in 46 micro-watersheds, was formed to improve water management through controlled use of water. The river parliament also explored improving soil, land and forest management; increasing agricultural productivity; seeking participation by women and generating self-employment and alternative livelihood options. Its social, economic and environmental impacts have generally been positive, and increased agricultural production has expanded livelihood opportunities. The Arwari River Parliament has provided a platform to resolve land, water and forest management disputes.[20]

To provide useful communication tools and systems for exchanging information, data and experiences, the Emilia-Romagna region in Italy developed a national forum on water conservation as a common platform for discussion and comparison of water conservation policies. The forum highlights the most modern, innovative policies for water saving and conservation at the national level. It is organized into thematic working groups (water saving in civil, agricultural and industrial sectors; drinking water losses in distribution systems; and communication). To expedite its work, the forum has a Website,[21] organizes an annual conference (held on World Water Day) and thematic workshops, produces newsletters and engages national and European experts.

Stakeholder engagement is important to improving water resources management through several channels, from direct participation in planning to expanding public awareness

Box 14.12 Public participation panels in the North American Great Lakes basin review of water quality

The United States and Canada established the International Joint Commission under the Boundary Waters Treaty of 1909. The two countries signed the Great Lakes Water Quality Agreement in 1972, with the goal of restoring and maintaining the chemical, physical and biological integrity of the Great Lakes basin ecosystem. The agreement called on the commission to conduct a study of pollution in the Great Lakes system from diffuse (non-point) sources in its drainage basin. The International Reference Group on Great Lakes Pollution from Land Use Activities (PLUARG) was established to undertake the study, focusing on three major questions:

- Are the Great Lakes being polluted from land drainage from non-point sources in the basin?
- If so, what is the extent of this pollution, what are its causes and where is the pollution occurring?
- What remedial measures can address these sources, and what would they cost?

The PLUARG study, which involved scientists, managers and policy-makers from both countries, concluded that non-point sources (particularly agricultural and urban runoff) were responsible for water quality problems from phosphorus, sediments, polychlorinated biphenyls, persistent pesticides, industrial organic chemicals and lead on either a lake-wide or localized basis. Non-point sources also contributed significant amounts of nitrogen, chloride, non-persistent pesticides and heavy metals, although these pollutants did not yet constitute a water quality problem.

Considering the many jurisdictions involved (see map), and the diversity of opinions to be considered, PLUARG recognized the need for public input to identify concerns and workable management strategies. PLUARG established nine public panels in the United States and eight in Canada, comprising industrialists, small business owners, farmers, labour representatives, educators, environmentalists, women's groups, sport and fishing associations, wildlife federations and elected and appointed government officials. Each panel met four times to discuss and make recommendations on the environmental, social and economic aspects of the study, with many expressing their goals for the Great Lakes. The panels reviewed and commented on a draft of the PLUARG report before its finalization and submission to the International Joint Commission. Each panel also submitted a report to PLUARG on its own views and recommendations on panel-identified problems, as well as suggested solutions. PLUARG also held numerous public meetings throughout the Great Lakes basin to gain additional perspectives. The input from the panels and the public meetings were a major contribution to the final report and the PLUARG technical report series.

Source: IJC 1978; PLUARG 1978.

Great Lakes provincial, state and county jurisdictional boundaries during 1978 PLUARG study

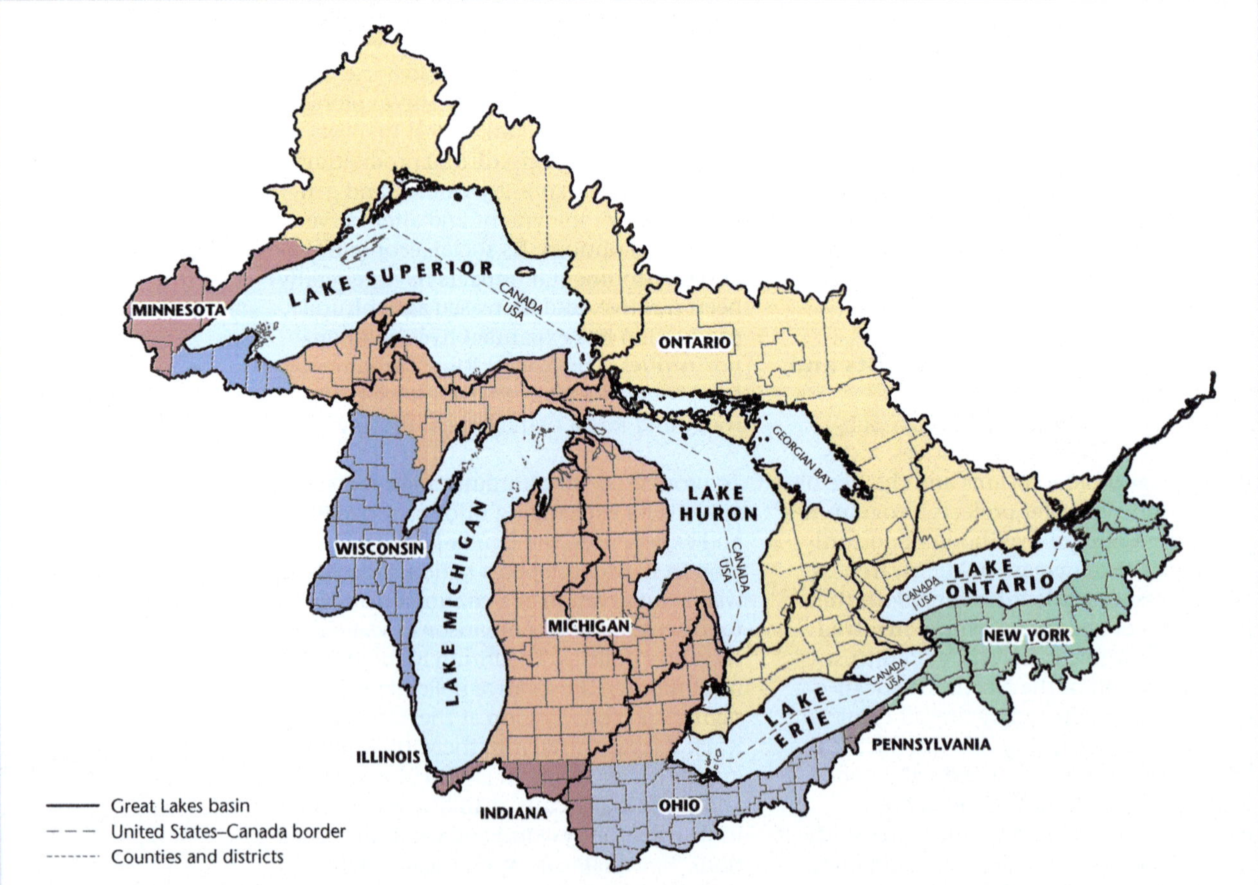

Source: Great Lakes Commission.

Box 14.13 Grass-roots water federations in Ecuador

In the highlands of Ecuador, as in other Andean countries, greater competition for declining water resources led to increasing conflicts and applications to register water rights. The provincial water agencies responsible for water allocations were understaffed, and water resources privatization, viewed as ineffective by many, was making national management and water conflict resolution more difficult. The decentralization of power to provincial authorities – rather than to user collectives – exacerbated perceptions of inequality and injustice in the water sector. In July 2005 thousands of water users demonstrated in Riobamba, demanding removal of the state water agency staff for inequitable treatment of indigenous peasant and female water users. Many community organizations also gathered to establish a provincial water user federation (*interjuntas*), to build capacity among water user organizations and foster discussion forums on water policy-making and law. This intersystem organization now facilitates participation by some 280 irrigation and drinking water users organizations, all with mostly indigenous and small-farmer household constituencies. The *interjuntas* also facilitates conflict management among water users and among associated systems, especially between poor groups and landlords and between indigenous rural peoples and the state, through its centre for the defence of rights and mediation of conflicts.

Other new water user organizations are starting to emerge. In Cotopaxi Province the Federation of Cotopaxi Irrigation Users now comprises 370 water user organizations composed of tens of thousands of *minifundio* (small land holding) water user families. Community water quantity and quality rights are represented at the national level through collaboration with the national water civil society platform, National Water Forum, and direct negotiations with the state. Federations such as *interjuntas* and the Federation of Cotopaxi Irrigation Users illustrate that equitable water distribution requires democratic decision-making and transparency of public investments and monitoring of government activities in the water sector.

Source: Dávila and Olazával 2006; Boelens 2008.

Box 14.14 Participatory irrigation management and the role of water user associations

Participatory irrigation management engages irrigation users at all levels and in all aspects of managing irrigation schemes. Based on the belief that water users are best suited to manage their water resources, participatory irrigation management allows considerable flexibility in water management methods.

India. Despite large investments, the irrigation schemes in the Indian state of Andhra Pradesh have been in serious trouble because of deteriorating infrastructure and low agricultural productivity. Policy reforms were introduced in 1996/97 to deal with irrigation concerns, including a threefold increase in water user charges, creation of water user associations and capacity building in water user associations across the state. Institutional reforms included the creation of farmer-government partnerships in irrigation operations and maintenance, consolidation of irrigation management transfer, new cost recovery methods, expenditure prioritization and capacity building for state agencies and water user associations.

Water user association board members are elected by local water users. Transferring management authority to user groups has created a strong sense of ownership and empowerment. Still, some studies criticize the reforms for being more a top-down government programme than a farmer-initiated effort, and some suggest that establishment of the water user associations resulted in needless proliferation of community organizations, when the village government could have handled the task (a view not shared by water users).

Some constraints to the new management system include limited power supply in rural areas, below design water discharge levels and continuing dependence on government funds in many cases. Nevertheless, there have been many positive results. Collection of water tariffs increased from 54% to 65% during the first year of implementation. Management of irrigation canals by water user associations has resulted in more effective water use, with an additional 52,361 hectares being irrigated in 1998 in the Tungabadra High Level Canal. Irrigation canal water-carrying capacity rose about 20%-30%, and agricultural productivity increased. There was also a dramatic reduction in farmer complaints.

Mexico. Most farming in Mexico is irrigated, with water services provided by the central government since the end of World War II. By the end of the 1980s the government was subsidizing more than 75% of the operation, maintenance and administration of irrigation districts – a non-sustainable outcome. As a result, farmers received relatively low-quality services for which they were reluctant to pay.

Under an extensive programme of agricultural reform, management of irrigation districts was transferred to water user associations, with responsibility for irrigation systems to be shared between the associations and the then newly formed National Water Commission (CNA). After the transfer of irrigation management responsibilities, the rate of tariff collection more than doubled in five years, peaking in 1997 at 72%. Water fees were raised, sometimes more than 100%. The expectation was that the water user associations would become financially self-sufficient, generating enough resources to cover the costs of CNA. That has not yet happened, however, so those costs are still being covered by a ministerial fund. Nevertheless, the irrigation systems have become financially more self-sufficient. In the lower Bajo Rio Bravo Irrigation District, for example, self-sufficiency rose from 36% in 1989 to 100% in 1994.

Turkey. Both the State Hydraulic Works and the General Directorate of Rural Services are responsible for managing soil and water resources in Turkey. The State Hydraulic Works is responsible for large-scale irrigation and water infrastructure; the General Directorate of Rural Services, for on-farm development and small irrigation schemes. Water scarcity has been a problem since the 1960s, and operation and maintenance of the country's irrigation systems was a financial and institutional burden for the government. Revenue collection was difficult, and water use was very high.

With support of the World Bank the central government began transferring irrigation schemes, even large ones, to water user associations in 1993, to reduce costs for the central agencies. Following Mexico's example, Turkey transferred 1,350,000 hectares of irrigated land to the water user associations by 1997, with 87% of irrigation projects transferred by 2007. This rapid transfer of irrigation

(continued)

Box 14.14 Participatory irrigation management and the role of water user associations (continued)

management was motivated by the rising costs of irrigation schemes to the central government, the availability of on-the-job training programmes in Mexico and the United States, the commitment of State Hydraulic Works staff and clearly defined goals and pilot projects. The transfer has resulted in more efficient use of water resources, increased investments in new technologies, higher water user tariff collection rates (from 42% in 1993 to 80% in 1997), lower energy costs (approximately 25%) and more equitable allocation of water resources.

The legal status of water user associations remains to be defined, and central agency obligations to provide technical and administrative assistance at the beginning of a transfer still need to be clarified. In 2004 the government abolished General Directorate of Rural Services as part of broader administrative reforms, delegating its responsibilities to the provincial governments.

Source: Jairath 2000; Raju 2001; Johnson 1997; Palacios 1999; Garces-Restrepo, Vermillion, and Muñoz 2007; Blue Plan, MAP, and UNEP 1999; Döker et al. 2003.

The participation requirements of article 14 of the EU Water Framework Directive are an attempt to launch a broader discussion about participatory approaches, as illustrated in box 14.15.

Addressing corruption and mismanagement in the water sector

Corruption can have enormous social, economic and environmental repercussions, particularly for poor people. Water-related construction projects such as aqueducts, sewer systems and basic sanitation and wastewater treatment plants have become magnets for corruption in many developing countries, which have limited oversight capacity for efficient use of public resources. Transparency International's *Global Corruption Report 2008,* prepared in collaboration with the Water Integrity Network, estimates that corruption in the water supply sector increases the investment costs of achieving the water supply and sanitation target of the Millennium Development Goals by almost $50 billion.[22]

In recent years preventing corruption has captured the attention of governments, private firms, civil society organizations and donors. A positive example is the formation of the Water Integrity Network, a global network promoting water integrity by coalition building. Several countries of the Southern Africa Development Community region have taken positive steps. Zambia established water watch groups in some cities to monitor relations between water regulators and service providers. South Africa set up telephone hotlines for consumer redress and complaints. Some districts in Malawi developed water board anticorruption policies aimed at improving water sector efficiency by preventing malpractice and addressing water consumer problems.[23] Box 14.16 provides additional examples of such efforts.

Capacity development for more effective action

Effective interactions between individuals, sector organizations and regulatory and administrative authorities are critical to advances in all areas. Administrative systems and sector policies in many countries are in need of considerable reform. This section offers some suggestions for governments and other water stakeholders to increase their capacity for effective action.

Assessing institutional and human capacities

A first step in improving services is to assess their ability to deliver more effective services and to prepare for future

Box 14.15 Public participation in water resources management

The RhineNet project highlights the value of public participation. The project covered floodplain restoration, construction of fish ladders, flood protection and recreational enhancement. Public involvement in projects increased public acceptance, even among those who might be considered 'losers' in the process.

The project plans for reactivating the Saar River floodplains in Hostenback, Germany, provide an example. The plans were presented to the mayor of Wadgassen and the general public. The project received widespread support, with citizens showing considerable flexibility over details of the required construction. One important agreement called for use of a much less used alternative path for the estimated 8,000 trucks that would be involved in the project's earth removal activities. Meetings and discussions were held regularly during the project, providing opportunities for citizens to voice their concerns, and events were reported in local bulletins, newspapers and electronic media. An important finding was that such efforts require considerable time for interviews and discussions with the individuals affected by the project – discussions that could not be replaced by media reports or press conferences.

Source: Lange 2008.

Box 14.16 Responding to corruption and mismanagement in the water sector – examples from Colombia, India and Lesotho

Development of anticorruption agreement with Colombian water pipe manufacturing companies. The Colombian Association of Environmental and Sanitary Engineers, whose affiliated water pipe manufacturing companies had a 95% share of the national market and a monopoly on bids in public tenders for water supply and sewer systems, undertook an anticorruption initiative as part of a sectoral antibribery agreement. The association, together with Transparency International-Colombia, worked to develop an agreement among pipe manufacturers based on Transparency International's Business Principles to Counteract Bribery. The agreement, signed in April 2005, resulted in substantial reductions in bid award prices, thereby reducing the scope for bribery.

The agreement was developed in reaction to the lack of transparency in the pipe business sector, particularly in public sector procurement, which resulted in overpriced products and substandard quality in public projects and utilities, creating an environment of mistrust. The situation eventually became untenable for the companies and the trade association, as transaction costs became unsupportable. Under the agreement each company prepared a general anticorruption policy and specific guidelines for each area specified in the Business Principles to Counteract Bribery (pricing and purchasing, distribution and sales schemes, implementation mechanisms, internal controls and audits, human resources management, protection of 'whistle blowers' and communications, internal reporting and consulting). The agreement also laid out the roles of an Ethics Committee and a Working Group tasked to supervise implementation, as well as extensive legal and economic powers for dealing with companies that fail to comply. A similar agreement was signed in Argentina in December 2005, and agreements also are being considered by Brazil and Mexico.

Citizen report cards for improved water services in Bangalore, India. To improve the quality of underperforming public water and sewer utilities in Bangalore, India, the national Public Affairs Centre established a system of benchmarks and citizen 'report cards'. These report cards triggered a series of reforms enhancing public sector accountability and responsiveness. Although the first report card in 1994 gave low ratings to all major city service providers, only a few service providers acknowledged the problems and took corrective actions. The second report card in 1999 indicated partial improvement in some services, while the third report card in 2003 revealed substantial improvements by almost all service providers, as well as a visible decline in corruption. Satisfaction levels rose dramatically, from 4% in 1994 to 73% in 2003.

Both supply and demand drivers of change contributed to this surprising turnaround. The trigger for public action seems to have been the public scrutiny and publicity attending the report cards, leading to important interventions on the supply side. A strategic decision of the state government was to establish a new public-private partnership forum to catalyse action and assist service providers in upgrading their services and responsiveness. The political support and commitment of the state's chief minister, the innovative practices resulting from the partnership forum, the active role of external catalysts (civil society groups and donors) and the learning experiences from initial responses contributed to the better performance. An open and democratic society is a prerequisite for the use of this monitoring and accountability tool.

Lesotho highlands project trials. The Lesotho Highlands Water project, the world's largest international water transfer project, transfers water from the Orange River to the Vaal River, providing extra water to Johannesburg, South Africa. Lesotho receives royalties for the water ($31 million in 2004, about 5% of its GDP). Under phase 1 of the project four dams and 110 kilometres of tunnels were completed at a cost of about $2 billion.

The first chief executive of the Lesotho Highlands Development Authority, charged with overseeing the project, was tried on bribery and fraud charges in 2001. His subsequent conviction was an important victory in the fight against corruption and illustrated what a determined government can do in fighting corruption. Recognizing that bribery has both a demand and supply component, the Lesotho government also brought charges against the multinational companies paying the bribes. Three major firms were convicted by the High Court of Lesotho, and all three were assessed fines. The World Bank also barred one company from bidding on future projects. The trials for future bribery prosecutions established several important precedents related to what the prosecution has to prove in regard to bribery, where the crime took place (determined by the location of the impacts of the crime) and what degree of financial transparency is required to prove the crime (a major breakthrough was the prosecution's access to Swiss banking records of the accused companies).

Source: Balcazar 2005; Stålgren 2006; Thampi 2005; Stålgren 2006; Earle and Turton 2005.

uncertainties. Such assessments can cover part of a sector (for example, management of river basins or sanitation) or focus on institutional architecture and capacity (for example, the education system, community management and the legal framework).

Improving a weak institutional environment is not a linear process.[24] It often requires efforts on several fronts, focusing on alleviating acute problems while creating the conditions for more favourable change over time. The United Nations Development Programme, for example, supported rapid water sector assessments in Bolivia, China, Ghana, Mali, Mexico and Peru, providing modest international support to local agencies to assess the challenges to their water sector and each country's capacities to devise new strategies with prioritized interventions.[25]

In China the Ministry of Water Resources and the Guizhou Provincial Administration worked on an institutional performance assessment, focusing on strengths and weaknesses in the economic and institutional aspects of water management, such as pricing, river basin management and stakeholder involvement in integrated water resources management. In Mexico the water initiative added to the National Water Commission's broader efforts to improve the water sector's overall performance. The Peruvian assessment guided a parliamentary debate on sectoral

Governments will need to rely more on an informed and capable civil society whose role in water management complements the work of government agencies

priorities, helping reform agencies and generating new skill mixes.[26] In Indonesia the 1998-2004 financial crisis triggered a deep institutional and administrative reform of the water sector, which is crucial to its economy, emphasizing decentralized decision-making, participatory irrigation management, cost recovery and cuts in staffing.

Strengthening institutional arrangements and capacity to support an agenda of change

Changes in society and in the environment call for regular adjustments of the institutional architecture of the water sector. Regular performance and capacity reviews can identify needed reforms and promote agreement on capacity development for implementing the reform. Uganda's National Water and Sewerage Corporation (NWSC) transformed itself from an organization plagued by unaccounted-for water, weak billing practices and high operating costs to one with steadily improving performance since 1998. NWSC separated operations and maintenance from performance monitoring and regulation. The capacity of utilities to work with this system was developed incrementally by multidisciplinary teams, beginning with simple performance contracts and upgrading to more complex arrangements. NWSC became a 'learning organization', creating an environment where managers must account for performance against goals and staff are expected to embrace innovation and tackle problems (see box 14.23 later in the chapter).

Engaging with civil society in developing its capacity

With the large numbers of water management stakeholders, governments are increasingly constrained in what they can achieve alone. They will need to rely more on an informed and capable civil society whose role in water management complements the work of government agencies.

As civil society becomes more water literate, it will come to a clearer understanding of the importance of water issues and lend solid support to water sector initiatives. That requires broad access to information and the capability to engage with governments on water service delivery issues – often requiring capacity-building efforts to be effective.

Many resources are available for community and civil society capacity development in water and sanitation. Examples of Internet-accessible resources include a comprehensive collection of capacity building materials on ecological sanitation emerging from the 4th World Water Forum in Mexico in 2006;[27] material on gender, water and capacity building provided by the Gender and Water Alliance;[28] an overview paper and guidelines for improving knowledge management at the personal and organizational level in the water and sanitation sector;[29] and a conceptual introduction to water sector alliances, case studies and lessons in scaling up innovations in water, sanitation and hygiene, both prepared by the International Water and Sanitation Centre.[30]

Stimulating professional knowledge

Water sector professionals require a sophisticated understanding of the hydrologic cycle and its variability. They also need a better understanding of the relationship between water use and sustainable economic development and water-related interactions in society, and of the needs of decision-makers. Non-professionals often require a better understanding of their interactions with and influences on water systems. This knowledge can come from research, the traditional domains of local communities and education, training and focused workshops. But the knowledge is often fragmented, held by a growing number of water stakeholders, each with part of the solution. Thus, communication is critical for building the knowledge base and institutional and human capacities needed to forge political consensus.

Networks are becoming increasingly important knowledge pools and mechanisms for knowledge dissemination (box 14.17), exchange and management. They must be structured, managed and funded according to their purpose (research, sharing of professional experience, training and so on). Networks are well suited to identifying and articulating large-scale, complex problems and to offering solutions and best practices tested in other places.

Learning alliances are groups of representative stakeholders with a focus on developing jointly owned approaches to problems that create a broader sense of ownership and lead to more rapid implementation. The Euro-Med Participatory Water Resources Scenarios (EMPOWERS), a regional pilot project led by Care International–UK from 2003 to 2007, sought to increase sustainable access to water for vulnerable populations in Egypt (where water demands exceed supplies), in Jordan (which has one of the world's lowest per capita water levels) and the West Bank and Gaza (where access to water is strictly controlled

Box 14.17 Networking to share water resources management experiences

Cap-Net is a network of 20 national and regional capacity-building networks and three global networks committed to capacity building in water resources management. Most of the member networks are informal voluntary associations of institutions and individuals committed to building capacity to address local needs and priorities. Cap-Net uses its global network structure to rapidly share international and regional knowledge. It facilitates knowledge development and exchange, supports the delivery of capacity development services to meet local priorities and brings expertise from networks and the international level to neglected areas. Having identified economic and financial instruments for implementing integrated water resources management as a neglected area, Cap-Net used its network experts to develop a training package structured to allow local adaptation. The training manual was tested, revised and translated into four languages within one year and has been implemented in Africa, Asia and Latin America through the partner networks.

The Regional Centre for Urban Water Management in Tehran is an organization of 13 countries and six international organizations whose mission is to transfer practical scientific knowledge and develop capacities in all dimensions of urban water management, to promote sustainable development and enhance human well-being and to facilitate integrated, transboundary water management. Its governing board comprises 10 water-related ministers (from Bangladesh, Egypt, India, Iran, Kuwait, Oman, Pakistan, Syria, Tajikistan and Yemen) and high-level representatives of three international organizations (the UNESCO Institute for Water Education, the International Water Academy and the International Water Association).

Source: Cap-Net 2008 and www.cap-net.org.

by an external authority). The International Centre for Water Hazard and Risk Management (ICHARM), an Asia-Pacific water knowledge hub of the Asian Development Bank, provides hazards mapping and tsunami training courses.

The Sustainable Water Management Improves Tomorrow's Cities Health (SWITCH) project is a major research partnership for innovation in integrated urban water management in implementing action-oriented, water demand-led research programs. The project encourages learning alliances to better define the research agenda and initiate research in aspects of the urban water cycle to help cities improve water sector integration and scaling-up impacts.[31] SWITCH has led to the establishment of learning alliances in 10 cities around the world.

In the Netherlands government institutions for surface water, groundwater and coastal protection were merged to improve their output. Innovation and research also were boosted in countries in Southern and Eastern Africa through WaterNet, linking some 50 university departments and institutions with common interests and expertise in water-relevant topics. Pooling their knowledge allows them to cover all the major aspects of water resources management (such as hydrology, environmental engineering, economics, law and water and sanitation technologies).

In India the Solution Exchange, which links people across the country and the globe, harnesses the power of communities of practice to leverage the knowledge and experience of multiple development practitioners for the common objective of problem-solving. In Madhya Pradesh, India, the UN-HABITAT Programme on Sustainable Cities works in four cities to improve and expand urban water supply, sewerage and sanitation, water drainage and solid waste management. The programme seeks to influence water use practices, policies and measures away from traditional approaches, which increase pressures on water resources and lead to overexploitation of groundwater, and toward innovative demand management to optimize available water in the municipal water supply system. The programme consolidated responses to a survey on experiences, suggestions and best practices from other parts of India, translated them into Hindi and circulated them at the Madha Pradesh State Parliamentarians Forum for Water.

Information and communication systems are mostly Internet-based, offering new tools for multistakeholder information sharing and communication. By providing access to scenario and forecasting tools and facilitating communication across administrative levels (from the local to the national and regional), these tools are well suited to facilitate negotiation.

Increasingly, water stakeholders are being linked through online knowledge networks and partnerships – communities of practice – linking researchers addressing

Increasingly, water stakeholders are being linked through online knowledge networks and partnerships linking researchers addressing similar issues

Water resources management plans can be difficult to fund and implement without the involvement of water users

similar issues. Simple tools, such as e-mail lists, are the preferred low-technology avenue for this activity.

Local stakeholders are usually the first to experience and address local problems and to find local solutions. Decision-makers can learn from this local knowledge and apply its lessons to building the capacity of local institutions and civil society (box 14.18). Informed decision-making requires a balanced combination of top-down (often larger-scale) and bottom-up (often smaller scale) approaches and procedures.

Stimulating public awareness

Increasing public awareness about water resources also facilitates sustainable use. Water resources management plans can be difficult to fund and implement without the involvement of water users.

One way to promote public knowledge is through science and education centres that compile, analyse and disseminate information on water resources. Such centres often focus on identifying and disseminating information on water system problems and sustainable use. They often consider the economic, ecosystem and cultural importance of specific water systems and their resources; direct and indirect uses and values and promising tools and strategies for management and lessons from case studies. They also provide valuable region-specific information, such as the public and media awareness campaign by the Yemen National Water Resources Authority, including the Yemen Water Awareness video.[32] Information centres such as the Lake Biwa Museum (Japan), Balaton Limnological Research Institute (Hungary) and Leahy Center of Lake Champlain (United States) focus their attention on lake basins.[33] The centres, sponsored by private foundations, corporations, government agencies, non-governmental organizations and academic institutions, demonstrate and promote the benefits of advancing scientific, educational and community development goals for water systems.

Public education and awareness raising can also be as simple as informing water stakeholders about changes in work routines or personal habits that can alleviate problems associated with unsustainable use of water resources (see chapter 2). Greater public awareness is facilitated by making details of water system problems and corrective programmes readily available to the public through the public education system, the information media and non-governmental organizations.[34]

The Emilia-Romagna Region in Italy launched a regional communication campaign on water conservation in 2004 called *"Acqua, risparmio vitale"* (Water, vital saving). The campaign was relaunched in 2008 with a new slogan, *"Mezzo pieno o mezzo vuoto? Comunque la pensi, l'acqua va risparmiata!"* (Half full or half empty? Whatever you think, save water!). This education strategy uses a combination of brochures for the general public, television and radio spots and advertisements on city buses.[35] Water managers can also use such communication activities to inform the key decision-makers working outside the water managers' domain (see chapter 15).

Developing appropriate solutions through innovation and research

Technological innovation covers a broad field, from technical issues to financial considerations, water service models and water governance issues (policy, sustainable financing, cultural values, political realities, law and so on). It can enable rapid and significant changes within the water sector. And it can improve existing systems (better hand pumps, for example) and develop new ways to address water issues.

Box 14.18 Incorporating local knowledge in networks – the Integrated Watershed Management Network of Eastern Africa

The Integrated Watershed Management Network (IWMNET), is a three-year (2007-10) capacity-building initiative for integrated watershed management in Eastern Africa involving German and Eastern African universities. Activities include specialized training on integrated water resources management and related issues and support for ongoing water sector reforms in Ethiopia, Kenya, Tanzania and Uganda. Water user associations, for example, are assisted in drafting and executing subcatchment management plans and in sharing their experience in drafting participatory catchment management plans. An online e-learning component makes all information accessible to students and professionals, even in rural areas.

Source: iwmnet-eu.uni-siegen.de and www.iwmnet.eu/index.html.

Developing a capability for innovation

Experience suggests that at least three key factors have contributed to the innovation driving the rapid economic transformation of many emerging market economies. One is investing heavily in basic infrastructure (roads, schools, water supplies, sanitation facilities, irrigation systems, health centres, telecommunications, energy resources and the like) as the foundation for technological learning. A second is developing and nurturing small and medium-size enterprises, which can supply local operational, repair and maintenance expertise and technicians. A third is developing and nurturing institutions of higher education (engineering and technological academies, professional engineering and technological associations, industrial and trade associations and the like).[36] Most of these efforts are outside the scope of water managers' responsibilities (some are discussed in chapter 15).

Business enterprises and non-governmental organizations can be a focus for learning.[37] An example is the innovative adaptation for use in Cambodia of ceramic filters, long used elsewhere to remove contaminants from water. Because most Cambodians lack access to microbiologically safe water at point of use, household-based water treatment is critical for protecting them from waterborne diseases. Most Cambodians must still collect water, store it for household use and treat it themselves to ensure safe drinking water. Potters for Peace, a non-governmental organization, developed an innovative use of ceramic filters, inventing a low-cost, on-the-spot production method using clay and sawdust treated with colloidal silver to increase bactericidal properties. Potters for Peace now trains others to operate filter-making facilities in Cambodia, Cuba, El Salvador, Ghana, Guatemala, Honduras, Indonesia, Kenya, Mexico, Sudan and Yemen.[38] Water management techniques also can be used in innovative ways to address a range of issues (box 14.19).

Because irrigation is the single largest water user on a global scale, improved irrigation technology is a good example of using technology to reduce water demands by improving water use efficiency (box 14.20).

Linking research and development in the developing world

Expanding demand-driven research capacity in developing countries is essential for reaching a critical mass of people engaged in research and development who can advance economic development.[39] The Paris Declaration on Aid Effectiveness of 2005, an international agreement affirmed by more than 100 ministers, heads of agencies and other senior officials, stressed that developing countries must become more capable of solving their own problems,

Most of the key factors that have contributed to the innovation driving the rapid economic transformation of many emerging market economies are outside the scope of water managers' responsibilities

Box 14.19 Using water management to preserve biodiversity and economic livelihoods – Kafue Flats, Zambia

Large dams can disrupt the natural water cycle and the ecosystems that rely on it. Kafue Flats in Zambia illustrates the use of technological innovation and cooperation to alleviate such disruptions.

Kafue Flats is a rich wildlife habitat occupying 6,500 square kilometres along the Kafue River, a major tributary of the Zambezi River. It sustains the livelihoods of local people engaged in hunting, fishing and cropping when floods recede on the flats at the end of the wet season. In 1978 the Itezhi-tezhi Dam was built upstream of the Kafue Gorge hydroelectric dam, Zambia's primary source of power, to store wet season peak flows to maximize hydropower production at the hydroelectric dam. The Itezhi-tezhi Dam ended the beneficial wet season flooding of Kafue Flats, adversely affecting the 300,000 people who rely on it for their livelihoods.

In 1999 the Zambian government and Zambia Electric Supply Corporation (ZESCO), in cooperation with WWF, local people and commercial farmers, initiated a project to restore a more natural flow to water releases from the Itezhi-tezhi Dam. An integrated water resources management study was undertaken in 2002, including development of a Kafue River basin hydrology model. The model was linked to real-time data from rainfall and river gauging stations in the catchment to predict water flows and reservoir levels. Based on this modelling exercise, agreement was reached among all the partners in 2004 to implement new dam operating rules. A major water flow mimicking the natural wet season flooding pattern was released for the first time in early 2007, and modules have been launched for wetlands rehabilitation, focusing on infrastructure development, tourism enhancement and community-based natural resources management.

The long-term results are expected to include improved ecological health for Kafue Flats and improved livelihoods for local people (particularly increased fish and pasture productivity), development of a wildlife-based tourism industry and sustained irrigation capacity. The hydroelectric production potential of the Kafue Gorge Dam is expected to be maintained or to increase. Discussions are under way on scaling up the environmental flows model to the other dams in the watershed (Kafue Gorge, Cahorra Bassa and Kariba) to extend benefits to the entire course of the rivers in Zambia and Mozambique. Preparations also are under way on a joint operations and management strategy for the three dams, involving the Zambezi River Authority, the Joint Operational Technical Committee for Cahorra Bassa and Kariba Dams, and the Southern Africa Development Community agreement for an integrated water resources management strategy for the Zambezi River under the auspices of its shared water protocol.

Source: WWF 2008.

Reliable and accurate water resources information and data, by reducing uncertainty about water resources, help decision-makers make more reliable and politically persuasive assessments of water risks

Box 14.20 Using irrigation technology to increase water use efficiency

Irrigation water can be delivered to crops through surface, sprinkler and drip irrigation systems. Although introducing new irrigation technology typically increases costs, it also increases water use efficiency, resulting in water savings. A California study, for example, found that water use efficiency ranged from 60%-85% for surface irrigation to 70%-90% for sprinkler irrigation and 88%-90% for drip irrigation. Potential savings would be even higher if the technology switch were combined with more precise irrigation scheduling and a partial shift from lower-value, water-intensive crops to higher-value, more water-efficient crops.

China introduced water-saving measures and irrigation system modernization in agriculture in the 1990s. China has some 400 large irrigation systems (each with an irrigated area of more than 20,000 hectares), which account for about a quarter of the total irrigated area of 56 million hectares. Modernization included the application of new materials and technologies to upgrade irrigation system structures and the application of modern irrigation concepts and institutions to improve irrigation management. Water conveyance and irrigation intervals have been shortened, and water losses have been reduced. Agricultural output in the programme area increased 46%, even though irrigation withdrawals have fallen from about 80% of total water withdrawals in 1980 to 60% today, a dramatic reduction.

Source: Cooley, Christian-Smith, and Gleick 2008; ICID 2008.

including through stronger research capacities that will enable them to absorb and use knowledge from other sources and countries as well as to advance knowledge (box 14.21).

Data and information needs

Reliable and accurate water resources information and data, by reducing uncertainty about water resources, help decision-makers make more reliable and politically persuasive assessments of water risks. More detailed and accurate information also guides better choices on needed infrastructure and makes public institutions more accountable for the impacts of their actions.

Water resources data and information

Water resources data include information on the quantity and quality of water resources and also on governance. Such data are crucial to a wide range of water resources stakeholders.

The World Bank conducted an assessment of the water supply and sanitation sectors in Ethiopia, including progress towards the Millennium Development Goals.[40] The amount and quality of data available for the assessment from different regions varied considerably, and some data were contradictory. The data thus could not be used for a sound public review of expenditures

Box 14.21 Research links in developing countries

The Water Law and Indigenous Rights programme (2001-07) began as an international action-research alliance to inform the debate on peasant, indigenous and customary rights in the Andean region and to facilitate action on local, national and international platforms. A major objective was to better understand local water rights and water management. The strategy was to focus on research and action, together with local, regional and international networks (both indigenous and non-indigenous), while training policy-makers, water professionals and grass-roots leaders. The programme deepened water policy debates on recognition of water rights, enabling better legislation and more democratic water governance and management policies. Network participants have extended their activities into new policy research and action networks on the plurality of water rights, multiscale water user organizations and ways to strengthen and recognize such processes through training and user-oriented intervention strategies. Concertación (2006-11) is an interdisciplinary research and capacity-building network concentrating on peasant empowerment and indigenous water management, with a focus on the Andean region. The Water Research Fund for Southern Africa, established in 1999, is available to researchers and institutions in any Southern Africa Development Community country. A peer review system ensures that high quality research proposals are selected for funding. A board consisting of researchers with different professional backgrounds and from different countries formulates the fund's research policy and defines priority areas. Although external donors are the main source of funding, rather than the research end users, the fund is exploring ways to better link societal demands to the research community and of reviewing its role in promoting this connection so that the supported research is responsive to the region's needs.

Source: Rap 2008; www.eclac.cl/drni/proyectos/walir/homee.asp; www.iwsd.co.zw/index.cfm.

in these sectors. The study revealed that most town water utilities do not have proper accounts, data on sanitation are absent or fragmentary and work by non-governmental organizations in the water supply and sanitation sectors is uncoordinated and undertaken without collaboration with responsible national authorities. The assessment also found that funding of these sectors, on both a sectoral and per capita basis, was low, resulting in poor water service throughout the country.

Water resources monitoring

Data on technical and scientific aspects of water resources management often come from monitoring activities, including measurements of chemical, biological and other parameters of the quantity or quality of water systems. Such data can illustrate the limits of the water resources, expose hard-to-see connections and contribute to innovative solutions to water resources problems, as illustrated in box 14.22.

Monitoring networks are inadequate and weakening (see chapter 13). Although modelling is not a substitute for accurate field measurements, modelling experience has advanced to the extent that some data can now be generated with models.

The Lake Naivasha Riparian Association in Kenya used modelling to investigate the potential impacts of water abstractions to supply intensive horticultural activities that began in the early 1980s. Association members were concerned that the horticultural activities were responsible for falling lake water levels. Horticultural interests denied the claim, pointing out that the lake was higher than it had been in the 1950s, prior to development of their industry. The modelling work considered the contributions of surface and groundwater to the lake levels, clearly illustrating that the declining lake levels coincided with commencement of horticultural activities in 1982. The results were ultimately accepted by all stakeholders, including the horticulturalists, resulting in their working more closely together to promote a stronger conservation ethic for the lake.[41]

Traditional knowledge and experiences of the people directly affected by a water system, whether as members of its shoreline community or as economic actors dependent on it for their livelihoods, are also important. Traditional knowledge can come from the legends and oral histories of indigenous people. In some areas such memories may be the only source of information on the history of a water system. The Ugandan government, for example, used local knowledge to identify and protect important fish breeding areas on the eastern shore of Lake Albert on the Uganda-Democratic Republic of Congo border.[42]

Financing

The need for sustainable financing is one of the most persistent concerns in water

The need for sustainable financing is one of the most persistent concerns in water resources decision-making

Box 14.22 Using monitoring information in water resources management

Information from monitoring activities can be used to expose the limits of water resources, reveal hidden connections and develop innovative solutions to water resources problems. Data on fishing intensity and gear were used in declaring temporary fishing moratoriums for Lakes Baringo and Naivasha in Africa and in setting restrictions on allowable fishing technologies for Lake Victoria. These measures contributed to the recovery or significant improvements of these fisheries.

Biophysical processes in water systems are complex and often expressed in small incremental changes that are not readily observable. Detailed measures and investigations at Lake Biwa, Japan, showed that declining snowfall over several decades, combined with a weakening water profile in the lake, had led to a decline in the dissolved oxygen levels in the lake's bottom waters, increasing lake eutrophication and pointing to the potential effects of global warming on the lake.

Field monitoring in the Lake Chad basin in Africa demonstrated that wet season conditions could be simulated by water releases from the Tiga and Challawa Dams and that artificial flooding of wetlands could be undertaken with the existing infrastructure.

Several years of monitoring data from Lake Ohrid in the Former Yugoslav Republic of Macedonia suggested that the phytoplankton and zooplankton communities in the lake were changing, consistent with its increasing eutrophication and making an unequivocal case for nutrient control.

Monitoring data collected from Lake Dianchi in China demonstrated the success of policies to reduce the external nutrient load from local enterprises. Such post-project monitoring can help marshal support for the often considerable outlays required for water resources management programmes and activities.

Source: ILEC 2005; World Bank 2005.

Small-scale water providers typically include entrepreneurs filling a void left by the market failure of the formal water sector and small companies providing services to municipalities under contract

resources decision-making. Ultimately, all financing in the sector comes from tariffs, the national budget and external aid. Decision-makers in the water sector do not control all of the factors influencing these financing sources, but they can help create a favourable investment climate and ensure good management of the financial resources at their disposal. Several key initiatives over the past five years have shaped the agenda of water financing, including the World Panel on Financing Water Infrastructure (chaired by Michel Camdessus), the Task Force on Financing Water for All (chaired by Angel Gurria) and the UN Secretary General's Advisory Board on Water and Sanitation (UNSGAB). This section offers examples of how some countries are tackling the problems and implementing the recommendations identified in these reports.

Water supply service providers manage financial resources well, creating favourable investment climates

Reforms have responded to the failure of public sector water providers to provide adequate levels of service and quality, often because of poor governance and inadequate investments. Reforms have also focused on transparency and accountability, low revenue collection, infrastructure deterioration and service breakdown. Privatization of water services is still an exception, and there is not enough experience to establish its long-term efficacy. Although the number of people served by the private water sector has grown from roughly 50 million in 1990 to about 300 million in 2002, most people in developing countries are not served by either private or partially private companies (see chapters 4 and 7 for additional discussion).[43]

The National Water and Sewerage Corporation (NWSC) in Uganda is a public sector utility that has assumed a more formal business structure. Within a short period this once poorly performing utility had improved service quality (expanding coverage from 12 towns in 1998 to 22 towns today) and strengthened financial viability through more efficient service provision and increased revenue collection. Its experience also illustrates the importance of good leadership, showing that substantial changes can occur rapidly under the right conditions.[44]

To improve efficiency and finances, NWSC introduced performance-based contracts for local utilities and separated operations and maintenance from performance monitoring and regulation. Results included a substantially larger service base, increased staff efficiency and a more viable financial base (box 14.23).

The Tunisian water authority SONEDE is another example of a water supply agency operating with good management practices. As indicated in box 14.3, a targeted pricing policy enables full cost recovery of operating the service, with tourist establishments paying the highest rates and households the lowest. Unaccounted for water in Tunis has been reduced below 10%.

Providing official development assistance for water resources in more user-friendly forms

In Kenya, output-based aid underpins the loans being made by a commercial microfinance agency. The international aid repays part of the investment cost once the work is completed. This approach differs from a normal guarantee of the lender in that it creates a strong incentive for the bank to ensure timely and effective completion of the project.

Small-scale local water providers

Small-scale water providers serve an estimated 25% of the urban population in Latin America and East Asia and about 50% in Africa and South-East Asia.[45] Small-scale water providers typically include entrepreneurs filling a void left by the market failure of the formal water sector and small companies providing services to municipalities or public authorities under contract.

The main obstacle to the expansion of small-scale water providers is lack of financing. Financing is generally from personal assets, profits from other businesses, community contributions and short-term credit from local banks or microcredit agencies. Options to increase access to financing for small-scale providers include microfinance schemes, access to local development and infrastructure banks, projects funded by non-governmental organizations or donors and investment subsidies (for example, in Cambodia grants from the International Development Association cover 50%-60% of investment costs). The Philippines has successfully used loan financing at market rates, while Colombia has used a middle path that preserves incentives and enhances financial sustainability by plotting increased tariffs over time and placing revenues in an investment fund.[46]

Box 14.23 **Addressing water supply and sanitation challenges in Uganda**

Uganda faced numerous challenges in addressing water supply and sanitation services related to urbanization and population growth pressures, including lack of harmony in water policies, inadequate governance, divergent stakeholder preferences, managerial inefficiencies and limited investment capital. To address these deficiencies, the government initiated a series of complementary activities:

Performance indicator	1998	2007
Service coverage (percent)	47	71
New water connections (number per year)	3,317	24,418
Total number of water connections	50,826	180,697
Number of employees	1,784	1,388
Staff per 1,000 water connections	37	7
Non-revenue water (percent of total)	51	32.5
Income (billions of Uganda shillings)	21.9	70.4
Profit after depreciation (billions of Uganda shillings)	-2.0	6.5

- Instituting timely, rational expansion of water production and network facilities.
- Using an optimal mix of technology-based approaches (such as yard taps, prepaid meters and public standposts).
- Introducing and subsidizing public access to water connections.
- Implementing innovative capital financing mechanisms (such as tariff indexation and conservation, 'soft' loan financing and bond issuance).
- Using output-based investment approaches to strengthen service targeting.
- Using community-based approaches and mainstreaming consumer preferences.
- Establishing stakeholder coordination forums, such as water councils, at different levels.
- Equitably providing water and sanitation infrastructure, with an emphasis on sanitation.

As a result of these activities, water delivery and service efficiency and profitability rose dramatically during 1998-2007 (see table). Coverage increased more than 50%, with total water connections more than 3.5 times higher in 2007 than in 1998. Employee efficiency also increased, and income rose more than 300%.

Several activities were especially helpful in improving Uganda's urban water supply and sanitation services:

- Using performance-contract arrangements that promote a private sector-like perspective, including market orientation, customer focus, incentive plans and accountability.
- Applying managerial subsidarity, keeping decision-making close to beneficiaries.
- Separating water supply and sanitation operations from monitoring and regulation to improve accountability.
- Incorporating sanitation improvement initiatives into water supply and sector plans based on appropriate technologies for affordability and social acceptability.
- Coordinating investment activities, operations management and user communities to ensure maximum impact.

Source: Mugisha and Sanford 2008.

Lack of legal recognition of small-scale providers is another obstacle to expansion. To strengthen legitimacy, local water authorities in Mauritania and Uganda enter into management contracts with private operators, which then provide water supplies to small towns.

Payment for environmental goods and services

Payment for environmental services is based on the recognition that the environment, such as wetlands and watersheds, provides a range of life-supporting goods and services, including potable water supply, irrigation water, food and fibre, wastewater treatment, flood control and aesthetic benefits (figure 14.1). Environmental goods and services, traditionally thought of as provided for 'free' by nature, lack a functioning market for pricing them. Further, for many environmental goods and services there is no direct link between the service provider and the consumer – for example, between an upstream provider (a land owner or resource manager) and a downstream user (public water supply, agriculture and industry). With the market for these services often poorly developed or non-existent, ecosystem managers have little economic incentive to improve their management efforts. The payment for environmental services concept attempts to address this problem by creating markets for environmental services, collecting money from water users and paying those providing the resource, thereby encouraging efficient and sustainable delivery of watershed services.

Experience with payment for environmental services is growing in several countries. In the United States New York City obtains its drinking water from watersheds in the Catskill Mountains north of the city. Water quality was traditionally very good, requiring little or no treatment. By the end of the 1980s, however, agricultural and other development in the Catskills threatened the water quality. New York City planners chose to work with upstream land owners in the Catskill watersheds to address potential water quality problems. The resultant plan included payments for both on-farm capital costs and pollution-reducing agricultural

Figure 14.1 **Ecosystems and some of the services they provide**

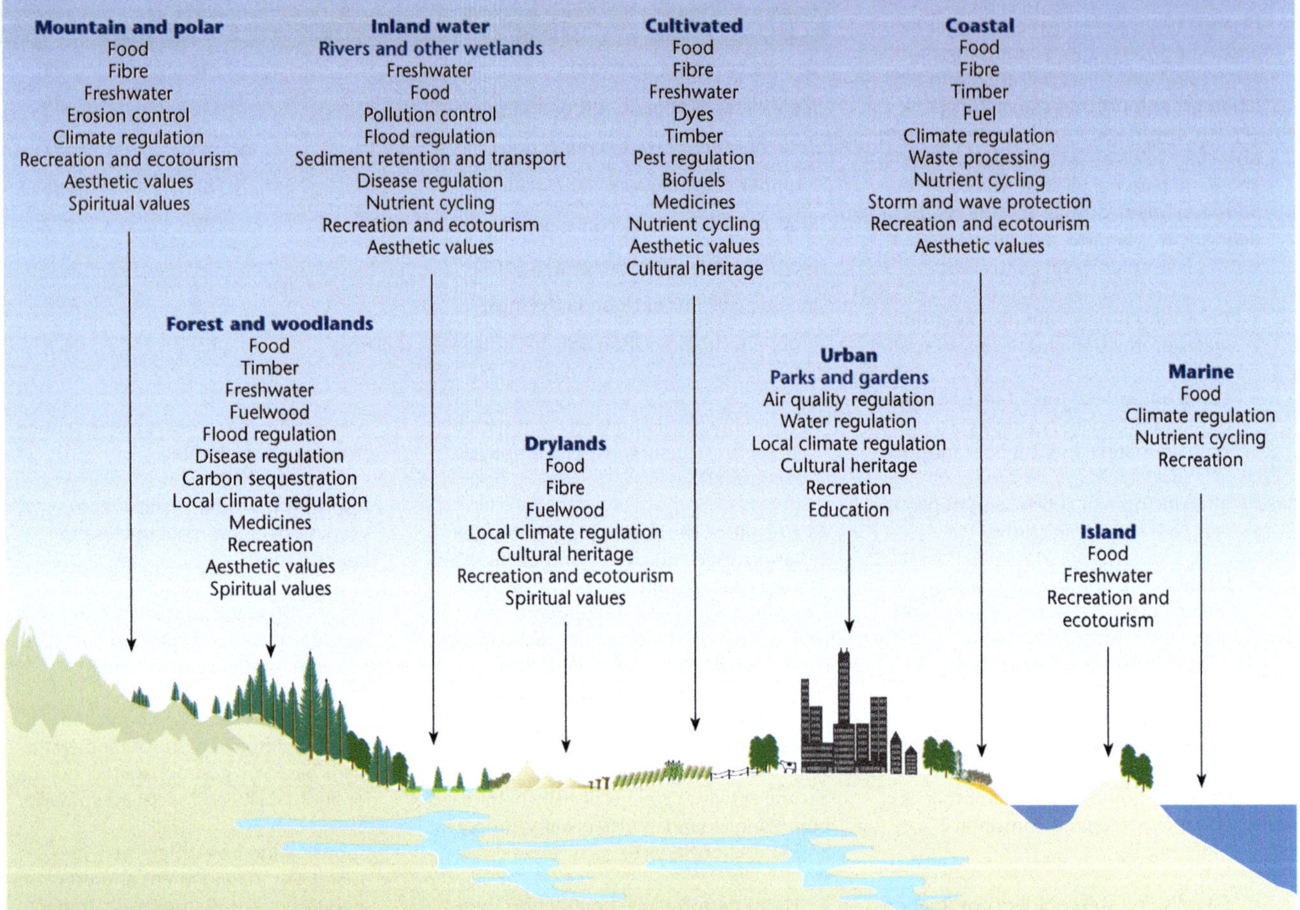

Note: Different combinations of services are provided to human populations from the various types of ecosystems represented here. Their ability to deliver the services depends on complex biological, chemical and physical interactions, which are in turn affected by human activities.
Source: Based on MEA 2005.

measures, which helped reduce the costs of conventional water treatment. This approach also protected the watersheds and the other environmental goods and services they provided (such as recreation and biodiversity conservation). The payments to water providers come directly from the revenues collected from water users in New York City.[47] A similar approach was used in Heredia, Costa Rica, which taxed its approximately 50,000 connected water users to pay farmers in the watershed for improved conservation measures.

Another example of payment for environmental services focuses on rewarding rural people for programmes that reduce the loss of topsoil from cropland in Kenya (box 14.24).

Some examples[48] suggest that, for payment for environmental services systems to create an enabling environment, infrastructure development may be necessary. This would be expected to increase the costs of implementing such systems.[49]

Box 14.24 Paying for environmental services – green water credits in Kenya

Over the last 25 years much of the cropland in Kenya has lost its topsoil. Meanwhile, the population has roughly doubled, increasing demand for food and power. The International Fund for Agricultural Development (IFAD)-supported green water credits initiative is an innovative mechanism for rewarding rural people for sustainable water management practices. The long-term goal is to empower upstream rural communities in the Tana River basin to better manage their land and water resources to improve food and water security and livelihoods both upstream and downstream in the watershed. It is also expected to decrease flood potential, improve blue water resources delivery downstream and provide diversified rural income sources.

Begun in the Tana River basin in 2006, this programme was also intended to demonstrate the feasibility of green water credits as an environmental service tool. The project builds on the difference between green water and blue water. Green water, which is soil moisture generated by rainfall that infiltrates the soil, is a fully consumptive water use, while some blue water (such as rivers and lakes) is used non-consumptively and can be reused.

The green water credits scheme includes quantifying current water fluxes in the watershed and estimating fluxes under the scheme, identifying potential participants (demand assessment) and developing a payment and reward mechanism for upstream land managers and downstream clients based on appropriate incentives. The main activities include upstream soil and water conservation techniques applied by farmers, leading to more water being available downstream. Results to date include lessened land degradation and improved quantity and quality of blue water supply. With the support of the Kenyan government, the green water credits concept is being scaled up to help improve food and water supplies nationwide. Increased power security has been another by-product.

Concurrent with implementation of this large scale, market-based scheme to improve land and water management, Kenya is reforming its entire water sector. To this end, the Tana River basin programme was strategically aligned with recent Kenyan water sector reforms, including the 2002 Water Act and the Kenyan National Water Resources Management Plan, which assign an economic value to water resources. These efforts reflect past success and future intentions by addressing water allocation issues along with financial considerations, to reward Kenya's land stewards, generating benefits for the rural poor and the country as a whole.

Source: Falkenmark 2003; ISRIC 2007; www.isric.org/UK/About+ISRIC/Projects/Current+Projects/Green+Water+Credits.htm

Notes

1. Examples include WCD 2000 and US EPA 2005.
2. Rogers and Hall 2002.
3. Bandaragoda 2006.
4. ILEC 2005; World Bank 2005.
5. UN-Water 2008.
6. UN-Water 2008.
7. See, for example, Eawag-WSSCC-SuSanA (2008); and IWA (2008).
8. Gyawali et al. 2006.
9. Bandaragoda 2006.
10. Cap-Net 2008.
11. Rast 1999.
12. ILEC 2005.
13. Bayoumi and Abumoghli 2007.
14. Phumpiu and Gustafsson 2009.
15. See WWAP 2006; Phumpiu and Gustafsson 2009; Caplan 2003; UNDP 2006.
16. Phumpiu and Gustafsson 2009.
17. Beccar, Boelens, and Hoogendam 2002.
18. Burchi 2005.
19. UNDP WGF-SIWI 2008.
20. Moench et al. 2003.
21. www.forumrisparmioacqua.it.
22. Transparency International 2008; UNDP and WGF-SIWI 2008.
23. Cap-Net 2008.
24. World Bank 2007.
25. UNDP 1997.
26. UNDP 1997.
27. www2.gtz.de/dokumente/bib/06-1322.pdf.
28. www.genderandwater.org/page/4208.
29. Visscher et al. 2006; www.irc.nl/page/29472.
30. Smits, Moriarity, and Sijbesma 2007; www.irc.nl/page/35887.
31. Butterworth et al. 2008.
32. www.youtube.com/watch?v=btWcXNSvOHw.
33. ILEC 2003.
34. ILEC 2005.
35. www.forumrisparmioacqua.it.
36. Conceição and Heitor 2003.
37. UN Millennium Project 2005; Juma and Agwara 2006.
38. WSP-UNICEF 2007; see also http://s189535770.onlinehome.us/pottersforpeace/?page_id=9.
39. UNDP 2006.
40. Watson et al. 2005.
41. ILEC 2005; World Bank 2005.
42. ILEC 2003.
43. UNDP 2006.
44. World Bank Institute 2005.
45. Dardenne 2006; McIntosh 2003.
46. Triche, Requena, and Kariuki 2006.
47. Pagiola and Platais 2002, 2007.
48. Muñoz Piña et al. 2008.
49. Pagiola and Platais 2007; World Bank 2007.

References

ANA (Brazil National Water Agency), GEF (Global Environment Facility), UNEP (United Nations Environment Programme), and OAS (Organization of American States). 2004. Strategic Action Program for the Integrated Management of the São Francisco River Basin and Its Coastal Zone – SAP. Brazil National Water Agency, Global Environment Facility, United Nations Environment Programme, and Organization of American States. www.gefweb.org/interior_right.aspx?id=238.

Balcazar, R. A. 2005. The Establishment of an Anti-Corruption Agreement with Pipe Manufacturing Companies: A Colombian Experience. Presentation at Seminar on Meeting International Targets without Fighting Corruption, World Water Week, 20-26 August 2005, Stockholm.

Bandaragoda, D. J. 2006. Status of Institutional Reforms for Integrated Water Resources Management in Asia: Indications from Policy Reviews in Five Countries. Working Paper 108. International Water Management Institute, Colombo.

Bayoumi, M., and I. Abumoghli. 2007. National Water Councils: Comparative

Experiences. United Nations Development Programme–Iraq, Amman.

Beccar, L., R. Boelens, and P. Hoogendam. 2002. Water Rights and Collective Action in Community Irrigation. In *Water Rights and Empowerment,* eds. R. Boelens and P. Hoogendam. The Netherlands: Gorcum Publishers.

Blue Plan, MAP (Mediterranean Action Plan), and UNEP (United Nations Environment Programme). 1999. *Participatory Irrigation Management Activities and Water User Organizations Involvement in Turkey.* Sophia Antipolis, France: Blue Plan.

———. 2002. Plan Bleu, CMDD (2002) Études de cas. Forum avancées de la gestion de la demande en eau en Méditerranée, Fiuggi, Italie, 3-5 Octobre 2002.

Boelens, R. 2008. The Rules of the Game and the Game of the Rules: Normalization and Resistance in Andean Water Control. Ph.D. diss., Wageningen University, The Netherlands.

Burchi, S. 2005. The Interface between Customary and Statutory Water Rights – A Statutory Perspective. Legal Papers Online No.45, Food and Agriculture Organization of the United Nations, Rome.

Butterworth, J. A., A. Sutherland, N. Manning, B. Darteh, M. Dziegielewska-Geitz, J. Eckart, C. Batchelor, T. Moriarty, C. Schouten, J. Da Silva, J. Verhagen, and P. J. Bury. 2008. Building More Effective Partnerships for Innovation in Urban Water Management. SWITCH Learning Alliance, Loughborough, UK. http://switchurbanwater.lboro.ac.uk/outputs/pdfs/WP6-2_PAP_Effective_partnerships_in_UWM_abridged.pdf.

Caplan, D. 2003. Investment in Water Infrastructure Benefits Orillia. Canada-Ontario Infrastructure Program, Infrastructure Canada, Toronto, Ontario.

Cap-Net. 2008. Performance and Capacity of River Basin Management Organizations: Cross-Case Comparison of Four RBOs. Cap-Net, Pretoria.

COHRE (Centre on Housing Rights and Evictions), AAAS (American Association for the Advancement of Science), SDC (Swiss Agency for Development and Cooperation), and UN-HABITAT (United Nations Human Settlements Programme). 2007. *Manual on the Right to Water and Sanitation.* Geneva: Centre on Housing Rights and Evictions. www.cohre.org/store/attachments/RWP - Manual_final_full_final.pdf.

Conceição, P., and M. V. Heitor. 2003. Techno-Economic Paradigms and Latecomer Industrialization. United Nations Educational, Social, and Cultural Organization Encyclopedia, Paris.

Cooley, H., J. Christian-Smith, and P. H. Gleick. 2008. *More with Less: Agricultural Water Conservation and Efficiency in California.* Oakland, CA: Pacific Institute. www.pacinst.org/reports/more_with_less _delta.

Dávila, G., and H. Olazával. 2006. *De la Mediación a la Movilización Social: Análisis de Algunos Conflictor por el Aqua en Chimborazo.* Quito: Abya-Yala.

Döker, E., D. Er, F. Cenap, H. Özlü, and E. Eminoglu. 2003. Decentralization and Participatory Irrigation Management in Turkey. Water Demand Management Forum: Decentralization and Participatory Irrigation Management, 2-4 February 2003, Cairo, Egypt.

Earle, A., and A. Turton. 2005. No Duck No Dinner: How Sole Sourcing Triggered Lesotho's Struggle against Corruption. Presentation at seminar on Can We Meet International Targets without Fighting Corruption? World Water Week, 20-26 August, 2005, Stockholm.

Eawag (Swiss Federal Institute of Aquatic Science and Technology)-Sandec (Department of Water and Sanitation in Developing Countries)-SuSanA (Sustainable Sanitation Alliance). 2007. Towards More Sustainable Sanitation Solutions. Swiss Federal Institute of Aquatic Science and Technology, Dübendorf, Switzerland. http://esa.un.org/iys/docs/Susana_backgrounder.pdf.

Falkenmark, M. 2003. Water Management and Ecosystems: Living with Change. TEC Background Paper No. 9, Global Water Partnership, Stockholm.

Garces-Restrepo, C., D. Vermillion, and G. Muñoz. 2007. *Irrigation Management Transfer. Worldwide Efforts and Results.* Water Reports No. 32. Rome: Food and Agriculture Organization of the United Nations.

Gyawali, D., J. A. Allan, P. Antunes, A. Dudeen, P. Laureano, C. L. Fernández, P. Monteiro, H. Nguyen, P. Novácek, and C. Pahl-Wostl. 2006. *EU-INCO Water Research from FP4 to FP6 (1994-2006) – A Critical Review.* Luxembourg: Office for Official Publications of the European Communities.

ICID (International Commission on Irrigation and Drainage). 2008. China Reduces Irrigation Water Withdrawals by 25 Percent. ICID Newsletter, International Commission on Irrigation and Drainage, New Delhi.

IJC (International Joint Commission). 1978. Great Lakes Water Quality Agreement, with Annexes and Terms of Reference, Between the United States of America and Canada, Signed at Ottawa, November 22, 1978. International Joint Commission, Washington, DC.

ILEC (International Lake Environment Committee Foundation). 2003. *World Lake Vision: A Call to Action.* Kusatsu, Shiga, Japan: World Lake Vision Committee Foundation.

———. 2005. *Managing Lakes and Their Basins for Sustainable Use: A Report for Lake Basin Managers and Stakeholders.* Kusatsu, Shiga, Japan: International Lake Environment Committee Foundation.

ISRIC (International Soil Research and Information Centre). 2007. Political, Institutional and Financial Framework for Green Water Credits in Kenya. Green Water Credits (GWC) Proof-of-Concept Report No.6, International Soil Research and Information Centre, Wageningen, The Netherlands.

IWA (International Water Association). 2008. Sanitation Challenges and Solutions. Sanitation 21 Task Force. International Water Association, London.

Jairath, J. 2000. Participatory Irrigation Management in Andhra Pradesh – Contradictions of a Supply Side Approach. Paper delivered at workshop on South Asia Regional Poverty Monitoring and Evaluation, 8-10 June 2000, New Delhi.

Johnson, S. H. 1997. *Irrigation Management Transfer in Mexico: A Strategy to Achieve Irrigation District Sustainability. Research Report 16.* Colombo: International Irrigation Management Institute.

Juma, C., and H. Agwara. 2006. Africa in the Global Knowledge Economy: Strategic Options. *International Journal of Technology and Globalisation* 2 (3/4) 218-31.

Kingdom, B., R. Liemberger, and P. Marin. 2006. The Challenge of Reducing Non-revenue Water (NRW) in Developing Countries. How the Private Sector Can Help: A Look at Performance-Based Service Contracting. Water Supply and Sanitation Board Discussion Paper Series 8, World Bank, Washington, DC.

Lange, J. 2008. A Guide to Public Participation, According to Article 14 of the EC Water Framework Directive. RhineNet Project Report, Freiburg, Germany.

McIntosh, A. 2003. *Asian Water Supplies. Reaching the Poor.* Manila: Asian Development Bank and London: International Water Association. www.adb.org/documents/books/asian_water_supplies/asian_water_supplies.pdf.

MEA (Millennium Ecosystem Assessment). 2005. *Living Beyond Our Means: Natural Assets and Human Well-Being,*

Statement from the Board. Washington, DC: World Resources Institute.

Moench, M., A. Dixit, S. Janakarajan, M. S. Rathore, and S. Mudrakartha. 2003. *The Fluid Mosaic: Water Governance in the Context of Variability, Uncertainty and Change.* Ottawa: Ottawa International Development Research Centre.

Moore, J. G., W. Rast, and W. M. Pulich. 2002. Proposal for an Integrated Management Plan for the Rio Grande/Rio Bravo. In *1st International Symposium on Transboundary Waters Management,* eds. A. Aldama, F. J. Aparicio, and R. Equihua. Avances en Hidraulica 10, XVII Mexican Hydraulics Congress, Monterrey, Mexico, Nov. 18-22, 2002.

Mugisha, S., and B. Sanford. 2008. State-Owned Enterprises: NWSC's Turnaround in Uganda. *African Development Review* 20 (2): 305-34.

Muñoz-Piña, C., A. Guevara, J. M. Torres, J. and Braña, J. 2008. Paying for the Hydrological Services of Mexico's Forests: Analysis, Negotiations and Results. *Ecological Economics* 65 (4): 725-36.

Palacios, E. V. 1999. Benefits and Second Generation Problems of Irrigation Management Transfer in Mexico. Water Conservation and Use in Agriculture Discussion Paper, Food and Agriculture Organization of the United Nations, Rome.

Phumpiu, P., and J. E. Gustafsson. 2009. When Are Partnerships a Viable Tool for Development? Institutions and Partnerships for Water and Sanitation Service in Latin Americ *Water Resources Management* 23 (1): 19-38.

PLUARG (Pollution from Land Use Activities Reference Group). 1978. Environmental Management Strategy for the Great Lakes System. Great Lakes Regional Office, International Joint Commission, Windsor, Ontario.

Raju, K. V. 2001. Participatory Irrigation Management in India (Andhra Pradesh) - IMT Case Study. International E-mail Conference on Irrigation Management Transfer, International Network on Partcipatory Irrigation Management, June-October 2001.

Rap, E. 2008. Interdisciplinary Research and Capacity Building Program on Water Policy and Water Management in Andean Countries. Background Paper for Knowledge on the Move Conference, 27-28 February 2008, The Hague.

Rast, W. 1999. Overview of the Status of Implementation of the Freshwater Objectives of Agenda 21 on a Regional Basis. *Sustainable Development International Journal* 1: 53-6.

Rogers, P., and A. W. Hall. 2002. Effective Water Governance. TEC Background Paper No. 7, Global Water Partnership, Stockholm.

Ruiz-Mier, F., and M. van Ginneken. 2006. Consumer Cooperative: An Alternative Institutional Model for Delivery of Urban Water Supply and Sanitation Services. Note No. 5, Water Supply & Sanitation Working Notes, World Bank, Washington, DC.

Stålgren, P. 2006. Corruption in the Water Sector: Causes, Consequences and Potential Reform. Policy Brief No. 4, Stockholm International Water Institute, Stockholm.

Thampi, G. K. 2005. Community Voice as an Aid to Accountability: Experiences with Citizen Report Cards in Bangalore. Presentation at seminar on Can We Meet International Targets without Fighting Corruption? World Water Week, Stockholm, August 2005.

Transparency International. 2008. *Global Corruption Report 2008: Corruption in the Water Sector.* Cambridge, UK: Cambridge University Press.

Triche, T., S. Requena, and M. Kariuki. 2006. Engaging Local Private Operators in Water Supply and Sanitation Services. Initial Lessons from Emerging Experience in Cambodia, Colombia, Paraguay, the Philippines and Uganda. Water Supply and Sanitation Working Notes No 12, World Bank, Washington, DC.

Treyer, Sébastien. 2004. Introducing Political Issues in the Debate on Water Resources Planning in Tunisia: A Necessity for the Implementation of Water Demand Management Policies. In *Proceedings of the Workshop on Water and Politics: Understanding the Role of Politics in Water Management, Marseille, France, 26-27 February 2004.* Marseille, France: World Water Council. www.worldwatercouncil.org/fileadmin/wwc/Library/Publications_and_reports/Proceedings_Water_Politics/proceedings_waterpol_full_document.pdf.

UNDP (United Nations Development Programme). 1997. *Capacity Building for Sustainable Management of Water Resources.* New York: United Nations Development Programme, Sustainable Energy and Environment Division.

———. **2006.** *Human Development Report 2006: Beyond Scarcity: Power, Poverty and the Global Water Crisis.* New York: Palgrave Macmillan.

UNDP (United Nations Development Programme) and WGF-SIWI (Water Governance Facility, Stockholm International Water Institute). 2008. *Water Adaptation: Freshwater in National Adaptation Programs of Action, NAPAs, and Climate Adaptation in Freshwater Planning.* Stockholm: United Nations Development Programme and Stockholm International Water Institute.

UNEP (United Nations Environment Programme). 2008. *Africa: Atlas of our Changing Environment.* Division of Early Warning and Assessment. Nairobi: United Nations Environment Programme. www.unep.org/dewa/africa/AfricaAtlas.

UN Millennium Project. 2005. *Investing in Development: A Practical Plan to Achieve the Millennium Development Goals.* New York: United Nations.

UN-Water. 2008. Status Report on IWRM and Water Efficiency Plans for CSD 16. Paper prepared for 16th session of the Commission on Sustainable Development in May 2008, UN-Water.

US EPA (Environmental Protection Agency). 2005. Handbook for Developing Watershed Plans to Restore and Protect Our Waters. Washington, DC: U.S. Environmental Protection Agency.

Velasco, H. G. 2003. *Water Rights Administration – Experiences, Issues and Guidelines.* Rome: Food and Agriculture Organization of the United Nations.

Watson, P. L., J. Gadek, E. Defere, and C. Revels. 2005. Assessment of Resource Flows in the Water Supply and Sanitation Sector. Ethiopia Case Study. World Bank, Washington, DC.

WCD (World Commission on Dams). 2000. *Dams and Development. A New Framework for Decision-Making.* London: Earthscan.

World Bank. 2005. Lessons for Managing Lake Basins for Sustainable Use. Environment Department, World Bank, Washington, DC.

———. **2007.** Promoting Market-Oriented Ecological Compensation Mechanisms: Payment for Ecosystem services in China. Policy Report, World Bank Analytical and Advisory Assistance Program, East Asia and Pacific Region, World Bank, Washington, DC.

———. **2008.** Climate Resilient Cities. 2008 Primer: Reducing Vulnerabilities to Climate Change Impacts and Strengthening Disaster Relief Management in East Asian Cities. Washington, DC: World Bank.

World Bank Institute. 2005. When Passionate Leadership Stimulates Enduring Change: A Transformational Capacity Development Anecdote from Uganda. Capacity Development Brief No. 13. World Bank Institute, Washington, DC.

WSP-AF (Water and Sanitation Programme – Africa). 2002. Rural Water Sector Reform in Ghana: A Major Change in Policy and Structure. Blue Gold Series, Field Note 2, Water and Sanitation Programme, Nairobi.

WSP (Water and Sanitation Programme)-UNICEF (United Nations Children's Fund). 2007. Use of Ceramic Water Filters in Cambodia. Field Note, Combating Waterborne Disease at the Household Level. Water and Sanitation Programme, Cambodia Field Office, Phnom Penh. www.wsp.org/UserFiles/file/926200724252_eap_cambodia_filter.pdf.

WWAP (World Water Assessment Programme). 2006. *The United Nations World Water Development Report 2. Water: A Shared Responsibility.* Paris: United Nations Educational, Scientific and Cultural Organization, and New York: Berghahn Books.

WWF. 2008. Kafue Flats, Zambia: Preserving Biodiversity through Water Management. *Innovation for Sustainable Development: Case Studies from Africa.* World Wildlife Fund, Gland, Switzerland.

Options inside the water box

Chapter 15

Options from beyond the water box

Authors: William Cosgrove and Joana Talafré

Contributors: Richard Connor, William Cosgrove, George de Gooijer, Gerry Galloway, Molly Hellmuth, Håkan Tropp and James Winpenny

Coordinator: Olcay Ünver (WWAP)

Facilitator: William Cosgrove

Key messages

- Dealing with risk and uncertainty has long been a routine challenge for water resources managers and policy-makers across sectors and the world. However, issues like climate change and demographic dynamics have made the risks greater and the task more complex. Risk management is now much more important – indeed essential – to analysis and decision-making.

- Drivers and policies outside the water sector have more impact on water management than do many policies championed and implemented by water-related ministries. Identifying trade-offs and synergies between water and other policy sectors can enhance policy impacts in all sectors and avoid some adverse effects on water.

- Because governments, civil society and business leaders make decisions every day that can affect water, it is important to identify where such decisions can also lead to improvements in water sector management and in water sector and environmental services.

- Examples of win-win situations abound – whether created by governments, communities or businesses – that point to promoting deliberate cooperation between water and non-water actors and integrating water issues into external decisions.

- International organizations, notably the UN system, can provide support and expertise to governments, help civil society build capacity and catalyse leadership in the private sector.

Chapter 14 shows that many actions defined within the water sector also require decisions by leaders and policy-makers outside the sector. Furthermore, external drivers and policies related to sectors that affect overall economic diversification – such as agriculture, trade, energy, health, housing, finance and social protection – have more impact on water management than do many policies championed and implemented by water-related ministries.[1] The preceding parts of this Report demonstrate the complex links between those external drivers and water.

Leaders in governments, civil society and the private sector all make decisions every day that affect water. It is important to identify how decisions motivated by economic or social objectives can also lead to improvements in water sector management and in water sector and environmental services. This chapter illustrates such win-win situations.

There are essentially two ways such win-win scenarios can unfold. First, policy decisions or reforms made outside the water sector can create space for change within

Curbing corruption in the water sector could save an estimated 20%-70% of invested resources

the water sector (discussed in the first part of this chapter). In many cases the success of reforms in the water sector has depended on broader enabling conditions – such as transparency and accountability in public institutions, sound trade conditions and access to information. Responses to some external drivers may therefore indirectly lead to improvements in water management.

The second way is more direct and involves avoiding the negative impacts of non-water sectoral interventions by taking water into account in decision-making (discussed in the second part of this chapter) or cultivating deliberate cooperation between water and non-water actors (discussed in the third part of this chapter).

The examples in this chapter provide useful insights into how programme and policy decisions can achieve multiple benefits when implemented so as to take impacts on water into account.

Promoting win-win scenarios by creating space for change

There are many paths to improving water resources management. Direct water interventions need to be supported by a broader enabling framework. Other chapters point to the need to increase investment in the water sector; to enhance water-related information, science and monitoring systems and to create durable financial mechanisms to enable better performance in the water sector. This section provides examples of decisions outside the 'water box' and their contribution to improving water services.

Box 15.1 Combating corruption

Donors are increasingly requesting anticorruption measures in their development assistance programming, to promote aid effectiveness. Transparency and accountability in government, along with clear rules and procedures, will go far in ensuring the stability needed to promote social and economic development. Efforts to promote integrity, transparency and accountability in one sector can pave the way for similar reforms in other sectors as they benefit from the successes and lessons learned.

Key elements of a successful anticorruption strategy include citizen participation and access to information, independent checks and balances and oversight mechanisms, public sector codes of conduct and attention to incentives to corruption in public administration.

Anticorruption strategies can be incorporated at all levels of project development and implementation, from procurement to monitoring and evaluation. For example, the Kecamatan Development Program in Indonesia, a project implemented by the Indonesian Ministry of Home Affairs (Community Development Office) and supported in part by the World Bank, encouraged village communities to participate in a water infrastructure project. Competition among villages was used to single out bad construction proposals and to strengthen social control in measuring quality throughout the project cycle.

Source: Martinez 2007; Olken 2005.

Improving governance: promoting transparency and accountability

Governance provides the context for making water management decisions, establishing partnerships, mobilizing and distributing resources and giving accounts. Decisions about governance imply complex political processes and are heavily influenced by history, culture and values.

The capacity to formulate and implement appropriate policies is affected by the inclusiveness of decision-making, the strength of institutions and attitudes towards corruption. Curbing corruption has particular relevance as corruption diverts resources, lowers effective investment and undermines public confidence and government decentralization efforts (see chapter 14).

Building integrity and accountability in public decision-making systems also involves promoting the role of civil society, non-governmental organizations and the private sector. This could be particularly relevant in the water services sector, where curbing corruption could save an estimated 20%-70% of invested resources.[2] Box 15.1 shows how government and donor efforts to combat corruption have also allowed for greater citizen participation in community water development, an essential part of water governance.

Citizen participation in decision-making at all levels is also central to promoting good governance – creating a climate of accountability and transparency. Organized interest groups, community-based associations and other groups with the capacity and legitimacy to express views are crucial. Promoting a culture of consultation within the water sector and increasing consultative and participatory capacity in general, will deliver benefits for collaborative water management.

Strengthening decision-making through decentralization

Decentralization involves complex processes to enable decision-making and promote the sharing of resources and responsibilities among various levels of government. Occasionally, it includes the devolution of some power and responsibilities to civil society. In the water sector decentralization of management places decision-making closer to the level at which services are provided.

Decentralization is typically part of a broader package of governance reform. The autonomy and resources accorded to local governments will vary. In some countries where decentralization is ongoing (for example, Uganda), local governments receive resources based on performance, and local officials are held accountable based on service and management standards.

Decentralized governance implies cooperation among a broader group of actors and stakeholders, including central and provincial government institutions and sometimes civil society, the private sector and resource users. 'When it works well, decentralization has many benefits: it can allow for a democratisation of decision-making through improved stakeholder inclusiveness, transparency and accountability. . . . It can also encourage the integration of traditional knowledge and practice with innovative technologies and science to promote fair and efficient management of water resources and services.'[3] The benefits of decentralization for water management depend on assigning adequate authority and resources to local governments (box 15.2).

Australia provides a model of such coordination. Its National Resource Management Ministerial Council develops policies and strategies for national approaches to conservation and sustainable use of Australia's natural resources, including land, vegetation, biological resources and water. The council's role is to facilitate, oversee and monitor implementation of national programmes and action plans and to evaluate outcomes, which would otherwise be difficult because of the division of constitutional powers and responsibilities among territorial, state and national government entities. The council is responsible for promoting community engagement in conserving natural resources and for coordinating with other ministerial councils and bodies.[4]

Promoting conducive macroeconomic policies

The effectiveness of water institutions depends on macroeconomic policies and the environment created by these policies. In the long term macroeconomic policies are so powerful and structurally influential that the best sectoral institutions cannot completely counteract their influence.

Many countries look to economic diversification – economy-wide and within sectors – to reduce economic risk and promote growth opportunities (box 15.3). Economic activity in most least developed countries is heavily concentrated in commodities, with agriculture and fisheries contributing more than 36% of GDP and 80% of export earnings.[5] A few commodities account for the bulk of each country's exports, with a single commodity often accounting for more than 60% of export earnings. More than 70% of the population of least developed countries is employed in primary commodity production.

Some diversification schemes have direct impacts on water use or water demand because they target a sector that uses water. Others have an indirect impact on water, for example, by altering patterns of energy use. Decisions about diversification are rarely made with water impacts in mind. Rather, they are made in response to a perceived economic vulnerability or opportunity (such as expanding biofuel markets), trends in the terms of trade, instability or price changes in commodity markets or the depletion of mineral or other natural resources.

The effectiveness of water institutions depends on macroeconomic policies and the environment created by these policies

Box 15.2 Decentralization in Uganda – making room for better environmental governance at the local level

Devolving decision-making powers over natural resources to publicly accountable local authorities is frequently advocated to achieve social development goals and enhance environmental management. Uganda's experience, however, suggests that to pave the way for better water governance and services, decentralization has to be accompanied by adequate legislative and regulatory powers, executive and enforcement authority for local authorities and adequate resources.

Under the Uganda Local Government Act of 1997 district and city councils can enact by-laws, provided the by-laws do not conflict with the national constitution or other laws. District councils are responsible for disease control, environmental sanitation, entomological services and vermin control, and forests and wetlands. Despite this transfer of authority, however, executive powers on environmental matters have not been effectively devolved. Instead, administrative deconcentration within line ministries in charge of particular resources has been attempted.

Local governments are required to help enforce environment-related rules and laws. Local councils below the subcounty level have the power to decide on penalties when rules are broken, but the lack of clarity regarding the powers of sanction and arbitration has undermined good environmental practices. In many cases responsibility has been transferred but resources have not, impeding local council efforts to promote better environmental management.

Despite these constraints, there have been positive achievements. In many areas supplemental resources have been mobilized from external aid sources to support local council efforts. Local governments have made significant progress in fulfilling their legal responsibility to protect wetlands and river banks by negotiating land and water use with the local communities.

Source: Bazaara 2003.

Measures to improve education, both access and quality, can improve access to water and sanitation

Box 15.3 Economic diversification in Mauritius helps promote access to water

In the late 1970s and early 1980s Mauritius launched a policy to diversify away from traditional exports (sugar and garments). In 1988 the government relaunched its tourism strategy, emphasizing low-environmental-impact, high-spending tourism. In 1970 agriculture accounted for 16% of global domestic product (GDP) and most exports (sugar), whereas manufacturing accounted for only 14% of GDP and mainly for the domestic market. The service sector, which made up the bulk of GDP, was largely non-tradable. By 1997 agriculture's share in GDP had fallen to 9% and less than a quarter of exports. Manufacturing's share had nearly doubled (to 25%). A flourishing tourism industry, which accounted for 15.7% of GDP and 26.8% of exports, had transformed services into a tradable sector. By 1999 manufactured goods accounted for 75% of merchandise exports.

Today, as an indirect result of these economic diversification efforts of the 1970s and 1980s, 94% of the population in both rural and urban areas has access to piped water and basic sanitation. While agriculture is still by far the greatest water user, accounting for 68%, the domestic sector is the second, at 30%. This is largely to support the tourism industry, the other economic mainstay of Mauritius, whose development was aided by significant investment in water supply infrastructure.

Source: Benevides and Pérez-Lucy 2001; FAO-AQUASTAT database; Zhang 2003; Bird, Chenje, and Medina 2002; Chang 2008; WHO/UNICEF Joint Monitoring Programme for Water Supply and Sanitation 2008, p. 48.

Economic diversification alone is not necessarily good for the environment or water systems. Its impacts depend on the trade-offs and choices that are being made, including which sectors are growing and what technology is being used. It is also increasingly recognized that longer-term considerations such as climate change (see chapters 5 and 11) should be factored into diversification schemes at local and national levels to ensure their sustainability and long-term economic viability.

Creating space through social policies for education and health

There are close and multiple links between education and water and sanitation. Provision of adequate sanitation in education facilities, particularly for girls, goes hand in hand with access to education. Similarly, interventions to increase access to basic education are often accompanied by efforts to provide drinking water in or around schools. In addition, education about health, water and sanitation, water in agriculture and environmental issues more broadly can lead to more sustainable performance (box 15.4).

Measures to improve education, both access and quality, can improve access to water and sanitation. Women with education have fewer children, which has beneficial consequences for the health of mothers and children and results in lower demand per family for food and water (see chapter 2, especially figure 2.3). Removing or reducing school fees could free household income to pay for other services – including food and water or agricultural inputs that could increase water productivity. Higher schooling achievement may contribute to better water management by increasing prospects for non-agricultural employment, potentially reducing the land under cultivation and, in turn, the amount of water used for irrigation. For land under cultivation to shrink, however, off-farm employment has to generate enough income to reduce a family's reliance on agriculture for its livelihood – still unlikely in most rural areas. In China, for example, estimates of the returns on investment in education show that the increase in rural income during the 1990s came mainly from off-farm employment. Studies have shown that government policies that increased investment in education and in infrastructure for non-farm

Box 15.4 More relevant schools lead to benefits for water

Efforts to increase the effectiveness of basic education programmes by integrating students' environment into the curriculum (school gardening in rural areas, for example) show that attendance increases and students learn appropriate ways of interacting with their environment, including in agriculture, land and water management. Schools are able to forge stronger links with their communities as service providers, including of agricultural information, making school more relevant.

Policy decisions that are essentially driven by education goals – increasing school attendance – can also create space for improvements in water management. The 2000 National Programme for Curriculum Development in Sri Lanka acknowledged that 'information regarding food habits and types of food . . . can be used in introducing innovative strategies in education. Outdoor activities using agricultural plots in the school and the home can also make learning more meaningful and hence attractive. These plots could be used for introducing concepts in mathematics, language and social studies, etc. . . . Since agriculture is the main occupation of the parents in the Sinhala Medium areas, every attempt must be made to help them to learn better practices. Well maintained agricultural plots in the school could . . . also serve as demonstration plots for the community.'

Source: Atchoarena and Gasperini 2003, pp. 180-85.

sectors in rural areas contributed significantly to rural non-farm employment and to household income.[6] In many cases this can make the agricultural sector more efficient, by generating income that can be used to provide better agricultural inputs.

There are indirect links between the loss of agricultural labour force to diseases such as HIV/AIDS and land degradation and unsustainable agricultural practices.[7] HIV-related disease and death cause agricultural productivity to decline. This creates or accelerates a feedback loop between poverty and pressures on natural resources, including water. There are similar links between agriculture and malaria, with integrated pest management or an ecosystem approach helpful in preventing malaria transmission as well as improving land and water productivity and food security.[8] An agricultural intervention in Kenya paved the way for improvements in natural resources management, including water, although its primary goal was malaria prevention (box 15.5).

Other interventions in the health sector also promote better water management or sanitation. For example, HIV/AIDS education and awareness programmes can be closely tied to education reform, including efforts to integrate the students' environment into education programmes (see box 15.4). Initiatives to limit the public's exposure to toxic chemicals (including agrochemicals) include efforts to limit chemical dispersal in the environment (air, land or water) and can have an impact on water quality. The United Nations Economic Commission for Europe and World Health Organization Euro Protocol on Water and Health, adopted in 1999, recognizes this mutually reinforcing relationship between health and water policies. The protocol is the first international treaty designed to reduce water-related deaths and diseases through improved water management and universal access to safe water and adequate sanitation. This goal is to be achieved through such measures as establishing systems of surveillance and response to water-related outbreaks and adopting national targets and mechanisms to ensure mutual support and compliance review.[9]

Clearing pathways towards win-win situations: avoiding negative impacts

Decisions at many levels and sectors can have large impacts on water, while water constraints can pose challenges to private entrepreneurship, economic development, poverty reduction and sustainable environmental services. Avoiding negative impacts on water is a matter not only of social responsibility, but also of the long-term sustainability of public and private investments.

Economic decisions rarely consider their potential environmental impacts, including possible adverse effects on water. And although water constraints can influence the outcomes of economic strategies, water managers and water actors rarely participate in economic decisions. But recognition is building that water use will need to be more sustainable to achieve economic growth agendas, especially where the economy is highly dependent on water. For example, in Djibouti, where water

Decisions at many levels and sectors can have large impacts on water, while water constraints can pose challenges to private entrepreneurship, economic development, poverty reduction and sustainable environmental services

Box 15.5 The ecohealth approach: combating malaria through agricultural practices in Kenya

A project in Kenya – under the umbrella of the International Development Research Centre Program Initiative on Ecosystem Approaches to Human Health and bringing together experts from the International Centre for Insect Physiology and Ecology and the International Water Management Institute as well as the local community – is examining links between agriculture and malaria using an ecosystem approach. The goal is to reduce the incidence of malaria through a range of agricultural practices:

- Reduce mosquito-breeding habitats by limiting the water used for rice cultivation.
- Reduce the length of time that paddies are wet, by changing flooding schedules or alternating rice cultivation with dryland crops such as soya. In addition to limiting the mosquitoes' habitat, planting soya could boost income, improve nutrition and save water.
- Introduce naturally occurring bacteria into stagnant water to kill mosquito larvae during the peak breeding season.

The heart of the ecosystem approach is to actively include community members in the research process, therefore helping researchers understand people's perceptions about the health and development problems in the area. This approach also seeks interventions and solutions that are transdisciplinary.

Source: Poverty-Environment Partnership 2008.

As the example of Ethiopia shows, all decisions about water use and management take place in a context of macro- and microeconomic policies and decisions

scarcity is severe, the 2004 Poverty Reduction Strategy Paper recognizes that 'the absence of a natural resources base, weak human capital, high costs of labour and production factors (energy, telecommunications and water), inadequate financial intermediation, and administrative constraints of all kinds are the key obstacles to economic growth and to the development of private initiative'.[10]

Tunisia, a country with very limited water resources, has succeeded in stabilizing its demands and has plans for more active demand management (see box 14.3 in chapter 14). Tunisia implemented a national water-saving strategy at an early stage for both urban and agricultural needs, in line with its long tradition as an 'oasis' culture with experience managing common but scarce water resources. Water demand for irrigation has been stable for more than six years, despite agricultural development, seasonal peaks in demand and unfavourable climate conditions (drought). The water demands of tourism (a source of foreign currency) and cities (a source of social stability) have been met.

Several principles underlie the Tunisian strategy:

- Abandoning isolated technical measures in favour of an integrated approach.
- Empowering users through a participatory approach (960 water user associations were created covering 60% of the irrigated public area).
- Introducing reforms gradually and adapting them to local conditions.
- Applying financial incentive mechanisms to promote use of water-efficient equipment and technologies (equipment purchases subsidized at 60%).
- Supporting farmers' incomes, to allow them to plan for agricultural investment and labour.
- Introducing a transparent and flexible pricing system, aligned with the national goals of food security and equity, leading to gradual recovery of costs. The drinking water pricing system supplies a basic-need level of services at low cost.[11]

Box 15.6 Measuring the water limits to growth in Ethiopia

Until recently, most policy and macroeconomic decisions in Ethiopia have been based on growth models that assume that rainfall is consistently distributed at historical average levels. These models do not take into account shocks to the economy caused by extreme water events, such as floods and droughts.

A World Bank study on Ethiopia estimated the magnitude of the impacts of high water variability on growth and poverty to assist the government in better managing water and other parts of the economy (trade, transport) and to reduce the impacts of water shocks. The study found that considering the effects of water variability reduced projected rates of economic growth 38% a year and increased projected poverty rates 25% over a 12-year period. It also found that investments in water infrastructure, such as irrigation systems, were needed to reduce vulnerability to rainfall variability and that transport infrastructure helped communities adjust to localized crop failures by enabling farms to sell their surplus in food-deficit areas. The analysis, conducted with the Ethiopian government, helped make water resources management a central focus of the government's national poverty reduction strategy.

Source: Sadoff 2006.

As the example of Ethiopia shows, all decisions about water use and management take place in a context of macro- and microeconomic policies and decisions (box 15.6). To be attractive to public and private actors, sustainable water management must also make economic sense. Even in the growing number of companies that promote social responsibility, economic viability remains the key driver of business decisions. But as the examples of the Coca-Cola Company (box 15.7) and Mondi (box 15.8) show, reducing the adverse effects of production on water is not only a matter of social responsibility – it can also make good business sense. Box 15.9 illustrates this point with several examples.

The Coca-Cola example shows that companies have much to lose in public image and consumer trust when they compete with traditional uses of water. In this case the damage to the company's image was potentially higher than the costs of taking preventive action. In the Mondi example another large company coupled its plans for expansion and profitability with measures designed to have a positive effect on the environment.

Further demonstration of rising private sector awareness of the centrality of sustainable water management is the adoption of the CEO Water Mandate by the members of the UN Global Compact as a voluntary platform for addressing sustainability in business operations and supply chains.[12] Launched 5-6 March 2008 by the UN Secretary-General and committed business leaders to help companies develop a comprehensive approach to water management, the mandate covers six areas: direct operations,

Box 15.7 The Coca-Cola controversy in Kala Dera, India – learning from mistakes

In 2000 the Coca-Cola Company began bottling operations in Kala Dera, a large village outside Jaipur, Rajasthan, India, where irrigated agriculture is the primary source of livelihood. Within a year the community noticed a rapid decline in groundwater levels, for which the plant was blamed. The community demanded closure of the plant, and the cause received some international support, mainly from university students in Canada, the United Kingdom and the United States, some of Coca-Cola's largest markets.

In late 2004 the company agreed to an independent enquiry into its water management practices by the Energy and Resources Institute (TERI), a non-profit international research organization based in New Delhi. Released in January 2008, the report was welcomed by both sides, each finding some support for its positions.

For the activists opposed to the bottling plant the report confirmed that the plant was operating in an 'overexploited' groundwater area and that the plant's operations were contributing to a worsening water situation and were a source of stress to the community. The assessment made four recommendations: the plant could transport water from the nearest unstressed aquifer, store water from low-stress seasons, relocate the plant to a water-surplus area or shut down this facility.

For Coca-Cola the report noted that the company was a very small user of water, tapping less than 1% of the area's available water. Nonetheless, the report recommended more consideration of community water needs in plant location and operations, arguing that 'assessment of water availability in the vicinity of a bottling operation should be from a perspective . . . wider than business continuity'. Coca-Cola India pledged to reach a net-zero balance in groundwater use by the end of 2009, installed a rainwater harvesting system for groundwater recharge and drafted guidelines for engaging with the community. To help farmers use water more efficiently, Coca-Cola, in partnership with local farmers and the government, has started drip irrigation projects with 15 farmers. Some 15 more projects are planned in 2008-09.

Assessments across various plant sites indicated that the state governments in India have not been able to value their water resources appropriately. The review report seems to move towards a view that holds corporations responsible for their environmental impact beyond the strictures of existing regulations. The report recommends that Coca Cola 'define a strategy wherein it is able to offset this anomaly through appropriate and commensurate interventions that . . . result in a stream of benefit flows to the community.'

Source: TERI 2008.

Box 15.8 Mondi South Africa achieves sustainable expansion

Mondi South Africa, a wholly owned subsidiary of Anglo American plc, produces pulp, paper, board, corrugated containers and mining support systems. Activities range from forestry operations to highly technical manufacturing and converting processes. To increase its competitiveness in a demanding international pulp market, Mondi implemented a 2.3 billion rand (approximately $24 million) expansion project, raising the mill's production capacity 25% and accommodating a 40% increase in timber supply from more than 2,800 small growers who form part of Mondi's Khulanathi timber growers scheme.

The project began in 2001 with an engineering study, followed by a detailed environmental impact assessment in 2002. Through improvements in energy supply infrastructure and targeted equipment and technology upgrades, the company achieved its expansion objectives, while realizing the following benefits:

- A 50% reduction in sulphur dioxide (2,177 tonnes).
- A 35% reduction in nitrous oxide (509 tonnes).
- A 50% reduction in carbon dioxide (297,121 tonnes).
- A 60% reduction in total sulphur.
- Total energy and water cost-savings of 38,678,843 rand (approximately $4.9 million).
- A reduction in water use of some 13,000 cubic metres per day.
- A more than 25% reduction in wastewater.

Source: WBCSD 2006.

supply chain and watershed management, collective action, public policy, community engagement and transparency. The endorsing members of the CEO Water Mandate include the chief executive officers of 32 companies working in all economic sectors around the world, from agrifood to textile and financial institutions.

Secretary-General Ban Ki-moon called on the business and international community during the 2009 Economic Forum meeting in Davos, Switzerland, to make water security one of the key topics for climate change adaptation and emphasized the importance of a Global Agenda Council on Water Security.[13]

Promoting win-win scenarios through cooperation and knowledge

In addition to avoiding negative impacts, win-win choices need to be cultivated in all aspects of social life to deal with the urgency of the water situation around the world. Examples of win-win scenarios can be found in all policy areas – from traditional water-using sectors to security policies – and implemented at various scales and levels.

Consultation and cooperative management

Citizen participation in decision-making is central to promoting good governance and creating a climate of accountability and

Box 15.9 Examples of clean production processes

- In Turkey a tyre-making factory in the Izmit region reduced its water consumption by nearly three-quarters, from 900,000 litres a day to 250,000 litres, thus also reducing its discharges into the community sewers. A detailed analysis made it possible to replace a cooling system with a closed-circuit system for an investment cost of $5,000 and a return time of two years.
- In Egypt one of the largest tinned-food manufacturers (Montazah, near Alexandria) underwent an eco-audit and introduced measures to reduce energy consumption: insulating steam pipes, replacing leaky parts, fitting a pressure regulator to the sterilizers and improving the recuperation system and boiler efficiency. Water consumption was reduced by implementing water-consumption hydrometer monitors, installing sprinklers (so that water flows only when needed) and improving the water collection and recycling system. The savings in water, steam and energy (nearly 40% savings in fuel consumption) made it possible to reduce discharges and amortize investments over 1-44 months.
- One of Croatia's biggest dairy companies, LURA in Zagreb-Lurat, undertook measures such as employee training, reducing the diameter of cleaning pipes and changing the hot-water circuit that reduced effluents by 286,000 cubic metres a year (or 27%) and drinking water by 280,000 cubic metres a year. These simple, low-cost measures (investment of €31,000) involved employees, saved water and energy (equivalent to €328,000 a year) and reduced effluents, with an investment amortization of less than one month.
- In Tunisia a manufacturer of car batteries identified 19 ways of preventing contamination and pollution (acids, lead scoria and wastewater) and saving lead and energy. The costs of the new measures were $522,500, while savings amounted to $1.5 million a year.
- In Mexico the General Motors de Mexico Ramos Arizpe Complex (2001 Stockholm Water Award recipient) employed physical, chemical and biological wastewater treatment processes to recover and reuse 70% of its industrial wastewater. It also promoted use of brackish well water by separating salts and increasing the useable amount withdrawn from 67% to 94%. It helped reduce pressure on the only source of water for a population of 40,000, a small confined and brackish aquifer.
- In India the Staple Fibre Division of Grasim Industries Ltd (2004 Stockholm Water Award recipient), a producer of viscose staple fibre, has reduced water consumption by 85%, process steam by 51% and electrical usage by 43% since 1980. The company was among the first to replace zinc – a pollutant particularly harmful to marine and aquatic life – with aluminium in its production process.
- In Australia, the driest inhabited continent in the world, Sydney Water (2006 Stockholm Water Award recipient) supplies water to 4.2 million people. As part of its operating licence, Sydney Water is required to reduce per capita consumption by 35% before 2011. Since the inception of its water conservation programme in 2001, Sydney Water has saved more than 20 million litres a day and more than 310 organizations have joined the initiative.
- In the United States, Orange County, California, (2008 Stockholm Water Award recipient) focused on reuse in its groundwater replenishment system, which diverts and purifies highly treated sewer water that was previously discharged into the ocean and returns the cleaned water to the groundwater basin. The system will provide enough additional water to meet the needs of 500,000 more people without diminishing groundwater resources for current users (2.3 million).

Source: Adapted from Benoit and Comeau 2005 and www.siwi.org/sa/node.asp?node=77.

transparency. Interventions to promote a culture of consultation and to increase consultative and participation capacity will help to deliver benefits in all areas, including collaborative water management.

In 1978 Québec, Canada, created the Bureau d'audiences publiques sur l'environnement (BAPE), an independent entity charged with consulting with citizens on projects that might have environmental implications (box 15.10). The BAPE example shows that an adequately mandated consultative process can legitimize policy decisions and identify multiple social, economic and environmental benefits around sound water management.

Creating institutions concerned with major water-related issues helps identify possibilities for collaborative initiatives in other areas of the economy as well. In the BAPE case the point of entry was the need for a new water policy that would answer the requirements of various users. Other points of entry can also be found. For example, technical investigations can pave the way to collaborative learning processes among scientists and decision-makers, as seen in the Upper San Pedro Partnership in Arizona in the United States (box 15.11).

Cooperative management can occur between different types of actors, including public-private partnerships. One example is payments for environmental services, using market-based tools to create incentives for maintaining a water resource or paying for watershed services (see chapter 14). In Costa Rica the water utility in the Heredia region pays landholders to protect forests on the hill slopes from which they derive their water, benefiting both landowners and municipal water customers.[14]

Win-win scenarios from sectoral interventions

Agriculture. Lasting win-win benefits for water and agriculture often result from explicitly recognizing and analysing trade-offs

Box 15.10 Québec's Bureau d'audiences publiques sur l'environnement consultation on water resources management

In 1998 the minister of the environment called on the Bureau d'audiences publiques sur l'environnement (Office for Public Hearings on the Environment, BAPE) to hold a public consultation on water management in Québec. The consultations – conducted through written comments, public hearings and discussions – helped develop recommendations on better water management.

From 15 March 1999 to 1 May 2000 the commission held 142 public meetings in the 17 administrative regions of Québec and heard 379 briefs. The BAPE report, *Water: A Resource to Be Protected, Shared and Enhanced,* was submitted to the minister of the environment on 1 May 2000. The report emphasized three areas:

- Improving governance through water management at the river basin level.
- Preparing regional portraits covering the public's expectations for water management and aquatic ecosystems.
- Reforming legislation and institutions needed to implement an integrated water and aquatic ecosystem policy.

The report addresses 16 themes, among them agricultural pollution abatement, hydroelectricity, integrated management of water and aquatic ecosystems at the watershed level, water exports and the special case of the St. Lawrence River.

The commission made 13 main recommendations for the short, medium and long terms. It also presented findings and observations for the government's reflection. In 2002 the government of Québec released its water policy, *Water: Our Life, Our Future.* The policy seeks to protect water resources as a unique heritage, manage water in a sustainable development perspective and better protect public health and the health of ecosystems.

The policy adopts most of the BAPE recommendations. It reaffirms that water is part of the collective heritage of the province and incorporates measures intended to:

- Implement integrated river basin management.
- Implement this form of management in the St. Lawrence River, by granting it a special status.
- Protect water quality and aquatic ecosystems.
- Continue water clean-up efforts and improve water services management.
- Promote water-related recreation and ecotourism activities.

Source: Cosgrove 2009; BAPE 2000; Québec Ministry of the Environment 2002.

Box 15.11 Collaborative learning processes – adopting an adaptive management framework

The Upper San Pedro Partnership in Arizona in the United States represents a diverse consortium of interests, including municipal, state and federal government institutions. From its initial goal of developing a definitive regional groundwater management plan in 1998, it evolved into a more complex, yet flexible, ongoing adaptive management planning process that endures a decade later.

An adaptive management approach allows actions with low risk or uncertainty to be taken early on. Member agencies realized that implementing certain water management strategies required substantial information through monitoring, research and modelling efforts as well as political assessments, while other projects represented relatively low-risk strategies and could be implemented sooner. By 2003 the partnership had identified more than 100 projects for implementation, including repair of leaky infrastructure, car-wash water recycling, voluntary retirement of agricultural pumping through conservation easements, recharge of treated effluent and reintroduction of beavers. In the 2004 Water Management and Conservation Plan the partnership prioritized additional projects – developing model codes and ordinances, establishing water-conservation surcharges for excessive use, exploring a transfer of development rights programme and other measures. Projects with greater uncertainty, higher political risks or significant costs were targeted for additional feasibility studies or evaluation using a decision-support system model. The partnership uses the model to evaluate combinations of management options (scenarios) such as the possible relocation of municipal wells, construction of additional recharge facilities and various water augmentation strategies. Unlike a simple annual 'bottom line' water-budget approach, the model considers spatial and temporal groundwater management concerns, essential to sustain the ecological values of the San Pedro Riparian National Conservation Area.

In San Pedro complex and controversial strategies such as water importation, the transfer of development rights and surcharges for excessive water use all had the potential to divide the community. To engage the community early on in planning processes, the partnership conducted a series of community meetings to provide citizens with an opportunity to thoughtfully consider issues and provide meaningful input.

Source: Holly Richter, Udall Center for Studies in Public Policy.

and factoring them into decision-making. Doing so may depend on the availability of information, collaborative decision-making and perceptions of available alternatives. In poor communities where survival is the main concern, people may have few choices about how they use land and water, or the perceived risks of alternatives could outweigh the potential benefits. This is why most successful integrated rural development initiatives are designed to help such communities reduce risks, develop alternatives and bring trade-offs to the forefront in decision-making (box 15.12).

Investment in water infrastructure and sanitation can reduce poverty by stimulating productive activity (see chapter 6). In rural areas there is no definitive line between 'household' water use and water use for productive purposes – watering plots for food and cash crop production,

Box 15.12 The Integrated Watershed Development Programme in Jhabua District, Madhya Pradesh, India

By the 1960s severe deforestation had seriously harmed the ecosystems and livelihoods of the people of Madhya Pradesh, India, resulting in widespread soil erosion, overgrazing and inappropriate land use, barren landscapes and seasonal migration of men in search of employment. Under a project implemented by the National Centre for Human Settlements and Environment and local communities with funds from the government of India, multiple interventions attempted to restore natural resources and improve the socioeconomic conditions of district residents. The project took an integrated management approach, based on community needs. Activities included:

- Protected afforestation on community land.
- Distribution of seedlings to encourage planting on private land.
- Soil and water conservation.
- Pasture improvement through planting pasture grasses.
- Water harvesting.
- Distribution of subsidized fuel and energy-saving devices.
- Integration of land use innovations with measures to improve community livelihoods.
- Promotion of alternative income-generating activities to reduce poverty and reduce seasonal migration.

In addition to immediate land productivity benefits, there were many positive ecosystem-wide benefits, including a marked increase in groundwater recharge and water supply from harvesting. Livelihoods improved as well. The model was subsequently adopted by neighbouring states.

Source: National Rural Development Institute n.d.; www.mprlp.in; www.nchse.org/projects.htm.

livestock, trade and other income-generating activities. The amount of water provided to poor households with access to land needs to be sufficient for these other income-generating purposes.

Multiple-use initiatives for water recognize the benefits to poor households of having adequate water for non-household, income-earning activities (box 15.13). But such initiatives are often not in line with water efficiency efforts. Surface irrigation is the most water-demanding form of irrigation, but the excess water can have other benefits: it enables aquaculture (as in rice fields in China) and washes off the salt accumulating in topsoil (avoiding salinization of cultivated lands). Efforts to save water by reducing water input would thus mean the loss of an income source (fish production) and potentially of cultivated lands, if salt accumulation becomes severe.

Integrated multiple-use systems are found worldwide – usually documented at the farm and field levels – but conflicting management objectives are just as common and create hurdles for the promotion of these systems. Multiple uses of water imply multiple interest groups whose water management objectives may not always be compatible. Despite much evidence of integrated water use at the farm level (such as rice-fish systems and irrigation-aquaculture systems), sectoral management at higher levels impedes true integration of water and irrigation with other sectors, including fisheries, forestry and sanitation. Furthermore, multiple uses and demands for water can generate opportunity costs and externalities, even when some uses are non-consumptive (fish farming in irrigation canals, for example). These problems are intensified by the seasonality of supply and the limited availability of irrigation water in semi-arid tropical countries as well as the common pool, open access nature of the resource.[15]

Box 15.13 Experience with multiple-use water services in Nepal

Since 2003 Winrock International Development Enterprises and local partners have worked with more than 70 communities in 14 districts in Nepal to develop multiple-use water systems serving some 1,600 households (more than 9,300 people). The approach combines multiple-use water services with supporting business services that allow the rural poor to realize health benefits as well as sustainable increases in income. Proven low-cost technologies (nearby low-flow sources that are often bypassed by conventional programmes) are used to provide cost-effective water services to small communities (15-50 households). Related business services focus on income-generating opportunities from productive water use activities, primarily high-value horticulture. Typically, such business services involve assessing rural enterprises (inputs, costs, demand, markets and prices) and identifying ways to add value through products and services.

An evaluation of the Nepal pilot programme found that, in addition to receiving water for drinking and domestic uses, households with multiple-use water services had increased their average gross incomes by $250 a year through the production and sale of horticultural products, made possible by reliable access to productive water. For 70% of the systems, gross margins covered costs within one year. All schemes are managed by water user associations, which are responsible for securing water rights, operating and maintaining the systems and managing finances. Households contribute to operation, maintenance and capital replacement of the system. About 75% of water user associations have savings and credit schemes, and many provide revolving microcredit for income-generating activities. Some 85% of households surveyed were highly satisfied with multiple-use water services, reporting greater quantities of water available within close proximity (less than 60 metres away, on average) for household use, irrigation and livestock.

Other important benefits included time saved from no longer hauling water, reduced incidence of diarrhoeal diseases, higher consumption of nutrient-rich vegetables and increased school attendance by girls (more than 65% of households). More than 60% of households had installed sanitary latrines, a combined effect of more water for sanitation, increased income and sanitation marketing.

Source: Development Vision – Nepal 2007.

Energy. Energy demand and water use are closely linked (see chapter 7). Energy costs and concerns about energy efficiency can be brought to bear in addressing water issues. In parts of South Asia and other regions where farmers use groundwater drawn from private wells, the price of electricity needed for pumping is a powerful instrument for controlling groundwater abstraction by influencing farmers' pumping behaviour.

However, electricity is underpriced in many areas. The arguments against raising electricity prices, and the practical problems in doing so, are similar to those for holding down irrigation water prices for small farmers. However, the rising cost of public subsidies for fuel and electric power has steeled the resolve of many governments to tackle this perverse incentive for overabstraction of water. The metering of electricity to farmers in North China shows the efficacy of economic pricing for resource management, given the right institutional structure.[16]

In the example in box 15.14 decisions about energy sources and consumption are motivated primarily by profitability and cost reduction, but they can also generate benefits for water.

Adequate water and energy are prerequisites for breaking the cycle of poverty. Using renewable energy in rural electrification, for example, is a win-win scenario at the local level. Greater availability of electricity would lessen dependence on fuelwood (still a significant source of energy in Africa, for example), reducing deforestation and its effects on the local water cycle, while providing an engine for growth.

In the Dominican Republic, where many people in rural areas have no access to electricity, the non-governmental organization Association for Solar Energy Development helped more than 5,000 households obtain home solar power systems, and numerous other public facilities have been powered by solar energy. The association estimates that 3-6 tonnes of carbon dioxide emissions are avoided for each 50 gigawatt panel that replaces kerosene lighting.[17] In addition to enabling communities to engage in alternative income-generating activities, the programme helped develop solar-powered potable water systems. Using solar energy to draw groundwater reduces the costs of access to water while providing access to higher-quality water, since surface waters are often polluted.

Local benefits vary by setting, but health benefits (from reductions in indoor air pollution), environmental benefits (from reduced deforestation and emissions) and economic benefits (from use of a renewable energy source) are all possible.

Tourism. With 898 million travellers worldwide in 2007 and 1.6 billion forecast for 2020, tourism is a growing sector of the economy in many developing countries.[18] In addition to direct employment and investment, tourism generates jobs and revenues in surrounding communities and stimulates infrastructure investment. Sustainable tourism, or ecotourism, can also provide the impetus for conservation efforts. Tourism depends on the availability of natural resources, landscapes and ecosystem services, chief among them water and biological diversity. But construction of tourism installations and infrastructure can also harm ecosystems through pollution, deforestation and overexploitation. Tourism also requires increased water supply and sanitation services, which can divert water from other uses. Antigua and Barbuda, whose economy is dependent on tourism, must purchase water from neighbouring Dominica to satisfy demand.

There is increasing evidence that win-win scenarios for tourism and water are possible, as well as for tourism and community development. In Tunisia higher water tariffs for tourist establishments cross-subsidize water supplied to Tunisians (see box 14.3 in chapter 14). With the recent rise in consumer awareness, tourism enterprises everywhere are trying to demonstrate corporate social and environmental responsibility. Governments are also working to promote more integrated tourism investment planning, sustainable use of protected

The rising cost of public subsidies for fuel and electric power has steeled the resolve of many governments to tackle this perverse incentive for overabstraction of water

Box 15.14 The drive for competitive advantage leads to water benefits

Eco-efficiency, a microeconomic objective, focuses on reducing the amount of water, energy, chemicals and raw materials used per unit of output. Eco-efficiency is motivated not only by environmental concerns but also by prospects of financial savings through reduced energy and water bills, less money spent on raw materials and fewer regulatory hurdles.

Swiss-based ST Microelectronics cut electricity use 28% and water use 45% in 2003 and reported saving $133 million. DuPont committed to keeping energy use flat no matter how much production increased, which has reportedly saved more than $2 billion over the past decade. Advanced Micro Devices tracks kilowatt hours per manufacturing index and reports a 60% reduction from 1999 to 2005.

Although motivated by the need to innovate and increase profitability, competitiveness and market shares, these measures have all had the added benefit of decreasing water use and limiting pollution.

Source: Worldwatch Institute 2008.

Water is emerging as a strategic resource that underpins many dimensions of security

areas, impact assessments and certification programmes. Ecotourism can help promote conservation, poverty reduction and sustainable water management when adequate enforcement and benefit-sharing mechanisms are in place (box 15.15).

Promoting multiple objectives through regional economic development

Regionally based development efforts at the subnational level usually involve a package of interventions in multiple and mutually reinforcing sectors. Because water resources are a cornerstone of economic development in many areas, regional economic development programmes are either organized around water interventions (as in the Southeast Anatolia example in box 15.16) or indirectly related to the water sector. As the example of the Tennessee Valley Authority in the United States shows, achieving multiple economic, social and environmental objectives simultaneously is not only possible, but also potentially easier than pursuing a single objective (box 15.17).

Promoting win-win scenarios through security and crisis prevention

There are multiple links between water and human security and, to some extent, national security – from individual food security, health, hygiene and sanitation, and economic security to regional conflicts among water users. Climate change, with its associated pressures on water resources, is expected to worsen human security everywhere – by undermining development efforts and by exposing more people to disasters resulting from extreme weather events and disrupted ecosystems.

Governments are increasingly concerned about security, including energy security, development and socioeconomic security and physical security. Political stability, a tenuous achievement in many countries, also depends on perceptions of economic vulnerability (as evidenced in early 2008 riots over rising food costs in such countries as Bangladesh, Haiti, Somalia and Yemen) and physical safety. The water-stability link is also stronger where water resources are scarce and competition among users is likely to occur on a large scale.[19]

Promoting human security. Water is emerging as a strategic resource that underpins many dimensions of security. And many interventions at the local, regional and global levels that are designed to address security concerns are linked to water resources management (box 15.18).

An example is the development of renewable energy to promote energy security. This can generate significant benefits for water – freeing it for other uses and for the protection of ecosystem integrity. Disaster preparedness and climate early

Box 15.15 Sustainable tourism, Çıralı, Turkey

Having experienced a surge in tourism, Turkey is predicted to join Greece and Croatia as a leading tourist destination in the Mediterranean by 2020. The Southwestern Anatolian coast of Turkey, identified by WWF as one of the most ecologically important in the Mediterranean, is also the most affected by mass tourism development. To avoid the destruction of its fragile coastline, the government of Turkey has adopted an integrated coastal management approach aimed at preserving not only the natural heritage but also social and cultural integrity.

Çıralı was chosen for a pilot project to implement this approach. Çıralı had shifted from an agricultural economy towards tourism in the late 1980s. The resulting construction of tourism facilities led to multiple problems. Pesticides from agricultural activities had already polluted soil and water sources, which were further threatened by the growth of tourist restaurants around the village's main spring. Illegal construction spread due to the lack of enforcement of land development regulations.

What precipitated a change was that Çıralı beach was a major nesting site for the endangered loggerhead sea turtle (*Caretta caretta*). WWF Turkey (formerly known as the Turkish Society for the Conservation of Nature) entered the scene initially to help protect the turtle. In 1997 the project was broadened from turtle protection to stimulating community awareness of its responsibility for the natural environment and to developing a model of sustainable tourism in Çıralı. WWF examined land resources management in the area and the potential to create diverse, sustainable and environment-friendly economic opportunities and to protect biodiversity.

With a new awareness of the need to become guardians of their natural heritage, the local community pressured for enforcement of the existing Coastal Law. The responsible ministry, with buy-in from local residents, moved the kiosks and restaurants that had been built too close to the shore in violation of the law. WWF and the ministry prepared further guidelines and recommendations for wise use of land resources, including a land use plan that defined the positioning of building infrastructure in the area and a land management plan. Ecotourism activities were started to generate support for conservation and create economic opportunities for the local community throughout the year, instead of just seasonally. As part of the venture locals were trained as nature guides, and trekking paths were identified – activities that attracted young people in the community. A move was also promoted from input-intensive agriculture, which polluted soil and water supplies, to organic agriculture. A cooperative was set up by the farmers to produce and market organic products and to create a brand for produce of Çıralı origin.

Source: peopleandplanet.net 2002; www.panda.org/news_facts/newsroom/news/index.cfm?uNewsID=13382&uLangID=1.

Box 15.16 Integrating multiple sectors in Southeast Anatolia, Turkey

Turkey's Southeastern Anatolia Project (GAP) is a multisectoral, socioeconomic development project designed to raise incomes in this less-developed region to narrow the regional income disparities by improving the economic structure, increasing productivity and employment opportunities in rural areas, enhancing the assimilative capacity of larger cities in the region and contributing to the national objectives of sustained economic growth, export promotion and social stability.

Begun as a water infrastructure development project, GAP evolved into a programme that supports sustainable social and economic development through additional investments in urban and rural infrastructure, agriculture, transport, industry, education, health, housing, tourism, environmental protection, gender equality and poverty reduction. The programme's total costs are split between investments in water-related projects (48%) and those in other sectors (52%).

The project area is home to 6.5 million people, or 10% of the country's population. The project will generate 27,470 gigawatt hours of electricity, boosting national hydroelectrical output by 70%, and add 1.7 million hectares to the 4.5 million hectares of irrigated land.

Governance of the project changed with the evolution of its philosophy. Management of the project began under the national water authority in the 1970s and shifted to the national planning organization in the 1980s. In 1989 the government established a mandated entity, the GAP Regional Development Administration, to coordinate special programmes and projects in social services (education, health), gender equity and empowerment, urban management, participatory irrigation management, agricultural productivity, resettlement, environmental conservation and protection, and institutional and community capacity-building. The GAP Higher Council, the highest decision-making body, is chaired by the prime minister and initially included ministers for planning, energy and natural resources, and public works. Ministers for agriculture and environment were added later. It has a coordinating body of line ministries operating in the project area, local governments, local chambers of industry and trade, local chambers of agriculture and regional universities.

The overall cost of the GAP is projected at $32 billion, of which $17 billion has been invested. Per capita farm incomes have tripled since irrigation was expanded. Rural electrification and accessibility reached 90%, literacy rates rose, infant mortality rates fell, and a more equitable land tenure system is in place. Urban settlements served with water supply quadrupled. The region's economy, once the least developed in the country, is now at the national average.

Source: Aysegül Kibaroglu, Department of International Relations, Middle East Technical University, Ankara, Turkey; Government of Turkey 1989.

Box 15.17 The Tennessee Valley Authority: economic and social transformation in a river basin and beyond

During the Great Depression of the 1930s the Tennessee River Valley, which stretches through seven southern states in the United States and drains a basin of 105,930 square kilometres, was a land of despair. More than 90% of valley residents had no electricity, and nearly 40% had no toilets or outhouses. Most of the power required for agriculture came from horses, and most of the residents were subsistence farmers working degraded soil or in areas experiencing serious and repeated flooding.

President Franklin Roosevelt launched many programmes designed to bring the economy out of depression and to improve the lot of people mired in poverty. With a strong interest in basin planning and in developing system-level solutions, he asked Congress to create 'a corporation clothed with the power of government but possessed of the flexibility and initiative of a private enterprise'. In 1933 Congress authorized the Tennessee Valley Authority (TVA).

Given a broad charter, the TVA focused on carrying out its mission on an integrated, watershed basis, dealing concurrently with hydropower production, navigation on the Tennessee River, flood control, health problems such as malaria prevention and resource challenges that included reforestation, erosion control and restoration of fisheries. It would have been labelled integrated water resources management, had the concept been identified in 1933.

By 1945 TVA had completed more than a dozen large dams and built a 1,050-kilometre navigation channel. It also had become the largest supplier of electricity in the United States and a major driver for rural electrification. Evaluations noted that the TVA was substituting order and design for haphazard, unplanned and unintegrated development in its programme for flood and navigation control, land reclamation and cheap electricity for light and power.

By the 1960s residents of the Tennessee Valley were sharing in unprecedented economic growth and a higher standard of living, with electricity, clean water and sanitation provided to most homes.

Today, the TVA provides electric power, recreation, improved water quality and a reliable supply of water to cool power plants and meet municipal and industrial needs. It generates more electricity than any other public utility, operates a system of 49 dams and reservoirs on the Tennessee River and its tributaries and manages 118,572 hectares of public land. It operates the river system as an integrated whole to provide year-round navigation. Barges on the Tennessee River carry some 50 million tonnes of goods annually. Structural and non-structural approaches prevent an estimated $230 million in flood damages annually. In one decade, through the development of water resources, the Tennessee Valley lifted itself out of poverty; in seven decades it has become a powerful economic and social force in the United States.

Source: Gerald Galloway, Department of Civil and Environmental Engineering, University of Maryland; www.tva.gov/heritage/hert_history.htm.

warning systems, increasingly used as preventive responses to climate variability, are directly linked to water management. In Djibouti, whose capital was severely affected by flooding in 2004, the World Bank supported the Flood Emergency Rehabilitation Project for the construction and rehabilitation of wadi banks, drainage infrastructure and flood control dikes. Although an immediate response to physical security concerns, these measures have a potentially long-term beneficial effect on

Box 15.18 Nepal: a community-led initiative to mitigate water-induced disasters

The Himalaya Mountain range is among the richest freshwater locations on Earth; the combined basins of nine major Asian rivers rise there and are home to some 1.3 billion people. It is also among the most fragile ecosystems, due primarily to the mountain system's propensity to mass-wasting (movement downslope of soil and rock under the force of gravity), seasonal monsoon precipitation fraught with extreme events and increasing population pressure. Nepal, located in the middle of the Himalaya range, has been subject to the risks associated with mass-wasting and flooding, which account for more than half the disaster-related deaths in the country. The situation has worsened in recent decades, with more intense and frequent extreme climate events associated with global climate change, such as torrential rains.

The community of Madhumalla in Morang District in southeastern Nepal is on the bank of the Mawa River, a small river with an upper watershed of about 20 square kilometres. This 25 kilometre long river has an average gradient of 4% in the upper reaches and 2% in the lower reaches, and its width varies from 200 metres to 700 metres. As do most rivers originating in the southern belt of Nepal, the Mawa River experiences unpredictable flooding caused mainly by monsoon rains, with sudden cloudbursts in the upper watershed generating torrents laden with debris, boulders and sediments. This causes rapid changes in river morphology, with a cycle of aggradation and degradation of the river bed, undercutting, erosion, overflowing of river banks and shifting of the entire river course. The local population is under constant threat of being washed away, and homes and crops are frequently damaged.

The Madhumalla community, led by then-Chairman Kashi Nath Paudyal, embarked on a remarkable mission some 14 years ago to address the threats posed by unpredictable and devastating floods. To defend against undercutting and erosion of the banks and degradation of floodplains, the community employed the bioengineering technique of planting a stratified greenbelt along the river bank using native trees, shrubs and grasses in conjunction with reinforcing materials. Structural measures, such as embankments and spurs made of gabion boxes, were used to protect the greenbelt from damage during the initial years.

The project was designed and implemented drawing on indigenous knowledge of locally available plant varieties and their ability to withstand the forces of river water and the community's understanding of the local physical environment and river morphology. The project is also preparing to sell forestry products from the plantation area, which are expected to generate hundreds of thousands of dollars annually within a few years – turning risks into rewards. The project mobilized substantial resources internally in the form of cash, labour and material assistance and received grants from several national and international donors totalling about $40,000.

The approach worked as intended and has been replicated in several other communities in the region, including nearby refugee camps. The project area is currently a training centre on bioengineering technology.

Source: Dinesh L. Shrestha, senior water and sanitation officer, United Nations Refugee Agency (UNHCR), Geneva, Switzerland, and Tako Ganai, consulting engineer, UNHCR, Jhapa, Nepal.

groundwater replenishment by controlling runoff from regular rainfall, thus contributing to the long-term water and food security of the city dwellers.

Rapid urban migration creates human security challenges. Many cities are ill-equipped to deal with a rapid influx of rural inhabitants, who frequently establish informal settlements. Governments adopt a variety of strategies to deal with these rapidly growing settlements; some invest in rural development and infrastructure to stem the flow of urban migrants.

Box 15.19 Resettlement in Rwanda

Rwanda's Imidugudu (group settlement) Policy was launched in 1997, after the civil war and genocide, to resettle returned refugees. By focusing on group settlement, the government hoped to solve the problem of land scarcity and more easily and efficiently provide social and economic services.

During the initial phases of resettlement many families had to leave the lowlands that provided ready access to water for higher areas where access to safe water was more difficult. In a 2002 status update Rwanda's Ministry for Infrastructure noted that while 22% of Rwandans had adopted the group settlement scheme, inadequate infrastructure development remained a major challenge, particularly access to clean water and adequate sanitation. Several supplementary initiatives were put in place to respond to this challenge, many of them funded by foreign assistance, to invest in rural water infrastructure and management.

Source: Rwanda Ministry of Infrastructure 2008; Japan International Cooperation Agency 2006.

Rural development and increased agricultural productivity may help to avoid slum formation by reducing migratory pressures. Like slum-upgrading initiatives, rural development initiatives are not entirely outside the domain of water managers, since they usually require some form of water intervention along with measures to increase basic social services and safety nets (food security and health care).

Crisis recovery. Economic and humanitarian crises, often brought about by prolonged droughts or other natural disasters, can quickly set back development progress. Crisis recovery efforts increasingly aim to build greater resilience to avoid future crises. For example, food aid programmes can be coupled with sustainable livelihoods initiatives and environmental rehabilitation. In most cases, as in Rwanda's group resettlement programme, crisis recovery and post-conflict reconstruction efforts involve interventions to restore damaged water services and infrastructure (box 15.19). In this sense they provide an opportunity to build stronger, more sustainable and more resilient water systems and to integrate risk management and climate change into planning.

Initiatives designed to address population growth, movement and geographic distribution alter water use and impose changes on the way water is managed. Land redistribution policies are currently being implemented in several countries. Their potentially controversial aspects aside, recent evidence has shown that effective land redistribution (in terms of its contribution to poverty reduction) also depends on adequate postsettlement support, including infrastructure development.[20] In Namibia, where most agricultural activity consists of livestock farming, many newly allocated farms remain underproductive due to lack of functioning water delivery points.[21]

Crisis recovery also often includes dealing with returning populations and with the movement of internally displaced people. As the examples of Rwanda (box 15.19) and Iraq (box 15.20) show, successfully reintegrating returning refugees requires an environment that allows them to meet their basic needs. Thus, returning a region to stability requires consideration of ecological services, particularly access to safe water and sanitation. Identifying win-win situations early on in crisis recovery efforts can prepare the way for sustainable peace and security.

In addition to promoting human security at the local level, interregional cooperation around shared waters can promote peace-building and trust among countries, as in the cases of the Nile River basin and the Senegal River (box 15.21). Although the causal pathways between water scarcity and conflict are still under debate, there is no doubt that international cooperation on shared waterways is occurring, even where tensions unrelated to water issues exist (for example, in Central Asia). The creation of international water institutions demonstrates that cooperation is possible and can be successful provided the rules are transparent and consensual. Box 15.22 also demonstrates the importance of institutions, including clear dispute resolution mechanisms, and the need to find solutions to water disputes, particularly in cases of increasing scarcity.

Tackling risk and uncertainty

Dealing with risk and uncertainty has long been a routine challenge for water resources managers and policy-makers across sectors and the world. However, issues like climate change and demographic dynamics are outside their control and have made the risks greater and their task more complex. Risk management is now much more important – indeed essential – to analysis and decision-making for water.

The creation of international water institutions demonstrates that cooperation is possible and can be successful provided the rules are transparent and consensual

Box 15.20 Rehabilitating the Iraqi Mesopotamian Marshlands for integration and stability

The Iraqi Mesopotamian Marshlands constitute part of the largest wetland ecosystem in the Middle East and Western Eurasia. They are a crucial part of intercontinental flyways for migratory birds, support endangered species and sustain freshwater fisheries and the marine ecosystem of the Persian Gulf. In addition to their ecological importance, these marshlands have been home to indigenous communities for millennia. The destruction of the marshlands, and the displacement of its indigenous Marsh Arab population, is one of the major humanitarian and environmental challenges facing Iraq. The competing roles of the marshlands as a transboundary water resource and a source of petroleum reserves have made the future of the marshlands region a priority on Iraq's reconstruction agenda while also presenting trade-offs for decision-makers.

Assessments in 2003 and 2004 reported that 85,000-100,000 Marsh Arabs reside within and near the remaining marshlands, with fewer than 10% of them leading a traditional subsistence existence. An estimated 100,000-200,000 Marsh Arabs remain internally displaced, and 100,000 are thought to be living as refugees outside Iraq, primarily in Iran. Other communities also reside within the marshlands region.

Following the collapse of the regime in mid-2003, local residents opened floodgates and breached embankments to bring water back into the marshlands. Satellite images analysed by the United Nations Environment Programme (UNEP) in 2003 revealed that some formerly dried-out areas have been re-inundated, helped by wetter than usual climate conditions. By April 2004 approximately 20% of the original marshland area was re-inundated, up from 5%-7% in 2003.

UNEP initiated the Iraqi Mesopotamia Marshlands project in 2004 through its International Environmental Technology Centre in Osaka and Shiga, Japan, with funds from the governments of Italy and Japan and in cooperation with the Centre for the Restoration of Iraqi Marshlands of the Ministry of Water Resources. Phase III commenced in 2007. The project has produced multiple benefits:

- Environmentally sound technologies on drinking water and sanitation provision and wetland restoration are being introduced and implemented, drawing on Iraqi expertise.
- Some 22,000 people in six pilot communities now have access to safe drinking water. Water treatment facilities (with a capacity of 750 cubic metres a day), 23 kilometres of water distribution pipes and 127 common distribution taps have been installed. Some displaced residents are returning to pilot site areas, partly because of the drinking water made available through the project. As stability returns, possibilities for rebuilding life in the marshlands grow.
- A sanitation system pilot project has been implemented in the Al-Chibayish community. Environmentally sound technology, using constructed wetlands, aims to serve approximately 170 inhabitants, who face health hazards from discharges of untreated wastewater to a nearby canal.

Source: http://marshlands.unep.or.jp.

The consequences of global climate change are manifested primarily through water – in glacial melt, floods, droughts and sea level

Box 15.21 Sustainable water institutions promote regional cooperation and stability – the case of the Senegal River

Created in 1972 by an agreement among the three riparian states (Mali, Mauritania and Senegal), the Organization for the Development of the Senegal River (OMVS) established the river as an international waterway and identified rational exploitation and management as a goal. In 2002 member states adopted a charter on using the river that sets out principles and modalities for distributing water resources among sectors and defines project approval criteria, environmental rules and participation modalities for broader public engagement.

The OMVS is a key institution through which countries agree on future projects (such as dams, hydropower and agricultural investments) and sectoral priorities. It is known as a water-focused institution that promotes collaboration on multiple uses of water and integrated water management. The OMVS also has an environmental observation arm that tracks the state of the resource and its associated ecosystems to ensure sustainability. Because of the river's central role in many activities in all three countries, cooperation has gradually extended to tripartite discussions at the central and local levels in other sectors, such as agriculture and enterprises, contributing to regional stability and integration.

Source: www.omvs.org.

Box 15.22 Allocation of water from the Rio Grande/Rio Bravo River between Mexico and the United States

In summer 2002 Texas farmers breathed a sigh of relief as the scarce water of the Rio Grande (called Rio Bravo in Mexico) began to flow again on their side of the border.

Under a 1944 treaty Mexico and the United States had agreed to share the waters of the Rio Grande and Colorado Rivers. Mexico agreed to assign 254,880 cubic metres of water from international storage, with the actual amount of water to be released from the Amistad and Falcon reservoirs determined by a complex accounting of reservoir inflows and water use by both countries that would ensure that neither country would run dry. But because of drought, Mexico had fallen in arrears on its water contributions.

A key part of the 2002 deal to resolve this dispute was infrastructure modernization, with the goal of increasing efficiency in water use. For the first time the two governments agreed to share costs and benefits by jointly investing in measures for water conservation, sustainability and efficiency. The agreement endorsed a bilateral financial package for water conservation initiatives.

Source: Moore, Rast, and Pulich 2002.

Climate change. The consequences of global climate change are manifested primarily through water – in glacial melt, floods, droughts and sea level rise. Planners can no longer rely on past hydrologic conditions to forecast future risks. Climate change increases the risk of failure or underperformance of structures and institutions. Developing countries are the most vulnerable to climate change because of their heavy dependence on climate-sensitive sectors, low capacity to adapt and poverty. Current climate variability and weather extremes already severely affect economic performance in many developing countries (see figure 5.2 in chapter 5).[22]

One way to cope with the uncertainty of climate change is through management measures that are flexible and robust under conditions of uncertainty. Such adaptive management principles, which involve a systematic process for improving management policies and practices by learning from previous strategies, are particularly relevant for decision-making under climate change.[23]

As this Report emphasizes throughout, the water sector must look beyond its traditional boundaries for solutions. Similarly, actors outside the water sector must be aware of links to risks in the water sector. For example, because natural disasters are often related to water, closer collaboration and coordination are needed between disaster risk-reduction agencies and those engaged in water resources management. Water managers and those outside the water sector can all contribute to innovative, integrated solutions. One innovation reducing risk is the provision of insurance linked to rainfall (box 15.23).

Several broader analytical tools offer promise in dealing with risk and uncertainty. Environmental impact assessments have long been used to promote a do-no-irreversible-harm approach to policy and business decisions. Strategic environmental assessments are also emerging as powerful tools to identify both potential damages and opportunities, which could help in identifying win-win scenarios. Their effective use

Box 15.23 Rainfall-related index insurance for farmers

Reducing climate risk for small farmers requires tools that can adjust to climate variability and underpin other adaptation strategies. Insurance schemes linked to the amount of rain (index insurance) provide security as well as productivity incentives.

Traditional insurance contracts that insure against crop failure have several shortcomings. They can create perverse incentives to allow crops to fail to collect the insurance. And they can result in a situation where less productive farmers buy insurance, while more productive farmers do not. This results in more payouts, leading to higher premiums, ultimately making this type of insurance too expensive for farmers in poor countries.

In one new type of insurance, contracts are written against a rainfall index. The index depends on an established relationship between lack of rainfall and crop failure, ideally verified by long historical records of rainfall and yield. If rainfall is below an agreed trigger point, farmers receive payouts. However, farmers still have the incentive to make the best productive choices, regardless of whether the insurance pays out. Index insurance is also cheaper to implement because the insurance company does not need to verify damage in the field. That also means that when rainfall is low enough to cause crops to fail, insurers will pay out to farmers within days or weeks. Thus farmers do not need to sell assets to survive, which can make them dependent on aid long after the drought has ended.

With rainfall-indexed insurance farmers might take greater risks with potentially higher returns, such as investing in fertilizers or sustainable land management practices. To address the impacts of climate variability and change, farmers must be able to take advantage of good years. By using index insurance to protect against massive losses during major droughts, farmers are able to put resources into being productive in good years instead of being limited by the low productivity of rare bad years.

There are also some disadvantages. If crops fail for reasons other than drought, or if rainfall at a farmer's field differs from that measured at the central rain gauge, the farmer receives no compensation. Index insurance systems are typically developed as a part of a larger risk-layering strategy in which index insurance is applied at the most appropriate point, and other tools – such as traditional insurance, government social safety nets and reinsurance – complete the package.

In Malawi farmers can now purchase index insurance to guarantee loans for agricultural inputs that would have otherwise been unavailable because of the threat of drought. Thousands of groundnut farmers in Malawi have purchased a packaged product that bundles index insurance with a loan giving them access to high-yielding groundnut seeds bred to be productive in the local climate. Index insurance packages and seed varieties can be designed to adapt together as risks change. In focus groups on climate change farmers in Malawi have volunteered that the index insurance is their primary mechanism to adapt to the changing climate. As climate change exacerbates risks, farmers can use the improved varieties and index insurance to take advantage of new opportunities.

Source: Dan Osgood, International Research Institute for Climate and Society (IRI); Molly Hellmuth, IRI, Columbia University.

depends on adequate enforcement and institutional capacity.

Knowledge and technology. Better knowledge about the state of resources and the impacts of decisions can also reduce uncertainty. Sound information on the state of water resources is needed not only for water planning but also for making policy choices in other sectors that are likely to affect water services or availability. As seen in the case of Ethiopia, information on water availability can help in making informed economic decisions (see box 15.6).

Building strong water monitoring networks is thus an essential part of water governance. In Australia the lead organization responsible for water monitoring relies on information from multiple water users, including individuals and corporations (box 15.24). However, monitoring networks are often beyond the reach of developing countries because they require technical means and financing that are unavailable (see chapters 13 and 14).

The principle of 'decentralized' collaboration, as applied in Australia, can also be found at different scales. For example, the United Nations Environment Programme Global Environment Monitoring System on Water (GEMS/Water), financed by the Canadian government, relies on data provided by countries and provides capacity development for water monitoring. Another example is World Water Monitoring Day™, an international education and outreach programme coordinated by the Water Environment Federation and the International Water Association. World Water Monitoring Day™ builds public awareness and involvement in protecting water resources around the world by engaging citizens to conduct basic monitoring of their local water bodies. It provides an easy-to-use test kit that enables everyone from children to adults to sample local water bodies for a core set of water quality parameters, including temperature, acidity, clarity (turbidity) and dissolved oxygen. Results are shared with participating communities around the globe through the World Water Monitoring Day™ Website.[24]

Sustaining change: changing habits through awareness

The important changes required to face the challenges of managing the world's water resources will be determined by changes in behaviour. The most profound changes in consumption patterns at a

Because of water's contributions to every aspect of development, funding for water need not compete with funding for other priorities

Box 15.24 Getting a complete picture of water resources through user contributions – a new role for Australia's Bureau of Meteorology

The Australian Bureau of Meteorology's role has expanded to transform Australia's water resources information. To secure a long-term water supply for all Australians, the government announced Water for the Future, a $12.9 billion water investment programme. The programme includes $450 million for the Improving Water Information Programme, administered by the Bureau of Meteorology and backed by the Commonwealth Water Act 2007 and key stakeholders.

The bureau will enhance the quality and utility of Australia's water information by producing a national water account, supported by a national water monitoring and data collection network. The programme includes development and maintenance of an integrated, national water information system accessible to the public.

Responsibilities of the bureau include:

- Issuing national water information standards.
- Collecting and publishing water information.
- Conducting regular national water resources assessments.
- Publishing an annual national water account.
- Providing regular water availability forecasts.
- Advising on matters relating to water information.
- Enhancing understanding of Australia's water resources.

A national system for water information storage, analysis and reporting requires an unprecedented level of collaboration with stakeholders. Under the new arrangements individuals, corporations and other organizations that possess, control or manage water are required by law to transmit information to the bureau for compilation and analysis.

Source: www.bom.gov.au/water/regulations/water-2008.shtml.

collective level are those that occur as the cumulative result of individual choices, when new patterns of behaviour enter the collective culture (see chapter 2).

One approach to changing behaviour is to use controls. Public institutions such as schools and hospitals can be mandated and enabled to provide sanitation and safe water, construction can be forbidden in fragile areas, and penalties can be imposed for the release of certain pollutants into air and water. Governments can also encourage or promote certain types of behaviour through subsidies, taxes and other incentives, such as payment for environmental services.

Awareness raising and social marketing are other tools for promoting behaviour change. Examples of social marketing campaigns around water issues can be found in almost all countries. In the Danube riparian countries educational institutions and the private sector created a Danube Educational Toolkit, called the 'Danube Box'. Problems and needs in several riparian countries were discussed and collected in national workshops, from which a prototype of a transboundary educational kit emerged. At least five Danube countries (Austria, Germany, Hungary, Romania and Serbia) are now promoting the Danube Box in various projects and activities. These efforts are being organized in close cooperation with the ministries for environment, water and education, along with pedagogical institutes and school authorities.

Ensuring sustainable financing

Because of water's contributions to every aspect of development – social, economic, political and environmental – funding for water need not compete with funding for other priorities. However, to realize maximum benefit, water's contributions should be recognized in every aspect of government and business planning. This may entail adopting new forms of costing or valuation. But it certainly means ensuring that water managers and stakeholders take part in decisions likely to affect water. Thus, this section explores ways to integrate water resources management in development planning. It also considers new ways of mobilizing resources for the water sector.

Integrating water resources management in development planning

Traditional funding mechanisms, including national budget allocations and development assistance, are likely to remain the main source of resources for water, environment and poverty reduction (see chapter 4). Evolution in development policies around the world points to the need for broader, more integrated platforms. Poverty reduction strategies provide a consensual basis for decisions on allocating national budgets and aid. Reliable information on water (including trends in availability and quality) could also be integrated into national budgets to inform economic objectives and investment packages that benefit water. As Zambia's experience shows, integrating water considerations in national development planning instruments can help leverage national and international financing and prepare the way for broad-based reforms (box 15.25).

International assistance for water provision will remain necessary, and some governments have demonstrated a commitment to financing water and sanitation. For example, the government of the Autonomous Community of the Basque Country (Spain) recently adopted a resolution reiterating its commitment to the Millennium Development Goals and proposing to dedicate 5% of its water fees to promote the sanitation target.[25]

International organizations, notably the UN system, can provide support and expertise to governments, help civil society build capacity and catalyse leadership in the private sector.

Cultivating innovations in financing

Innovative mechanisms are being devised in many countries for mobilizing and channelling resources to the water sector. Public-private partnerships, payments for ecosystem services and similar schemes are being explored for generating financing for the water sector. Such schemes also offer innovative modalities of cooperation at all levels (box 15.26).

Managing the multiple ecosystem services of water use to optimize benefits and minimize adverse impacts is another promising approach. Using urban wastewater for agricultural production, for example, avoids discharging pollutants directly into waterways and safeguards water quality for domestic users.[26] This role of farmers as environmental stewards should be recognized and valued through economic incentives from other sectors. This is the aim of programmes such as green water credits in Kenya (see box 14.24 in chapter 14). Cross-sectoral incentives can also be applied at a broader level, as in Ecuador, where the water sector is financed through a tax on telecommunications (box 15.27).

* * *

Government, the private sector and civic organizations – having faced and dealt with water issues – provide many examples of the way forward. The examples in this chapter show how government policy, business decisions and civic initiatives can benefit water services and achieve multiple benefits. They show that sustainable development is possible when links

The examples in this chapter show how government policy, business decisions and civic initiatives can benefit water services and achieve multiple benefits

Box 15.25 Zambia's experiences linking integrated water resources management with national development plans

Zambia still has adequate water resources, but declining rainfall patterns and increased water use are causing increased water stress. Water is used by households and for industrial production, as well as for hydropower production, which earns Zambia on average some $10 million a year in exports. Agriculture is also a key sector for development, depending on water for irrigation, livestock watering and other uses.

During the past 10 years Zambia has grown rapidly in response to economic and administrative reforms. Increasing pressure on water resources for economic expansion led the government to undertake broad water sector reforms. As early as 1994 the government introduced the National Water Policy, following up with reforms of water supply and sanitation and then water resources management. The reforms were planned through the 2001 Water Resources Action Programme, which included an institutional and legal framework and water resources demand, supply and infrastructure components. Reforms included a new water policy, legislation and a new institution to manage and develop water resources.

In the Fifth National Development Plan 2006-10 the government of Zambia expressed concern with ensuring that the benefits of a healthier economy reach the rural and urban poor. Recognizing the importance of water for development, the government integrated the water sector reforms, including the new integrated water resources management plan, and the National Development Plan. Linking these was seen as fundamental to poverty reduction and achieving all the Millennium Development Goals.

In formulating the Fifth National Development Plan, the government used a participatory approach, involving 17 sector advisory groups including water and a government-led high-level forum to improve sector coordination and advise the government on water sector reform, performance, and monitoring and evaluation. The forum includes representatives from key institutions and stakeholders inside and outside the water sector. This process secured political support for the integrated water resources management plan, which was selected as the instrument for implementing the water-related programmes of the development plan.

The 1994 water policy had dealt mainly with reforms within the sector. The new integrated water resources management policy is intent on integrating water management throughout all sectors in Zambia. As a result of this integration of water and national development planning, many donors have incorporated water-related investment in their assistance packages to Zambia.

Source: Mike Muller, visiting professor, Graduate School of Public and Development Management, Witwatersrand University, based on Nyambe and Feilberg n.d.

are forged between water managers and decision-makers in other sectors. Because water issues are central to every aspect of development, actors within and outside the water sector need to strengthen their collaboration to create and promote appropriate mechanisms for collecting and sharing information, financing priority investments and building responsive institutions. The political will to address water issues worldwide remains crucial, along with a willingness to consider innovative ways to approach local, regional and international cooperation.

Box 15.26 Payments for ecosystem services help curb climate change and conserve biodiversity while protecting water resources

The highland forests and *paramos* (treeless plateaus) of the Andes mountain chain provide a variety of ecosystem services for human and natural communities in northern South America. In the centre of the Colombian Andes the Chingaza and Sumapaz National Parks cover more than 225,000 hectares of these fragile ecosystems, which contain habitats for a variety of threatened flora and fauna and provide crucial water for downstream human populations. The area is the watershed for Bogota, which has more than 7 million residents. In recent years the water supply has declined considerably due to human activities such as agriculture and livestock raising, which have led to degradation of the high Andes ecosystems. The Bogota Water and Sewerage Company (EAAB) estimates that water demand will rise substantially by 2020. Immediate measures are necessary to protect the watershed in order to meet that anticipated demand.

Recently EAAB, together with the Colombian government and the support of Conservation International, implemented a pilot project for financing and protecting these critical watersheds. The project provides multiple benefits. Conceived to provide climate change mitigation benefits as well as reliable freshwater supplies, it provides clean, renewable energy while protecting the watershed and supplying funding both to EAAB and the Colombian national parks.

EAAB's Santa Ana small-scale hydroelectric power plant generates electricity without the greenhouse gas emissions associated with traditional fossil fuel plants. The dam is recognized as a Clean Development Mechanism project by the United Nations Framework Convention on Climate Change. It generates about 23,000 certified emission reduction units each year with an approximate value of $450,000. EAAB has committed to giving half this revenue to the Colombian Park Service, which uses the resources to consolidate and expand Chingaza National Park. Increased funds for park management will slow deforestation and ecosystem degradation, by reducing another source of greenhouse gases while conserving biodiversity and improving water supplies.

In the next phase of the project Conservation International and partners are supporting an initiative to consolidate a regional biological corridor to connect Chingaza with Sumapaz National Park, the eastern Andean mountain slopes and the San Rafael water basin. A variety of financing mechanisms will generate income based on the water and climate mitigation services the area provides. Through a large-scale reforestation effort on degraded lands in the corridor, the project will expand its carbon sequestration and water provision, providing benefits to other communities that depend on the water from threatened high Andean ecosystems.

Source: www.conservation.org/explore/regions/south_america/andes/Pages/andes.aspx.

Box 15.27 A water services concession with public funding in Guayaquil, Ecuador

Guayaquil, Ecuador's economic capital, is home to 2.4 million people, or a third of the national urban population. When a concessionaire took over management of water services in 2001, water coverage lagged far behind the national average: only 60% of residents had household connections in 2000, compared with the national urban average of 81% in 1998. The gap was smaller for sewerage, with coverage of 56% compared with a national urban average of 61%. The concession is the second largest in Latin America by population served after Buenos Aires.[1]

The concession rapidly expanded access to piped water through household connections. Starting with 245,000 connections in 2000, it installed 160,000 new connections in the first five years of operation – a more than 10% annual increase and three times the contractual target of 55,000. Those gains brought water coverage in the city up to 82% in 2005 and benefited about 800,000 people, most of them living in poor neighbourhoods not previously served by the network. Over the same period progress in extending urban water access at the national level stagnated. The concessionaire's performance in improving sewer access was more modest, rising from 56% to 62%.

The good performance on water access was made possible by a special tax transfer mechanism that the central government introduced in the 1980s to subsidize new water connections. Revenues from a 10% tax on telephone bills is transferred to utilities to support expansion of the water network in uncovered urban areas. New water connections are provided free of charge in urban areas not previously covered by the water network (areas where most of the population is poor). Part of the cost of expanding the network is also subsidized. Sewer connections were ineligible for these tax transfers and subsidies, explaining the more modest progress.

Note

1. Santiago de Chile (5.3 million customers) is larger but operated under a divestiture scheme (private ownership of infrastructure).

Source: Yepes 2007.

Notes

1. World Bank 2007.
2. ADB and OECD 2001.
3. WWAP 2006, p. 75.
4. See Bayoumi and Abumoghli 2007 for further examples.
5. CFC and Benevides and Pérez-Lucy 2001; Zhang 2003.
6. Fan, Zhang, and Zhang 2004; De Brauw et al. 2002.
7. See, for example, Hlanze, Gama, and Mondlane 2005; www.fao.org/hivaids/publications/farming_en.htm.
8. Poverty-Environment Partnership 2008.
9. UNECE 1999.
10. Djibouti Government 2004, p. 5.
11. Blue Plan, MAP, and UNEP 2003, 2007.
12. United Nations Global Compact 2008.
13. www.un.org/apps/sg/sgstats.asp?nid=3682.
14. Worldwatch Institute 2008, p. 119.
15. Palanisami and Easter 2000.
16. Shah et al. 2004.
17. GEF Small Grants Facility 2003; Smith 2000.
18. World Tourism Organization (www.unwto.org).
19. Trondalen 2008.
20. See for example, Binswanger et al. 2008.
21. Werner and Kruger 2007.
22. AfDB et al. 2003; Stern 2006.
23. Pahl-Wostl 2007; Pahl-Wostl, Kabat, and Möltgen 2007.
24. www.worldwatermonitoringday.org/About/About_Main.html.
25. www.lehendakaritza.ejgv.euskadi.net/r48-11912/en/contenidos/informacion/organismos_multilaterales/en_multilat/presentacion.html.
26. Ragab and Koo-Oshima 2006.

References

ADB (Asian Development Bank) and OECD (Organisation for Economic Co-operation and Development). 2001. *Progress in the Fight against Corruption in Asia and the Pacific.* Manila: Asian Development Bank; and Paris: Organisation for Economic Co-operation and Development.

AfDB (African Development Bank), ADB (Asian Development Bank), DFID (Department for International Development), DGD-EC (Directorate-General for Development, European Commission), BMZ (German Federal Ministry for Economic Cooperation and Development), BuZa (Dutch Ministry of Foreign Affairs), OECD (Organization for Economic Cooperation and Development), UNDP (United Nations Development Programme), and the World Bank. 2003. Poverty and Climate Change: Reducing the Vulnerability of the Poor through Adaptation. Consultative draft presented at the Eighth Conference of Parties to the United Nations Framework Convention on Climate Change in New Delhi, 23 October-1 November 2002.www.unpei.org/PDF/Poverty-and-Climate-Change.pdf.

Atchoarena, David and Lavinia Gasperini, eds. 2003. *Education for Rural Development: Towards New Policy Responses.* Rome: Food and Agriculture Organization of the United Nations, and Paris: United Nations Educational, Scientific, and Cultural Organization.

BAPE (Bureau d'audiences publiques sur l'environnement). 2000. *Water: A Resource to Be Protected, Shared and Enhanced.* Québec City: Government of Québec.

Bayoumi, Mohammed, and Iyad Abumoghli. 2007. National Water Councils: Comparative Experiences; Report submitted to UNDP, February 2007.

Bazaara, Nyangabyaki. 2003. Decentralization, Politics and Environment in Uganda. Environmental Governance in Africa. Working Paper 7, World Resources Institute, Washington, DC.

Benavides, D. D., and E. Pérez-Ducy. 2001. *Tourism in the Least Developed Countries.* Madrid: United Nations Conference on Trade and Development, World Tourism Organization.

Benoit, G., and A. Comeau, eds. 2005. *A Sustainable Future for the Mediterranean: The Blue Plan's Environment and Development Outlook.* London: Earthscan.

Binswanger, Hans P., Roland Henderson, Zweli Mbhele, and Kay Muir-Leresche. 2008. Accelerating Sustainable, Efficient, and Equitable Land Reform: Case Study of the Qedusizi/Besters Cluster Project, Africa Region Working Paper Series 109. World Bank, KwaZulu Natal, South Africa. www.worldbank.org/afr/wps/wp109.pdf.

Bird, Geoffrey, Jacquie Chenje, and Sarah Medina. 2002. *Africa Environment Outlook: Past, Present and Future Perspectives.* Hertfordshire, UK: EarthPrint. www.unep.org/ dewa/Africa/publications/aeo-1/159.htm.

Blue Plan, MAP (Mediterranean Action Plan), and UNEP (United Nations Environment Programme). 2003. Results of the Fiuggi Forum on Advances of Water Demand Management in the Mediterranean: Findings and Recommendations. Document prepared for the Next Mediterranean Commission for Sustainable Development Meeting, Sophia Antipolis, France, United Nations Environment Programme/Mediterranean Action Plan, Athens.

———. **2007.** Water Demand Management, Progress and Policies: Proceedings of the 3rd Regional Workshop on Water and Sustainable Development in the Mediterranean. Zaragoza, Spain, 19-21 March. MAP Technical Reports Series 168. Athens: United Nations Environment Programme.

CFC (Common Fund for Commodities) and UNCTAD (United Nations Conference on Trade and Development). 2001. Enhancing Productive Capacities and Diversification of Commodities in LDCs and South-South Co-operation. Workshop. 21-23 March. Common Fund for Commodities and United Nations Conference on Trade and Development, Geneva.

Chang, Philip. 2008. Armenia. In *Asian Development Outlook 2008.* Manila: Asian Development Bank.

Cosgrove, William. 2009. Public Participation to Promote Water Ethics and Transparency. In *Water Ethics: Marcelino Botin Water Forum 2007*, eds. M. Ramon Llamas, L. Martinez Cortina, Aditi Mukherji. London: Taylor and Francis Group.

De Brauw, A., Huang J., S. Rozelle, Zhang L., and Zhang Y. 2002. The Evolution of China's Rural Labor Markets during the Reforms. Department of Agricultural and Resource Economics, University of California Davis Working paper 02-003. University of California, Davis.

Development Vision – Nepal. 2007. Final Report on Evaluation of UJYALO Program. Volume 1, Main Text. Development Vision – Nepal (P.) Ltd., Kathmandu. http://pdf.usaid.gov/pdf_docs/PDACK585.pdf.

Fan, S., Zhang L., and Zhang X. 2004. "Reforms, Investment, and Poverty in Rural China." *Economic Development and Cultural Change* 52 (2): 395-421

FAO (Food and Agriculture Organization of the United Nations). 2005. Toward Multi-Sectoral Responses to HIV/AIDS: Implications for Education for Rural People (ERP). Paper presented as Ministerial Seminar on Education for Rural People in Africa: Policy Lessons, Options and Priorities, 7-9 September 2002, Addis Ababa.

GAP (Southeastern Anatolia Project). 1989. Southeastern Anatolia Project GAP Master Plan Study. Nippon-Koei and Yüksel Joint Venture, Ankara, Turkey.

GEF Small Grants Programme. 2003. Solar Rural Electrification via Micro-Enterprises, Dominican Republic. Latin America & Caribbean: Dominican Republic-3. Global Environment Facility, Washington, DC, and United Nations Development Programme, New York. http://sgp.undp.org/download/SGP_DominicanRepublic3.pdf.

Government of Djibouti. 2004. Poverty Reduction Strategy Paper, Djibouti.

Hlanze, Zakhe, Thanky Gama, and Sibusiso Mondlane. 2005. The Impact of HIV/AIDS and Drought on Local Knowledge Systems for Agrobiodiversity and Food Security. LinKS Project on Gender, Biodiversity and Local Knowledge Systems for Food Security Report 50. FAO-LinKSSwaziland, Mbabane.

IMF (International Monetary Fund). 2004. *Djibouti: Poverty Reduction Strategy Paper.* IMF Country Report No. 04/152. Washington, DC: International Monetary Fund. www.imf.org/external/.

ISRIC (International Soil Research and Information Centre). 2008. Green Water Credits. World Soil Information, Wageningen, The Netherlands. www.isric.org/UK/About+ISRIC/Projects/ Current+Projects/Green+Water+Credits.htm.

Japan International Cooperation Agency. 2006. Project on Improvement of Water Supply and Sanitation in the Southern Part of the Eastern Province (Pura-Sani Project). 1-1 Details of the Project. Japan International Cooperation Agency, Tokyo. http://project.jica.go.jp/rwanda/0605427/english/01/01.html.

Martinez, Grit. 2007. The Role of Donors in Tackling Sector Development Corruption. Power Point presentation 26 September 2007, Water Integrity Network, Accra. www.waterintegritynetwork.net/content/download/2038/34265/file/3_The%20Role%20 of%20donors.pdf.

Moore, J. G., W. Rast, and W. M. Pulich. 2002. Proposal for an Integrated Management Plan for the Rio Grande/Rio Bravo. In *1st International Symposium on Transboundary Waters Management*, eds. A. Aldama, F. J. Aparicio, and R. Equihua. Avances en Hidraulica 10, XVII Mexican Hydraulics Congress, Monterrey, Mexico, Nov. 18-22, 2002.

National Rural Development Institute. n.d. Integrated Watershed Development Project in Madhya Pradesh. National Rural Development Institute, Hyderabad, India. http://nird.ap.nic.in/clic/water_madhya.html.

Nyambe, Imasiku A., and Miriam Feilberg. n.d. Zambia – National Water Resources Report for WWDR3. Ministry of Energy and Water Development, Lusaka.

Olken, Benjamin. 2005. *Monitoring Corruption: Evidence from a Field Experiment in Indonesia.* NBER Working Paper 11753. Cambridge, MA: National Bureau of Economic Research.

Palanisami, K., and K.W. Easter. 2000. *Tank Irrigation in the 21st Century – What Next?* New Delhi: Discovery Publishing House.

Pahl-Wostl, C. 2007. Transition towards Adaptive Management of Water Facing Climate and Global Change. *Water Resources Management* 21(1): 49-62.

Pahl-Wostl, C., P. Kabat, J. Möltgen, eds. 2007 *Adaptive and Integrated Water Management: Coping with Complexity and Uncertainty.* New York: Springer Publishing.

Peopleandplanet.net. 2002. A Showcase for Sustainable Tourism in Turkey. Planet 21, London. www.peopleandplanet.net/doc.php?id=1723.

Poverty-Environment Partnership. 2008. *Poverty, Health and Environment: Placing Environmental Health on Countries' Development Agenda.* Joint Agency Paper. www.unpei.org/PDF/Pov-Health-Env-CRA.pdf.

Québec Ministry of the Environment. 2002. *Water. Our Life, Our Future: Québec Water Policy.* Québec: Bibliothèque Nationale du Québec.

Ragab, Ragab, and Sasha Koo-Oshima. 2006. *Proceedings of the International Workshop on Environmental Consequences of Irrigation with Poor Quality Waters: Sustainability, Management and Institutional, Water Resources, Health and Social Issues.* 12 September 2006, Kuala Lumpur, Malaysia. New Delhi: International Commission on Irrigation and Drainage.

Rwanda, Ministry of Infrastructure. 2008. Habitat and Urbanism Status in Rwanda. Kigali. www.mininfra.gov.rw/docs/HABITAT%20AND%20URBANISM%20IN%20RWANDA.pdf.

Sadoff, Claudia. 2006. Can Water Undermine Growth? Evidence from Ethiopia. In *Agriculture and Rural Development Notes.* Issue 18. World Bank, Washington, DC. http://siteresources.worldbank.org/INTARD/Resources/Note18_Ethiopia_web.pdf.

Shah, Tushaar, Christopher Scott, Avinash Kishore, and Abhishek Sharma. 2004. *The Energy-Irrigation Nexus in South Asia: Groundwater Conservation and Power Sector Viability.* Colombo: International Water Management Institute.

Smith, Julie, A. 2000. Solar-Based Rural Electrification and Microenterprise Development in Latin America: A Gender Analysis. NREL/SR-550-28955. National Renewable Energy Laboratory, Golden, CO. www.nrel.gov/docs/fy01osti/28995.pdf.

Stern, Nicholas. 2006. Stern Review: The Economics of Climate Change. Cambridge, UK: Cambridge University Press.

TERI (The Energy and Resources Institute). 2008. Third Party Assessment of Coca-Cola Facilities in India. New Delhi: The Energy and Resources Institute. www.teriin.org/cocacola_report_toc.php.

Trondalen, Jon-Martin, Simon Mason, and Adrian Muller. 2008. Water and Security: Entangled and Emerging Issues. Working paper. Compass Foundation, Geneva.

UNECE (United Nations Economic Commission for Europe). 1999. Protocol on Water and Health to the 1992 Convention on the Protection and Use of Transboundary Watercourses and International Lakes. United Nations Economic Commission for Europe, London. www.unece.org/env/documents/2000/ wat/mp.wat.2000.1.e.pdf.

United Nations Global Compact. 2008. *CEO Water Mandate: An Initiative by Business Leaders in Partnership with the International Community.* New York: UN Secretary-General.

WBCSD (World Business Council for Sustainable Development). 2006. Anglo American/Mondi: Improved Energy and Efficiency. Case Study. World Business Council for Sustainable Development, Geneva. www.wbcsd.org/DocRoot/6NDZbCwuNIJpaFU9g0i6/mondi_energy_efficiency_full_case_web.pdf.

Werner, Wolfgang, and Bertus Kruger. 2007. Redistributive Land Reform and Poverty Reduction in Namibia: Livelihoods after Land Reform. Country Paper. Desert Research Foundation of Namibia, Windhoek.

WHO (World Health Organization)/UNICEF (United Nations Children's Fund) Joint Monitoring Programme for Water Supply and Sanitation. 2008. Progress on Drinking Water and Sanitation: Special Focus on Sanitation. New York: United Nations Children's Fund, and Geneva: World Health Organization.

World Bank. 2007. Making the Most of Scarcity: Accountability for Better Water Management Results in the Middle East and North Africa. MENA Development Report. Washington, DC: World Bank.

Worldwatch Institute. 2008. *State of the World: Innovations for a Sustainable Economy.* New York: Norton Institute.

WWAP (World Water Assessment Programme). 2006. *The United Nations World Water Development Report 2: Water: A Shared Responsibility.* Paris: United Nations Educational, Scientific, and Cultural Organization, and New York: Berghahn Books.

Yepes, G. 2007. *Los subsidios cruzados en los servicios de agua potable y saneamiento.* IFM Publications. Washington, DC: Inter-American Development Bank, Infrastructure and Financial Markets Division.

Zhang Le-Yin. 2003. Economic Diversification in the Context of Climate Change. Background Paper for UNFCCC Workshop on Economic Diversification, 18-19 October 2003, Teheran. United Nations Framework Convention on Climate Change, Tehran. http://unfccc.int/files/meetings/workshops/other_meetings/application/pdf/bgpaper.pdf.

Options from beyond the water box

Chapter 16

The way forward

Authors: Andrew Bullock and William Cosgrove

Contributors: Fatma Attia, Gerry Galloway, Jack Moss, Uri Shamir and Albert Wright

Coordinator: Olcay Ünver (WWAP)

Facilitator: William Cosgrove

Key messages

- Water and water systems must be managed to achieve social and economic development objectives and to sustain development. Water resources, properly managed, are critical to the survival and well-being of individuals. They can ensure equity and security in water and sanitation for families, businesses and communities. And they can ensure adequate water for food, energy and the environment as well as protection from floods and droughts.

- Decision-making on water requires seeking synergies and selecting appropriate trade-offs. It also requires distinguishing between short-term 'fire-fighting' – responding to the urgent issues of the day – and long-term strategic development. Developing multipurpose water schemes and reusing water wherever feasible can lessen the need for trade-offs by enabling the same volumes of scarce water to deliver multiple outcomes.

- The donor community can incorporate water into the broader frameworks of development aid and focus assistance on areas where it is needed most – in sub-Saharan Africa, in Asian and Latin American slums and in states recovering from conflict. Recent G-8 efforts in this direction are promising.

- The chief executives of the UN agencies, following the example of their joint discussions of and collective responses to climate change, can convene to examine the role of water, water systems and water management in development and environmental services, providing direction to agencies and advice to member countries.

- The World Water Assessment Programme and its partners are working to help reduce uncertainty, facilitate decision-making and accelerate investment by highlighting the links between socioeconomic development and investment in water management capacity and infrastructure in other sectors.

- The challenges are great, but unsustainable management and inequitable access to water resources cannot continue. We might not have all the information we would like to have before acting, but we do know enough now to begin to take significant steps. Actions must include increased investment in water infrastructure and capacity development. Leaders in the water domain can inform the processes outside their domain and manage water resources to achieve agreed socioeconomic objectives and environmental integrity. But leaders in government, the private sector and civil society will determine the direction that actions take. Recognizing this responsibility, they must act now!

Good water resources management can transform social systems for the better and sustain development

Making water an integral part of all planning and management decisions

Good water resources management can transform social systems for the better and sustain development. It can alleviate the anxiety and fear arising from concerns about too little water – or too much. It can respond to humanity's concerns for natural and social changes and to our growing understanding of uncertainties, thus helping avoid political instability in fragile states. Action today is more important than ever – for the poor as for those better off.

Meeting the many objectives of water resources management with limited human, financial and institutional resources will require more integrated and collaborative procedures and stronger water management institutions and capacity. It will require better monitoring, data analysis and information products to fill the gaps in knowledge of how water is used and managed today and over time – and of the health of ecosystems that supply it. And it will require securing sustainable financing to meet the costs of service provision using existing infrastructure and to invest in essential new infrastructure, both physical and institutional.

The challenges are daunting. But they are surmountable. Managers and professionals in the water sector can work with leaders and decision-makers in all sectors to meet the challenges. They need to act within a framework that:

- Ensures accountability and transparency in planning and implementation, particularly through greater stakeholder participation, with the appropriate incentives and disincentives.
- Integrates gender-sensitive and equitable approaches in water issues.
- Provides options for decision-makers outside the 'water box', informed by clear messages substantiated by sound analyses from a water community that understands its role in the broad development agenda.
- Develops solutions that strike a convincing balance between objectives and alternative means of achieving them, through consultation, innovation and research.
- Provides data and information to reduce uncertainty.
- Secures financing for investment in physical and institutional water infrastructure, using a variety of financing instruments.
- Develops capacity to improve the effectiveness and scales of progress.
- Recognizes that nature also needs water, to ensure the continuing delivery of life-supporting environmental goods and services.

But leaders in the water sector, acting alone, risk being overlooked in broader decision-making for social and economic development. And leaders outside the water sector risk making uninformed and suboptimal development decisions. The most valuable evolution of integrated water resources management could be its extension into dialogue and partnerships with water-using sectors whose policies and strategies are governed by many factors outside the water sector.

Chapter 14 offers practical examples of solutions within the water domain. Options that show promise involve:

- Institutional and human capacity development, to prepare institutions for current and future water and related challenges.
- Water law, both formal and customary, including regulations in other sectors that affect water resources management.
- Consultation with stakeholders and accountability in planning, implementation and management to build trust, as effective management involves pluralistic governance, transparency and interactions among parties with different interests.
- Use of financial options and economic instruments to support the reliability and quality of the services provided.
- Innovation and research to develop appropriate realistic and sustainable solutions.
- Payment for environmental services as an incentive for improving water management efforts and for supporting sustainable ecosystems and water security.
- Creation by water sector decision-makers of a favourable investment climate

based on sound water management and accountability.

Working through practical partnerships

For decades hydrologists and water professionals have been writing about and advocating for the essential role of water resources management in addressing social and economic development, pointing to the need to balance objectives under financial, human and institutional constraints. But decisions on water development and management objectives, and the allocation of human, financial and environmental resources to meet them, are made by leaders in government, the private sector and civil society, not by water professionals alone. Therefore, water professionals need to inform the decisions of these leaders outside the water domain on such issues as spatial and development planning, demographic planning, health, education, agriculture, industry, energy, economic development and the environment.

As government, business and community leaders respond to the needs and initiatives of water users, they should be guided by sound information about the economic and environmental trade-offs in providing these services. The international development community can provide guidance by collecting evidence on successful and failed interventions and introducing this information into national planning. Decision-makers can learn from the examples in chapters 14 and 15.

This Report takes as a starting point that decisions and actions are most effective when implemented at the lowest institutional level qualified to do so. Where individuals are capable of meeting a challenge, their efforts should be supported. Where individual efforts are not enough, community members, including the professional and business community, can come together to meet the challenge, often working with local governments. To facilitate this possibility, a higher level of government may need to step in to delegate authority and provide the needed technical or financial support.

Other actions may be beyond the capacity of individual communities and local governments. Controlling and allocating surface water and groundwater on a basin and aquifer scale and establishing pollution control standards require the involvement of the water authority or ministry. For national well-being, the president or prime minister is the ultimate water manager, supported by the cabinet as the water management team.

International organizations, especially the UN agencies, can provide support for the member states, capacity assistance for civil society and catalytic guidance for the private sector in their efforts to incorporate water in decision-making processes within and outside the water domain. The leadership of the UN system and intergovernmental organizations can provide counsel, drawing on their experience worldwide. As in the joint discussion and collective response on climate change, the chief executives of the UN agencies could meet on the role of water and water management in socioeconomic development processes, environmental sustainability and achievement of the Millennium Development Goals. Donor governments and philanthropic organizations can enter this discussion, in ways compatible with their mission and objectives.

Municipalities, decentralized bodies and local administrations are all engaged in delivery service and managing water resources. These responsibilities have typically been decentralized by the central governments, but often without transferring the necessary financial and human resources. These local bodies face difficult choices in managing water systems, water resources and water and sanitation systems – on whether to regulate through concessions and contracts with private partners and on how to engage with non-governmental organizations. They can inform their decisions by drawing on a growing base of water management experience worldwide. All groups with a stake in water resources management can work together to craft national development plans for water and sanitation, guided by poverty reduction and environmental strategies and international guidelines.

Obtaining and sharing information

Reducing environmental and social risks today and preparing for a future of increasing climate variability and climate extremes require information about the availability and variability of water resources today and tomorrow and about trends in demand. Mechanisms for gathering this information are needed even as countries proceed with infrastructure construction. Information and tools for decision-making under uncertainty will help to avoid making decisions for the short term that have irreversible harmful environmental and social effects over the long term.

As government, business and community leaders respond to the needs and initiatives of water users, they should be guided by sound information about the economic and environmental trade-offs in providing these services

Because the demands on water are many and diverse, water management needs to be guided by broader social and economic development objectives that clarify expected outcomes

Hydrology has moved out of catchments and basins. While the volume of water in circulation on our planet remains constant, climate change models suggest some significant movements in water resources and changes in local ecosystems in response to human use and misuse. Data on water resources – where the water is, where and how it flows – are essential for understanding these global changes. Yet the necessary data are not being collected. As the need for information is increasing, attention to collecting environmental data is waning, even though technologies for collecting data, democratizing data access and gathering are becoming more affordable. We must invest in these technologies and in often-neglected local data-gathering systems to enhance our common understanding of water systems, water resources and water management.

Equally important is information about how much water is being used, by whom and for what purposes, the ability and willingness of water users to pay, cost recovery rates and the investments needed to achieve socioeconomic objectives.

Water synergies, balances and trade-offs

Because the demands on water are many and diverse, water management needs to be guided by broader social and economic development objectives that clarify expected outcomes. 'Integrated' development plans – such as national development strategies, poverty reduction strategies, rural development strategies and regional, district and city development plans – should identify the full set of expected outcomes for water managers.

Because drivers, demands and what can reasonably be achieved given limited resources differ, there will necessarily be trade-offs. Where water is abundant, trade-offs may have little detrimental impact on concerned parties or the natural environment. As water becomes increasingly scarce, trade-offs will be harsher, requiring sound leadership to guide decision-making.

A key area of decision-making concerns economic and environmental trade-offs, a highly political process. It is important to differentiate between short-term 'fire-fighting' – responding to the urgent issues of the day – and strategic development. Multipurpose schemes and water reuse can reduce the need for trade-offs by enabling the same volumes of scarce water to deliver multiple outcomes.

In negotiating trade-offs, interest groups strive to protect the interests of their members. Industry generally lobbies for self-regulation rather than control. Governments are concerned with enforcing laws and regulations. Local pressures and international regulators sometimes encourage price controls, which can have negative as well as positive impact, including preventing harmful price manipulations by speculators. Some international groups advocate for global public goods and services. Nongovernmental organizations advocate for water as a human right, charging governments with responsibility for service provision and users and beneficiaries for use that ensures long-term social, economic and environmental sustainability.

Choices about water uses

Options depend on social, economic and environmental conditions, the availability of water over space and time, and the threat of droughts and floods, all of which vary around the world.

Where water is scarce, the challenge is to select the development path that attains the best social, economic and environmental outcomes. Such decisions shift the trade-offs away from water resources alone to broader concerns of environmental, economic and social benefits. Making decisions about water in this context can sometimes introduce inefficiencies in other development activities. For example, importing food rather than producing it domestically may permit water to be used for higher value outputs, but many farmers will then need to find other ways to earn a living.

Scales of space and time

Actions are constrained by the time horizon for which they are planned (box 16.1). Politicians and water managers can find long-term planning difficult, because the

Box 16.1 Timescales for long-term planning

Planning and response timescales may be categorized as follows:

- Responses to crises (such as droughts, floods, civil strife) (1-2 years).
- Human resources changes (2-3 years).
- Political horizons (3-5 years).
- Small infrastructure horizons (3-5 years).
- Outcomes horizons (5-8 years).
- Behavioural change horizons (10 years).
- Large infrastructure horizons (10-20 years).
- Development horizons (15- 20 years).
- Long-term capacity and intra-generational equity horizons (25+ years and beyond, depending on the level of the plan).

Source: Authors' compilation.

plans and objectives often exceed the time horizon of their tenure. Overcoming this requires frameworks and incentives that support long-term planning.

Status of economic development

The stage of economic development and the financial and human resources available affect which water management options are feasible. For example, when financial resources are limited, choices may have to be made on allocating funds to sectors with the highest economic rate of return or to those that provide basic services. Lack of resources may require finances from sources outside the budget and using resources from outside the country.

High-income countries are experiencing water management problems that are very different from those of poor countries. While high-income countries can afford to pay more attention to the environment and to long-term water system sustainability, developing countries prioritize eliminating poverty and raising the overall level of health and well-being, sometimes at the expense of environmental sustainability.

The challenge is to build a new dialogue between developed countries and developing countries on water management and its role in sustainable development. Developed countries and developing countries must work together to identify socioeconomic priorities and to invest in and use water to power the engines of growth. They must break cycles of poverty while avoiding the harmful environmental and health consequences of unbridled development experienced in many developed countries. Cooperation between developed countries and developing countries can build mitigation, adaptation, avoidance and no-regret measures into decision-making, to avoid incurring the costs of neglected environmental management later.

Working towards better development outcomes

A nation's water resources are used and managed most effectively when they are linked to broader development objectives. What are the objectives, for example, for feeding the population, for providing power for industry, commerce and households, for job creation and incomes, and for child education and health? What are the relations between these objectives, and water and water systems? How should water be managed to achieve these objectives?

To incorporate these and other water considerations, countries need strategic policies and plans. Such plans must incorporate continuity despite changes in personnel and avoid loss of direction when administrations change or key individuals leave. Development partners can consult these plans to stay informed of government intentions and to direct water-related investments into actions responsive to the country's needs.

Actions in a world of change, risk and uncertainty

Risk and uncertainty are part of decision-making. What level of protection is affordable (complex dams or simple refuges)? What are the trade-offs between investments that provide direct benefit (hospitals, schools) and those that protect against possible extreme events (flood protection)? How much should be invested in research where the stakes are high, but the results uncertain? Perception of risk is not fixed but is coloured by socioeconomic conditions, culture and religion, as well as environmental realities.

Risk management encompasses more than managing extremes such as floods and droughts. It entails the use of a structured approach to manage uncertainty regarding these events. Decision-makers must take into account multiple uncertainties, including those associated with limited or low-quality data and information and the inherent unpredictability of climate and other environmental factors. A promising approach for dealing with climate risk is to integrate management of current climate variability and extremes with adaptation to climate change.

The world faces major choices in meeting the challenge of climate change and its potential environmental and socioeconomic impacts. Public policy, so far dominated by mitigation, could benefit from a better balance between mitigation and adaptation. Carbon is a measure of the anthropogenic causes of climate change – water is a measure of its impacts. The international community also has to balance investing for tomorrow's likely problems of greater climate variability and global warming against investing for today's problems of climate variability to prevent losses from droughts and floods. While both are vital, focusing on today's problems can also create greater resilience for dealing with tomorrow's problems.

Because of uncertainty, decisions on current problems should leave the way open

Developed countries and developing countries must work together to identify socioeconomic priorities and to invest in and use water to power the engines of growth

Informed decisions – based on expected results and the consequences of failing to act – are needed now

for future options. No-regrets strategies – actions that would significantly reduce the adverse impacts of change but would not cause harm if projections of impacts of change are wrong – are important in responding to climate change. In contrast, failure to act carries risks because the situation may deteriorate if no action is taken.

The World Water Assessment Programme and its partners are working to reduce uncertainty, facilitate decision-making and accelerate investment by identifying the links between socioeconomic development, environmental sustainability, water management capacity and investments in water-related infrastructure and other sectors. Work has begun, in concert with UN-Water, to identify indicators and supporting databases to guide decisions on water policies and actions by leaders inside and outside the water sector. It will expand the information base on options tested in many contexts. The World Water Assessment Programme is also working on scenarios linking external drivers to the water sector that could be applied globally, regionally and nationally. The next *United Nations World Water Development Report* will include the fruits of that work, along with additional examples of how challenges are being addressed on the ground.

Targeting official development assistance and philanthropic aid

For members of the international community there are choices between pursuing their traditional regional financial and political interests and focusing on areas where aid of all types is needed most. The greatest gaps in access to such water-related services as drinking water supply and sanitation are in sub-Saharan Africa, Asian and Latin American slums and countries recovering from conflict. Inefficient water use in agricultural production also is a continuing problem in many countries around the world, both developed and developing.

The 2008 World Economic Forum in Davos led to calls for policies to encourage a minimum water impact alongside a minimum carbon footprint; the 2009 forum included appeals to fight water scarcity. At the 2009 G-8 meeting in Italy the G-8 leaders are scheduled to review the 2003 G-8 Evian Water Action Plan and discuss strategies with their African partners for enhancing its implementation.

Against a background of political and strategic alignments, the international community must look for ways to support the construction of the infrastructure required to provide a range of direct and indirect services provided by water systems, including water supply and sanitation, production of food and energy and adaptation and mitigation of climate variability. National and local governments can optimize their investments by identifying actions that will produce the greatest socioeconomic and environmental benefits. Supporting countries already on track to achieve the Millennium Development Goals at the expense of countries lagging behind can only entrench global divisions. sub-Saharan Africa in particular suffers from lack of development of its water resources infrastructure for multiple uses.

Deciding – and acting!

Informed decisions – based on expected results and the consequences of failing to act – are needed now. Lagging investment in water leaves hundreds of millions of people exposed to the risks of environmental degradation and water-related disasters and susceptible to political unrest. Billions of the world's people suffer from water-related diseases and hunger. Acting to reduce such exposure can improve their health and well-being and provide access to healthcare for millions more. Country examples indicate that proper water management could increase gross domestic product by 5% to 14% – an impact that may be unachievable through any other intervention.

The challenges are great, but the unsustainable management and inequitable access to water resources cannot continue – because the risks of inaction are even greater. We might not have all the information we would like to have before acting, but we know enough to begin to take significant actions. Some leaders are already acting, showing the way. Others are ready to act. Leaders inside and outside the water domain have critical, complementary roles. Leaders in the water domain can inform the processes outside this domain and manage water resources to achieve agreed socioeconomic and environmental objectives. But leaders in government, the private sector and civil society determine the direction that actions will take. Recognizing this, they must act now!

Appendixes

Appendix 1

World Water Development Report indicators

Author: Mike Muller

Even as we pay more attention to the current state, uses and impacts of water resources and identify the challenges that the global community faces in managing them, the flow of information to support this work is drying up rather than growing.

The first edition of *The United Nations World Water Development Report,* published in 2003, included an extensive compilation of information, drawn from multiple sources, documenting the state of water, the resource and its uses. Many agencies and individuals opened their formal and informal archives to share information from their knowledge bases. These enthusiastic contributions established an important baseline from which to move forward.

In all, more than 160 indicators were reported on, ranging from the global quantum of water available and withdrawals for human use to compliance with water quality standards for key pollutants and governance mechanisms to support water management.

The first report also explicitly recognized the need for further work, notably in collecting biogeophysical and socioeconomic data as well as data on environmental protection and investment in water. It highlighted the danger of data availability driving the selection of indicators, which results in a 'data-rich, but information-poor' syndrome, in which plenty of data are produced but they are not tailored to information needs

The second edition of *The United Nations World Water Development Report* discussed the consequences of poor data availability:

> Data on almost every subject related to water issues is usually lacking, unreliable, incomplete or inconsistent. We have learned that merely collecting data is not enough. It must be brought together, analysed and converted into information and knowledge, then shared widely within and between countries and stakeholders to focus attention on water problems at all scales. It is only when the data has been collected and analysed that we can properly understand the many systems that affect water (hydrological, socio-economic, financial, institutional and political alike), which have to be factored into water governance.

The number of indicators presented in the second edition of the report declined to 62 because there was no systematic process for updating the data used for most of the indicators presented in the first report. Water supply and sanitation has been an exception: the World Health Organization and United Nations Children's Fund Joint Monitoring Programme has systematically addressed the challenge, investing to ensure a regular flow of updated information on this subsector.

Three years later the production team for the third *United Nations World Water Development Report* is in a similar situation to its predecessors. During preparations for the

Mike Muller is co-chair, World Water Assessment Programme Expert Group on Indicators, Monitoring and Data/Metadata Bases.

report, a survey of data providers suggested that new data would be available for only some of the indicators used in the second report. At press time 30 indicators had been updated. Because some indicators included in the second report were identified as not useful by the source agency, 58 indicators are now listed in table A1.1. The profile sheets describing these indicators are available on the World Water Assessment Programme Website (www.unesco.org/water/wwap/).

While the indicators on the water resources situation in the first *United Nations World Water Development Report* provided policy-makers at the national, regional and global levels with a critical overview of the situation, insight into the trends of key indicators is vital in a rapidly changing world.

In most cases it has not proved possible to provide this insight. No new global estimates of available water resources or of volume abstracted by major sectoral users are available. So while this third edition of the report again contains much important information, it remains impossible to provide information on the evolution of key indicators. (One notable exception is a new indicator, Status of progression on Agenda 21, which has been updated and included in this report.)

Several actions are planned to address this gap: UN-Water has created the Task Force on Indicators, Monitoring and Reporting to address the challenge of producing key global indicators of the state of water resources to meet the needs of policy- and decision-makers at all levels. And the World Water Assessment Programme has established the Expert Group on Indicators, Monitoring and Data/Metadata Bases to support this work, specifically by promoting a dialogue between indicator users and data providers/interpreters about the feasibility of providing data for the key indicators on a sustainable, ongoing basis. The expert group will also propose strategies to improve data collection and interpretation.

It is hoped that the next *United Nations World Water Development Report* will be able to report some substantive progress and answer key questions about whether and how changing water resources endowments affect countries and regions, whether the efficiency of water use for national socioeconomic development is improving and whether degradation of the water environment has been slowed. At the very least, it should be able to report on steps taken to improve the flow of data and information needed to establish and monitor key trends.

Table A1.1 List of *United Nations World Water Development Report* indicators and location of detailed data

Topic	Indicator	Category in cause-effect approach[a]	Type of indicator[b]	Location[c]: In *World Water Development Report 2*	Location[c]: In *World Water Development Report 3*
Level of stress on the resource	Index of non-sustainable water use	Driving force, Pressure, state	Key	Section 1	na
	Rural and urban population	Pressure, state	Basic	Section 1	Map 2.1 Figure 2.1
	Relative Water Stress Index	Pressure, state	Key	Section 2	na
	Sources of contemporary nitrogen loading	Pressure, state	Key	Section 3	na
	Domestic and industrial water use	Pressure, state	Basic	Section 3	Table 7.1 Figure 7.1
	Impact of sediment trapping by large dams and reservoirs	Pressure	Key	Section 4	na
	Coefficient of variation for the Climate Moisture Index	State	Key	Section 4	na
	Water Reuse Index	Pressure, state	Key	Section 4	Figure 8.6
Governance	Access to information, participation and justice	Response	Developing	Table 2.2 Table 2.3	na
	Assessing progress towards achieving the integrated water resources management target	Response	Key	Table 2.1	na

(continued)

Table A1.1 List of United Nations World Water Development Report indicators and location of detailed data (continued)

Topic	Indicator	Category in cause-effect approach[a]	Type of indicator[b]	Location[c]: In *World Water Development Report 2*	Location[c]: In *World Water Development Report 3*
Settlements	Index of performance of water utilities	State	Developing	na	na
	Urban water and sanitation governance index	State	Developing	na	na
	Slum profile in human settlements	Pressure	Developing	na	na
State of the resource	Total actual renewable water resources	State	Key	Table 4.3	*
	Precipitation	Driving force	Basic	Table 4.3	Table 10.1 Map 11.1 **
	Total actual renewable water resources per capita	State	Developing	Table 4.3	**
	Surface water as share of total actual renewable water resources	State	Developing	Table 4.3	na
	Overlap as share of total actual renewable water resources	State	Developing	Table 4.3	na
	Inflow from other countries as share of total actual renewable water resources[d]	State	Developing	Table 4.3	**
	Outflow to other countries as share of total actual renewable water resources	State	Developing	Table 4.3	na
	Total use as share of total actual renewable water resources[e]	State	Developing	Table 4.3	**
	Groundwater development as share of total actual renewable water resources	State	Key	Table 4.3	na
Ecosystems	Fragmentation and flow regulation of rivers	State, impact	Key	Map 5.3 Figure 5.4	Figure 8.2
	Dissolved nitrogen (nitrates + nitrogen dioxide)	State	Key	Map 5.2	*
	Trends in freshwater habitat protection	State, response	Key	Fig. 5.7	na
	Freshwater species population trends index	State	Key	Fig. 5.2	Figure 8.1
Health	Disability-adjusted life year	Impact	Key	Table 6.3	Table 6.3
	Prevalence of stunting among children under age 5	Impact	Developing	na	Map 6.2
	Mortality rate of children under age 5	Impact	Developing	Table 6.2	*
	Access to safe drinking water	Impact	Key	Map 6.1	Figure 7.3
	Access to basic sanitation	Impact	Key	Map 6.2	Figure 7.4
Food, agriculture and rural livelihoods	Percentage of undernourished people	State	Key	Map 7.2 Figure 7.10 Figure 7.11	*
	Percentage of poor people living in rural areas	State	Key	na	*
	Agriculture GDP as share of total GDP	State	Key	na	*
	Irrigated land as a percentage of cultivated land	Pressure, state	Key	Map 7.1	Map 7.5
	Agriculture water withdrawals as share of total water withdrawals	Pressure	Key	na	Table 7.1
	Extent of land salinized by irrigation	State	Key	na	na
	Groundwater use as share of total irrigation	Pressure, state	Key	na	Figure 7.1

(continued)

Table A1.1 **List of United Nations World Water Development Report indicators and location of detailed data** (continued)

Topic	Indicator	Category in cause-effect approach[a]	Type of indicator[b]	Location[c]	
				In *World Water Development Report 2*	In *World Water Development Report 3*
Industry and energy	Trends in industrial water use	Pressure	Key	Figure 8.1	na
	Water use by major sector	State	Key	Figure 8.3	Table 7.1 Figure 7.1
	Organic pollution emissions (biochemical oxygen demand) by industrial sector	Impact	Key	Figure 8.4	*
	Industrial water productivity	Response	Key	Table 8.4	Figure 7.8
	Trends in ISO 14001 certification	Response	Key	Table 8.2	Figure 8.7
	Electricity generation by energy source	State	Key	Figure 9.1	Figure 7.11
	Total primary energy supply by source	State	Key	Figure 9.2	*
	Carbon intensity of electricity generation	Impact	Key	Table 9.4	na
	Volume of desalinated water produced	Response	Key	Table 9.1	Box 9.5
	Access to electricity and water for domestic use	Pressure	Key	Table 9.5	*
	Capability for hydropower generation	State	Key	Table 9.6	Map 7.6 *
Risk assessment	Disaster Risk Index	State	Key	Box 10.4	na
	Risk and policy assessment indicator	Response	Key	Figure 10.7	na
	Climate Vulnerability Index	State	Key	Map 10.3	na
Valuing and charging for the resource	Water sector share in total public spending	Response	Developing	na	na
	Ratio of actual to desired level of public investment in drinking water supply	Response	Developing	na	na
	Ratio of actual to desired level of public investment in basic sanitation[f]	Response	Developing	na	na
	Rate of cost recovery[g]	Driving force, response	Developing	na	na
	Water charges as percentage of household income[h]	Driving force, response	Developing	Figure 12.5	na
Knowledge base and capacity	Knowledge Index	State	Developing	Map 13.2	*

na designates that the indicator is not used in the report, although for many of these indicators updated information is provided online (see table note).

Note: An Indicator profile sheet with a detailed definition and explanation of how the indicator is computed (as well as data tables for some indicators) is available for most indicators at www.unesco.org/water/wwap/wwdr/wwdr3/indicators. Exceptions are subindicators for 'Total actual renewable water resources'.

*See table accompanying the online Indicator profile sheet.
**See table accompanying the online Indicator profile sheet for Total actual renewable resources.

a. The categories are based on the DPSIR (driving forces, pressures, state, impact, response) framework. For details, see www.unesco.org/water/wwap/wwdr/wwdr1/pdf/chap3.pdf and www.unesco.org/water/wwap/wwdr/wwdr2/pdf/wwdr2_ch_1.pdf (pp. 33-38).
b. Basic indicators provide fundamental information and are well established and widely used; data are generally widely available for all countries. Key indicators are well defined and validated, have global coverage and are linked directly to policy goals. Developing indicators are in a formative stage and may evolve into key indicators following refinement of methodological issues or data development and testing.
c. Because of updates to data and sources, data may not match across reports.
d. Now called 'Dependency ratio'.
e. Now called 'Millennium Development Goal water indicator'.
f. Proposed for *United Nations World Water Development Report 3*.
g. Now called 'Rate of operation and maintenance cost recovery for water supply and sanitation'.
h. Now called 'Water and sanitation charges as percentage of various household income groups'.

Source: Compiled by Engin Koncagül and Akif Altundaş.

Appendix 2

Water-related goals and objectives of major conferences and forums, 1972-present

Table A2.1 **Water-related goals and objectives of major conferences and forums, 1972-present**

International forum	Agreed goals and objectives
UN Conference on the Human Environment, Stockholm, Sweden, 1972	• The main issues of the conference were preservation and enhancement of the human environment. • The declaration of the conference acknowledged that 'a point has been reached in history when we must shape our actions throughout the world with a more prudent care for their environmental consequences.'
UN Conference on Water, Mar del Plata, Argentina, 1977	• The main objective of this first global-scale conference on water was to promote greater awareness nationally and internationally of global problems related to water and to assess water resources and water use efficiency through an integrated approach to water resources management. • Led to declaring the 1980s International Drinking Water Supply and Sanitation Decade, with the objective of providing drinking water and sanitation for all people by 1990.
International Drinking Water Supply and Sanitation Decade, 1981-90	'The goal of the Decade was that, by the end of 1990, all people should possess an adequate water supply and satisfactory means of excreta and sullage disposal. This was indeed an ambitious target as it has been estimated that it would have involved the provision of water and sanitation services to over 650,000 people per day for the entire ten year period. Although major efforts were made by government and international organisations to meet this target, it was not achieved.' (C. Choguill, R. Francys and A. Cotton, 1993, Planning for Water and Sanitation)
Global Consultation on Safe Water and Sanitation for the 1990s, New Delhi, India, 1990	• The main issues were safe drinking water and environmental sanitation. • The New Delhi Statement declared: 'Safe water and proper means of waste disposal . . . must be at the center of integrated water resources management.' (Environment and health, New Delhi Statement)
World Summit for Children, New York, United States, 1990	• The main issues were health and food supply. • The World Declaration on the Survival, Protection and Development of Children stated: 'We will promote the provision of clean water in all communities for all their children, as well as universal access to sanitation.'
International Decade for Natural Disaster Reduction, 1990-2000	Recognized the increased vulnerability of people and property to natural disasters and sought 'to reduce through concerted international action, especially in developing countries, the loss of life, property damage and social and economic disruption caused by natural disasters.' (Resolution 44/236 of the UN General Assembly)
International Conference on Water and Environment, Dublin, Ireland, 1992	The most important achievement was the development of the Dublin Principles: • Freshwater is a finite and vulnerable resource, essential to sustain life, development and the environment. • Water development and management should be based on a participatory approach, involving users, planners and policy-makers at all levels. • Women play a central part in providing, managing and safeguarding water. Water has an economic value in all its competing uses and should be recognized as an economic good.

(continued)

Table A2.1 **Water-related goals and objectives of major conferences and forums, 1972-present** (continued)

International forum	Agreed goals and objectives
UN Conference on Environment and Development , Rio de Janeiro, Brazil, 1992	Agenda 21, chapter 18, 'Protection of the quality and supply of freshwater resources: application of integrated approaches to the development, management and use of water resources', dealt with the basis for action, objectives and activities concerning: • Integrated water resources development and management. • Water resources assessment. • Protection of water resources, water quality and aquatic ecosystems. • Drinking water supply and sanitation. • Water and sustainable urban development. • Water for sustainable food production and rural development. • Impacts of climate change on water resources.
Ministerial Conference on Drinking Water Supply and Environmental Sanitation, Noordwijk, The Netherlands, 1994	• The main issues were drinking water supply and sanitation. • The Programme of Action identified 'assign[ing] high priority to programmes designed to provide basic sanitation and excreta disposal systems to urban and rural areas.'
UN International Conference on Population and Development, Cairo, Egypt, 1994	• The Programme of Action highlighted 'ensur[ing] that population, environmental and poverty eradication factors are integrated in sustainable development policies, plans and programmes.' (Chapter III – Interrelationships between population, sustained economic growth and sustainable development)
World Summit for Social Development, Copenhagen, Denmark, 1995	• The main issues were poverty reduction, water supply and sanitation. • The outcome was the Copenhagen Declaration on Social Development.
UN Fourth World Conference on Women, Beijing, People's Republic of China, 1995	• The main issues were gender, water supply and sanitation. • The outcome was the Beijing Declaration and Platform for Action.
Second UN Conference on Human Settlements (Habitat II), Istanbul, Turkey, 1996	• The main issues were sustainable human settlements development in an urbanizing world. • The outcome was the Habitat Agenda.
World Food Summit, Rome, Italy, 1996	• The main issues were food, health, water and sanitation. • The outcome was the Rome Declaration on World Food Security.
1st World Water Forum, Marrakech, Morocco, 1997	The main issues were water and sanitation, management of shared waters, preserving ecosystems, gender equity and efficient use of water 'to recognize the basic human needs to have access to clean water and sanitation, to establish an effective mechanism for management of shared waters, to support and preserve ecosystems, to encourage the efficient use of water.' (Marrakech Declaration)
International Conference on Water and Sustainable Development, Paris, France, 1998	The outcome was the Paris Declaration on Water and Sustainable Development, whose objective was 'to improve co-ordination between UN Agencies and Programmes and other international organizations, to ensure periodic consideration within the UN system. [To] emphasize the need for continuous political commitment and broad-based public support to ensure the achievement of sustainable development, management and protection, and equitable use of freshwater resources, and the importance of civil society to support this commitment.' (Paris Declaration)
Millennium Declaration, New York, United States, 2000	The Millennium Development Goals include the following water-related targets: • 'To halve, by the year 2015, the proportion of the world's people whose income is less than one dollar a day and the proportion of people who suffer from hunger and, by the same date, to halve the proportion of people who are unable to reach or to afford safe drinking water.' • 'To stop the unsustainable exploitation of water resources by developing water management strategies at the regional, national and local levels, which promote both equitable access and adequate supplies.'

(continued)

Table A2.1 **Water-related goals and objectives of major conferences and forums, 1972-present** (continued)

International forum	Agreed goals and objectives
2nd World Water Forum, The Hague, The Netherlands, 2000	The Ministerial Declaration identified the following main challenges: • Meeting basic needs – access to safe and sufficient water and sanitation. • Securing the food supply, particularly of the poor and vulnerable. • Protecting ecosystems – ensure the integrity of ecosystems through sustainable water resources management.ASharing water resources, by peaceful cooperation between water users at all levels. • Managing risks from floods, droughts, pollution and other water hazards. • Valuing water – managing water to reflect its economic, social, environmental and cultural values. • Governing water wisely, including involving the public and the interests of all stakeholders.
International Conference on Freshwater, Bonn, Germany, 2001	• The main issues were governance, finance mobilization, capacity building and knowledge sharing. • Identified water as key to sustainable development. • The outcome was the Ministerial Declaration Recommendations for Action: 'Combating poverty is the main challenge for achieving equitable and sustainable development, and water plays a vital role in relation to human health, livelihood, economic growth as well as sustaining ecosystems. . . . The conference recommends priority actions under the following three headings: governance, mobilising financial resources, capacity building and sharing knowledge.' (Ministerial Declaration Recommendations for Action)
World Summit on Sustainable Development, Johannesburg, South Africa, 2002	The summit dealt with the following freshwater-related issues: • Decentralization of governance. • Community empowerment. • Service provision: rural and urban challenges. • Information management. • Integrated water resources management. • Education and awareness. • Financial and economic mechanisms. • Regional challenges were particularly recognized and identified.
3rd World Water Forum, Kyoto, Japan, 2003	The outcomes included: • A water and climate dialogue, including agreed action points. • A water and poverty dialogue, including agreed action points. • A final report on financing water infrastructure. • Outcomes from the Dialogue on Food, Water and Environment. • A detailed document on Water Actions.
G-8 Evian Summit, Evian, France, 2003	One outcome was a G-8 Action Plan on Water: • Promoting good governance. • Using all financial resources. • Building infrastructure by empowering local authorities and communities. • Strengthening monitoring, assessment and research. • Reinforcing engagement of international organizations.
Water for Life Decade, 2005-15	Launched by the United Nations System, the aim is to promote efforts to fulfil international commitments made on water and water-related issues by 2015, with special emphasis on the involvement and participation of women.
The World Conference on Disaster Reduction, Kobe (Hyogo), Japan, 2005	Adopted the 'Hyogo Framework for Action 2005-2015: Building the Resilience of Nations and Communities to Disasters', recognizing the importance of water-related disaster risk reduction.

(continued)

Table A2.1 **Water-related goals and objectives of major conferences and forums, 1972-present** (continued)

International forum	Agreed goals and objectives
4th World Water Forum, Mexico City, Mexico, 2006	The ministers at the forum reaffirmed commitments made at the UN Conference on Environment and Development, World Summit on Sustainable Development, and the Commission on Sustainable Development during 2005, emphasizing: • Expediting implementation in water, sanitation and human settlements. • Enhancing the sustainability of ecosystems. • Applying innovative practices such as rainwater management and development of hydropower projects in some regions. • Involving relevant stakeholders, particularly women and youth, in planning and management. • They also expressed support for relevant UN water-related activities, including the coordinating role of UN-Water.
5th World Water Forum, Istanbul, Turkey, 2009	The theme is Bridging Divides for Water.

Source: www.un.org/esa/sustdev and www.worldwatercouncil.org.

Abbreviations, data notes and units of measure

Abbreviations

ANA	National Water Agency (of Brazil)
AQUAREC	Integrated Concepts for Reuse of Upgraded Wastewater
AQUASTAT	FAO database on water and agriculture
BAPE	Bureau d'audiences publiques sur l'environnment (Office for Public Hearings on the Environment, Québec, Canada)
BOD	Biochemical oxygen demand
CALM	Circumpolar Active Layer Monitoring observational network
CNA	National Water Commission of Mexico (Comisión Nacional del Agua)
DOC	Dissolved organic carbon
EAAB	Water and Sewerage of Bogotá (Empresa de Acueducto y Alcantarillado de Bogotá)
EIONET	European Environment Information and Observation Network
FAO	Food and Agriculture Organization of the United Nations
FRIEND	Flow Regime from International Experimental Data
GAP	Southeastern Anatolia Project (Güneydoğu Anadolu Projesi)
GCM	General circulation model
GDP	Gross domestic product
GEO	*Global Environment Outlook*
GLAAS	Global Annual Assessment of Sanitation and Drinking-Water
GNI	Gross national income
GOCE	Gravity Field and Steady-State Ocean Circulation Explorer
GRACE	Gravity Recovery and Climate Experiment satellites
GRAPHIC	Groundwater Resources Assessment under the Pressures of Humanity and Climate Change programme
GW-MATE	World Bank Groundwater Management Advisory Team
IFAD	International Fund for Agricultural Development
IPCC	Intergovernmental Panel on Climate Change
MDG	Millennium Development Goal
NAFTA	North American Free Trade Agreement
NEWS	Global Nutrient Export from Watersheds models

NGO	Non-governmental organization
NWSC	National Water and Sewerage Corporation of Uganda
OAS	Organization of American States
OECD	Organisation for Economic Co-operation and Development
OMVS	Organization for Development of the Senegal River (Organisation pour la mise en valeur du fleuve Sénégal)
PDSI	Palmer Drought Severity Index
PLUARG	International Reference Group on Great Lakes Pollution from Land Use Activities
POC	Particulate organic carbon
R&D	Research and development
RADWQ	Rapid Assessment of Drinking Water Quality survey method
SAGUAPAC	Cooperative for Urban Water and Sanitation Services in Santa Cruz
TVA	Tennessee Valley Authority
UN	United Nations
UNEP	United Nations Environment Programme
UNESCO	United Nations Educational, Scientific and Cultural Organization
UNFCCC	United Nations Framework Convention on Climate Change
UN-HABITAT	United Nations Human Settlements Programme
UNICEF	United Nations Children's Fund
UNSGAB	UN Secretary-General's Advisory Board
WBCSD	World Business Council for Sustainable Development
WHO	World Health Organization

Data notes and units of measure

$	All dollars are current US dollars unless otherwise indicated
Billion	Billion is 1,000 million
Water	Refers to freshwater unless otherwise indicated

Btu	British thermal unit (1,054.35 joules)
exajoule	Unit of energy equal to 10^{18} joules
Gt	Gigatonne
kcal	Kilocalorie
kg	Kilogram
km^2	Square kilometre
km^3	Cubic kilometre
m^2	Square metre
m^3	Cubic metre
mm	Millimetre
terawatt	Unit of energy equal to 1 trillion watts (10^{12})

List of boxes, figures, maps and tables

Boxes

Figures

Maps

Tables

Index

For Product Safety Concerns and Information please contact our EU
representative GPSR@taylorandfrancis.com Taylor & Francis Verlag GmbH,
Kaufingerstraße 24, 80331 München, Germany

Printed and bound by CPI Group (UK) Ltd, Croydon, CR0 4YY
14/09/2026
02216248-0002